Weeds

WALTER CONRAD MUENSCHER
Late Professor of Botany, New York State
College of Agriculture at Cornell University

 Weeds SECOND EDITION

FOREWORD AND APPENDIXES BY
PETER A. HYYPIO
Curator of the Herbarium and Extension Botanist,
L. H. Bailey Hortorium, Cornell University

Comstock Publishing Associates   A DIVISION OF
Cornell University Press   ITHACA AND LONDON

COPYRIGHT, 1935, 1955 BY THE MACMILLAN COMPANY
COPYRIGHT © 1980 BY CORNELL UNIVERSITY

All rights reserved. Except for brief quotations in a review, this book, or parts thereof, must not be reproduced in any form without permission in writing from the publisher. For information, address Cornell University Press, Sage House, 512 East State Street, Ithaca, New York 14850.

Second edition 1955

Reissued 1980 with a new foreword and appendixes by Cornell University Press.

First printing, Cornell Paperbacks, 1987.

This edition is reprinted by arrangement with Macmillan Publishing Co., Inc.

International Standard Book Number 0-8014-1266-8 (cloth)
International Standard Book Number 0-8014-9417-6 (paper)
Library of Congress Catalog Card Number 79-40017

Printed in the United States of America

Librarians: Library of Congress cataloging information appears on the last page of the book.

Cornell University Press strives to utilize environmentally responsible suppliers and materials to the fullest extent possible in the publishing of its books. Such materials include vegetable-based, low-VOC inks and acid-free papers that are also either recycled, totally chlorine-free, or partly composed of nonwood fibers.

Paperback printing   10   9   8   7   6   5   4

To My
FATHER and MOTHER

The labors of the agriculturist are a constant struggle; on the one hand, by presenting the most favorable conditions possible, he endeavors to make certain plants grow and produce to their utmost capacity; and on the other hand, he has to prevent the growth of certain other plants that are ready to avail themselves of these favorable conditions.—William Darlington. 1859.

Foreword to the Reissue

Twenty-five years after the appearance of the second edition, *Weeds*, by Walter Conrad Muenscher, continues to be an outstanding book on the identification and control of weeds. It is still the most comprehensive survey of the common weeds of the northern United States and Canada. Except for changes in scientific names, most of the basic botanical and ecological information in the book is as sound as ever. Muenscher provided, in careful detail, specific methods for nonchemical control of weeds. By omitting all reference to chemical weed control, he avoided the pitfalls of including rapidly changing recommendations and restrictions in the use of herbicides. In my own experience as an extension botanist during the last fifteen years *Weeds* has been an indispensable tool and a ready source of valuable information. It has been the one weed book I have recommended above all others. Unfortunately, it has been out of print too long.

In this reissue of the second edition, typographical errors have been corrected, and three appendixes have been added. Appendix I lists the changes in botanical names, including the changes in the authorities, that have been published since 1955 in floras, manuals, monographs, and other taxonomic literature in this country and abroad. The changes in nomenclature have occurred as the result of continuing advances in taxonomic research, of revisions in systematic concepts, and out of concern for strict adherence to the Rules and Recommendations of the International Code of Botanical Nomenclature. Since most of our weeds are European in origin, many of the changes in their names have come from Tutin et al., eds., 1964–1976, *Flora Europaea*. Many of these changes, especially at specific and subspecific levels, have already been incorporated in recent American floras. However, at the generic level we tend to be more conservative, and I have chosen to follow the generic concepts recognized in recent American works.

Appendix II is a list of the plants found in the second edition arranged alphabetically by their botanical names followed by their corresponding "standardized" common names. Common names, unlike scientific ones, are not governed by any formal code of nomenclature. However, many technical workers dealing with plant and vegetation control in agriculture and wildlife management have been aware of the need for standardization of common names for weeds and other plants for use in technical and popular publications. The Weed Science Society of America recognized this need by appointing a Subcommittee on Standardization of Common and Botanical Names of Weeds as a unit of their Terminology Committee. Most of the common names appearing in Appendix II were taken from the 1971 report of this subcommittee. Names not found in the report were taken from the American Joint Committee on Horticultural Nomenclature, 1942, *Standardized Plant Names*, whenever possible. A few others were located in the floras and manuals listed in Appendix III. I was compelled to devise only one common name—"hispid marshcress"—for *Rorippa palustris* subsp. *hispida*.

Appendix III is a bibliography of the literature I used in compiling the information in Appendix I and Appendix II.

PETER A. HYYPIO

ITHACA, NEW YORK
November, 1979

Preface to First Edition

In the preparation of this book the primary aim has been to make more available the information on the identification and control of weeds. Part I is devoted to a consideration of those characteristics and habits of weeds by which they affect other plants or interfere with man's activities, and also to the methods employed for their eradication and control. In Part II is brought together data concerning the names, duration, reproduction, dissemination, habitat, range, source, recognition and control of the commonest weeds of the northern United States and Canada. Particular emphasis has been given to identification and control.

For the purpose of this book it has been necessary to make a somewhat arbitrary distinction between weeds and other plants. In addition to the common weeds there have been included some weeds of local distribution and also recent introductions that are yet limited to restricted areas but have shown aggressiveness and promise of becoming more widely scattered. Some native American plants that may be considered weeds under certain conditions have been omitted because descriptions of these can be found by consulting manuals or floras of the regions in which they occur.

The principal manuals covering areas within the United States are: [1] Robinson and Fernald, 1907, *Gray's New Manual of Botany,* the area from Missouri east and north; Britton and Brown, 1913, *An Illustrated Flora of the Northern United States, Canada and the British Possessions;* Small, 1933, *Manual of the Southeastern Flora;* Rydberg, 1932, *Flora of the Prairies and Plains of Central North America;* Coulter and Nelson, 1909, *New Manual of Botany of the Central Rocky Mountains;* Tidestrom, 1925, *Flora of Utah and Nevada;* Jepson, 1925, *A Manual of the Flowering Plants of Cali-*

---

[1] Abrams, *Illustrated Flora of the Pacific States* vol. 1 has been printed in a new edition in 1940; and volumes 2 and 3 are now ready and were used in the revision of *Weeds.* Jepson, *A Manual of the Flowering Plants of California* was reprinted in 1951. Gray's *Manual of Botany* was revised by M. L. Fernald in 1950 (the 8th edition) and Henry Allen Gleason in 1952 revised Britton and Brown under the title, *The New Britton and Brown Illustrated Flora of the Northeastern United States and Adjacent Canada.*

fornia; Abrams, 1923, vol. I, *Illustrated Flora of the Pacific States.*

The following general books on weeds have been found helpful for their original contributions and literature references: Darlington, 1859, *American Weeds and Useful Plants;* Clark and Fletcher, 1906, *Farm Weeds of Canada;* Pammel, 1911, *Weeds of Farm and Garden;* Georgia, 1914, *A Manual of Weeds;* Brenchley, 1920, *Weeds of Farm Land* (England); Long, 1929, *Weeds of Arable Land* (England); Korsmo, 1930, *Unkräuter im Akerbau der Neuzeit* (Norway, Germany).

Many of the more recent contributions to our knowledge concerning the habits, distribution or control methods of weeds in the United States and Canada are referred to in the text and cited in the list of references. This list is not intended to be a complete bibliography but supplies the principal leads to the published source material on weeds. Additional references to special weeds will be found in many of the publications cited.

The weeds are arranged in alphabetical order, under their scientific names, by families. The name used for a species is generally the one employed in most of the botanical manuals and floras. Where there is a difference in usage, synonyms are included. The English, or so-called common names of weeds, are frequently numerous and confusing. Many weeds are known by different names in different regions, and the same name is often used for several weeds. When a weed has more than one English name, that in most general use or deserving preference because it avoids confusion with other weeds immediately follows the scientific name. The statements of distribution for the various species are not necessarily their botanical ranges but the approximate area or areas in which they occur as weeds. The months during which a weed usually blossoms in the northern United States are indicated for each kind. This period must be considered only as an approximation since the time of flowering may be affected by many factors.

The key to the species is based on 500 weeds, most of which are found in the northern United States. It is based primarily on flowers and fruits which are more constant than the vegetative organs and habits of weeds. Botanical terminology has been employed because it permits greater accuracy in the necessarily brief statements in the key and descriptions. The reader who is unfamiliar with this terminology should find the glossary helpful.

Preface to First Edition

I desire to acknowledge the coöperation and helpful criticism of my colleagues and to express my particular indebtedness for specific help received in the preparation of this book. Most of the illustrations were drawn from the plants by Mrs. Helen Hill Craig. The descriptions of the weed seeds are largely based on measurements and observations made by Mrs. Antoinette Boesel on the specimens in the seed collection of Cornell University. Mr. Alton A. Lindsey supplied the data in Table V on viability of weed seeds. Professor Charles C. Chupp verified the data in Table VII on the causal organisms of diseases common to weeds and certain crops. Dr. W. Blauvelt verified the data in Table VIII concerning insect pests common to weeds and crop plants. Minnie Worthen Muenscher typed the final manuscript and assisted with the proof reading and preparation of the index.

W. C. MUENSCHER

ITHACA, NEW YORK,
February 14, 1935.

Preface to Second Edition

This revision of "Weeds" contains the original five hundred weeds and an additional seventy-one. Besides these five hundred seventy-one which are described, with mechanical control methods suggested for each, a number of other weeds of local significance are given brief mention. There are twelve new plates picturing nineteen additional kinds of plants.

The nomenclature has been brought up-to-date, following in general the eighth edition of Gray's *Manual of Botany*. Wherever the names have been changed from those in the first edition of "Weeds," the former name is given in parentheses. Species names have been decapitalized, following the recommendation in the *International Code of Botanical Nomenclature*.

The descriptions of the original five hundred plants are the same as in the first edition except for a few changes in distribution data. The Keys (pages 65 to 105) have been rewritten to include the seventy-one additional species. The number after each plant in the Key refers to the number given that plant in Part 2. New plants have been added to several of the tables in Part 1 and the section on noxious weeds has been rewritten. Otherwise, except for material on chemical control of weeds, Part 1 is essentially the same as in the first edition.

It is acknowledged that chemicals may often offer the best and, for many weeds, the most economical method of control. Each year brings forth new and improved chemicals which are superior to the older ones for specific weed problems. Methods of application, too, change with increased knowledge of the problems involved. Also such factors as temperature, rainfall, humidity, soil type, and soil moisture greatly influence the results. Because of these factors and of the great advances which have been made in this field in the past ten years it seems unwise to consider them in this book. Omission of all mention of chemical control does not mean it is not considered important, but rather that it is too important to be treated in a book dealing primarily with the identification of weeds. Therefore, the chapter on chemical control and the mention of specific methods of control by chemicals have been omitted. Instead, a list of selected references on chemical control is included in Chapter 3.

Throughout the work on this revision my colleagues, students, and former students have been most helpful. I wish to acknowledge my indebtedness to them all; and especially to Dr. Oren Justice, Seed Analyst for the United States Department of Agriculture, for the section on Noxious Weeds; to Dr. Stanford N. Fertig, Professor of Agronomy at Cornell University, for the selected references on Chemical Weed Control; to Dr. George Schumacher, Professor of Botany at Harpur College, for the revision of the Keys; and to Dr. Robert F. Thorne, Professor of Botany at the University of Iowa, for consultive help with Part 2 and for reading the manuscript. Also to Miss Elfriede Abbe, Cornell University, for the additional plates. My wife, Minnie Worthen Muenscher, has helped invaluably with both clerical and editorial work. Our daughters, Joanne Droppers, Helen Tryon, and Elizabeth De Velbiss, have also given varied assistance which is gratefully acknowledged. I should like, also, to express my appreciation to Miss Carol H. Woodward, Editor of Outdoor Books at Macmillan, for her courteous assistance.

W. C. MUENSCHER

ITHACA, NEW YORK,
February, 1955.

Figures 44, 46, 47, 49, 51, 57, 67, 69, 70, 128, 133 and 134 are taken from Cornell University Extension Bulletins 168, 191, 192 and 195. Figures 4, 5, 6, 11, 14, 15, 16, 17, 18, 31, 33, 34, 39, 40, 41, 42, 64, 65, 66, 72, 74, 75, 76, 77, 78, 80, 83, 84, 95, 96, 106, 114, 115, 120 and 126 are published by courtesy of Cornell University, with the permission of the Dean of the New York State College of Agriculture. Figure 79 is in part adapted from Revision of the *North American and West Indian Species of Cuscuta* by T. G. Yuncker.

Contents

PART I

GENERAL: WEEDS AND THEIR CONTROL   PAGE 1

CHAPTER I

DISSEMINATION AND IMPORTANCE OF WEEDS	3
CHARACTERISTICS AND HABITS OF WEEDS	3
Habits of growth	3
Habits of seed production	4
THE REPRODUCTION OF WEEDS	9
THE DISSEMINATION OF WEEDS	14
Ornamentals escaping	14
Impurities in agricultural seeds	16
Hay and feed-stuffs	16
Ballast from freight cars and boats	17
Threshing machines and hay balers	17
Cultivating implements	17
Manure	18
Packing materials	18
Wind as an agent of dissemination	18
Water as an agent of dissemination	18
Animals as agents of dissemination	20
THE SOURCES OF WEEDS	21
LOSSES CAUSED BY WEEDS	24
Reduction in yield of crops	24
Increase of cost of operation of farms	24
Injurious effects of weeds	25
BENEFITS DERIVED FROM WEEDS	32

CHAPTER II

WEEDS OF SPECIAL HABITATS	36
WEEDS OF LAWNS AND TURFS	36
WEEDS OF PASTURES	38
WEEDS OF HAY FIELDS AND MEADOWS	41
WEEDS OF CULTIVATED FIELDS AND GARDENS	43

	PAGE
WEEDS OF GRAIN FIELDS	44
WEEDS OF CRANBERRY BOGS	45
WEEDS OF RICE FIELDS	46

CHAPTER III

THE CONTROL OF WEEDS — 48

METHODS FOR PREVENTING THE SPREAD OF WEEDS INTO NEW AREAS — 49
1. Use of clean seed — 49
2. Avoiding the scattering of weed seeds with farm products and machinery — 49
3. Prevention of seed production by weeds in nearby waste areas — 50

METHODS FOR DESTROYING THE TOPS OF WEEDS — 50
4. Hand pulling — 50
5. Hand hoeing — 51
6. Cultivation — 51
7. Plowing — 52
8. Harrowing — 52
9. Disking — 53
10. Pasturing and grazing — 53
11. Mowing — 54
12. Spudding — 54
13. Burning — 55
14. Steam — 55
15. Biological methods of weed control — 55
16. Chemicals — 56

METHODS FOR DESTROYING THE UNDERGROUND PARTS OF WEEDS — 57
17. Hand digging — 57
18. Clean cultivation — 57
19. Summer fallow — 58
20. Rotation — 58
21. Drainage — 59
22. Smother crops — 59
23. Straw mulch — 60
24. Mulch paper — 60
25. Chemicals — 61

METHODS FOR DESTROYING WEED SEEDS IN THE SOIL — 61
26. Harrowing and shallow cultivation — 62
27. Deep plowing — 62

PART II

WEEDS ARRANGED ACCORDING TO FAMILY, TOGETHER
WITH KEY — 63

	PAGE
KEY TO THE GROUPS AND SPECIES OF WEEDS	65
EQUISETACEAE (Horsetail Family)	106
OSMUNDACEAE (Flowering Fern Family)	108
POLYPODIACEAE (Fern Family)	110
GRAMINEAE (Grass Family)	111
CYPERACEAE (Sedge Family)	148
COMMELINACEAE (Spiderwort Family)	152
JUNCACEAE (Rush Family)	152
LILIACEAE (Lily Family)	154
MYRICACEAE (Sweet Gale Family)	162
URTICACEAE (Nettle Family)	162
POLYGONACEAE (Buckwheat Family)	164
CHENOPODIACEAE (Goosefoot Family)	180
AMARANTHACEAE (Amaranth Family)	192
NYCTAGINACEAE (Four-o'clock Family)	196
PHYTOLACCACEAE (Pokeweed Family)	197
AIZOACEAE (Carpet-weed Family)	199
PORTULACACEAE (Purslane Family)	199
CARYOPHYLLACEAE (Pink Family)	201
RANUNCULACEAE (Crowfoot Family)	217
BERBERIDACEAE (Barberry Family)	226
PAPAVERACEAE (Poppy Family)	226
CAPPARIDACEAE (Caper Family)	228
CRUCIFERAE (Mustard Family)	229
RESEDACEAE (Mignonette Family)	262
CRASSULACEAE (Orpine Family)	262
ROSACEAE (Rose Family)	264
LEGUMINOSAE (Pulse Family)	276
ZYGOPHYLLACEAE (Caltrop Family)	291
OXALIDACEAE (Wood Sorrel Family)	291
GERANIACEAE (Geranium Family)	292
EUPHORBIACEAE (Spurge Family)	296
ANACARDIACEAE (Cashew Family)	306
MALVACEAE (Mallow Family)	310
GUTTIFERAE (St. Johns-wort Family)	314
VIOLACEAE (Violet Family)	316
PASSIFLORACEAE (Passion-flower Family)	317

xviii *Contents*

	PAGE
CACTACEAE (Cactus Family)	317
LYTHRACEAE (Loosestrife Family)	318
ONAGRACEAE (Evening Primrose Family)	319
UMBELLIFERAE (Parsley Family)	321
ERICACEAE (Heath Family)	331
PRIMULACEAE (Primrose Family)	334
APOCYNACEAE (Dogbane Family)	336
ASCLEPIADACEAE (Milkweed Family)	339
CONVOLVULACEAE (Morning-glory Family)	344
POLEMONIACEAE (Phlox Family)	354
HYDROPHYLLACEAE (Waterleaf Family)	354
BORAGINACEAE (Borage Family)	356
VERBENACEAE (Vervain Family)	365
LABIATAE (Mint Family)	368
SOLANACEAE (Nightshade Family)	383
SCROPHULARIACEAE (Figwort Family)	391
BIGNONIACEAE (Bignonia Family)	404
OROBANCHACEAE (Broom-rape Family)	406
PLANTAGINACEAE (Plantain Family)	407
RUBIACEAE (Madder Family)	411
CAPRIFOLIACEAE (Honeysuckle Family)	415
VALERIANACEAE (Valerian Family)	415
DIPSACACEAE (Teasel Family)	417
CUCURBITACEAE (Gourd Family)	418
CAMPANULACEAE (Bluebell Family)	419
LOBELIACEAE (Lobelia Family)	421
COMPOSITAE (Composite Family)	422
GLOSSARY	507
READY REFERENCE DATA	518
LITERATURE REFERENCES	519

APPENDIXES:
  I. CHANGES IN BOTANICAL NOMENCLATURE — 533
  II. WEEDS MENTIONED IN THIS BOOK (ARRANGED ALPHABETICALLY BY SCIENTIFIC NAME FOLLOWED BY STANDARDIZED COMMON NAME) — 537
  III. BIBLIOGRAPHY FOR CURRENT NOMENCLATURE AND COMMON NAMES — 557

INDEX — 559

PART I
GENERAL: WEEDS AND THEIR CONTROL

CHAPTER I

Dissemination and Importance of Weeds

The word weed suggests a useless, ugly or harmful plant that persists in growing where it is not wanted. Many weeds possess one or more, but not necessarily all, of these characteristics. A plant may be useful, beautiful or harmless and still be a weed under certain conditions. Weeds are those plants, with harmful or objectionable habits or characteristics, which grow where they are not wanted, usually in places where it is desired that something else should grow. Strictly speaking, there are no species of weeds. Whether a plant of a given species is considered a weed depends not only on its characteristics and habits but also on its relative position with reference to other plants and man.

Characteristics and Habits of Weeds

Among the most striking characteristics that enable plants to become weeds are those relating to their habits of growth and seeding.

Habits of growth.

a. Many weeds are capable of thriving under adverse conditions as well as under those favorable for the growth of crop plants.

b. Many weeds, especially perennials, are able to regenerate lost parts. If the crown of a dandelion or buckhorn plantain is removed, the roots send up new shoots.

c. A number of perennial weeds can spread by vegetative methods even though they are prevented from producing seeds. The rootstocks of quack-grass and the roots of perennial sow thistle may creep for several yards in one season. The roots or rootstocks of such weeds are often scattered far and wide during cultivation.

d. Many weeds are able to grow under adverse conditions because they have much reduced or modified leaves and other aerial parts and thus conserve moisture. This enables certain weeds to win out in the competition when water is a limiting factor.

e. The flowers of many, not all, weeds are very small and inconspicuous. Such flowers often mature seeds before their presence is suspected.

f. Many weeds contain substances which give them a disagreeable taste or odor; others are covered with sticky materials, stiff hairs, spines or thorns. All these devices tend to protect such weeds against injury by natural enemies or domestic animals.

Habits of seed production.

a. It is generally known that weeds mature enormous numbers of seeds every year. A single individual of some of the worst kinds may produce enough seeds in one season to cover an entire acre of ground if they all developed into plants in the next season. For example, single plants of some common weeds growing about Ithaca, New York, matured the following numbers of seeds during the season of 1924.

Hedge mustard, *Sisymbrium altissimum*	511,208 seeds
Black mustard, *Brassica nigra*	58,363 "
Tumble weed, *Amaranthus albus*	180,220 "
Amaranth pigweed, *Amaranthus retroflexus*	196,405 "
Fleabane, *Erigeron canadensis*	243,375 "
Wild lettuce, *Lactuca scariola*	52,700 "
Nightshade, *Solanum nigrum*	178,000 "
Jimson-weed, *Datura stramonium*	23,400 "
Purslane, *Portulaca oleracea*	193,213 "
Poison-hemlock, *Conium maculatum*	38,000 "

One plant of hedge mustard produced over a half million seeds, or enough, if evenly scattered, to sow eleven seeds on every square foot in an acre of land or enough to sow 3,200 seeds on every acre of a 160-acre farm. Some weeds produce over a million seeds on a single plant in one season. The numbers of seeds produced by many of the more common weeds have been reported by Stevens.[1]

b. Buried weed seeds may remain alive for many years. Experiments with buried seeds conducted by the Michigan Experiment Station[2] and also by the United States Department of Agriculture,[3] over periods of twenty to forty years, have revealed a remarkable longevity in many common weed seeds. It is not uncommon for weed seeds to retain their germinating power ten, twenty or even forty years after they have been buried in the soil.

[1] Stevens, 1932.   [2] Darlington, 1922.   [3] Goss, 1924.

TABLE I

THE PERCENTAGE OF GERMINATION OF WEED SEEDS BURIED IN THE SOIL FROM ONE TO FIVE YEARS *

Species	Seed Stored Dry			Seed Buried in Sandy Soil					Seed Buried in Clay Soil				
	1–2 Weeks	4–5 Mo.	5 Years	1 Year	2 Years	4 Years	5 Years		1 Year	2 Years	4 Years	5 Years	
24. Bromus secalinus	88	85	76	0	0	0	0		0	0	0	0	
58. Setaria glauca	24	90	74	62	39	35	9		72	55	25	5	
111. Rumex obtusifolius	85	83	93	21	51	40	38		18	62	46	29	
118. Chenopodium album	6	32	51	8	48	23	18		6	53	36	20	
136. Amaranthus albus	24	54	82	15	41	20	13		8	20	23	17	
146. Portulaca oleracea	13	42	88	5	26	25	28		8	32	35	26	
147. Agrostemma githago	98	97	62	0	0	0	0		0	0	0	0	
164. Silene noctiflora	97	95	82	9	43	24	10		8	60	14	15	
188. Barbarea vulgaris	70	83	33	12	25	22	31		7	41	28	27	
192. Brassica kaber	38	74	83	31	64	63	37		13	45	47	48	
208. Erucastrum gallicum	12	66	32	5	35	28	26		10	51	42	33	
212. Lepidium campestre	12	65	26	0	0	0	0		0	0	0	0	
218. Raphanus raphanistrum	38	57	20	16	42	28	19		14	53	22	12	
226. Thlaspi arvense	8	65	1	10	35	40	25		14	25	56	34	
324. Daucus carota	20	77	71	25	44	52	37		5	43	45	30	
408. Solanum carolinense	6	74	—	30	17	—	—		3	14	—	—	
471. Arctium lappa	81	89	94	34	42	48	36		38	47	42	54	
519. Galinsoga ciliata	0	19	28	6	40	25	29		5	28	31	27	
529. Hieracium aurantiacum	43	60	66	9	27	29	15		5	34	24	26	
535. Hypochoeris radicata	0	82	71	15	26	31	32		18	38	18	41	
560. Sonchus arvensis	15	56	53	11	32	37	12		12	29	41	2	

* The seeds were harvested as they matured in the autumn of 1928. They were buried on Sept. 30, 1928, at a depth of 6 inches.

TABLE II

Number of Viable Weed Seeds in One Cubic Decimeter of Soil Taken 1 Dm. below the Surface of the Soil

Source of soil sample	Meadow on stony clay	Meadow on clay	Abandoned field, clay	Old meadow, clay	Old field, fine gravel	Grass along canal bank	Old field, clay	Old meadow, sand	Grassy road-side, sand	Meadow, stony clay	Old meadow, clay	Old field, stony clay
Years since soil was disturbed	3	3	8	6	9	7	3	8	10	4	10	7
Species of weed												
Acalypha rhomboidea—Three-seeded mercury												4
Amaranthus retroflexus—Amaranth pigweed	1											
Ambrosia artemisiifolia—Ragweed		4	3	34	3		11	1	8	14	2	3
Barbarea vulgaris—Winter-cress									4	19		
Brassica kaber—Wild mustard			3						12			
Cerastium vulgatum—Mouse-ear chickweed	11	6		2	4	2		4	4	4	2	
Chenopodium album—Lambs-quarters		2	3	1		15				2		
Chrysanthemum leucanthemum—Ox-eye daisy		2				21			1	6	5	2
Daucus carota—Wild carrot	6	16		6	15	8	13	6	46	17	8	
Erigeron strigosus—Rough daisy fleabane				2	7		15		2	8	1	
Hieracium aurantiacum—Orange hawkweed				7	26		1				27	2
Hypericum perforatum—St. Johns—wort				66		9					1	
Linaria vulgaris—Butter-and-eggs				2						1		
Medicago lupulina—Black medic				4		20						
Melilotus alba—White sweet clover						6			3			
Nepeta cataria—Catnip												
Panicum capillare—Old witch-grass	3	7	3	8		5	6	3		6	1	
Plantago lanceolata—Buckhorn plantain	1	2		1	3				21	13	49	
Plantago major—Broad-leaved plantain	3	1								2	9	
Polygonum aviculare—Knotweed	3				1							
Polygonum convolvulus—Black bindweed				12	6		2	1				
Portulaca oleracea—Purslane						8				3	1	

TABLE II—Continued

NUMBER OF VIABLE WEED SEEDS IN ONE CUBIC DECIMETER OF SOIL TAKEN 1 DM. BELOW THE SURFACE OF THE SOIL—Continued

Source of soil sample	Meadow on stony clay	Meadow on clay	Abandoned field, clay	Old meadow, clay	Old field, fine gravel	Grass along canal bank	Old field, clay	Old meadow, sand	Grassy roadside, sand	Meadow, stony clay	Old meadow, clay	Old field, stony clay
Potentilla norvegica—Cinquefoil				28		2				1	1	1
Potentilla recta—Sulfur cinquefoil				8	29						5	
Prunella vulgaris—Heal-all											9	
Ranunculus acris—Tall field buttercup												
Rumex acetosella—Sheep sorrel				5	2						3	
Rumex crispus—Curly dock	2			10	9	19		5				
Setaria glauca—Yellow foxtail	14	18	10	13		9	12		3	15		
Solidago graminifolia—Narrow-leaved goldenrod												
Stellaria media—Chickweed				1								
Taraxacum officinale—Dandelion		2										
Verbascum blattaria—Moth mullein								2	18	4		3
Verbascum thapsus—Mullein					19				21	8		
Veronica officinalis—Speedwell												
Veronica peregrina—Purslane speedwell		2				5		3			3	
Veronica serpyllifolia—Thyme-leaved speedwell			2						4		1	

Wild or black mustard (*Brassica nigra*), pigweed (*Chenopodium album*), purslane (*Portulaca oleracea*) and amaranth pigweed (*Amaranthus retroflexus*) germinated after having been buried from twenty to forty years. Canada thistle (*Cirsium arvense*), yellow foxtail (*Setaria glauca*), small ragweed (*Ambrosia artemisiifolia*), Jimsonweed (*Datura stramonium*), ox-eye daisy (*Chrysanthemum leucanthemum*) and nightshade (*Solanum nigrum*) germinated after having been buried twenty years. Chickweed (*Stellaria media*), buckhorn plantain (*Plantago lanceolata*) and hedge mustard (*Sisymbrium altissimum*) germinated after having been buried ten years.

In 1928 an experiment was started at the New York State College of Agriculture to determine the longevity of the seed of the common weeds of New York. The seeds were buried under field conditions at a depth of 6 inches in sandy soil and clay soil. Seeds were taken up after having been buried one, two, four and five years and their viability was determined. The results of the first five years of this experiment are recorded in Table I. These results show that in all but three of the species tested many seeds were still able to germinate after having been buried in the soil for five years. This is longer than the period covered by the usual system of crop rotation. Chess (*Bromus secalinus*), corn cockle (*Agrostemma githago*) and hoary pepper-grass (*Lepidium campestre*), did not germinate after having been buried for one year.

Under natural conditions the soil of a field may become foul with weed seeds that have been plowed under. An accurate tabulation of all the weed seedlings appearing between June 1 and July 30, 1928, on an area 10 by 10 feet in an old clover field near Ithaca, New York, revealed 4,000 individual weeds, representing 28 families, 65 genera and 82 species. This is an unusually large number of species but the number of individuals is not unusual.

In 1930, 1 cubic decimeter soil samples were taken 1 decimeter below the soil surface from several counties in western New York where wild carrots were abundant. These soils had been undisturbed from three to ten years; nevertheless, every sample contained viable seeds of wild carrot as well as several other common weeds of the vicinity (see Table II).

c. Many weeds can mature seeds after they are removed from the soil. Purslane, galinsoga and chickweed may continue to bear flowers and to mature seeds for days afterwards. The tops of perennial sow

Dissemination and Importance of Weeds

thistle, dandelion and many of the mustards frequently ripen seeds from flowers and unripe seeds after they have been mowed off.

d. Some weeds ripen their seeds at the same time or just before the crop, among which they grow, is mature. Corn cockle, chess and spurry ripen about the same time as winter wheat. These weed seeds are shattered out before harvest or are harvested with the grain. Wild radish and Canada thistle frequently ripen about the same time as oats.

e. Some weed seeds are difficult to detect or to separate from crop seeds because they are very similar in size, shape and weight. Such seeds are often associated as impurities with certain crop seeds because they can be separated out only by special processes or equipment. Red clover seed frequently contains buckhorn plantain, night-blooming catchfly, dodder and green foxtail seeds. White clover and alsike clover seed often contain sheep sorrel and witch-grass.

f. The seeds or fruits of many weeds are provided with special structures which aid in dissemination. The seeds of milkweeds and the achenes of many weeds of the Composite family, such as the dandelion and sow thistle, are provided with hairy appendages so that they are scattered for long distances by the wind. The seeds of sandbur, burdocks, and cockleburs are surrounded by a covering with hooks or spines so that they can cling to animals and are carried from place to place.

The Reproduction of Weeds

One of the most important habits of weeds is their ability to reproduce their kind very freely. If this were not so, it would be much easier to get rid of weeds. The methods by which a weed reproduces must be taken into account in planning its eradication or control.

Weeds are generally classified into four groups: annuals, winter annuals, biennials and perennials. The annuals start from seed in the spring or summer, develop into mature plants, ripen their seeds and die in the same season. The winter annuals start from seed in autumn, develop a low rosette of leaves before winter, produce flowers and ripen their seeds in the following spring or summer and then die. Biennials start from seed in the spring and produce a rosette or crown of leaves from a somewhat fleshy tap-root in the first year. In the following spring the over-wintered tap-root sends forth a flowering shoot which dies, after the seeds are ripened, at the end of the second

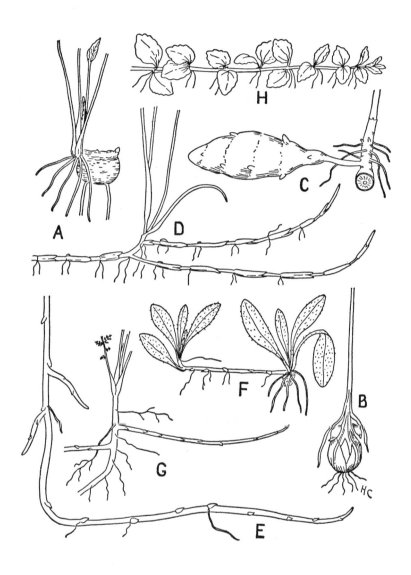

Fig. 1—Vegetative propagation of weeds by modified stems. **A**, corm, *Ranunculus bulbosus*. **B**, bulb, *Allium vineale*. **C**, tuber, *Helianthus tuberosus*. **D**, rhizome, *Agropyron repens*. **E**, rhizome, *Convolvulus sepium*. **F**, stolon, *Hieracium pratense*. **G**, rhizome, *Achillea millefolium*. **H**, creeping stem, *Veronica serpyllifolia*. Reduced to about ⅓ natural size.

Dissemination and Importance of Weeds

season. Perennial weeds may flower and produce seed in their first season from seed or more often not until one or more years later. The aerial parts of perennial weeds usually die but the underground parts may continue to live and send up a new growth forming flowering shoots and produce a crop of seeds, year after year.

In brief, then, this classification groups the weeds according to whether they live one season, one year, two years or more than two years. Unfortunately, this behavior is not constant but the duration of the weeds may be determined, to a large extent, by climatic factors. Many weeds that are annuals or biennials in very severe climates may act as biennials or perennials in milder climates or in mild winters.

An examination of 500 weeds described in this book shows that, according to their usual duration over the greater part of their range in the northern United States, 45 per cent (including 4.8 per cent shrubs) are perennial, 34 per cent annual, 7 per cent biennial. In addition, the duration of 8 per cent may be annual or winter annual; 4 per cent may be annual or biennial; and 2 per cent may be biennial or perennial (see Table III). This variation is probably largely due to the differences in the extreme climatic conditions between the various parts of the country in which weeds grow, or partly also to variations in the severity of certain seasons in a given locality. Concerning the seed production, this classification shows that most weeds produce seeds in summer or autumn, the annuals, winter annuals and biennials only once and the perennials during several years. As a matter of fact, with very few exceptions, all weeds produce seeds, the exact time of maturity depending on a number of conditions. In this respect most weeds are alike.

The impression is sometimes held that many weeds, especially those

TABLE III

DURATION OF FIVE HUNDRED WEEDS OF NORTHERN UNITED STATES

Duration	Number of Species	Percentage of Species
Annual	170	34
Annual or winter annual	40	8
Annual or biennial	20	4
Biennial	35	7
Biennial or perennial	10	2
Perennial (herbs)	201	40.2
Perennial (shrubs)	24	4.8
Total	500	100

that have depended on vegetative propagation for some time, have lost their ability to produce seed. An examination of the seed habits of the common weeds does not support this belief. Only a very few weeds in the northern United States do not produce seed. In this group belong those among the ferns and fern allies (Pteridophytes) which are non-seed plants, reproducing only by spores and by vegetative propagation. Among the seed plants a few cultivated forms which are propagated vegetatively have escaped as weeds in some localities; among these the orange day-lily (*Hemerocallis fulva*), and apparently the spearmint (*Mentha spicata*), and peppermint (*Mentha piperita*) have never been reported to produce seed in the United States. Several weeds such as the wild garlic (*Allium vineale*), and the leafy spurge (*Euphorbia lucida*), rarely produce seeds in the northern United States.

From the standpoint of eradication and control of weeds it is important to know not only how and when they reproduce by seed but also how they propagate and spread by other means. It is the ability of weeds to produce modified stems or roots capable of storing reserve foods and of producing new plants that enables them to resist many of the control treatments intended to kill them by mechanical injury or starvation. If weeds are examined as to their ability to propagate their kind by vegetative methods, they fall into four general groups as follows.

A CLASSIFICATION OF WEEDS

(Based on their usual method of vegetative reproduction)

1. Weeds in which neither roots nor stems develop new plants (no vegetative propagation).
   a. Those producing seeds during one season only.
      This group includes many annuals, most winter annuals and biennials.
   b. Those producing seeds during several seasons.
      This group includes such strong tap-rooted perennials as *Phytolacca americana*.
2. Weeds in which stems develop new plants (by formation of roots).
   a. Aerial stems develop roots.
      This group contains such annuals as *Galinsoga ciliata, Digitaria, Stellaria media,* and such perennials as *Lysimachia nummularia, Glechoma hederacea* and *Veronica serpyllifolia*.
   b. Subterranean stems develop roots.
      This group includes a number of perennials as *Agropyron repens* and *Convolvulus sepium*.

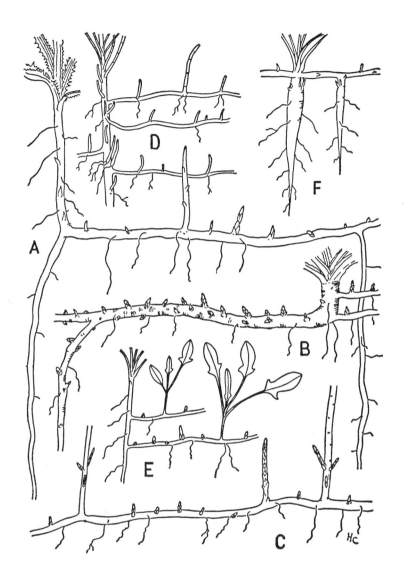

FIG. 2—Vegetative propagation of weeds by roots. **A**, *Cirsium arvense*. **B**, *Sonchus arvensis*. **C**, *Euphorbia esula*. **D**, *Linaria vulgaris*. **E**, *Rumex acetosella*. **F**, *Campanula rapunculoides*. Reduced to about ⅓ natural size.

3. Weeds in which roots develop new plants (by formation of shoots). This group includes such perennials as *Sonchus arvensis* and *Euphorbia esula*.
4. Weeds in which both stems and roots may develop new plants. This group contains such perennials as *Convolvulus arvensis*, *Cirsium arvense*, *Rorippa austriaca* and *Rorippa sylvestris*.

THE DISSEMINATION OF WEEDS

Most weeds are good travelers. They have no power of locomotion but use various forces or agents to transport and scatter them from place to place. Of all the agents by which weeds are disseminated, man, wind, water and animals play the most important roles.[4] These agents are so effective that many weeds have been scattered to parts of the world where they did not occur formerly. Man frequently carries the seeds of weeds across natural barriers such as large bodies of water, mountains or deserts. After the weeds have become established in a new locality, their spread is aided by the wind, water and animals. Most weeds have modifications of some kind which adapt them for dissemination by one or more of these agents.

Ornamentals escaping.

A number of weeds, more or less common today in many sections of the United States, were originally introduced as ornamentals or for other uses, were well taken care of in gardens or fields until they became established and then were allowed to escape. The following example, taken from many that might be mentioned, illustrates that this is still going on. In 1919 the writer was asked to view a highly prized and rare introduced plant growing in a choice location among other rare exotic species in a rock-garden in western Washington. The plant was the tall field buttercup (*Ranunculus acris*). In 1929 the farmers in the immediate vicinity were speculating as to the source of the yellow-flowered weed which was over-running their pastures. In 1932 a botanist reported *Ranunculus acris* as an addition to the flora of Washington. In 1933 lowland pastures, covering many square miles, were yellow with this new pest, unknown in the region only a few years previously.

Common weeds that were originally introduced into the United States as ornamental or other "useful" plants are listed in Table IV.

[4] Ridley, H. N., 1930, in his comprehensive treatment on "The dispersal of plants throughout the world," records many data that are of interest in the study of weed dissemination.

Dissemination and Importance of Weeds 15

TABLE IV
SOME INTRODUCED CULTIVATED PLANTS THAT HAVE ESCAPED AND BECOME WEEDS IN SOME SECTIONS OF THE UNITED STATES

ANNUALS

o		488	Centaurea cyanus—	Bachelors-button
o		121	Chenopodium botrys—	Jerusalem-oak
o		360	Ipomoea purpurea—	Morning-glory
o		133	Kochia scoparia—	Summer-cypress
o		404	Nicandra physalodes—	Apple-of-Peru
v		219	Raphanus sativus—	Wild radish

BIENNIALS

h		321	Carum carvi—	Caraway
d		181	Chelidonium majus—	Celandine
d		323	Conium maculatum—	Poison-hemlock
o	d	417	Digitalis purpurea—	Foxglove
v		327	Pastinaca sativa—	Wild parsnip

PERENNIALS

h		473	Artemisia absinthium—	Wormwood
o		479	Bellis perennis—	English daisy
o	s	179	Berberis vulgaris—	Barberry
o		460	Campanula rapunculoides—	Bellflower
v		120	Chenopodium bonus-henricus—	Good King Henry
v		499	Cichorium intybus—	Chicory
o	s	257	Cytisus scoparius—	Scotch broom
o		292	Euphorbia cyparissias—	Cypress spurge
o	s	259	Genista tinctoria—	Dyers broom
o		305	Hibiscus trionum—	Flower-of-an-hour
d		536	Inula helenium—	Elecampane
o		456	Knautia arvensis—	Field scabious
d		388	Leonurus cardiaca—	Motherwort
o		453	Lonicera japonica—	Japanese honeysuckle
o		334	Lysimachia nummularia—	Moneywort
o		306	Malva moschata—	Musk mallow
h		391	Marrubium vulgare—	Horehound
h		395	Mentha spicata—	Spearmint
h		396	Nepeta cataria—	Catnip
o		383	Glechoma hederacea—	Ground ivy
o		83	Ornithogalum umbellatum—	Star-of-Bethlehem
o		405	Physalis alkekengi—	Chinese lantern-plant
o		95	Polygonum cuspidatum—	Japanese knotweed
o	s	244	Rosa eglanteria—	Sweet brier
h		105	Rumex acetosa—	Sour dock
o		156	Saponaria officinalis—	Bouncing Bet
o		229	Sedum acre—	Stonecrop
o		230	Sedum purpureum—	Live-for-ever
d		376	Symphytum officinale—	Comfrey
d		563	Tanacetum vulgare—	Tansy
o		429	Veronica filiformis—	Veronica

```
        d = drug plant       o = ornamental
        h = herb             v = vegetable
          o s = ornamental shrub
```

Impurities in agricultural seeds.

The ease with which weeds are scattered with crop seeds has been recognized for a long time. Over a century ago Schweinitz [5] had already reported more than fifty of the common weeds treated in this book as introduced fortuitously with agricultural seeds in New York. Because of the similarity of the habits of certain weeds and crop plants, they grow together and, if they have similar seeds, such weeds frequently become associated with certain crops. Agricultural seeds such as the grasses, clovers, alfalfa, and to a less extent cereals, frequently contain weed seed impurities. The seeds of some weeds have about the same size, shape and weight as some crop seeds so that it is difficult to separate them without special machinery. Seeds of the buckhorn plantain, field pepper-grass and night-flowering catchfly are commonly found in cheaper grades of alfalfa or red clover seed. Sheep sorrel seed is a common impurity in white clover seed. Such weeds are scattered widely and occur wherever conditions are favorable for growing clovers or alfalfa.

Nearly every state now has a "seed laboratory" in which one or more seed analysts are engaged in analyzing seeds for impurities and testing them for germination. Farmers and seed-growers may have seed samples analyzed free or for a small fee. Most states now require that seeds offered for sale be tagged with a label indicating, among other things, the names of noxious weed seeds contained in them (see pages 31-32).

Hay and feed-stuffs.

Many mixed and ground feeds contain weed screenings separated from wheat or other grains prior to milling. These weed seeds are mixed with bran or other feeds with or without grinding. The grinding process usually destroys only the larger weed seeds. Many of the smaller sorts escape injury from the grinding and, unless the feed is very finely ground or treated by heating, the seeds may retain their viability. Certain coarsely ground or mixed feeds have been found to contain as many as 1,000 to 20,000 viable weed seeds to a pound.[6, 7, 8, 9] Many screenings are heated to destroy the viability of weed seeds contained in them before mixing with other feeds. The

---

[5] Schweinitz, 1832. [6] Hills and Jones, 1907. [7] Jenkins, 1909.
[8] Beach, 1908. [9] Korsmo, 1930.

weed seeds most prevalent in mixed and ground feeds are those of species commonly growing in grain fields. A list of these is given on page 45.

Ballast from freight cars and boats.

The effectiveness of this means of spreading seeds for long distances is illustrated by an examination of the ballast grounds of any large port or yard where freight cars are cleaned. In such a locality one can find plants from almost any part of the world with which it has been in direct communication. To be sure, most of the seed immigrants on their arrival do not find the climatic or soil conditions favorable so that they never develop into mature plants and, being unable to reproduce, fail to establish themselves and soon disappear. However, seeds that have been brought to a new region from other parts of the world where conditions were similar often establish themselves and spread. They become naturalized and may even drive out some of the native plants.

Threshing machines and hay balers.

Such equipment as threshing machines and hay balers, which are frequently used by several farmers in a community, if moved from place to place without thorough cleaning, sometimes scatter seeds of certain weeds from one farm to another. In certain states regulations governing the operation of threshing machines require that the operator remove all weed seeds from the machine before it is taken from the farm.

Cultivating implements.

Often a noxious perennial makes its first appearance on a farm or in a field as a single plant or as a small patch. If such weeds are prevented from producing seed and the soil is left undisturbed, they may spread very little from year to year. However, if the isolated patch is plowed and harrowed, the roots or rhizomes of such weeds may be scattered through a much larger area or throughout an entire field. Canada thistle, leafy spurge, bindweed and perennial sow thistle are sometimes held in check by frequent mowing to prevent seed formation in a limited area until it is plowed and dragged. Then suddenly, instead of one infestation, there are many which need to spread very little to overrun the entire field.

Manure.

Before the days of the automobile, the common practice of hauling manure from the cities to farms provided a continuous introduction of new weed seeds. Unless the manure was thoroughly composted before spreading on the soil, a variety of weeds, limited only by what went into the manure, could always be expected.[10] This is not now such an important source of weed infestations.

Packing materials.

Hay and straw have been used extensively for packing materials. Since nearly always the poorer grades are employed, they often contain weed seeds. Certain states now have laws prohibiting the use of such packing materials as a precaution against spreading agricultural pests. Nursery materials from weedy land, especially trees with "balls of soil" around them, may be a means of scattering weeds to new areas. Not only seeds but pieces of roots and rhizomes may be transported with the soil. Quack-grass, sow thistle, bindweed, yellow cress and cypress spurge have been known to make their appearance in certain localities with the soil about nursery stock.

Wind as an agent of dissemination.

Many weeds have modifications or adaptations which aid them in becoming scattered by the wind.[11, 12] Seeds or small fruits with tufts of hair or wing-like appendages are carried by the wind over long distances. The achenes of dandelions or thistles floating in the air on a windy day are a common sight when these weeds are shedding seeds. When the milkweed pod bursts open, its seeds, each provided with a tuft of hair, are blown far and wide by the slightest breeze. The "tumble weeds," of which the Russian thistle and winged pigweed are examples, have a much branched and spreading stem which breaks loose from the soil and the whole plant rolls or tumbles over the fields, scattering seeds as it goes. A few weeds, such as the brake fern and horsetail ferns and other Pteridophytes, reproduce by minute spores which are carried by the wind.

Water as an agent of dissemination.

Many weed seeds are light or are covered with an oily film so that they float on the surface of water. Such seeds are frequently washed

[10] Oswald, 1908.   [11] Bolley, 1895.   [12] Ridley, 1930.

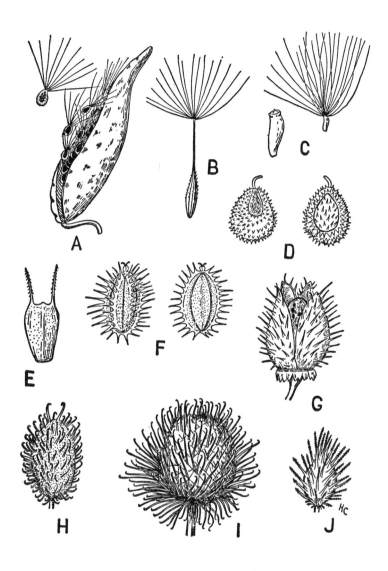

FIG. 3—Fruits and seeds of some weeds showing modifications for dissemination. A, *Asclepias syriaca,* reduced to ½ natural size. B, *Taraxacum officinale* (× 2). C, *Cirsium arvense* (× 1 and 4). D, *Cynoglossum officinale* (× 2). E, *Bidens vulgata* (× 2). F, *Daucus carota* (× 4). G, *Datura stramonium* (× ½). H, *Xanthium orientale* (× 1). I, *Arctium lappa* (× 1). J, *Cenchrus pauciflorus* (× 2).

into streams by surface run-off during heavy rains or they are picked up by overflowing streams and are carried to other fields lower down the valley. The various mustards, winter-cress, wild parsnips, sedges, cocklebur, ragweeds and marsh elder are a few examples of weeds whose seeds are frequently carried long distances by water without losing their viability.

In the irrigated regions of the western states, one of the most important agents for scattering weed seeds is the irrigation water. Rank weeds are often allowed to grow along the banks of canals. They ripen seeds and shed them into the water to be carried down stream and through the laterals to conditions ideal for their growth. Investigations have shown that irrigation ditch banks are sometimes a more prolific source of weed infestation than roadsides. Several million weed seeds have been found to pass a given point on a twelve-foot ditch within twenty-four hours.[13] In Colorado (1935) the weed seeds most commonly found in irrigation water were *Amaranthus graecizans, A. retroflexus, Chenopodium album, Polygonum convolvulus, P. persicaria, Rumex crispus, Taraxacum officinale* and *Iva xanthifolia*. The prevention or removal of weeds along irrigation ditches by grazing or mowing will keep large numbers of weed seeds from being distributed and thus reduce the weed problem on the irrigated fields.

Animals as agents of dissemination.

Many weed seeds are disseminated by animals. The burs, such as burdocks and cockleburs, and fruits or nutlets with hook like or spiny appendages, such as beggar-ticks, stick-tights and houndstongue, may be carried about by clinging to the wool or fur of larger animals. Seeds are carried short distances by rodents and ants, and are stored for future use, where they are sometimes left to germinate.

Many birds eat the seeds or small fruits of weeds. The seeds with hard coats may be expelled in a viable condition after passing through the digestive tract or they may be regurgitated in pellets. Poison ivy, poison sumac, choke-cherries and barberries are frequently scattered in this manner. Small weed seeds sticking to the feathers, or with mud to the feet of birds, may be carried from place to place, although this method of dissemination does not seem as important as was formerly believed.

Domestic animals feeding on hay or coarse feed containing weed

[13] Egginton and Robbins, 1920.

seeds usually pass many of the smaller ones in an undigested form capable of germinating.[14, 15, 16, 17] If manure from such animals is spread on the soil without first composting it for two or three months, many of these weed seeds will grow.[18] Table V shows that when a number of weed seeds common in weedy hay or coarsely ground grain were fed to a horse, cow, swine and sheep, with the exception of the larger sorts, some seeds of most of the species passed through the alimentary tract without losing their viability.[19]

The Sources of Weeds

Many of the common weeds of North America were brought from other parts of the world. Josselyn, as early as 1672, reported that within a few years after the English colonists had settled in New England such introduced weeds as quack-grass, dandelion, shepherds-purse, sow thistle, knotweed, chickweed, docks, plantains, groundsel, mullein and others had made their appearance and were flourishing. The subsequent westward migration of these weeds has been almost unlimited except in those areas in which climatic or soil conditions were unfavorable.

The opinion is frequently held that noxious weeds have come from the Old World where they have grown in cultivated land so long that they have become adjusted or adapted to such an environment and are, therefore, more noxious and more persistent than native species which have not had this long period for adaptation. In those regions of North America which were originally covered by forests, most of the noxious weeds are introduced. However, where the original vegetation consisted of the open grassland, prairie or plains type, many native species have persisted and spread as weeds. When the timber was removed from the forested regions and the land was cleared for agricultural purposes, usually only a few weeds were present to contend with during the first few years. The changing of the forest to an open field produced a new environment. The native woodland species that were accustomed to the reduced sunlight, moist humus cover, higher atmospheric humidity and other factors and conditions associated with a woodland environment, usually were unable to persist very long after the land had been cleared. Soon,

[14] Korsmo, 1930.
[16] Ridley, 1930.
[18] Oswald, 1908.
[15] Atkeson, Hulbert and Warren, 1934.
[17] Harmon and Keim, 1934.
[19] Lindsey, 1933.

TABLE V
The Effect of Passing Through the Digestive Tract of Some Domestic Animals on the Viability of Weed Seeds

Species	Horse	Cow	Swine	Sheep
Agropyron repens	+	+	−	+
Bromus secalinus	−	−	−	−
Digitaria sanguinalis	−	+	+	−
Panicum capillare	+	+	+	+
Setaria viridis	+	+	+	+
Polygonum convolvulus			−	−
Polygonum persicaria	+		+	−
Rumex crispus	+	+	+	+
Chenopodium album	+	+	+	+
Amaranthus retroflexus	+	+	+	+
Silene noctiflora	+	+	+	+
Spergula arvensis	−		+	+
Stellaria media	+	+	+	+
Portulaca oleracea	−	−	+	+
Ranunculus acris	+	+	+	+
Barbarea vulgaris	+	+	+	+
Brassica kaber	−	+	+	+
Erucastrum gallicum	+	+	+	+
Raphanus raphanistrum	−	−		
Potentilla recta	+	+	+	+
Medicago lupulina	+	−		
Euphorbia cyparissias	−	−	−	−
Hypericum perforatum	+	+		
Oenothera biennis	−	+		
Daucus carota	+	−	−	−
Lithospermum officinale			−	−
Verbena hastata	−	+		
Nepeta cataria	+	+	+	+
Solanum carolinense	+	+	+	+
Verbascum blattaria			+	+
Verbascum thapsus	−	+		
Plantago indica	+	+	+	+
Plantago major	−	+		
Galium aparine	−	−	−	−
Dipsacus sylvestris	+	+	+	+
Lobelia inflata	−	+	+	+
Achillea millefolium	+	+	+	+
Chrysanthemum leucanthemum	+	+	+	+
Cirsium arvense	−	−	+	+
Taraxacum officinale	−	−	+	+

+ = germinated after passing.
− = did not germinate after passing.
blank = not fed to that animal.

however, exotic weeds, species native to other parts of the world where the original habitat was similar to that produced by the removal of the forest, arrived, established themselves and frequently multiplied and spread rapidly. The explanation of the large number of introduced species of weeds in certain parts of North America may be found, at least in part, in the change of environment from woodland to open fields, rather than by assuming that the introduced species have had a long time to change their habits and become "adapted' to cultivated land.

The prairie farmer who has succeeded in establishing a woodlot has observed that the native plants accustomed to the open prairie soon disappear and woodland species accustomed to the new environment arrive, establish themselves and thrive. Likewise, it might be expected when a woodland is changed to an open field, the most successful species would be those which have come from regions whose environment corresponds most nearly to the new.

The sources of five hundred weeds treated in this book (mostly those occurring in the northern United States) are indicated in the treatment of each species, and tabulated by continents in Table VI.

TABLE VI

THE SOURCE OF FIVE HUNDRED WEEDS OF THE NORTHERN UNITED STATES

Origin	Number of Species	Percentage of Species
Native to North America	196	39.2
Widespread	51	10.2
Eastern	95	19
Western	42	8.4
Southern United States	8	1.6
Tropical America	15	3
North America and Europe	13	2.6
North America and Eurasia	16	3.2
Europe	177	35.4
Asia	12	2.4
Eurasia	66	13.2
Africa and Eurasia	3	.6
Doubtful	2	.4
Total	500	100

The place of origin indicated is the continent, or in the case of North America the section, in which each species is considered native.

Losses Caused by Weeds

Reduction in yield of crops.

Weeds compete with crops for water, nutrients and light. Like the crop plants among which they grow, they require water, mineral nutrients and sunlight. If any one of these factors is lacking or reduced below a certain minimum, neither crops nor weeds can survive. In many fields the crop plants must compete with the weeds for water, mineral nutrients and sunlight.[20] The weeds are usually better adapted to obtain the larger share of whatever water is available and of the fertilizers supplied so that, unless they are destroyed, they soon outgrow the crops and shade them. The more water and fertilizers are used by weeds, the less remains for the crop plants which become stunted and do not yield as well as if growing without the competition of weeds.

Some weeds, such as the dodders, grow as parasites upon the stems of such crops as clover and alfalfa. These weeds have no chlorophyll and no true roots so that they are dependent on the host plant, which they rob for their nourishment.

Many weeds, especially those that have climbing or twining stems, pull down certain crops so as to cause a loss by ordinary methods of harvesting. Grains are sometimes caused to lodge so that much of the crop is destroyed or lost by ordinary methods of harvesting.

Increase of cost of operation of farms.

Tillage operations, such as plowing, harrowing, disking and cultivating, are time-consuming and expensive. They often represent from one-tenth to one-third of the total value of the crop produced. Much of this outlay could be avoided if it were not necessary to control weeds to permit the growth of crop plants. Cultivation experiments [21] with a number of vegetable crops at Ithaca and on Long Island,[22] New York, indicated that the chief advantage derived from cultivation was through the control of weeds. Tests conducted in several states show that the chief benefit from cultivation derived by corn plants is the removal of weeds.[23] If weeds were not present, most of the cultivation could be eliminated without reducing the yield of crops.

[20] Pavlychenko and Harrington, 1934.   [21] Thompson, 1927.
[22] Thompson, Wessels, and Mills, 1931.   [23] Cates and Cox, 1912.

Dissemination and Importance of Weeds 25

Grain crops sometimes contain a large bulk of weeds such as mustards, wild radish or smartweeds. The additional cost of harvesting and threshing these weeds increases the cost of materials and labor required for the production of each bushel of grain, so that the crops may yield a loss instead of a profit. Factory peas sometimes contain so many weeds as to interfere with the proper cutting and handling. Potato fields are sometimes so weedy that diggers cannot be used effectively. Under these conditions, weeds not only injure the crop but also place an added wear and tear on machinery and increase the power required to operate it.

Certain weeds, such as the smartweeds, remain green longer and dry much more slowly than the grain with which they grow. When such weeds are present in grain they delay the time required for its drying and curing. If such grain is stacked it usually heats and delays the threshing. If threshing is done in the field, unripe weed seeds in the grain may cause it to heat in storage and to become moldy or unfit for milling or seed purposes. The common live-for-ever of meadows delays the curing of hay and remains green and moist long after the hay is cured, causing it to mold in the shock or even in the hay-mow.

In some farming sections the climate, soil and marketing facilities make it advantageous to follow a certain system of crop rotation in order to obtain the greatest return for the labor and at the same time maintain the soil fertility. The appearance of certain persistent perennial weeds, such as quack-grass, wild morning-glory or perennial sow thistle, frequently makes it necessary to modify such a planned rotation or abandon it entirely for other practices designed primarily to eradicate, or at least control, these noxious weeds.

Certain weeds grow in ditches, especially at times when the water is low. Such weeds frequently clog ditches so that they must be cleaned out at least once a year. Other weeds grow along the banks of ditches and fall into them and are carried from field to field at times of high water. The roots of some weeds stop up drain tiles.

Injurious effects of weeds.

A large number of weeds act as hosts for fungi, bacteria and mosaics which are also transferred to crop plant hosts. In some diseases the same causal organism attacks both crop plants and closely related weeds. A number of the parasitic fungi attack certain crop plants and spend a part of their life cycle on botanically unrelated weed hosts. In either case the presence of the weed hosts makes it much more difficult to control crop diseases and to reduce the losses caused by them. Table VII gives examples of common diseases attacking both weeds and crop plants.

TABLE VII

SOME COMMON WEED HOSTS WHICH HARBOR DISEASES OF COMMON CROP PLANTS

Weed Host	Disease	Cause	Crop Host
Brassica kaber—Field mustard	Club-root	Plasmodiophora brassicæ	Cabbage
Agropyron spp.— Quack-grass Avena fatua—Wild oats	Black stem-rust	Puccinia graminis	Wheat, oats, barley
Agropyron spp. Quack-grass	Ergot	Claviceps purpurea	Rye, barley
Berberis vulgaris— European barberry	Black stem-rust	Puccinia graminis	Wheat, oats, barley
Potentilla spp.	Leaf-spot	Mycosphærella fragariæ	Strawberry
Solanum carolinense —Horse nettle	Septoria blight	Septoria lycopersici	Tomato
Atriplex spp. Chenopodium spp.	Spinach mildew	Peronospora effusa	Spinach, Swiss chard
Daucus carota—Wild carrot	Carrot blight	Macrosporium carotae	Carrot
Solanum spp.	Verticillium wilt	Verticillium albo atrum	Eggplant
Galinsoga ciliata Rudbeckia serotina Veronica peregrina	Yellows	Virus	Lettuce, salsify, aster
Phytolacca americana —Pokeweed Physalis spp.— Groundcherry Asclepias syriaca— Milkweed Nepeta cataria—Catnip Leonurus cardiaca— Motherwort	Cucumber mosaic	Mosaic	Cucumber
Physalis spp.— Groundcherry Solanum spp.— Nightshade	Tomato mosaic	Mosaic	Tomato
Solanum spp.— Nightshade	Potato mosaic	Mosaic	Potato

Dissemination and Importance of Weeds

Many common pests of crops live on weeds during certain stages of their development or until crop plants are available. Some insect pests attack many different kinds of weeds, but others live on a very limited number, usually those species closely related to the crop plants which they attack. Table VIII records some examples of common insect pests attacking both weeds and crop plants.

TABLE VIII
SOME COMMON WEED HOSTS WHICH HARBOR INSECTS OF COMMON CROP PLANTS

Weed Host	Insect Disease	Organism	Crop Host
Brassica spp.—Mustards	Cabbage root-maggot	Hylemyia brassicæ	Cabbage, cauliflower, radish, turnips
Brassica spp.—Mustards	Spinach aphis	Myzus persicæ	Spinach, eggplant, cabbage, etc.
Echinocystis lobatus—Wild cucumber	Striped cucumber beetle	Diabrotica vittata	Cucumber, melons, squashes
Sedum purpureum—Live-for-ever	Melon aphis	Aphis gossypii	Melons, cucumber, squash
Stellaria media—Chickweed			
Plantago spp.—Plantain			
Brassica spp.—Mustard			
Convolvulus sepium—Wild morning-glory			
Daucus carota—Wild carrot	Carrot rust-fly	Psila rosæ	Carrot, parsnip, celery
Pastinaca sativa—Wild parsnip			
Daucus carota—Wild carrot	Carrot weevil	Listronotus latiusculus	Carrot
Daucus carota—Wild carrot	Tarnished plant-bug	Lygus pratensis	Celery, dahlia, nursery stock
Ambrosia artemisiifolia—Ragweed			
Amaranthus retroflexus			
Plantago spp.—Plantain	Rosy apple aphis	Aphis Sorbi	Apple
Rumex crispus—Dock	Rhubarb curculio	Lixus concavus	Rhubarb
Echinocystis lobatus—Wild cucumber	Squash vine-borer	Melittia satyriniformis	Squashes, cucumber, pumpkin, melon
Solanum spp.—Nightshades	Potato stalk-borer, also Colorado potato beetle	Trichobaris trinotata	Potato
Physalis subglabrata		Leptinotarsa decimlineata	Potato
Datura stramonium			

TABLE VIII—*Continued*

SOME COMMON WEED HOSTS WHICH HARBOR INSECTS OF COMMON CROP PLANTS

Weed Host	Insect Disease	Organism	Crop Host
Ambrosia spp.—Ragweed Brassica spp.—Mustards Solanum spp.—Nightshade	Onion thrips	Thrips tabaci	Onion
Phytolacca americana—Pokeweed Solanum spp.	Greenhouse red-spider mite	Tetranychus tetaricus	Rose, dahlia, carnation, etc.
Arctium—Burdock Cirsium—Thistle Ambrosia—Ragweed	Burdock borer	Papaipema cataphracta	Tomato, corn, dahlia, aster
Ambrosia spp.—Ragweed Arctium spp.—Burdock Chenopodium spp.—Lambs-quarters	Common stalk-borer	Papaipema nitela	Tomato, corn, dahlia, aster, lily, peony, etc.

Several common weeds are poisonous when eaten by stock. These contain poisonous compounds or compounds which under certain conditions yield poisonous substances. The losses of livestock, due to eating poisonous weeds, are relatively few in the eastern United States; but in certain of the western states, where natural vegetation is utilized for grazing to a much greater extent, losses are sometimes very considerable. A few weeds, like the poison ivy, are poisonous to the touch and are the source of considerable discomfort to many individuals. Table IX records some weeds which are poisonous, at least under certain conditions.

TABLE IX

POISONOUS WEEDS AND THE PARTS THAT MOST COMMONLY CAUSE POISONING

NAME	INJURIOUS PART
1. Equisetum arvense—Field horsetail	Tops
2. Equisetum hyemale—Scouring-rush	Tops
3. Equisetum palustre—Horsetail	Tops
4. Equisetum sylvaticum—Wood horsetail	Tops
8. Pteridium aquilinum—Brake fern	Leaves and stems
41. Holcus lanatus—Velvet-grass	Leaves and stems
42. Hordeum jubatum—Squirrel-tail-grass	Awns cause mechanical injury
46. Lolium temulentum—Darnel	Seeds containing fungus
61. Sorgum halepense—Johnson-grass	Tops
83. Ornithogalum umbellatum–Star-of-Bethlehem	Entire plant

Dissemination and Importance of Weeds

TABLE IX—Poisonous Weeds—Continued

NAME	INJURIOUS PART
84. Veratrum viride—American hellebore	Tops and rootstocks
85. Zigadenus venenosus—Meadow death camas	Bulbs and tops
88. Urtica dioica—Stinging nettle	Hairs on leaves and stems cause irritation
89. Urtica gracilis—Stinging nettle	Hairs on leaves and stems cause irritation
144. Phytolacca americana—Pokeweed	Entire plant, mainly the roots
147. Agrostemma githago—Corn cockle	Seeds
156. Saponaria officinalis—Bouncing Bet	Entire plant
157. Saponaria vaccaria—Cow cockle	Entire plant
170. Delphinium menziesii—Low larkspur	Tops
171. Delphinium strictum—Larkspur	Tops
172. Ranunculus abortivus—Small-flowered buttercup	Fresh tops
173. Ranunculus acris—Tall field buttercup	Fresh tops
175. Ranunculus bulbosus—Bulbous buttercup	Fresh tops
176. Ranunculus repens—Creeping buttercup	Fresh tops
177. Ranunculus sceleratus—Cursed crowfoot	Fresh tops
181. Chelidonium majus—Celandine	Tops
192. Brassica kaber—Wild mustard	Seeds
243. Prunus virginiana—Choke-cherry	Wilted leaves
256. Crotalaria sagittalis—Rattle-box	Tops
264. Lupinus spp.—Lupines	Tops and seed pods
291. Euphorbia spp.—Spurges	Tops
302. Rhus radicans—Poison ivy	Leaves, bark, etc., poisonous to touch
303. Rhus vernix—Poison sumac	Leaves, bark, etc., poisonous to touch
309. Hypericum perforatum—St. Johns–wort	Tops
322. Cicuta maculata—Water-hemlock	Roots and tops
323. Conium maculatum—Poison-hemlock	Roots and tops
327. Pastinaca sativa—Wild parsnip	Leaves poisonous to touch
330. Kalmia angustifolia—Sheep laurel	Leaves
331. Kalmia latifolia—Mountain laurel	Leaves
332. Kalmia polifolia—Swamp laurel	Leaves
333. Anagallis arvensis—Scarlet pimpernel	Tops
343. Cynanchum nigrum—Black swallow-wort	Tops
344. Cynanchum vincetoxicum—White swallow-wort	Tops
403. Datura stramonium—Jimson-weed	Seeds, unripe fruits and leaves
408. Solanum carolinense—Horse nettle	Tops and berries
409. Solanum dulcamara—European bittersweet	Tops and berries
411. Solanum nigrum—Black nightshade	Tops and berries
412. Solanum rostratum—Buffalo bur	Tops and berries
417. Digitalis purpurea—Foxglove	Leaves and stem
462. Lobelia inflata—Indian tobacco	Tops

TABLE IX—Poisonous Weeds—*Continued*

NAME	INJURIOUS PART
517. Eupatorium rugosum—White snakeroot	Leaves
524. Helenium autumnale—Sneezeweed	Leaves and stems
525. Hellenium nudiflorum—Sneezeweed	Leaves and stems
526. Helenium tenuifolium—Bitterweed	Leaves and stems
570. Xanthium orientale—Cocklebur	Germinating seeds and very young seedlings

A number of common weeds possess hard sharp structures such as awns, spines or thorns, which produce mechanical injury to horses, cattle and sheep.[24, 25] The wounds in and about the mouth and about the eyes of an animal not only may be painful but may act as channels through which bacterial infections may start, often resulting in swelling and pus formation. Such animals may starve as a result of blindness or because their eyes are swollen shut so that they cannot find forage. These structures may also produce serious wounds or obstructions in the alimentary tract. Some of the more common mechanical injury-producing weeds are wild oats (*Avena fatua*), downy brome-grass (*Bromus tectorum*), sandbur (*Cenchrus longispinus*), squirrel-tail-grass (*Hordeum jubatum*), Russian thistle (*Salsola kali*), puncture vine (*Tribulus terrestris*) and species of Stipa, Aristida, Xanthium, Cirsium and Centaurea.

Several common weeds of pastures and meadows have a very strong flavor which is imparted to milk from animals feeding upon them. Such dairy products always have a reduced value and often cannot be disposed of at any price. The more common weeds producing undesirable flavors in dairy products are wild garlic (*Allium vineale*), wild onion (*A. canadense*), wild leek (*A. tricoccum*), all found in the eastern states, and various other wild onions native in the western states. Wild carrot, ragweeds, marsh elder (*Iva axillaris*) and several mustards are very widespread. In the southern states bitterweed (*Helenium tenuifolium*), and in the northwestern states and western Canada, fanweed (*Thlaspi arvense*), are commonly the cause of undesirable flavors in dairy products.

In the arid regions where irrigation is practiced, many common weeds find their most favorable conditions for growth along the canals and laterals through which the water is supplied to the crops. In such places weeds often interfere seriously with the proper distribution

[24] Fleming and Peterson, 1919.   [25] Pammel, 1929.

Dissemination and Importance of Weeds 31

of the water. Additional labor and expense is required to clear the weeds from the ditches and their banks in order to provide a proper flow of water and to prevent their seeds from falling in and becoming scattered over the fields.

The presence of weeds in agricultural seeds decreases their sale value. Timothy, clover, alfalfa, and other seeds sometimes contain so many weed seeds that they command a very low price or cannot be marketed at all unless they are first put through an expensive cleaning process. Every state in the Union now has a law regulating, to some extent at least, the sale of agricultural seeds. Each of these laws contains a provision that defines certain weeds as "noxious weeds." In most states the noxious weeds are grouped into two categories: (a) primary noxious or prohibited weeds and (b) secondary noxious weeds. The most commonly prohibited weed seeds, listed in the order of the number of states prohibiting them, are as follows: *Convolvulus arvensis, Cirsium arvense, Agropyron repens, Cardaria draba* including var. *repens, C. pubescens, Centaurea picris, Euphorbia esula, Sonchus arvensis, Cyperus rotundus, Sorgum halepense, Alhagi camelorum* and *Solanum elaeagnifolium.*

The state laws permit secondary noxious-weed seeds in agricultural and vegetable seed provided the label shows the name and the rate of occurrence, on a pound or ounce basis, of each secondary noxious-weed seed. The number of permitted secondary noxious-weed seeds is frequently limited. The seeds classed as noxious in the various states are changed from time to time by legislative action so that no table of them is included in this revision. The 1935 edition of "Weeds" included a table of 33 species considered noxious in two or more states and listed 28 additional weeds which were declared noxious by only one state. In January 1954, there were a total of approximately 150 species declared noxious in at least one state— over 80 being declared noxious in two or more states.

The following species have been declared noxious in five or more states as of January 1954: *Agropyron repens, Agrostemma githago, Alhagi camelorum, Allium canadense, A. vineale, A.* spp., *Ambrosia psilostachya, Avena fatua, Brassica hirta, B. juncea, B. kaber, B. nigra, B.* spp., *Bromus secalinus, Cardaria draba, C. pubescens, Centaurea picris, C. solstitialis, Chrysanthemum leucanthemum, Cirsium arvense, Convolvulus arvensis, C. sepium, C.* spp., *Cuscuta epilinum, C. epithymum, C. indecora, C. pentagona, C. planiflora,*

C. suaveolens (*C. racemosa* var. *chiliana*), *C.* spp., *Cynodon dactylon*, *Cyperus esculentus*, *C. rotundus*, *Daucus carota*, *Euphorbia esula*, *Franseria discolor*, *Helianthus ciliaris*, *Hypericum perforatum*, *Iva axillaris*, *Lactuca pulchella*, *Lolium temulentum*, *Oryza sativa*, *Plantago aristata*, *P. lanceolata*, *Raphanus raphanistrum*, *Rorippa austriaca*, *Rumex acetosella*, *R. altissimus*, *R. crispus*, *R. obtusifolius*, *R.* spp., *Solanum carolinense*, *S. elaeagnifolium*, *S. rostratum*, *Sonchus arvensis*, *Sorgum halepense*, *Thlaspi arvense*, *Tribulus terrestris*.

Of the 33 weeds considered noxious in 1935 in several states only one, *Cichorium intybus*, is no longer regarded as noxious. Ten of those declared noxious at that time by one state were not so regarded by any state in 1954. An equal number, however, is now regarded as noxious by several states. For instance, *Solanum elaeagnifolium* was then considered noxious by California and is now outlawed by 20 states; and *Helianthus ciliaris* is now outlawed by 12 states instead of only Texas as was the case in 1935.

In August 1939, a Federal Seed Act was passed to regulate interstate and foreign commerce in seed. The act defines the following as noxious in imported seed: *Cardaria draba*, *C. draba* var. *repens*, *C. pubescens*, *Cirsium arvense*, *Cuscuta* spp., *Agropyron repens*, *Sorgum halepense*, *Convolvulus arvensis*, *Centaurea picris*, *Sonchus arvensis*, *Euphorbia esula*. The act does not designate the species of weeds that shall be considered as noxious in seed moving in interstate commerce but accepts, for this purpose, the noxious weeds and other labeling requirements of the particular states into which the seed is shipped.

BENEFITS DERIVED FROM WEEDS .

After a consideration of the harmful characteristics of weeds, it may not be out of place to take up briefly those attributes of weeds which have been beneficial.

Weeds, when plowed under or returned to the soil in some other way, add some humus and nutrients. Most weeds contain considerable nitrogen and ash constituents,[26, 27] but since the weeds originally obtained the minerals from the soil, they really return only what they took and do not add to the soil fertility. If the soil did not support weeds, they would not draw anything from it. A weed crop for one or

[26] Ince, 1915.   [27] Campbell, 1924.

Dissemination and Importance of Weeds

two years proved more beneficial than a legume crop or bare fallow on the yield and acre value of tobacco crops in parts of Maryland.[28] Weeds sometimes form a cover which is useful in that it may prevent the soil from being blown about by wind or eroded by water from rain or floods. This is especially important in regions of sudden downpours on sloping soils or of strong winds on dry sandy soils. However, except for abandoned fields, more desirable grasses or other soil-binding plants might be substituted to serve this purpose even better.

A number of weeds furnish forage for animals when more palatable plants are scarce. This is frequently the case in late autumn or during a drought, when the more desirable vegetation is dried up but some of the deep-rooted perennial weeds may still be green. Some weeds, like the Russian thistle,[29] pigweeds, sunflowers, ragweeds and others, when young and tender may be used for silage. At best, these emergency foods form a poor substitute, especially when more nutritious and palatable crop plants can be grown.

Many common weeds produce seeds or small fruits that are eaten by certain birds. Species of Amaranthus, Chenopodium and Polygonum, which retain many of their seeds on the plant until winter, often form an important food supply for seed-eating birds when the ground is covered with snow.

Several of the commonest weeds of gardens, fields or waste places are edible. Their young tender shoots, when cooked for greens or pot-herbs, are as palatable as spinach and many other garden greens. Some of them have a characteristic flavor or texture which supplies a welcome change from spinach. The most popular of these edible weeds are as follows:

Pteridium aquilinum—	Brake fern
Urtica gracilis—	Nettle
Rumex crispus—	Curly dock
Rumex acetosa—	Sour dock
Chenopodium album—	Lambs-quarters
Phytolacca americana—	Pokeweed
Portulaca oleracea—	Purslane
Barbarea vulgaris—	Winter-cress
Brassica kaber—	Wild mustard
Asclepias syriaca—	Milkweed
Taraxacum officinale—	Dandelion

[28] Brown and McMurtrey, 1934.   [29] Powell, 1933.

Weeds

A number of weeds have medicinal properties and are used in the preparation of certain medicines or drugs.[30] The more common weeds having medicinal value, together with the parts thereof used, are recorded in Table X. The demand for these, while reasonably constant, in most cases is rather limited.[31]

TABLE X
WEEDS USED AS MEDICINAL PLANTS

Name	Part Used
10. Agropyron repens—Quack-grass	Rootstocks
84. Veratrum viride—American hellebore	Rootstocks
86. Comptonia peregrina—Sweet-fern	Leaves and tops
108. Rumex crispus—Curly dock	Roots
119. Chenopodium ambrosioides—Wormseed	Fruits and tops
144. Phytolacca americana—Pokeweed	Roots, berries
181. Chelidonium majus—Greater celandine	Entire plant
190. Brassica hirta—White mustard	Seeds
193. Brassica nigra—Black mustard	Seeds
283. Geranium maculatum—Wild cranesbill	Roots
312. Passiflora incarnata—Passion-flower	Tops
320. Angelica atropurpurea—Purple-stem angelica	Roots
323. Conium maculatum—Poison-hemlock	Fruits
335. Apocynum androsaemifolium—Spreading dogbane	Roots
336. Apocynum cannabinum—Indian hemp	Roots
340. Asclepias syriaca—Milkweed	Roots
341. Asclepias tuberosa—Butterfly-weed	Roots
376. Symphytum officinale—Comfrey	Roots
378. Verbena hastata—Blue vervain	Tops
383. Glechoma hederacea—Ground ivy	Tops
384. Hedeoma pulegioides—American pennyroyal	Leaves and flowering tops
389. Lycopus americanus—Water horehound	Entire herb
391. Marrubium vulgare—Horehound	Leaves and tops
394. Mentha piperita—Peppermint	Leaves and flowering tops
395. Mentha spicata—Spearmint	Leaves and flowering tops
396. Nepeta cataria—Catnip	Leaves and flowering tops
403. Datura stramonium—Jimson-weed	Leaves and seeds
408. Solanum carolinense—Horse nettle	Ripe berries
409. Solanum dulcamara—European bittersweet	Young branchlets
417. Digitalis purpurea—Foxglove	Leaves
426. Verbascum thapsus—Mullein	Flowers and leaves
462. Lobelia inflata—Indian tobacco	Leaves and tops
463. Achillea millefolium—Yarrow	Tops
471. Arctium lappa—Great burdock	Roots
472. Arctium minus—Smaller burdock	Roots
473. Artemisia absinthium—Wormwood	Leaves and tops
516. Eupatorium perfoliatum—Boneset	Leaves and flowering tops
523. Grindelia squarrosa—Gumweed	Leaves and flowering tops

[30] Henkel, 1917. [31] Sievers, 1930.

TABLE X—Continued
WEEDS USED AS MEDICINAL PLANTS—Continued

Name	Part Used
536. Inula helenium—Elecampane	Roots
541. Lactuca scariola—Wild lettuce	Leaves
563. Tanacetum vulgare—Tansy	Leaves and flowering tops
565. Taraxacum officinale—Dandelion	Roots
569. Tussilago farfara—Coltsfoot	Leaves and roots

CHAPTER II

Weeds of Special Habitats

Some weeds seem omnipresent because they grow along roadsides, in waste places, or wherever there is an abandoned or undisturbed piece of land, be it large or small. Fortunately, many of these weeds are not adapted by any special modifications to persist among crops. They soon succumb when subjected to the special treatments employed in the tillage and harvesting of certain crops, or other treatments such as are received by meadows, pastures or lawns, where the tops of the weeds are removed by the periodic harvesting of the crop or the continuous cutting of the grass.

However, certain weeds are common and most troublesome under particular conditions or among certain crops. These weeds, of what might be termed special habitats, are among the most noxious and difficult to control as long as the grower insists on continuing to raise the same type of crop on the infested land. Weeds which have a life cycle similar to that of certain crops, when once established, tend to become more abundant as long as those crops are grown. The conditions favorable for the growth of the crop often are also beneficial to the weeds associated with it. The secret of the control of weeds of special crops is a system of crop rotation—a change of habitat.

WEEDS OF LAWNS AND TURFS

Most lawns and turfs contain weeds. Sometimes these become so abundant that they crowd out most of the lawn grasses. While lawn weeds vary somewhat in different localities and climatic conditions, the common ones are few and widely distributed.[1, 2, 3]

The weeds that persist and thrive under the treatment ordinarily given to a well-kept lawn do so by virtue of certain characteristics and habits of growth. These consist of two distinct types, those which produce a very short stem or crown with a rosette of leaves close to the ground, and those which make a mat of rootstocks just below the

[1] Durrell, 1926.   [2] Fiske, 1931.   [3] Westover, 1931.

Weeds of Special Habitats 37

surface or runners on top of the soil. Both of these types are adapted to withstand, and to grow in spite of, the frequent close cuttings of the lawn or turf. Successful lawn weeds have one habit in common, their growing points are close to the surface of the ground where they are not injured by frequent mowing. Many annuals and other weeds with a more erect habit and with higher growing points are common in newly seeded lawns, but these usually disappear in the second season as they cannot stand close cutting.

The sources of weeds appearing in a new lawn are many. Most of these may be traced to the following: impurities in the lawn grass seed; impurities in the soil, manure or mulch used for top-dressing; wind-blown seeds of such weeds as dandelions carried from infested to clean areas. An individual land owner may prevent weeds from entering his premises by using clean seeds and top-dressings, but to prevent the entrance of wind-blown weed seeds from nearby lands requires the united efforts of the entire community.

The following list of the commonest weeds in American lawns shows that nearly all of the most persistent are perennials; a few, mostly grasses, are annuals; biennial weeds are rare in a well-kept established lawn.

THE COMMONEST WEEDS OF LAWNS AND TURFS

ANNUALS

Capsella bursa-pastoris—Shepherds-purse
Dactyloctenium aegyptium—Egyptian-grass
Digitaria ischaemum—Small crab-grass
Digitaria sanguinalis—Crab-grass
Eleusine indica—Goose-grass
Euphorbia maculata—Milk purslane
Poa annua—Annual spear-grass
Polygonum aviculare—Knotweed
Scleranthus annuus—Knawel
Setaria glauca—Yellow foxtail
Stellaria media—Chickweed
Veronica peregrina—Speedwell
Veronica persica—Speedwell

BIENNIALS

Daucus carota—Wild carrot
Malva neglecta—Round-leaved mallow
Verbascum thapsus—Mullein

PERENNIALS

Achillea millefolium—Yarrow
Bellis perennis—English daisy
Cerastium arvense—Field chickweed
Cerastium vulgatum—Mouse-ear chickweed
Chrysanthemum leucanthemum—Ox-eye daisy
Cichorium intybus—Chicory
Galium mollugo—Cleavers
Galium verum—Yellow bedstraw
Geranium molle—Cranesbill
Glechoma hederacea—Ground ivy
Hieracium aurantiacum—Orange hawkweed
Hieracium pilosella—Mouse-ear
Hieracium pratense—Yellow paintbrush
Hydrocotyle sibthorpioides—Lawn pennywort
Hypochoeris radicata—Gosmore
Leontodon autumnalis—Fall dandelion
Plantago lanceolata—Buckhorn plantain
Plantago major—Broad-leaved plantain
Plantago rugelii—Rugels plantain
Potentilla argentea—Silver cinquefoil
Prunella vulgaris—Heal-all
Ranunculus acris—Tall field buttercup
Ranunculus repens—Creeping buttercup
Rumex acetosella—Sheep sorrel
Rumex crispus—Curly dock
Rumex obtusifolius—Broad-leaved dock
Sherardia arvensis—Blue field-madder
Taraxacum officinale—Dandelion
Thymus serpyllum—Creeping thyme
Trifolium dubium—Little hop clover
Veronica officinalis—Speedwell
Veronica serpyllifolia—Thyme-leaved speedwell

WEEDS OF PASTURES

Pastures, whether they consist of native grasslands, hay fields that have been used for grazing, or land that is too poor or otherwise unsuited for anything but grazing, frequently have certain weeds associated with them. The appearance of many weeds in pastures is usually due to overgrazing, too extensive or undergrazing, lack of proper drainage, or depletion of one or more of the soil nutrients, such as nitrogen or phosphorous, necessary for the growth of the pasture. Many of the commonest pasture weeds are plants that are un-

Weeds of Special Habitats

palatable, distasteful, or even poisonous to stock. Others have spines, thorns or modifications which cause stock to avoid them. Other weeds grow in pastures in spite of a low soil fertility or in poorly drained lands.[4]

As a result of overgrazing, especially on areas of native vegetation, introduced weeds often invade pastures and ranges and crowd out the more desirable forage plants. When the pastures are grazed too extensively, as where only a few animals have access to a large acreage, native herbs and shrubby species often encroach on the pasture, and not infrequently transform it into a bramble patch or thicket. The early stages of such a change may consist largely of "poverty-grasses," foxtail-grass, wild asters, and goldenrods. In New York, and to a certain extent in many of the northeastern states, brambles, thorn-apples, sweetbrier, spireas and sweet-fern are the commonest shrubby weeds invading pastures. In western Washington and Oregon, Canada thistle, smartweeds, bog rush (*Juncus effusus*), hardhack (*Spiraea douglasii*), and sweetbriar (*Rosa eglanteria*) are frequently very troublesome, especially in lowland pastures; the bracken fern is the most troublesome weed on upland pastures.

In general, pasture weeds consist of those species which, on account of some special taste or some mechanical adaptations, are avoided or cannot be attacked by stock, or those weeds which can thrive in spite of frequent or continuous grazing by stock.

THE COMMON PASTURE WEEDS

FERNS AND FERN ALLIES

Dennstaedtia punctilobula—Hay-scented fern
Equisetum arvense—Field horsetail
Pteridium aquilinum—Brake fern

GRASSES AND RUSHES

Agropyron repens—Quack-grass
Andropogon virginicus—Beard-grass
Bromus tectorum—Downy brome-grass
Danthonia spicata—Wild oat-grass
Hordeum jubatum—Squirrel-tail-grass
Juncus effusus—Bog rush
Juncus tenuis—Path rush
Panicum capillare—Old witch-grass
Setaria glauca—Yellow foxtail

[4] Johnstone-Wallace, 1933.

SHRUBS

Artemisia spp.—Sagebrush
Comptonia peregrina—Sweet-fern
Crataegus spp.—Thorn-apple
Rosa eglanteria—Sweet brier
Spiraea alba—Meadow-sweet
Spiraea douglasii—Hardhack
Spiraea tomentosa—Steeplebush

ANNUALS

Gnaphalium spp.—Cudweed
Hedeoma pulegioides—American pennyroyal
Lobelia inflata—Indian tobacco
Polygonum spp.—Smartweed

BIENNIALS

Cirsium vulgare—Bull thistle
Cynoglossum officinale—Hounds-tongue
Daucus carota—Wild carrot
Echium vulgare—Blue thistle
Verbascum blattaria—Moth mullein
Verbascum thapsus—Mullein

PERENNIALS

Achillea millefolium—Yarrow
Anaphalis margaritacea—Pearly everlasting
Antennaria spp.—Ladies tobacco
Asclepias syriaca—Milkweed
Aster spp.—Wild aster
Chrysanthemum leucanthemum—Ox-eye daisy
Eupatorium perfoliatum—Boneset
Hieracium aurantiacum—Orange hawkweed
Hieracium florentinum—King devil
Hieracium pratense—Yellow paintbrush
Hypericum perforatum—St. Johns-wort
Inula helenium—Elecampane
Lithospermum officinale—Gromwell
Mentha arvensis—Wild mint
Penstemon digitalis—Beard-tongue
Phytolacca americana—Pokeweed
Plantago lanceolata—Buckhorn plantain
Plantago major—Broad-leaved plantain
Potentilla simplex—Cinquefoil
Prunella vulgaris—Heal-all

Weeds of Special Habitats 41

Ranunculus acris—Tall field buttercup
Ranunculus repens—Creeping buttercup
Rumex acetosella—Sheep sorrel
Silene cucubalus—Bladder campion
Solidago graminifolia—Narrow-leaved goldenrod
Thymus serpyllum—Creeping thyme
Verbena hastata—Blue vervain
Veronica officinalis—Speedwell

WEEDS OF HAY FIELDS AND MEADOWS

Meadows and hay fields are sometimes very weedy. A new meadow is sometimes started with impure grass seed or on land foul with weed seeds. If such a new seeding does not have ideal conditions for the growth of the grass or leguminous forage plants, the weeds are bound to get a head start and certain ones may continue until they finally drive out most of the meadow grasses. Certain weeds, as indicated in the following list, have seeds which appear as common impurities in clover and alfalfa seed. These weeds are usually associated with new seedings of these crops but seldom cause much trouble in good stands after the second year.

Many common weeds do not make their appearance until the meadow is old or run down by frequent cutting and removal of the crop, without the addition of any fertilizers to the soil. Many of these weeds mature their seeds about the time the hay is cut. In certain localities, especially where weather conditions are less favorable for making hay earlier in the season, weedy hay is not cut until many of the weeds have ripened their seeds. The use of such hay for feeding or bedding purposes supplies a constant source of weed seed contamination of manure and the soils on which it is distributed. The seeds of most meadow weeds are often passed through the digestive tract of horses, cattle and sheep without losing their viability (see page 22).

THE COMMONEST WEEDS OF HAY FIELDS AND MEADOWS

ANNUALS OR BIENNIALS

    Ambrosia artemisiifolia—Ragweed
x Cuscuta spp.—Dodder
    Daucus carota—Wild carrot
    Eleusine indica—Goose-grass
    Erigeron annuus—White-top

Weeds

    Erigeron canadensis—Fleabane
    Erigeron strigosus—Rough daisy fleabane
x Lepidium campestre—Field pepper-grass
x Silene noctiflora—Night-flowering catchfly

PERENNIALS

    Achillea millefolium—Yarrow
    Agropyron repens—Quack-grass
x Anthemis arvensis—Corn chamomile
    Anthoxanthum odoratum—Sweet vernal-grass
x Barbarea vulgaris—Winter-cress
    Chrysanthemum leucanthemum—Ox-eye daisy
    Cichorium intybus—Chicory
    Cirsium arvense—Canada thistle
    Equisetum arvense—Field horsetail
    Erigeron philadelphicus—Philadelphia fleabane
    Erigeron pulchellus—Robins plantain
    Galium mollugo—Cleavers
    Hieracium aurantiacum—Orange hawkweed
    Hieracium pratense—Yellow paintbrush
    Holcus lanatus—Velvet-grass
x Hordeum jubatum—Squirrel-tail-grass
    Hypericum perforatum—St. Johns-wort
    Linaria vulgaris—Butter-and-eggs
x Lychnis alba—White cockle
    Lychnis flos-cuculi—Ragged robin
    Malva moschata—Musk mallow
    Onoclea sensibilis—Sensitive fern
    Origanum vulgare—Wild marjoram
    Penstemon digitalis—Beard-tongue
    Physalis subglabrata—Ground-cherry
x Plantago lanceolata—Buckhorn plantain
    Plantago rugelii—Rugels plantain
    Potentilla simplex—Cinquefoil
    Potentilla recta—Sulfur cinquefoil
    Ranunculus acris—Tall field buttercup
    Ranunculus bulbosus—Bulbous crowfoot
    Rudbeckia serotina—Black-eyed Susan
    Rumex acetosa—Sour dock
    Rumex acetosella—Sheep sorrel
    Rumex crispus—Curly dock
    Sedum purpureum—Live-for-ever
x Silene cucubalus—Bladder campion
    Solanum carolinense—Horse nettle

x = most troublesome weeds in new seedings of clover, alfalfa and grass.

Weeds of Special Habitats

WEEDS OF CULTIVATED FIELDS AND GARDENS

The weeds of cultivated crops and gardens are mostly annuals or perennials introduced from sections of the world that were originally non-wooded. These weeds respond to cultivation much like the cultivated crops, but do not persist long in competition with grasses or on land that has been left undisturbed.

The annual weeds of cultivated fields and gardens usually produce large numbers of seeds. These seeds frequently germinate very unevenly and some of them may germinate after each time the crop is cultivated. Many come up after the last cultivation and mature a new crop of seed before the first frost. Some of these annual weeds, like purslane, chickweed, galinsoga and ragwort, take root very freely after cultivation as long as the plant is left in contact with the soil, even under ordinary weather conditions.

Among the most troublesome weeds of cultivated crops are those perennials which reproduce and spread, not only by seeds but also by underground stems and creeping roots, every small piece of which may send up a new shoot. Many of these weeds have underground parts so deep in the soil that they are not disturbed by ordinary plowing and cultivation. The common weeds of cultivated fields and gardens are relatively few, but very generally distributed throughout the northern United States.

SOME COMMON WEEDS OF CULTIVATED FIELDS AND GARDENS

ANNUALS

Abutilon theophrasti—Velvet-leaf
Amaranthus albus—Tumble weed
Amaranthus hybridus—Green amaranth
Amaranthus retroflexus—Amaranth pigweed
Arenaria serpyllifolia—Thyme-leaved sandwort
Brassica kaber—Wild mustard
Capsella bursa-pastoris—Shepherds-purse
Chenopodium album—Lambs-quarters
Chenopodium paganum—Pigweed
Digitaria sanguinalis—Crab-grass
Echinochloa crusgalli—Barnyard-grass
Erysimum cheiranthoides—Wormseed mustard
Euphorbia helioscopia—Sun spurge
Galinsoga ciliata—Galinsoga
Lactuca scariola—Wild lettuce

Lamium amplexicaule—Dead nettle
Panicum capillare—Old witch-grass
Polygonum convolvulus—Black bindweed
Polygonum persicaria—Ladys-thumb
Portulaca oleracea—Purslane
Raphanus sativus—Radish
Senecio vulgaris—Groundsel
Setaria glauca—Yellow foxtail
Setaria viridis—Green foxtail
Sisymbrium officinale—Hedge mustard
Sonchus asper—Spiny-leaved sow thistle
Sonchus oleraceus—Sow thistle
Spergula arvensis—Spurry
Veronica peregrina—Purslane speedwell
Veronica persica—Speedwell

PERENNIALS

Agropyron repens—Quack-grass
Cardaria draba—Hoary cress
Convolvulus arvensis—Bindweed
Convolvulus sepium—Wild morning-glory
Cyperus esculentus—Yellow nut-grass
Cyperus rotundus—Nut-grass
Linaria vulgaris—Butter-and-eggs
Oxalis europaea—Yellow wood sorrel
Sonchus arvensis—Perennial sow thistle

Weeds of Grain Fields

Certain weeds are common in grain fields, such as wheat, oats, barley and rye. Most of these weeds are annuals or winter annuals having a life cycle very similar to that of the grain crop among which they grow, germinating in autumn or spring and maturing their seed just before or about the time that the grain ripens. The seeds of these weeds are scattered before the crop is harvested or they may be harvested with the crop and become common impurities in grain seed. The actual number of species of weeds common in grain fields, while varying somewhat in different parts of the United States, is not so very large in any one region. Most of the weeds of grain fields cause the greatest trouble in the western and northwestern states, especially in regions where grain is grown as a continuous crop. The following list of the most common weeds of American grain fields shows that most of them are annuals; the few perennials which do occur are among the most persistent and the most difficult to control.

Weeds of Special Habitats

THE COMMONEST WEEDS OF GRAIN FIELDS

ANNUALS

Agrostemma githago—Corn cockle
Ambrosia artemisiifolia—Ragweed
Avena fatua—Wild oats
Brassica kaber—Field mustard
Brassica juncea—Indian mustard
Bromus secalinus—Chess
Camelina microcarpa—False flax
Capsella bursa-pastoris—Shepherds-purse
Centaurea cyanus—Bachelors-button
Centaurea melitensis—Napa thistle
Chenopodium album—Lambs-quarters
Cnicus benedictus—Blessed thistle
Conringia orientalis—Hares-ear mustard
Echinochloa crusgalli—Barnyard-grass
Erucastrum gallicum—Dog mustard
Helianthus annuus—Sunflower
Lithospermum arvense—Gromwell
Neslia paniculata—Ball mustard
Polygonum convolvulus—Black bindweed
Polygonum hydropiper—Smartweed
Polygonum lapathifolium—Pale smartweed
Raphanus raphanistrum—Wild radish
Raphanus sativus—Radish
Salsola kali—Russian thistle
Saponaria vaccaria—Cow cockle
Setaria glauca—Yellow foxtail
Silene gallica—English catchfly
Sonchus oleraceus—Sow thistle
Spergula arvensis—Spurry
Thlaspi arvense—Fan-weed

PERENNIALS

Agropyron repens—Quack-grass
Allium vineale—Wild garlic
Anthemis arvensis—Corn chamomile
Cirsium arvense—Canada thistle
Convolvulus arvensis—Bindweed
Sonchus arvensis—Perennial sow thistle

WEEDS OF CRANBERRY BOGS

A cranberry bog represents a very specialized habitat. The cranberry plants take several years to become established, but when

46 Weeds

once fully so, they may be expected to thrive for many years without cultivation. However, if the weeding or the proper flooding of a cranberry bog is neglected, it will soon be invaded by numerous native bog plants and almost any of the native shrubs and trees that grew on the soil before the cranberry bog was established.[5, 6, 7, 8]

SOME COMMON WEEDS OF CRANBERRY BOGS

Equisetum arvense—Field horsetail
Equisetum telmateia—Giant horsetail
Osmunda cinnamomea—Cinnamon fern
Woodwardia virginica—Chain fern
Holcus lanatus—Velvet-grass
Calamovilfa brevipilis—Reed-grass
Carex bullata—Cut-grass
Carex spp.—Sedges
Dulichium arundinaceum—Dulichium
Leersia oryzoides—Rice cut-grass
Scirpus americanus—Shore rush
Scirpus cyperinus—Wool-grass
Eriophorum spp.—Cotton-grass
Juncus spp.—Rushes
Smilax spp.—Green brier
Salix spp.—Willows
Geum aleppicum var. strictum—Yellow avens
Potentilla anserina—Goose-grass
Rubus spp.—Blackberry and dewberry
Spiraea douglasii—Hardhack
Trifolium fimbriatum—Clover
Amphicarpum purshii—Double-seeded millet
Gaultheria shallon—Salal
Chamaedaphne calyculata—Leather-leaf
Lysimachia terrestris—Loosestrife
Epilobium angustifolium—Fireweed
Hypochoeris radicata—Gosmore, false dandelion

WEEDS OF RICE FIELDS

Because of the frequent flooding of rice fields, the weeds consist of species that naturally grow in water or marshy areas that are subject to periodic inundations. The source of these weeds may be seeds in the soil of the seed-bed, impurities in the rice seed, seeds brought in by the irrigation water, or wind-blown seeds carried from plants

[5] Beckwith and Fiske, 1925. [6] Brown, 1927.
[7] Crowley, 1929. [8] Darrow, 1924.

Weeds of Special Habitats

growing along nearby irrigation ditches or marshy places. Several species are very troublesome in rice fields [9, 10, 11] and the seeds of these weeds are often found as impurities in seed rice.[12]

SOME COMMON WEEDS OF RICE FIELDS

Typha latifolia—Broad-leaved cat-tail
Echinochloa colonum—Jungle rice
Echinochloa crusgalli—Water-grass or barnyard-grass
Panicum dichotomiflorum—Spreading witch-grass
Paspalum distichum—Joint-grass
Paspalum floridanum
Phalaris brachystachya—Canary-grass
Phalaris paradoxa—Canary-grass
Oryza sativa—Red rice
Leptochloa fascicularis—Scale-grass
Cyperus spp.—Umbrella plant
Eleocharis palustris—Spike rush
Fimbristylis autumnalis
Scirpus spp.—Bulrush
Ammania coccinea—Red stem
Sesbania exaltata—Indigo weed
Aeschynomene virginica—Curly indigo

[9] Kennedy, 1922. [10] Prince, 1927.
[11] Jones, 1924. [12] Bellue, 1932.

CHAPTER III

The Control of Weeds

In deciding on the most effective or practicable method of attacking a particular weed, it is necessary to consider its habits, habitat, and distribution. A knowledge of the habits of a weed supplies information concerning: the time and conditions under which its seeds germinate; how much time it requires to mature seeds; whether it dies at the end of one or two years, or lives several years; whether it reproduces and spreads only by seeds or also by vegetative propagation. The habitat of a weed is the place in which it is at home, that is, the kind of a place in which it thrives best and where it may be expected to be troublesome. Many weeds are not particular in their choice of habitat, but others require a particular set of conditions in order to grow well enough to become troublesome and present control problems. Such weeds may be restricted to wet or dry soils, sandy or clay, rich or poor soils, grasslands or cultivated lands, open fields or shady places, acid, neutral, or somewhat alkaline soils. A knowledge of the distribution of a weed is useful as it supplies information concerning the extent of the area over which it grows at a certain time, whether it is limited to a few small patches or is scattered over the whole countryside.

The three fundamental objectives of the various methods of combating weeds are prevention, eradication, and control. The main object of prevention measures is the exclusion of weeds from areas not yet infested or preventing their spread from infested to clean fields. The aim of eradication measures is the elimination of a weed after it has become established in an area. This involves the complete destruction of a weed, including tops, underground parts, and seeds in the soil. Such measures are most practicable if employed soon after a weed becomes established in a field or region and while it is still restricted to a small area. After a weed has spread over large areas, its complete eradication becomes very difficult and often impossible or economically impractical. Control measures are used where prevention and eradication were neglected or have failed.

The Control of Weeds

Control measures interfere with the growth and reproduction of weeds and lessen their injury to the crops; they do not, as a rule, eradicate the weeds, but they make it possible to raise crops in spite of their presence. Sometimes after control measures have been applied to a weed consistently for several years, it may be reduced in numbers, area, or vigor so as to make its complete eradication practicable.

From a practical viewpoint the various methods of prevention, eradication, and control of weeds fall into three general groups depending on whether their primary aim is to prevent the introduction or spread of weeds to new areas, to destroy the tops of weeds, or to destroy the underground parts.

METHODS FOR PREVENTING THE SPREAD OF WEEDS INTO NEW AREAS

1. *Use of clean seed.*

Many weeds have been brought into localities free from them with agricultural seeds. Cheap seed is often the most expensive because it may contain one or more kinds of noxious weed seeds. In the end it is always better to buy and sow recleaned seed which tests high in purity and germination than to take a chance on cheap seed. The Federal Seed Act of 1939 requires that seed offered for sale in interstate commerce be provided with a label containing a statement of the percentage of pure seed and weed seed contained by them. The label should always be examined to determine whether the seed contains noxious weeds. This precaution may save work and expense later on. Most seeds sold today by reputable dealers possess a high degree of purity; however, not all seeds in the market are fit to sow. Home-grown seed grain, such as is sometimes bought from or exchanged with neighbors, is frequently planted without recleaning or just as it comes from the thresher. Such seed sometimes contains many weeds, and crops grown from them are more expensive to produce than if clean seed had been purchased at a higher price.

2. *Avoiding the scattering of weed seeds with farm products and machinery.*

Much can be done to prevent weeds from entering clean fields by using certain precautions in the transporting of farm products and materials containing weed seeds. If noxious weeds are known to be present in hay, straw, grain, or coarsely ground feeds containing weed seed screenings, they should be used or treated in a manner which will destroy the seeds before they enter the soil. Manure containing waste material, litter, and bedding originating from these products should be thoroughly composted for four or five months. This will kill most of the weed seeds before they are spread over the fields. Top-soil, gravel, dirt about nursery stock, and packing materials consisting of hay or straw, should be examined, either before using or afterwards, to make sure that no noxious weeds get started from seeds or vegetative parts that might be transported with them. Threshers and hay presses should be cleaned before they are removed from a farm known to be infested with noxious weeds to make sure that none is carried to the next farm.

3. *Prevention of seed production by weeds in nearby waste areas.*

Weeds with wind-borne seeds or seeds capable of being carried by water, if allowed to mature in waste places, along roadsides, fences, ditches or in swales or stony areas unfit for cultivation, may scatter seeds to adjoining clean fields. Weeds in such areas should be mowed at least twice a year, or as often as necessary to prevent them from maturing seed. Sometimes it is practicable to fence such areas so that the weeds can be kept grazed closely by sheep.

METHODS FOR DESTROYING THE TOPS OF WEEDS

The consistent destruction of the tops of weeds will prevent them from producing seed, and this treatment also gradually starves and weakens the roots and underground stems of perennial species. Most annuals and biennials are killed by the destruction of their tops since, under ordinary conditions, their roots do not produce new shoots. The efficiency of the several methods of destroying the tops of weeds depends on the kind and the conditions under which they are growing.

4. *Hand pulling.*

This method may be employed to destroy the seedlings of any kind of weed. It is also effective for destroying annuals, biennials and some perennials after they have reached considerable size.

The destruction of weeds growing within the rows of garden

plants, truck crops or within the rows and hills of cultivated crops where they cannot be reached by cultivators, depends largely on hand weeding. When these weeds are still small they can be easily removed before they rob the crop plants of light, water and mineral nutrients.

Annual and biennial weeds occurring in small patches or scattered among grass or grain can be pulled after a rain when the soil is loose without disturbing the crop plants. Such weeds should be pulled as soon as they are large enough so that they can be grasped but before they produce seed. The pieces of root remaining in the soil will not sprout again. The tops of some perennials can also be pulled but in many species their roots will send out new shoots. In order to kill the roots it is necessary to remove the tops several times.

5. *Hand hoeing.*

Weeds can be destroyed readily while they are still small by hand hoeing, either by cutting off their tops or by stirring the surface soil so as to expose the seedlings to the drying action of the sun. However, this method is too laborious except for small areas. It is useful in small gardens and in rows of cultivated crops and for supplementing machine cultivation late in the season when the crop plants are large or to destroy scattered weeds missed by the cultivator. Scattered weeds in pastures and grasslands can be destroyed with a hoe by cutting their crowns well below the ground line so as to prevent them from sending out new shoots.

6. *Cultivation.*

Cultivation is one of the oldest and most widely practiced methods of controlling weeds. It not only destroys the seedlings and the tops of large kinds, but frequently when the soil is stirred other weed seeds in it are induced to germinate and can then be destroyed. Of the several cultivators available for destroying weeds in cultivated crops, those with a knife or "duck-foot" type of blades are the most effective for cutting roots and rootstocks. Cultivation should be started early, as soon as the crop plants are up enough so that the rows are visible and while the weeds are still small so that they succumb easily and have not yet stunted the growth of the crop plants. Hill crops should be planted in check rows so that the cultivator can be run in both directions of the field and thus eliminate most of the hand hoeing. Crops in drilled rows often require hand hoeing or hand weeding after cultivation.

7. *Plowing.*

The tops of annual, biennial and many perennial weeds are killed by plowing them under. Such a treatment is a very effective method of disposing of a luxuriant weed growth in a field in which the crop has been overrun by weeds because of an unfavorable season or poor seed. Most plowing is done either in autumn or spring while the crop is out of the way. Autumn plowing brings many of the roots and rootstocks of perennial weeds to the surface where they are killed by exposure to the action of freezing or alternate freezing and thawing. Autumn plowing in some localities or on certain soils makes it possible to prepare the seed-bed earlier in the spring so that one or two crops of weed seedlings can be destroyed before the crop is planted. Land on which weeds have been allowed to produce a large crop of seeds should not be plowed under until after the surface soil has been stirred several times by harrowing or shallow disking so as to induce the seeds to germinate. These seedlings will be destroyed by subsequent harrowing or by freezing. If the seeds are plowed under, many will lie dormant until they are brought near the surface by subsequent plowing or cultivation. A field on which a crop has been kept under clean cultivation for one or two seasons is fairly free from weed seeds in its upper layers of soil. If such a field is plowed, viable weed seeds are brought from the deeper soil to the surface where conditions are favorable for germination. A grain or meadow crop following a clean cultivated crop is much freer from weeds if it is sowed after harrowing or shallow disking of the field without first plowing it.

8. *Harrowing.*

Shallow surface tillage with a spike-tooth harrow is an effective method for the wholesale destruction of small weeds before the crops are planted, after they are up, and after grains are harvested.

Grain fields seeded with grass or clover frequently develop a crop of annual or biennial weeds after the grain is harvested. One or two harrowings of the stubble field will destroy many of these weed seedlings and give the meadow plants a better chance.

Small weed seedlings in cultivated crops like potatoes, corn and sugar beets are easily killed by harrowing the field with a light harrow just as the weeds are coming through the soil surface.

Many annual weeds, such as mustards, can be killed in very young stages by harrowing a field when the grain is about four to six inches high. If a light harrow is used, very little damage is done to the grain plants. This method can be employed to advantage only in fields where the surface of the soil is well finished, rolled, and free from large stones, roots of perennial weeds, or other obstructions which might clog the harrow and cause the grain plants to be dragged out.

9. *Disking.*

The disking of fields in autumn will destroy annual and biennial weeds and also the new top growth of perennials appearing after the crop is harvested. By setting the disk cutters at an angle the top growth can be cut and turned under the soil. Autumn plowed land often forms a crust over the surface by spring. Disking in the spring destroys the top growth of any weeds present and brings seeds in the upper soil to the surface so that a new crop of seedlings appears in a few days. These can be destroyed before the crop is planted.

10. *Pasturing and grazing.*

The continued removal of the tops of weeds by grazing animals, like close mowing, prevents seed formation and also gradually weakens the underground parts. Sheep, goats, hogs and cattle are effective in destroying many kinds of weeds. Many farms would profit by maintaining a flock of sheep for cleaning up weeds in pastures, fields and along fence-rows. Grazing animals should not be turned on pastures too early in the spring or before the grass has made a good growth. If pastures are subdivided into two or more lots, permitting animals to be turned from one to another in rotation, more weeds will be cleaned up than if they are allowed to roam over a large area and choose the more desirable forage and leave the weeds to become rank and unpalatable. Geese have been recommended for controlling weeds in such places as berry patches [1] and asparagus beds. Imported beetles have been used to control St. Johns-wort, *Hypericum perforatum*.[2]

[1] Magill, 1952.
[2] Huffacker and Kennett, 1952.

11. *Mowing.*

Weeds that have been allowed to become too large to destroy by cultivation or those growing in places where cultivation is impracticable or impossible can be prevented from producing seed by mowing close to the ground. On large areas of even ground a mower can be employed, but on rough or stony ground a scythe must be used. Perennial weeds usually require several cuttings before the underground parts exhaust their reserve food supply. If only a single cutting can be made, the best time is when the plants begin to blossom. At this stage the reserve food supply in the roots has been nearly exhausted, and new seeds have not yet been produced. Weedy meadows should be mowed early to prevent the spreading of weed seeds with hay.

In fields where it is not practicable to plow, such as a new meadow or new seeding of alfalfa or clover, many weeds can be controlled by mowing them about the time the blossoms appear or at least before seeds are matured. This cutting can be done with a scythe or mowing machine. The sickle bar of the mowing machine should be set as high as possible and yet cut off the lowest flowers from the weed stalks. This will prevent seeds from forming before the first crop of hay is cut. If the stems of the weeds are cut high above the ground, any new shoots that may appear will be high enough so that the mower will cut them when the first cutting of hay is made. If the first cutting of weeds is not made, they may ripen and scatter seeds before hay is cut. If the weeds are cut low the first time, not only may the hay crop be set back, but the weeds may form new flowering shoots so close to the ground that the mower will not cut them off in the first cutting of hay.

12. *Spudding.*

The spud is a tool with a long chisel-like blade designed for the removal of weeds with long tap-roots from lawns, pastures and meadows without disturbing the sod. Spudding may be resorted to at any time, but it is most effective before the weeds have become very large. The roots should be cut as far below the crown as possible. This can be done with the least effort in the spring or after a heavy rain when the soil is soft and the roots lift easily.

The Control of Weeds

13. *Burning.*

Noxious weeds which have been allowed to mature seed are sometimes cut and burned on the spot to prevent the spreading of their seeds. Patches of dodder (Cuscuta) growing in clover and alfalfa fields can be prevented from spreading to other parts of a farm by burning them without moving. Weedy fields, pastures, and meadows are sometimes burned with the hope of destroying weeds. Burning is no more effective in destroying the underground parts of perennial weeds than any other method of removing their tops. Burning the tops of weeds also destroys grass and other organic matter which should normally be returned to the soil by plowing it under or by other methods. Several types of weed-burners are now available in the market. These are sometimes employed by railroads for keeping their rights-of-way free from vegetation. Small portable weed-burners are used for keeping paths, driveways, ditch banks, fence-rows and similar places free from weeds. These burners are effective and easily manipulated for killing weeds while they are still small. They do not kill the underground parts.

14. *Steam.*

Super-heated steam has been used by some railroads for killing the tops of weeds along their rights-of-way. This method is not practicable for killing weeds under agricultural conditions because the equipment required to supply the necessary flow of steam would be too heavy to operate except on a highway or railroad track.

15. *Biological methods of weed control.*

The most perfect method of controlling weeds is one in which their natural enemies are set to attack them. There has been considerable interest in this phase in recent years, but so far this method is still in the experimental stage. Many weeds have insect enemies or parasitic fungi or bacteria which attack them and, under certain conditions, kill or destroy them. The chief danger in the introduction and utilization of these enemies of weeds is that most noxious weeds have closely related species which are important crop plants. The introduction of the enemies of the weeds might mean the destruction of related crop plants as well.

A few attempts at biological weed control might be mentioned.

Early in this century, there appeared on the farm of the New York State College of Agriculture at Ithaca, New York, a fungus disease on the leaves of live-for-ever (*Sedum triphyllum*). For several years diseased shoots of the live-for-ever were sent to farmers of New York state who inoculated the live-for-ever plants in their meadows, with the result that they became diseased and died. The fungus disease was so successful that all the live-for-ever plants in the diseased patch on the college farm were killed and the fungus disappeared before it had been isolated so that its identity could be determined. About forty years ago the New Jersey Agricultural Experiment Station offered to send cultures of Canada thistle (*Cirsium arvense*), diseased by a rust (*Puccinia suaveolens*), to farmers for the purpose of inoculating healthy Canada thistles. The Canada thistle rust is very common and widespread today wherever its host is present. While the rust stunts the growth of the thistles somewhat, they seem to grow and spread in spite of its attack.

16. *Chemicals.*

Since methods of chemical weed control vary with the weeds to be treated and the associated crops, and the weeds vary in different sections of the United States and Canada, it is wise to consult with the State Experiment Station or Agricultural College. Their bulletins may be comprehensive like *Oregon Weed Control Recommendations,* Virgil Freed, Readers Weeders No. 28, Oregon State College, Corvallis, Oregon, 1953, and *1954 Weed Control in Field Crops,* Stanford N. Fertig, Cornell Extension Circular No. 821, Revised February 1954, Ithaca, New York; or more specific like *Control of Early Weed and Grass Seedlings in Cotton,* H. E. Rea, Texas Agricultural Experiment Station Progress Report No. 1508, November 14, 1952.

A list of *Selected References* follows. Some include comprehensive bibliographies. Readers are especially urged to study such books as *Principles of Weed Control* by Ahlgren, Klingman, and Wolf; and *Weed Control* by Robbins, Crafts and Raynor, 2nd. Edition.

SUGGESTED REFERENCES

1. AHLGREN, GILBERT A., GLENN C. KLINGMAN, and DALE E. WOLF. Principles of Weed Control. New York: Wiley. 1951. 368 pp.
2. Bibliography of Weed Investigations. Division of Weed Investigations. U.S. Dept. of Agr. Vol. 1, 1951+ Reprint from Weeds.

The Control of Weeds

3. Eastern Section National Weed Committee of Canada Proceedings. Published annually. Vol. 1, 1947+
4. North Central Weed Control Conference Proceedings. Published annually. Vol. 1, 1944+ Also, Research Report North Central Weed Control Conference.
5. Northeastern Weed Control Conference Proceedings. Published annually. Vol. 1, 1947+
6. ROBBINS, W. W., A. S. CRAFTS and R. N. RAYNOR. Weed Control. New York: McGraw-Hill. 2nd ed. 1952.
7. Southern Weed Conference Proceedings. Published annually. Vol. 1, 1948+
8. Weeds, Journal of the Association of Regional Weed Control Conferences. Geneva, New York: W. F. Humphrey Press, Inc. Published quarterly. Vol. 1, 1951+
9. Western Canadian Weed Control Conference Proceedings. Published annually. Vol. 1, 1947+
10. Western Weed Control Conference Proceedings. Published annually. Vol. 1, 1938+

METHODS FOR DESTROYING THE UNDERGROUND PARTS OF WEEDS

The underground structures of perennial weeds, either roots, rootstocks or both, store food and in many cases also serve as organs for propagation. The destruction of these underground organs is necessary to prevent the growth, reproduction and spreading of weeds.

17. *Hand digging.*

The removal of rootstocks by hand digging is a slow but sure way of destroying certain perennial weeds. The work must be thorough to be effective. Weeds in which the roots are capable of sending forth new shoots are more difficult to destroy by digging because every piece of root that breaks off and remains in the soil may produce a new plant. This method cannot be used on large areas but it is practical for preventing the spreading of noxious weeds from gardens, small patches in fields, and from about trees and shrubs where other methods cannot be used.

18. *Clean cultivation.*

Repeated removal of the tops of perennial weeds by clean cultivation of crops gradually weakens and destroys their roots and rootstocks. The cultivation must be thorough and at short intervals so that no green shoots are allowed to appear above the soil surface, where they can manufacture foods to replenish the roots. Crops planted in check rows allow for thorough cultivation in both directions of the field. Cultivators with knife or "duck-foot" types of blades are more efficient in cutting the rootstocks and roots than ordinary shovel or tooth cultivators which permit many of the weeds to slip by unharmed. Some of the more persistent deep-rooted perennial weeds require two or more years of clean cultivation before the roots are killed by starvation.

19. *Summer fallow.*

Perennial weeds are sometimes so abundant that it is useless to try to eradicate them and raise a crop at the same time. Under such conditions clean cultivation without growing a cultivated crop—summer fallow or bare fallow—is the most efficient method of destroying their underground parts. However, this method is expensive because it requires a large amount of labor and, since it is carried on throughout the growing season, the farmer loses the use of the land for a year. The field must be plowed shallow or disked at short intervals throughout the summer so that no green shoots appear; otherwise the work is done in vain. The summer fallow method of weed control has been employed extensively in dry land grain sections where rotation is not feasible because of the difficulty of growing cultivated crops without irrigation. In dry-farming regions the summer fallow is also considered effective in the absorption of precipitation by the soil and in the retention of absorbed moisture.[3]

A modification of the summer fallow method, beginning about midsummer after the infested field has been pastured or has produced a grain or hay crop, has been found more practicable in humid regions than fallowing throughout the growing season. Such a method starts the fallowing at a time when most perennial weeds are depleted of much of the foods in the underground organs and are less resistant to the treatment.

20. *Rotation.*

Many weeds thrive best and cause the most trouble when the same crop is grown year after year. Some weeds are associated with certain crops or grow only in special habitats (see Chapter II). Crop rotation or changing the habitat interferes with the normal life cycle of many weeds. On the average general farm in the northern United States

[3] McCall and Wanser, 1924.

The Control of Weeds 59

the four common types of weed habitats are cultivated crops, grain fields, grasslands and meadows and pastures. Various systems of crop rotation have been employed from time to time. Many are useful for controlling weeds but the practicability of any specific method for a particular locality must be determined by considering such factors as climate, rainfall, suitability of soil, availability of markets and opportunity for utilizing or disposing of the crops. In general, a rotation over a four-year period is a useful standard to follow. This may be shortened by one year or increased by one or two years to meet local conditions. This permits the land to be occupied for one year by each of the four general types of crops:

> First year—clean cultivated crop
> Second year—grain crop
> Third year—clover crop
> Fourth year—grassland

The fourth year may be used for a hay crop or for pasture. If alfalfa is sown in the third year, the rotation may be increased for several years or as long as a good clean stand of alfalfa remains.

The success of a system of rotation, as far as the control of weeds is concerned, depends largely on the thoroughness and persistence with which the cultivated crop is kept free from weeds rather than on the kind of crop.

21. *Drainage.*

Most land on which cultivated crops grow well is drained sufficiently so that any improvement of the drainage would have little effect on the weeds. Pastures and meadows are frequently poorly drained and are filled with native weeds which thrive in wet soil, swales and sloughs. Various species of hardhack, buttercups, rushes, horsetails and many others are characteristic weeds of poorly drained, under-grazed pasture lands. By improving the drainage of such land, conditions are made less favorable for these weeds and more favorable for the grasses and legumes so that they can compete with, and finally drive out, many weeds.

22. *Smother crops.*

A thick stand of a rapidly growing crop competes with the weeds for water and light to such an extent as to prevent their top growth and weaken their roots. The smother crops most commonly used are alfalfa, sorghum, sweet clover, buckwheat, rye, rape, millet, soybean and sunflower. The type of smother crop is determined largely by the soil and climatic conditions and the opportunity for utilizing the product. Alfalfa, where it succeeds well, is an excellent smother crop because of its extensive root system, which enables it to compete with weeds for water, and its dense top growth which has a smothering effect on the new growth of weeds. The frequent cutting of the alfalfa destroys the tops of the weeds before they have had a chance to ripen seeds. In the control of weeds the principal value of smother crops is that they weaken the underground parts so that the weeds succumb more easily to the cultivation that follows.

23. *Straw mulch.*

When perennial weeds are restricted to a very limited area, it is possible to kill them, or at least to prevent them from spreading, by covering the area with a thick layer of straw, hay or manure. The effect of such a mulch is to exclude the light from the tops of the weeds until the reserve foods in the roots and rootstocks are withdrawn and the weeds are starved. This method destroys many weeds, but one of the most noxious deep-rooted perennials, bindweed (*Convolvulus arvensis*), has been observed to grow through a layer of straw four feet deep. The roots of Austrian field cress (*Rorippa austriaca*) have been found alive after having been buried under the middle of a straw stack for two years. If a straw mulch is used, it should extend well beyond the edge of the weed area; otherwise shoots will grow out around the edges.

24. *Mulch paper.*

Paper as a mulch for the control of weeds was first developed on an extensive scale in the cane plantations of Hawaii.[4] Its success soon led to the extensive use of paper for the control of weeds in the Hawaiian pineapple fields [5] so that by 1935 a large per cent of the Hawaiian pineapple crop was grown under mulch paper.[6] Since about 1925 a number of workers in various parts of the United States have used mulch paper on a number of crops, chiefly vegetables.[6, 7, 8, 9, 10, 11, 12]

In general, the results of these investigators show that the mulch

[4] Eckhart, 1923.   [5] Hartung, 1926.   [6] Hutchins, 1933.
[7] Flint, 1928.   [8] Flint, 1929.   [9] Magruder, 1930.
[10] Thompson and Platenius, 1932.   [11] Edmond, 1929.
[12] Smith, 1931.

paper is very effective in the control of weeds. Mulch paper, under certain conditions, indirectly also produces beneficial effects by increasing the soil temperature or by holding it more uniform at night, by conserving soil-moisture, and by increasing nitrification. The beneficial effect may increase the yield of certain crops but this is usually offset by the added expense of the paper and the labor of laying it. It appears, therefore, that the chief benefit derived from its use on most crops is the control of weeds and a reduction in the cost of cultivation. In a few warm-season crops of high acre value such as tomatoes, peppers, muskmelons, when grown in the northern states, the increase in temperature and moisture may not only increase the total yield, but it may force the crops to mature earlier in the season when the prices are higher, thus raising the income sufficiently to make it worth while to use paper.

Mulch paper is black or dark, impervious to water, and can be obtained in various weights and widths, with or without perforations. The paper is sometimes laid in strips with a row of vegetables planted between. The usual method, however, is to lay the paper and sow the seeds in hills, or transplant the plants, in the holes along the middle of the strip of paper.

The chief disadvantage in the use of mulch paper is the difficulty with which it is held in place. For this purpose laths, steel rods, wire and staples, stones and covering the edges of the paper with soil have been advocated. In 1925 the writer tried various methods at Ithaca, New York, but found that the only practical means of anchoring the paper was by covering the edges with soil. The other methods are too expensive or laborious to make their use economical except for very special crops. There have been machines developed for laying and covering the edges of the paper with soil in one operation.[13, 14] However, in humid regions some of the lighter grades of paper when covered with soil will decay before the end of the growing season.

25. *Chemicals.*

See suggestions under 16. *Chemicals* (page 56).

METHODS FOR DESTROYING WEED SEEDS IN THE SOIL

It is evident that very little progress can be made toward cleaning a field of weeds by destroying the plants unless measures are taken

[13] Hutchins, 1933. [14] Musselman, 1929.

also to destroy the seeds already in the soil. These seeds, which may have been produced during one or more previous seasons, will continue to come up and produce new crops of weeds during several following seasons unless they are destroyed.

26. *Harrowing and shallow cultivation.*

By harrowing or disking the stubble in the autumn as soon as the grain crop is harvested, many seeds may be induced to germinate in time to be killed by frosts or autumn plowing before they have had an opportunity to produce seeds. Early shallow cultivation should follow in the spring. This will induce many seeds to germinate so that they may be destroyed by the preparation of the seed-bed a week or two later. If the field is badly infested with seeds, it may even be more desirable to fit the land a second time even if the crop must be planted later. It is easier to deal with the weeds before, rather than after, the crop is planted. All seeds that are induced to germinate and all weeds that are destroyed before the crop is planted will not have to be contended with later when the crop is in the way. After two or more seasons of this treatment, combined with other methods if necessary, depending on how badly the soil is infested, the weed plants should be reduced to the point where they can be pulled or hoed by hand.

27. *Deep plowing.*

Land on which the surface soil is infested with viable weed seeds should not be plowed deeply until after it has been harrowed or disked several times so as to induce the seeds to germinate. The seeds of some weeds retain their viability for many years if they are buried under six or eight inches of soil. On being brought to the surface again they germinate. The seeds of a few weeds such as chess *(Bromus secalinus),* and corn cockle *(Agrostemma githago),* are usually destroyed by plowing them under to a depth of six to eight inches and leaving the soil undisturbed for two years. Such weeds seldom cause trouble by starting from seeds remaining in the soil except when a grain crop is grown on the same field year afer year. They are most abundant where their seeds are sowed as impurities with the grain.

PART II
WEEDS ARRANGED ACCORDING TO FAMILY TOGETHER WITH KEY

KEY TO THE GROUPS AND SPECIES OF WEEDS

	PAGE
1. Stems woody, persistent; weed a shrub................A	65
1. Stems herbaceous, dying to the ground; weed an herb.......AA	67
2. Plant a pteridophyte (producing spores)................B	67
2. Plant a spermatophyte (producing flowers)............BB	67
3. Leaves parallel-veined; weed a monocotyledon..........C	67
4. Weed a sedge..	68
4. Weed a grass..	69
3. Leaves netted-veined; weed a dicotyledon.............CC	73
5. Ovary superior...................................D	73
6. Petals absent, or not unlike sepals..................E	73
6. Petals present, unlike the sepals................EE	79
7. Corolla polypetalous, petals separate............F	79
7. Corolla gamopetalous, petals united...........FF	88
5. Ovary inferior................................DD	96
8. Flowers in involucrate heads......................	96
8. Flowers not in involucrate heads...................	104

A. Weeds with woody stems, shrubs.
  1. Leaves compound.
    2. Plants without spines or prickles.
      3. Leaves opposite; flowers orange to red.... *Campsis radicans*, 436 *
      3. Leaves alternate; flowers whitish, green or yellow.
        4. Leaflets 3–10 cm. long; flowers small, green or whitish; fruit a white or yellowish drupe.
          5. Leaves with 3 leaflets................ *Rhus radicans*, 302
          5. Leaves with 5 or more entire leaflets...... *Rhus vernix*, 303
        4. Leaflets 1–2 cm. long; flowers showy, yellow.
          6. Leaflets 5 or more, silky; flowers regular; branchlets brownish; fruit an achene........ *Potentilla fruticosa*, 237
          6. Leaflets 3, glabrous; flowers irregular; branchlets green; fruit a legume.................... *Cytisus scoparius*, 257
    2. Plants with spines or prickles.
      7. Stems perennial; stipules grown to the base of the petiole; flower producing a cup with many dry seed-like fruits.
                                        Rosa eglanteria, 244
      7. Stems biennial; stipules free from the base of the petiole; flower producing a fleshy aggregate fruit....... *Rubus spp.*, 245

* These numbers refer to the descriptions of the species beginning on page 106.

1. Leaves simple.
    8. Branches, at least some of them, with spines or thorns.
        9. Leaves reduced to spine-like petioles; flowers yellow, irregular.
                                                        Ulex europaeus, 272
        9. Leaves broader, not spine-like.
            10. Leaves entire; flowers irregular; fruit a legume.
                                                        Alhagi camelorum, 251
            10. Leaves toothed or lobed; flowers regular.
                11. Spines branched, or 3 in a cluster; flowers yellow.
                                                        Berberis vulgaris, 179
                11. Thorns unbranched; flowers pink or white.
                                                        Crataegus spp., 232
    8. Branches without spines or thorns.
        12. Margins of leaves entire.
            13. Stems climbing or trailing; leaves opposite; fruit a berry.
                                                        Lonicera japonica, 453
            13. Stems erect.
                14. Foliage leathery, persistent; flowers pink, rose, or white; fruit a 5-valved capsule.
                    15. Leaves distinctly petioled, green on both surfaces.
                        16. Leaves mostly opposite or whorled; capsules in lateral clusters.................*Kalmia angustifolia,* 330
                        16. Leaves mostly alternate; capsules in terminal clusters..........................*Kalmia latifolia,* 331
                    15. Leaves nearly sessile; whitish on the lower surface.
                                                        Kalmia polifolia, 332
                14. Foliage thin, deciduous; flowers yellow.
                    17. Leaves alternate; flowers irregular; fruit a legume.
                                                        Genista tinctoria, 259
                    17. Leaves opposite, with pellucid dots; flowers regular; fruit a 3–5 valved capsule....*Hypericum spathulatum,* 310
        12. Margins of leaves lobed or toothed.
            18. Margins of leaves lobed.
                19. Leaves silvery-white, wedge-shaped, with 3–5 terminal teeth or lobes....................*Artemisia tridentata,* 476
                19. Leaves not silvery-white.
                    20. Leaves pinnately lobed, sweet-scented; flowers in catkins; fruit dry.................*Comptonia peregrina,* 86
                    20. Leaves with one deep basal lobe on each side, with disagreeable odor; flowers purple, in cymes; fruit a red berry.
                                                        Solanum dulcamara, 409
            18. Margins of leaves toothed.
                21. Petiole with a pair of glands near the upper end; flowers white, producing stone-fruits.........*Prunus virginiana,* 243
                21. Petiole without glands; flowers producing dry fruits.
                    22. Leaves woolly on the lower surface; flowers pink or rose.
                        23. Seed-pods hairy; eastern........*Spiraea tomentosa,* 250

Key to the Weeds 67

    23. Seed-pods glabrous; western....... *Spiraea douglasii*, 248
   22. Leaves not woolly on the lower surface.
    24. Leaves finely toothed, narrow; stems yellowish-brown;
       flowers white........................*Spiraea alba*, 247
    24. Leaves coarsely toothed, broad; stems reddish-brown;
       flowers mostly pink or rose-colored.. *Spiraea latifolia*, 249
AA. Weeds without woody stems, herbs.
  B. Reproduction by spores borne in sporangia, Ferns and Fern Allies (Pteridophytes). (See also BB, page 67.)
    1. Leaves reduced to small scales, whorled, on jointed hollow stems; sporangia borne in terminal cones.
      2. Plants producing two kinds of stems, fertile ones appearing first, not green, terminated by a cone, sterile ones appearing later, green and much branched.
        3. Fertile stems simple; sterile stems with erect or spreading branches; leaves, "teeth," black........*Equisetum arvense*, 1
        3. Fertile stems with 1 or 2 whorls of branches; sterile stems with drooping branches; leaves, "teeth," brown.
                            Equisetum sylvaticum, 4
      2. Plants producing one kind of stem, green, unbranched or with a few simple branches, bearing a cone at the apex.
        4. Stems perennial, evergreen..........*Equisetum hyemale*, 2
        4. Stems annual, dying at the end of the season.
                            Equisetum palustre, 3
    1. Leaves (fronds) prominent, large and compound, alternate, from underground stems (rootstocks); sporangia borne on the fronds.
      5. Fronds of two kinds; fertile ones bearing sporangia in globular segments, sterile ones green and 1-pinnate.
        6. Rootstock creeping; fronds scattered, without hairs.
                          Onoclea sensibilis, 7
        6. Rootstock not creeping; fronds clustered, hairy.
                        Osmunda cinnamomea, 5
      5. Fronds all alike, scattered, green and 2–3 times pinnate.
        7. Fronds long and narrow; sporangia in scattered sori on the lower surface of pinnae........*Dennstaedtia punctilobula*, 6
        7. Fronds broadly triangular; sporangia in a continuous sorus along the margin of pinnae.
          8. Underside of fronds pubescent or tomentose.
                    Pteridium aquilinum var. *pubescens*, 8
          8. Underside of fronds glabrous.
                    Pteridium aquilinum var. *latiusculum*, 9
BB. Reproduction by seeds borne in flowers, vegetative propagation often present (Spermatophytes).
  C. Leaves mostly parallel-veined; flower-parts in 3's; cotyledons 1. (Monocotyledons). (See also CC, on page 73.)
    1. Herbage with the odor of onion or garlic (stem bulbous at base, bulblets often present at tip of stem).

68 Weeds

  2. Leaves elliptic-lanceolate, 2–5 cm. wide........ *Allium tricoccum*, 79
  2. Leaves linear, less than 1 cm. wide.
    3. Stems leafy at least to the middle............. *Allium vineale*, 80
    3. Stems leafy only near the base.............. *Allium canadense*, 78
1 Herbage without odor of onion or garlic.
  4. Leaves broad.
    5. Plants prostrate, rooting at the nodes; flowers irregular.
          Commelina communis, 73
    5. Plants erect; flowers regular.
      6. Stems leafy only at the base.
        7. Leaves numerous, in 2 ranks; flowers large, over 5 cm. in
            diameter, orange or yellow.......... *Hemerocallis fulva*, 82
        7. Leaves 2–3; flowers small, about 1 cm. in diameter,
            white............................. *Convallaria majalis*, 81
      6. Stems leafy to the top; flowers greenish..... *Veratrum viride*, 84
  4. Leaves narrow, grass-like or rush-like.
    8. Perianth-segments 6.
      9. Plant with bulb; perianth-segments white or yellowish-white;
          leaves flattened.
        10. Flowers in terminal umbel... *Ornithogalum umbellatum*, 83
        10. Flowers in racemes............... *Zigadenus venenosus*, 85
      9. Plants without bulb; perianth-segments green or brownish;
          leaves wiry.
        11. Inflorescence lateral .... *Juncus effusus*, 75
        11. Inflorescence terminal.
          12. Plants annual, low .... *Juncus bufonius*, 74
          12. Plants perennial, tall.
            13. Rhizomes slender, elongate and horizontal; leaf
                sheaths extending above lower fourth of stem.
                              Juncus gerardii, 76
            13. Rhizomes fibrous and short; leaf sheaths covering
                only lower fourth of stem .... *Juncus tenuis*, 77
    8. Perianth absent (or reduced to bristles); flowers in spikelets;
        fruit a grain or achene.
      14. Leaves in 3 ranks; stems mostly solid; fruit an achene
          (Sedges).
        15. Scales of the spikelet in 2 ranks.
          16. Plants annual; forming neither tubers nor corms.
            17. Scales of spikelet marked with dark brown or pur-
                ple; styles 2-cleft................ *Cyperus diandrus*, 66
            17. Scales of spikelet yellow-brown; styles 3-cleft.
                              Cyperus iria, 68
          16. Plants perennial; with tubers or corm-like bases.
            18. Plants with tuber bearing rootstocks.
              19. Spikelets light brown or straw-colored.
                          Cyperus esculentus, 67

Key to the Weeds

  19. Spikelets dark chestnut-brown or purple.
                Cyperus rotundus, 69
  18. Plants without tubers but with corm-like bases.
                Cyperus strigosus, 70
 15. Scales of spikelet in 3 ranks.
  20. Bristles barbed, about as long as achene, rudimentary or lacking .... *Scirpus atrovirens,* 71
  20. Bristles smooth, rust-colored and much exceeding the achene.
                Scirpus cyperinus, 72
14. Leaves in 2 ranks; stems mostly hollow; fruit a grain (Grasses).
  21. Spikelets (1–5 together) inclosed in a globular spiny bur-like involucre............................. *Cenchrus longispinus,* 27
  21. Spikelets not inclosed in a spiny bur-like involucre.
    22. Flowers in spikelets 1 (rarely 2–3, the lower sterile).
      23. Arrangement of spikelets in 2 rows along an axis.
        24. Spikelets in 2 rows on one side of a flattened axis.
          25. Plant with long creeping rootstocks.
                Cynodon dactylon, 28
          25. Plant without creeping rootstocks.
            26. Second flowering glume thin, with hyaline flat margins; annuals.
              27. Stems erect or ascending, rachis narrowly winged.
                  Digitaria filiformis, 31
              27. Stems often rooting at lower nodes, decumbent or spreading; rachis broadly winged.
                28. First empty glume wanting, or minute and scarious, second as long as spikelet; pedicels nearly terete, glabrous or nearly so; leaves 3–6 mm. wide.
                    Digitaria ischaemum, 32
                28. First empty glume minute, herbaceous, second one-half as long as the spikelet; pedicels angled and scabrous; leaves 4–10 mm. wide.
                    Digitaria sanguinalis, 33
            26. Second flowering glume thick, with firm inrolled margins; perennials.
              29. Spikelets mostly solitary on rachis, rounded and glabrous....................... *Paspalum laeve,* 55
              29. Spikelets mostly in pairs.
                30. Spikelets ovate to orbicular, glabrous or minutely pubescent.............. *Paspalum ciliatifolium,* 53
                30. Spikelets ovoid and acute, villous.
                  Paspalum dilatatum, 54
        24. Spikelets 2–3 at a node, in 2 rows on opposite sides of the solitary terminal spike.
          31. Flowering glume of middle spikelet pedicelled; empty glumes ciliate................. *Hordeum murinum,* 43

70 Weeds

31. Flowering glume of middle spikelet sessile; empty glumes scabrous.
32. Empty glumes not all alike, first empty glume of lateral spikelets and both empty glumes of middle spikelets much broader than the others............*Hordeum pusillum*, 45
32. Empty glumes of 3 spikelets at a node all alike, awn-like.
   33. Flowering glumes of lateral spikelets long-awned.
Hordeum jubatum, 42
   33. Flowering glumes of lateral spikelets awnless or short-awned..........................*Hordeum nodosum*, 44
23. Arrangement of spikelets in panicles or spike-like racemes.
  34. Fertile flower solitary in spikelet (not subtended by sterile flowering glume).
    35. Spikelets in a dense cylindrical panicle; flowering glume with a dorsal awn........................*Alopecurus geniculatus*, 12
    35. Spikelets in an open narrow panicle; flowering glume with a terminal awn or awnless.
     36. Awns straight, unbranched, or absent.
      37. Erect slender plants; flowering glume awnless, loosely enveloping the grain.
       38. Annual; spikelets 2.5–3 mm. long.
Sporobolus neglectus, 62
       38. Perennial; spikelets 1.5–2 mm. long.
Sporobolus poiretii, 63
      37. Decumbent plants; flowering glume awned or acuminate, closely enveloping the grain.
       39. Empty glumes minute; scaly rootstock not apparent.
Muhlenbergia schreberi, 48
       39. Empty glumes at least half the length of the flowering glume; rootstock obviously scaly.
Muhlenbergia frondosa, 47
     36. Awns twisted or branched.
      40. The awns branched.
       41. Lateral awns much shorter than the middle awn, 3–6 mm. long......................*Aristida dichotoma*, 17
       41. Lateral awns not much shorter than the middle awn, 3–7 cm. long....................*Aristida oligantha*, 18
      40. The awns twisted, 1–2.5 dm. long.
       42. Awn slender, fragile; empty glumes 2–2.8 cm. long.
Stipa comata, 64
       42. Awn rigid; empty glumes 2.8–3.5 cm. long. *Stipa spartea*, 65
  34. Fertile flower subtended by 1 or 2 sterile flowering glumes.
    43. Sterile flowering glumes 2.
     44. And not reduced...............*Anthoxanthum odoratum*, 16
     44. And much reduced, unawned; empty glumes equal.
Phalaris canariensis, 56

Key to the Weeds

43. Sterile flowering glumes 1.
  45. First empty glume larger than the second; spikelets with long silky hairs, spikelets in pairs or 3's, one sessile and fertile, the others pedicelled and sterile or rudimentary.
    46. Spikelets in 3's, 1 fertile and 2 sterile or rudimentary.
Sorgum halepense, 61
    46. Spikelets in pairs, 1 fertile and 1 sterile.
      47. Raceme solitary............*Andropogon scoparius*, 14
      47. Racemes in fascicles of 2–6.
        48. Pedicelled spikelet staminate.
Andropogon gerardii, 13
        48. Pedicelled spikelet rudimentary.
Andropogon virginicus, 15
  45. First empty glume smaller than the second; spikelets all fertile, without silky hairs.
    49. Spikelets surrounded by a bristly involucre, in spike-like clusters.
      50. Bristles 5 or more about each spikelet, tawny; spikelets 3 mm. long....................*Setaria glauca*, 58
      50. Bristles 1–3, green or purple; spikelets 2.5 mm. long.
        51. Barbs on bristles pointing upward.
Setaria viridis, 60
        51. Barbs on bristles pointing downward.
Setaria verticillata, 59
    49. Spikelets without involucre, in open panicles.
      52. Branches of the panicle one-sided.
        53. Spikelets very bristly; coriaceous flowering glume acute; nodal hairs of panicle short.
Echinochloa pungens, 35
        53. Spikelets not conspicuously bristly; coriaceous flowering glume obtuse; nodal hairs long.
Echinochloa crusgalli, 34
      52. Branches of the panicle not one-sided.
        54. Plant hairy.
          55. Spikelets acuminate, 2–4 mm. long.
Panicum capillare, 49
          55. Spikelets obtuse or apiculate, 1.3–2.2 mm. long.
Panicum gattingeri, 51
        54. Plant glabrous.
          56. Annual, much branched, decumbent.
Panicum dichotomiflorum, 50
          56. Perennial, strict, erect, unbranched.
Panicum virgatum, 52
22. Flowers in spikelets 2-many (sometimes upper flower sterile).
  57. Spikelets arranged in 2 rows along axis.

58. Arrangement of spikelets in 2 rows on one side of a flattened axis (spikes 1-several, usually digitate).
59. Rachis of spike extending beyond spikelets.
Dactyloctenium aegypticum, 29
59. Rachis not extending beyond spikelets...... *Eleusine indica*, 36
58. Arrangement of spikelets in 2 rows on opposite sides of the main stem, spike solitary, terminal.
60. Spikelets solitary at each joint of rachis.
61. The spikelets placed edgewise to rachis.
Lolium temulentum, 46
61. The spikelets placed flatwise to the rachis.
Agropyron repens, 10
60. Spikelets 2-3 at each joint of rachis, placed edgewise to rachis (lateral spikelets reduced to bristles)......*Hordeum jubatum*, 42
57. Spikelets in panicles or racemes.
62. Empty glumes extending to the tip of the lowest flowering glume or beyond, awns on the back of or between the teeth of the flowering glume, or absent.
63. Rachilla prolonged behind the upper flower.
Aira caryophyllea, 11
63. Rachilla not prolonged behind the upper flower.
64. Awn from between 2 apical teeth........*Danthonia spicata*, 30
64. Awn attached to the back of the glume.
65. Plant velvety; awns minute.............*Holcus lanatus*, 41
65. Plant not velvety; awn 1 cm. long, or more.
Avena fatua, 19
62. Empty glumes not extending to the tip of the lowest flowering glume; awns at the tip of, or rarely just below the tip, or between the teeth, or absent.
66. Flowering glumes 3-nerved.
67. Spikelet 2-4 flowered.................*Eragrostis frankii*, 37
67. Spikelet 5-many flowered.
68. Width of spikelet 0.8-1.5 mm......*Eragrostis pectinacea*, 39
68. Width of spikelet 1.8-3 mm.
69. Flowering glumes densely imbricated; joints of the rachilla not visible........... *Eragrostis megastachya*, 38
69. Flowering glumes loosely imbricated; joints of the rachilla visible................*Eragrostis poaeoides*, 40
66. Flowering glumes 5-many-nerved.
70. Spikelets less than 5 mm. long; flowering glumes keeled, often with cobwebby hairs at base............*Poa annua*, 57
70. Spikelets more than 10 mm. long, exclusive of awns; flowering glumes rounded on the back, with no cobwebby hairs.
71. Awn much longer than the body of the narrow-lanceolate flowering glume; lower empty glume 1-nerved, upper one 3-nerved.

Key to the Weeds

  72. Panicle dense, ovoid, erect.
             Bromus rubens, 23
  72. Panicle spreading, drooping.
   73. Length of awn about 1.5 cm.; flowering glume villous-strigose; panicle dense . . . . *Bromus tectorum,* 26
   73. Length of awn 2–3 cm.; flowering glume scabrous; panicle open.
             Bromus sterilis, 25
 71. Awn shorter, or not much longer, than the body of the broadly elliptical flowering glume; lower empty glume 2–5-nerved, upper one 5–9-nerved.
  74. Sheaths glabrous . . . *Bromus secalinus,* 24
  74. Sheaths pubescent.
   75. Panicle dense, erect.
             Bromus mollis, 21
   75. Panicle open.
    76. Panicle long, drooping; flowering glume 9–10 mm. long.
            Bromus commutatus, 20
    76. Panicle short, erect; flowering glume less than 7 mm. long.
            Bromus racemosus, 22
CC. Leaves mostly netted-veined; flower-parts usually in 4's or 5's; stem bundles in a ring; cotyledons 2 (Dicotyledons).
 D. Ovary superior. (See also DD on page 96.)
  E. Corolla absent or resembling calyx. (See also EE on page 79.)
   1. Plant covered with whitish branched hairs.
            Eremocarpus setigerus, 290
   1. Plant not covered with whitish branched hairs.
    2. Ovary 2-many-celled or ovaries several.
     3. Cells of ovary 10; flowers white, in racemes.
           Phytolacca americana, 144
     3. Cells of ovary 3 or separating into 3 1-celled carpels.
      4. Leaves whorled; sap not milky.
           Mollugo verticillata, 145
      4. Leaves alternate or opposite (the upper sometimes whorled).
       5. Flowers with a calyx; sap watery.
        6. Lower surface of leaves with stellate pubescence.
           Croton capitatus, 289
        6. Pubescence not stellate or wanting.
         7. Leaves large and palmate, plant with rather stiff stinging hairs . . . . *Cnidoscolus stimulosus,* 288
         7. Leaves not lobed.

8. Pistillate flower with a 5–9 lobed leafy bract.
Acalypha rhomboidea, 286
8. Pistillate bract 9–15 lobed............*Acalypha virginica*, 287
5. Flowers without calyx; a pistillate flower and several staminate flowers surrounded by a cup-shaped involucre; sap milky (Euphorbia).
  9. Flowers (involucres) axillary; leaves opposite, serrate; glands 4, appendaged; annuals.
    10. Stems ascending or erect, nearly smooth; leaves 1.5–3.5 cm. long, with a purple spot; capsule broadly ovate, sharply angled, smooth; seeds ash-colored..............*Euphorbia maculata*, 297
    10. Stems prostrate, hairy; leaves 1.5 cm. long or less.
      11. The stems reddish or greenish, hirsute; leaves bright green; capsule smooth; seeds ash-colored.
Euphorbia vermiculata, 301
      11. The stems flesh-color, puberulent; leaves dull, often with a purple spot; capsule hairy; seeds reddish-flesh-colored.
Euphorbia supina, 300
  9. Flowers (involucres) in terminal, simple or compound umbels, or the inflorescence dichotomous; leaves serrate, dentate, or entire.
    12. Upper leaves white-margined..........*Euphorbia marginata*, 298
    12. Upper leaves not white-margined.
      13. Appendages of the involucral glands large, showy, white; leaves entire, linear-elliptic; perennials.
Euphorbia corollata, 291
      13. Appendages absent; leaves linear or obovate.
        14. Leaves toothed; annuals.
          15. The leaves alternate below, obovate, finely serrate, glabrous; seeds sculptured........*Euphorbia helioscopia*, 295
          15. The leaves opposite, coarsely dentate, hairy; seeds minutely tubercled.................*Euphorbia dentata*, 293
        14. Leaves entire, the lower ones alternate.
          16. Shape of leaves obovate; seeds sculptured; annuals.
Euphorbia peplus, 299
          16. Shape of leaves linear to oblong; seeds smooth; perennials.
            17. Primary bracts 15–20 mm. wide; leaves 4–25 mm. wide; glands horned.
              18. Leaves 15–25 mm. wide, bright green; pods wrinkled.......................*Euphorbia lucida*, 296
              18. Leaves 4–9 mm. wide, glaucous, pods minutely granular or smooth................*Euphorbia esula*, 294
            17. Primary bracts 4–7 mm. wide; leaves 1–2 mm. wide; glands scarcely horned; pods minutely granular.
Euphorbia cyparissias, 292

Key to the Weeds

2. Ovary 1-celled, solitary.
   19. Seeds in ovary several.................*Cerastium vulgatum,* 151
   19. Seeds in ovary 1.
      20. Stipules sheathing.
         21. Sepals 4–5, nearly equal; stigmas not tufted (Polygonum).
            22. Aquatic; floating leaves long-petioled, glabrous with rounded to cordate bases.........*Polygonum coccineum,* 93
            22. Terrestrial; without floating leaves.
               23. Stems twining; leaves broadly ovate; cordate at base.
                  24. Calyx sharply angled; achenes dull black.
                      Polygonum convolvulus, 94
                  24. Calyx wing-angled; achenes glossy black.
                      Polygonum scandens, 103
               23. Stems not twining; leaves linear to ovate, or cordate.
                 25. Flowers axillary.
                    26. Plant usually prostrate; calyx in fruit 2–3.5 mm. long, margin whitish or pink.
                        Polygonum aviculare, 92
                    26. Plants erect or ascending.
                       27. Leaves linear to elliptical; fruiting calyx 1.5–5 mm. long.
                          28. Fruiting calyx 3–5 mm. long; achene reticulate and dull...............*Polygonum erectum,* 96
                          28. Fruiting calyx 1.5–2 mm. long; achene minutely dotted and glossy....*Polygonum argyrocoleon,* 91
                     27. Leaves broadly ovate, 8–15 cm. long, 5–12 cm. wide; fruiting calyx 8–9 mm. long.
                        Polygonum cuspidatum, 95
               25. Flowers in terminal spikes.
                  29. Sheaths ciliate with a row of bristles.
                      30. Sepals glandular-dotted; stamens 6.
                        Polygonum hydropiper, 97
                      30. Sepals not glandular-dotted.
                        31. Leaves broadly ovate; spikes nodding, stout; flowers 3–5 mm. long; stamens 7; annual.
                            Polygonum orientale, 100
                        31. Leaves lanceolate; spikes erect.
                          32. Annual; flowers dull, pale, greenish-purple or greenish-white.....*Polygonum persicaria,* 102
                          32. Perennial; flowers white or pink.
                             33. Sheaths with bristles 5–15 mm. long.
                               Polygonum setaceum, 104
                             33. Sheaths with bristles 1–7 mm. long; spikes often interrupted at base.
                               Polygonum hydropiperoides, 98
                  29. Sheaths not ciliate, except rarely the uppermost.

76 Weeds

 34. Perennial with creeping rootstock.
              Polygonum coccineum, 93
 34. Annual without creeping rootstock.
  35. Peduncle glandular-pubescent; spikes thick, erect; stamens 8............*Polygonum pensylvanicum,* 101
  35. Peduncle glabrous or but slightly glandular; spikes slender, drooping; stamens 6.
             Polygonum lapathifolium, 99
21. Sepals 6, the 3 inner ones much enlarged in fruit (except *Rumex acetosella*); stigmas tufted.
 36. Plants dioecious; leaves hastate or sagittate.
  37. Sepals not enlarged in fruit; leaves hastate; low plants with running rootstocks....................*Rumex acetosella,* 106
  37. Sepals enlarged in fruit; plants erect, 3–15 dm. high.
   38. Leaves oblong to broadly lanceolate and arrow-shaped; valves of fruit 4–6 mm. wide............*Rumex acetosa,* 105
   38. Leaves lanceolate to hastately 3-lobed; valves of fruit 2–4 mm. wide.......................*Rumex hastatulus,* 109
 36. Plants with perfect flowers or imperfectly monoecious; leaves not hastate or sagittate.
  39. Tubercles on fruiting calyx 3; lower leaves rounded or cordate at base.
   40. Leaves dark green, crenate, crisped; pedicel jointed about one-fourth its length from base..........*Rumex crispus,* 108
   40. Leaves pale green, not crisped, nearly entire; pedicel jointed about one-tenth its length from base.
             Rumex mexicanus, 110
  39. Tubercles on fruiting calyx 1.
   41. Valves with long prominent teeth; lower leaves round or cordate at base...................*Rumex obtusifolius,* 111
   41. Valves entire or slightly dentate; leaves tapering or acuminate.
    42. Valves 6–9 mm. long; leaves dark green with crispness often apparent....................*Rumex patientia,* 112
    42. Valves 4–6 mm. long; leaves pale.....*Rumex altissimus,* 107
20. Stipules not sheathing, often absent.
 43. Style or stigma solitary.
  44. Plant without stinging hairs; flowers 3–5 in a spreading 5-lobed involucre.........................*Mirabilis nyctaginea,* 143
  44. Plant with stinging hairs; flowers without involucre.
   45. Annuals...............................*Urtica urens,* 90
   45. Perennials, with rootstocks.
    46. Leaves thin, ovate; plant dioecious.........*Urtica dioica,* 88
    46. Leaves firm, ovate-lanceolate or lanceolate; plant usually monoecious..........................*Urtica gracilis,* 89
 43. Styles or stigmas 2–5.

Key to the Weeds

47. Leaves palmately cleft or compound............Cannabis sativa, 87
47. Leaves pinnately lobed, toothed or entire.
   48. Flowers in scarious-bracted and often spiny clusters.
      49. Leaves with stipular spines.............Amaranthus spinosus, 142
      49. Leaves without stipular spines.
         50. Flowers in thick axillary clusters (in *A. lividus* a flowering terminal spike appears as well); plants prostrate or low and diffuse.
            51. Bracts much longer than sepals; stems whitish; plant erect, low and diffuse....................Amaranthus albus, 135
            51. Bracts scarcely exceeding sepals or shorter, stems not whitish.
               52. Bracts equal to or barely longer than sepals; sepals 3–5; plant prostrate..........Amaranthus graecizans, 136
               52. Bracts about half as long as sepals; sepals 2; plant procumbent or erect...............Amaranthus lividus, 138
         50. Flowers in prominent terminal spikes or panicles (small axillary inflorescences often occur as well).
            53. Plants dioecious; bracts greatly exceeding flowers; leaf tips tapering.......................Amaranthus palmeri, 139
            53. Plants monoecious.
               54. Sepals 1.5–2 mm. long; bracts 2–4 mm. long; branches of panicle slender, about 5 mm. wide above the middle.........................Amaranthus hybridus, 137
               54. Sepals 3 mm. or longer, bracts 4 mm. or longer; branches of panicle stiff and stout.
                  55. Stamens 5; sepals with rounded tips.
                                            Amaranthus retroflexus, 141
                  55. Stamens 3 (5); sepals with acute tips.
                                            Amaranthus powellii, 140
   48. Flowers without scarious bracts.
      56. Foliage reduced and awl-shaped.
         57. Leaves alternate, spine-like.
                                   Salsola kali var. tenuifolia, 134
         57. Leaves opposite, not spine-like.........Scleranthus annuus, 158
      56. Foliage not reduced and not awl-shaped.
         58. Flowers imperfect, the pistillate inclosed by 2 bracts.
            59. Bracts fused at the base; pistillate flowers without sepals.
               60. Leaves gray-mealy; fruiting bracts often warty; leaves mostly rhombic-ovoid...............Atriplex rosea, 115
               60. Leaves not gray-mealy.
                  61. Shape of leaves lanceolate...........Atriplex patula, 113
                  61. Shape of leaves hastate...Atriplex patula var. hastata, 114
            59. Bracts separate; pistillate flowers with 3–4 sepals.
                                   Axyris amaranthoides, 116
         58. Flowers mostly perfect.

62. Calyx 5-parted, winged in fruit.
  63. Mature sepals with incurved spike at tip; stems whitish; leaves short, flat, oblong linear.................*Bassia hyssopifolia,* 117
  63. Mature sepals without incurved spikes; stems light green to purple; leaves not as above.
    64. Leaves lanceolate oblong, sinuate-toothed.
Cycloloma atriplicifolium, 131
    64. Leaves linear, lanceolate, or spatulate and entire.
      65. Flowers solitary or paired in clusters; leaves with ciliate margins...................................*Kochia scoparia,* 133
      65. Flowers 3–5 in a cluster; leaves somewhat glaucous and fleshy.
Halogeton glomeratus, 132
62. Calyx 2–5 parted or absent, not winged in fruit.
  66. Calyx of a solitary sepal or absent; leaves linear.
Corispermum hyssopifolium, 130
  66. Calyx 3–5 parted; leaves broader.
    67. Foliage glandular (strong-scented when crushed).
      68. Leaves coarsely toothed; panicle broad.
Chenopodium ambrosioides, 119
      68. Leaves pinnately-lobed; panicle narrow, the branches one-sided...........................*Chenopodium botrys,* 121
    67. Foliage not glandular; often mealy.
      69. Plants perennial...........*Chenopodium bonus-henricus,* 120
      69. Plants annual.
        70. Seeds vertical or the terminal ones sometimes horizontal.
          71. Calyx fleshy in fruit; leaves not white beneath.
            72. Mature flowers bright red, berry-like; seeds dull-black with narrow rim-like margin; leaves triangular and often hastate..............*Chenopodium capitatum,* 122
            72. Mature flowers not bright red or berry-like; seeds a glossy-brown with rounded margin; leaves rhombic-ovoid, tapering at tip and base.
Chenopodium rubrum, 128
          71. Calyx not fleshy in fruit, green; leaves white beneath.
Chenopodium glaucum, 123
        70. Seeds all horizontal.
          73. Sepals thick, more or less keeled, closed in fruit; leaves and flowers mealy.
            74. Seed 1.5–2 mm. in diameter; leaves yellow-green or dark green................*Chenopodium paganum,* 126
            74. Seed 1.2 mm. in diameter; leaves glaucous-green.
Chenopodium album, 118
          73. Sepals thin, not keeled, somewhat open in fruit; leaves and flowers rarely slightly mealy.
            75. Axillary flower-clusters shorter than or equal to the leaves.

Key to the Weeds

      76. Leaves ovate to rhombic-ovate, coarsely dentate or wavy margined; seeds dull-black.
            Chenopodium murale, 125
      76. Leaves oblong to ovate, entire; seeds glossy-black.
            Chenopodium polyspermum, 127
   75. Axillary flower-clusters, at least the upper, longer than the leaves.
      77. Panicles narrow, racemose, leafy; leaves oblong, coarsely toothed; seeds glossy.
            Chenopodium urbicum, 129
      77. Panicles very loose and open, nearly naked; leaves broad, large with few large teeth or lobes.
            Chenopodium hybridum, 124

EE. Corolla present, unlike the calyx.
  F. Corolla polypetalous (See also FF on page 88).
    1. Stamens perigynous.
      2. Flowers regular.
        3. Sepals 2; pistil 1; plant prostrate; leaves fleshy.
            Portulaca oleracea, 146
        3. Sepals 5, or calyx 5–6 lobed; leaves not fleshy.
          4. Stems climbing by tendrils.........*Passiflora incarnata,* 312
          4. Stems not climbing by tendrils.
            5. Fruits inclosed in a dry receptacle; pistils 1–4.
              6. Leaves opposite or whorled; flowers purple.
                  Lythrum salicaria, 314
              6. Leaves alternate; flowers yellow or green.
                7. Receptacle top-shaped, ribbed, with hooked spines; flowers yellow... *Agrimonia gryposepala,* 231
                7. Receptacle 4-angular, naked; flowers greenish.
                    Sanguisorba minor, 246
            5. Fruits not inclosed in the receptacle; petals yellow; carpels several, separate.
              8. Receptacle enlarged in fruit, fleshy and red when mature......................*Duchesnea indica,* 233
              8. Receptacle not enlarged in fruit, dry when mature.
                9. Styles in fruit elongated and hooked.
                  Geum aleppicum, 234
                9. Styles in fruit not elongated.
                  10. Flowers solitary at the nodes of runner-like stems.
                    11. Leaves glabrous.........*Potentilla reptans,* 241
                    11. Leaves pubescent beneath.

Weeds

12. First flower from the node above the second or third well-developed internode.
          Potentilla simplex, 242
12. First flower from the node above the first well-developed internode.
         Potentilla canadensis, 236
 10. Flowers in cymose clusters.
  13. Leaves with 3 leaflets........*Potentilla norvegica,* 239
  13. Leaves with 5–9 leaflets.
   14. Leaflets green beneath.
    15. Flowers in compact almost leafless cymes.
           Potentilla recta, 240
    15. Flowers in open leafy cymes.
          Potentilla intermedia, 238
   14. Leaflets white-tomentose beneath.
          Potentilla argentea, 235
2. Flowers irregular; carpel 1.
 16. Calyx 6-lobed; placenta axile...............*Cuphea petiolata,* 313
 16. Calyx of 5 sepals or 5-lobed; placenta parietal.
  17. Foliage simple ....*Crotalaria sagittalis,* 256
  17. Foliage compound.
   18. Leaves with 3 leaflets.
    19. Leaflets entire.
     20. Pod 1-seeded; leaflets not stipellate.
          Lespedeza violacea, 263
     20. Pod 2-several-seeded; leaflets stipellate.
         Desmodium canadense, 258
    19. Leaflets serrulate.
     21. Pods curved or coiled; flowers in heads.
          Medicago lupulina, 265
     21. Pods straight.
      22. Flowers in heads; stamens attached to corolla.
       23. Plants silky-hairy; corolla pink.. *Trifolium arvense,* 269
       23. Plants not silky-hairy; corolla yellow.
        24. Leaflets all sessile; plants erect.
           Trifolium agrarium, 268
        24. Terminal leaflet stalked; plants spreading.
         25. Head 20–40-flowered.. *Trifolium procumbens,* 271
         25. Head 3–20-flowered.......*Trifolium dubium,* 270
      22. Flowers in racemes; stamens free from the corolla.
       26. Corolla white; legume reticulate....*Melilotus alba,* 266
       26. Corolla yellow; legume with transverse ridges.
          Melilotus officinalis, 267
   18. Leaves with more than 3 leaflets, rarely only 2.
    27. Leaves palmately compound..........*Lupinus perennis,* 264
    27. Leaves pinnately compound.
     28. Leaves usually bearing tendrils.

Key to the Weeds

29. Style flat, bearded down inner face.
    30. Leaflets 2; flowers yellow.... *Lathyrus pratensis*, 262
    30. Leaflets 6–10; flowers purple.. *Lathyrus palustris*, 261
29. Style filiform, bearded at apex only.
    31. Flowers sessile, 1–3 in axils of leaves.
        Vicia angustifolia, 273
    31. Flowers in peduncled racemes.
        32. Leaves with 2–5 pairs of leaflets; flowers less than 5 mm. long; legume 4 seeded.
        Vicia tetrasperma, 275
        32. Leaves with 4–12 pairs of leaflets; flowers 1 cm. long or longer; legume 6–12 seeded.
        Vicia cracca, 274
28. Leaves without tendrils, flowers yellow.
    33. Flowers irregular; stamens diadelphous.
        Glycyrrhiza lepidota, 260
    33. Flowers regular; stamens separate.
        34. Flowers in axillary racemes.
            35. Leaflets 10–18, lance-oblong; pods flattened, 8–11 mm. wide............*Cassia marilandica*, 253
            35. Leaflets 4–6, obovate; pods 4-angled, 3–4 mm. wide.........................*Cassia tora*, 255
        34. Flowers solitary or in small axillary clusters.
            36. Flowers 4–8 mm. broad; anthers 5.
            Cassia nictitans, 254
            36. Flowers 2.5–4 cm. broad; anthers 10.
            Cassia fasciculata, 252
1. Stamens hypogynous; flowers mostly regular.
    37. Stamens monadelphous or diadelphous (at least at base).
        38. Stem climbing by tendrils; leaves alternate, simple.
        Passiflora incarnata, 312
        38. Stem not climbing by tendrils.
            39. Ovary 1-celled; placenta free-central; leaves opposite.
                40. Plant creeping; corolla yellow...*Lysimachia nummularia*, 334
                40. Plant prostrate but not creeping; corolla red or white.
            Anagallis arvensis, 333
            39. Ovary 4-many-celled; placenta axile; leaves alternate.
                41. Stamens 10, united only at the base.
                      42. Plants producing rhizomes; stipules minute or absent.
                          43. Pedicels in fruit reflexed; capsule mostly 10–15 mm. long..............................*Oxalis florida*, 278
                          43. Pedicels in fruit not reflexed; capsule mostly 5–12 mm. long.......................*Oxalis europaea*, 277
                      42. Plants not producing rhizomes; stipules prominent, oblong; capsule mostly 15–25 mm. long.
            Oxalis stricta, 279

41. Stamens numerous, united more than half their length.
  44. Carpels 5, united at maturity, forming a 5-celled dehiscent capsule or separating.
    45. Seeds in carpel 1; involucre none........ *Sida spinosa,* 308
    45. Seeds in carpel several—many; involucre present.
      Hibiscus trionum, 305
  44. Carpels 10–20, separating from the central axis when mature.
    46. Involucre present; carpels 1-seeded, indehiscent.
      47. Stems procumbent; flowers axillary; leaves with shallow crenate lobes.................*Malva neglecta,* 307
      47. Stems erect; flowers clustered toward the end of branches; leaves deeply lobed......*Malva moschata,* 306
    46. Involucre absent; carpels 2-seeded, dehiscent.
      Abutilon theophrasti, 304
37. Stamens separate (in fascicles in *Hypericum*).
  48. Carpels 3-many, separate.
    49. Flowers irregular, spurred.
      50. Stems low, with few leaves..........*Delphinium menziesii,* 170
      50. Stems taller, leafy.................*Delphinium simplex,* 171
    49. Flowers regular.
      51. Carpels 5; leaves not lobed, very fleshy.
        52. Leaves flat and broad; corolla purple; low, spreading.
          Sedum purpureum, 230
        52. Leaves cylindric to ovate; corolla yellow; erect.
          Sedum acre, 229
      51. Carpels many; leaves, at least the upper, lobed, scarcely fleshy.
        53. Flowers orange or red, solitary, terminal; leaves with narrow linear dissections...................*Adonis annua,* 160
        53. Flowers yellow, solitary or in corymbose clusters, leaves 3-lobed, parted or deeply divided.
          54. Achenes with two lateral vesicles, tomentose; sepals tomentose..................*Ranunculus testiculatus,* 178
          54. Achenes without vesicles, glabrous.
            55. Flowers small, 1 cm. in diameter, or less.
              56. Basal leaves lobed or parted; annual or winter annual.
                57. Achenes 4.5–10 mm. long with numerous spines.
                  Ranunculus arvensis, 174
                57. Achenes 1 mm. long, granular or ridged.
                  Ranunculus sceleratus, 177
              56. Basal leaves not lobed; biennial.
                Ranunculus abortivus, 172
            55. Flowers large, 1.5–2.5 cm. in diameter.
              58. Plant low, stoloniferous.......*Ranunculus repens,* 176
              58. Plant erect, not stoloniferous.

Key to the Weeds 83

        59. Leaves with terminal division stalked; base of stem corm-like................*Ranunculus bulbosus,* 175
        59. Leaves with all divisions sessile; base of stem not corm-like....................*Ranunculus acris,* 173
48. Pistil solitary.
  60. Ovary 1-celled.
    61. Leaves opposite.
      62. Leaves punctate; placenta parietal.
                                      Hypericum perforatum, 309
      62. Leaves not punctate; placenta free-central.
        63. Sepals united.
          64. Calyx with scaly bracts or small leaves at base; styles 2.
                                          Dianthus armeria, 152
          64. Calyx without bract, naked.
            65. Styles 5, rarely 4.
              66. Styles opposite the petals.....*Agrostemma githago,* 147
              66. Styles alternate with the petals.
                67. Flowers perfect; petals 4-lobed.
                                        Lychnis flos-cuculi, 155
                67. Flowers imperfect.
                  68. Flowers fragrant, opening in evening.
                                          Lychnis alba, 153
                68. Flowers inodorous, open in daytime.
                                          Lychnis dioica, 154
            65. Styles 2-3.
              69. Styles 2; calyx obscurely nerved.
                70. Calyx terete; perennial.....*Saponaria officinalis,* 156
                70. Calyx 5-angled; annual......*Saponaria vaccaria,* 157
              69. Styles 3; calyx 5-many-nerved.
                71. Perennials or biennials; glaucous, flowers in panicles.
                    72. Calyx much inflated; seeds 1-1.5 mm. long.
                                          Silene cucubalus, 161
                    72. Calyx but slightly inflated; seeds 0.6-1 mm. long........................*Silene cserei,* 160
                71. Annuals; not glaucous.
                    73. Stem glabrous, each joint with a glutinous band....................*Silene antirrhina,* 159
                    73. Stem pubescent and viscid.
                      74. Flowers in racemes.
                        75. Racemes branched; leaves lanceolate.
                                          Silene dichotoma, 162
                        75. Racemes simple; leaves spatulate.
                                            Silene gallica, 163
                    74. Flowers in cymes..........*Silene noctiflora,* 164
        63. Sepals separate, more or less spreading.
          76. Scarious stipules present.

77. Leaves whorled; pistil with 5 styles, 5-valved.
   Spergula arvensis, 165
77. Leaves fascicled; pistil with 3 styles, 3-valved.
   Spergularia rubra, 166
  76. Scarious stipules absent.
    78. Petals entire .... *Arenaria serpyllifolia,* 148
    78. Petals 2-parted.
      79. Capsule splitting into valves; plant glabrous, or stems with rows of hairs.
        80. Stems and flower-stalks pubescent; leaves ovate.
          Stellaria media, 168
        80. Stems and flower-stalks glabrous; leaves lanceolate.
          Stellaria graminea, 167
      79. Capsule opening by a row of teeth at apex; plant hairy.
        81. Perennials.
          82. Petals much longer than the sepals.
            Cerastium arvense, 149
          82. Petals equal to, or shorter than, the sepals.
            Cerastium vulgatum, 151
        81. Annuals (pedicel usually less than 1 cm. long).
          Cerastium viscosum, 150
 61. Leaves alternate.
   83. Leaves, ovary, and capsule spiny .... *Argemone mexicana,* 180
   83. Leaves, ovary, and capsule not spiny.
    84. Styles 3 .... *Reseda lutea,* 228
    84. Style 1.
      85. Herbage with colored or milky juice; stipules absent (flowers regular; stamens many).
        86. Juice milky; stigma discoid .... *Papaver rhoeas,* 182
        86. Juice orange-red, stigma not discoid.
          Chelidonium majus, 181
      85. Herbage without colored or milky juice.
        87. Leaves trifoliate, without stipules.
          88. Pod sessile or with a short stipe above the calyx, clammy; stamens 6–12 .... *Polanisia graveolens,* 184
          88. Pod with a stipe above the calyx about 1 cm. long, not clammy-pubescent; stamens 6 .. *Cleome serrulata,* 183
        87. Leaves not trifoliate, with prominent stipules; stamens 5 .... *Viola arvensis,* 311
 60. Ovary 2-many-celled.
  89. Leaves opposite.
    90. The leaves simple, entire, punctate; stamens many in clusters; ovary ripening into a capsule .... *Hypericum perforatum,* 309
    90. The leaves pinnately compound; stamens 8–10; ovary ripening into a spiny bur .... *Tribulus terrestris,* 276
  89. Leaves alternate or basal, entire, lobed or divided, not punctate.

Key to the Weeds

91. Sepals 5; petals 5.
   92. Leaves pinnately veined; stamens with anthers 5.
Erodium cicutarium, 280
   92. Leaves palmately veined; stamens with anthers usually 10.
      93. Flowers large; petals about 15 mm. long; perennial.
Geranium maculatum, 283
      93. Flowers small; petals 12 mm. long or less; annual or biennial.
         94. Lobes of ovary hairy.
            95. Sepals awned; ovary hirsute; seeds pitted.
Geranium carolinianum, 281
            95. Sepals awnless; ovary puberulent; seeds smooth.
Geranium pusillum, 285
         94. Lobes of ovary glabrous or nearly so.
            96. Pedicels 0.5–2 cm. long; petals 4–5 mm. long; sepals awnless............................*Geranium molle,* 284
            96. Pedicels 3–9 cm. long; petals 8–10 mm. long; sepals awned.
Geranium columbinum, 282
91. Sepals 4; petals 4.
   97. Length of fruit not more than 3 times width.
      98. Flowers white (rarely greenish).
         99. Fruit wider than long, 2-lobed, each lobe 1-seeded, separating as a rugose nutlet................*Coronopus didymus,* 201
         99. Fruit not as above.
            100. Fruit flattened parallel with a broad partition (petals 2-cleft).
               101. Seeds many in each cell; fruit glabrous.
Draba verna, 206
               101. Seeds 4–8 in each cell; fruit stellate-pubescent.
Berteroa incana, 189
            100. Fruit flattened at right angles to a narrow partition.
               102. Seed 1 in each cell.
                  103. Cauline leaves clasping.
                     104. Plant annual or winter annual.
                        105. Lower leaves simple; plant soft-downy.
Lepidium campestre, 212
                        105. Lower leaves compound or finely divided; plant glabrous or nearly so.....*Lepidium perfoliatum,* 214
                   104. Plant perennial.
                     106. Stems hairy; fruits heart shaped, often inflated.
Cardaria draba, 198
                     106. Stems puberulent; fruits obovoid, strongly inflated...................*Cardaria pubescens,* 199
                 103. Cauline leaves not clasping.
                     107. Petals present; embryo accumbent.
Lepidium virginicum, 216
                     107. Petals absent; embryo incumbent.

108. Basal leaves bipinnatifid; racemes loose and simple; plant usually with offensive odor.
Lepidium ruderale, 215
108. Basal leaves toothed or pinnatifid, racemes dense and numerous...........*Lepidium densiflorum,* 213
 102. Seeds several in each cell.
  109. Fruits obcordate.............*Capsella bursa-pastoris,* 196
  109. Fruits orbicular or ovate.
   110. Lower cauline leaves clasping; seeds smooth.
Thlaspi perfoliatum, 227
   110. Lower cauline leaves not clasping; seeds wrinkled.
Thlaspi arvense, 226
98. Flowers yellow.
 111. Leaves entire or toothed.
  112. Fruit winged, 1-celled, 1-seeded...........*Isatis tinctoria,* 211
  112. Fruit not winged, 2-celled, several-seeded.
   113. Shape of fruit globose, indehiscent, wrinkled.
Neslia paniculata, 217
   113. Shape of fruit obovoid, dehiscent, smooth.
    114. Annuals....................*Camelina microcarpa,* 195
    114. Perennials.....................*Rorippa austriaca,* 220
 111. Leaves pinnately lobed.................*Rorippa islandica,* 221
97. Length of fruit 4 to many times width.
 115. Fruit indehiscent.
  116. Flowers yellow, rarely white, with dark veins; fruit moniliform, breaking into short pieces.....*Raphanus raphanistrum,* 218
  116. Flowers purple or whitish; fruit thick and fleshy, not breaking into pieces........................*Rhaphanus sativus,* 219
 115. Fruit dehiscent.
  117. Petals rose-colored, pink or white.
   118. Perennial; glabrous; petals 3 times as long as sepals.
Cardamine pratensis, 197
   118. Annual or biennial; with or without hairs; petals not 3 times as long as sepals.
    119. Leaves lyrate-pinnatifid; cotyledons accumbent.
Sibara virginica, 223
    119. Leaves not pinnatifid; cotyledons incumbent.
     120. Leaves oblong to spatulate, margins entire or shallowly toothed.....................*Arabidopsis thaliana,* 186
     120. Leaves reniform or deltoid, margins coarsely toothed.
Alliaria officinalis, 185
  117. Petals yellow or yellowish.
   121. Apex of fruit beyond valves 2 mm. long or more; petals 6–20 mm. long.
    122. Petals with dark veins..................*Eruca sativa,* 207
    122. Petals without dark veins.

Key to the Weeds

123. Racemes leafy-bracted................*Erucastrum gallicum*, 208
123. Racemes not leafy-bracted.
    124. Seeds in 2 rows in each cell; fruit flat.
        Diplotaxis tenuifolia, 205
    124. Seeds in 1 row in each cell; fruit not flat.
        125. Apex of fruit beyond valves less than 4 mm. long; leaves semi-succulent, glabrous.
            126. Pedicels slender..................*Barbarea vulgaris*, 188
            126. Pedicels as thick as the fruit..........*Barbarea verna*, 187
        125. Apex of fruit beyond valves more than 4 mm. long, or if shorter the leaves thin and often hairy.
            127. Upper cauline leaves clasping, glaucous and glabrous.
                Brassica rapa, 194
            127. Upper cauline leaves tapering at base.
                128. Beak of fruit flattened or angular, usually with 1 seed in basal part; valves 3-nerved.
                    129. Fruiting pedicels about 1 cm. long; beak flattened; fruit hispid; seed yellowish........*Brassica hirta*, 190
                    129. Fruiting pedicels about 0.5 cm. long, beak angular; fruit glabrous or hispid; seed dark brown or black............*Brassica kaber* var. *pinnatifida*, 192
                128. Beak of fruit conical; valves 1-nerved or nearly so.
                    130. Fruits appressed, 1-2 cm. long; plant hairy.
                        Brassica nigra, 193
                    130. Fruits spreading, 2.5-7 cm. long; plant glaucous.
                        Brassica juncea, 191
121. Apex of fruit beyond valves 1 mm. long or less; petals 3-10 mm. long.
    131. Leaves entire or dentate.
        132. Leaves oval-oblong, entire, clasping, glaucous.
            Conringia orientalis, 200
        132. Leaves linear-lanceolate, dentate, not clasping.
            133. Fruits about 2 cm. long; leaves scarcely dentate.
                Erysimum cheiranthoides, 209
            133. Fruits 4-10 cm. long; leaves repand-dentate.
                Erysimum repandum, 210
    131. Leaves pinnate or pinnatifid.
        134. Leaves waxy, terminal lobe, if present, rounded.
            135. Pedicels slender.....................*Barbarea vulgaris*, 188
            135. Pedicels as thick as fruit...............*Barbarea verna*, 187
        134. Leaves not waxy, terminal lobe, if present, not rounded.
            136. Valves of fruit with 1-3 veins; cotyledons incumbent.
                137. Pubescence of stem with branched hairs; valves 1-nerved.
                    138. Seed pods linear; seeds in one row.
                        139. Leaves bi-tripinnate; fruits about 2 cm. long.
                            Descurainia sophia, 204

88   Weeds

139. Leaves pinnatifid or bipinnatifid; fruits 0.5–1.5 cm.
long..................Descurainia richardsonii, 203
138. Seed pods club-shaped; seeds in 2 rows.
Descurainia pinnata, 202
137. Pubescence lacking or composed of simple hairs; valves 1–3 nerved.
140. Fruit awl-shaped, 1–2 cm. long, appressed.
Sisymbrium officinale, 225
140. Fruit cylindrical, 6–10 cm. long, spreading.
Sisymbrium altissimum, 224
136. Valves of fruit without veins; cotyledons accumbent.
141. Fruits, 1–2.5 cm. long; stem prostrate or ascending; a perennial plant creeping by subterranean shoots.
Rorippa sylvestris, 222
141. Fruits 5 mm. long or less; stem mostly erect; a variable species, plants either annual or biennial.
Rorippa islandica, 221
FF. Corolla gamopetalous.
1. Plants without chlorophyll, parasitic, usually straw-colored or reddish.
2. Erect, branched plants, slightly fleshy; parasitic on roots.
Orobanche ramosa, 437
2. Slender twining vines; parasitic on stems and leaves of green plants.
3. Stigma elongated; capsule with circumscissile dehiscence; flowers 5-merous.
4. Style with stigma, longer than the ovary.
5. Calyx-lobes triangular-ovate, acute at apex, not overlapping......................Cuscuta epithymum, 350
5. Calyx-lobes broadly ovate, fleshy at apex, overlapping.
Cuscuta planiflora, 355
4. Style with stigma, not longer than the ovary; calyx-lobes obtuse, overlapping..................Cuscuta epilinum, 349
3. Stigmas capitate; capsule without circumscissile dehiscence; flowers 4–5-merous.
6. Corolla lobes obtuse and either erect or spreading.
Cuscuta gronovii, 352
6. Corolla lobes acute.
7. Bracts 3–5; sepals free.............Cuscuta glomerata, 351
7. Bracts absent; sepals fused.
8. Flowers mostly 4-merous.
9. Calyx lobes acute; corolla inflexed; seeds dark brown.........................Cuscuta coryli, 348
9. Calyx lobes ovate-obtuse; corolla erect; seeds yellow brown...................Cuscuta polygonorum, 356

Key to the Weeds

8. Flowers mostly 5-merous.
   10. Flowers 1–2 mm. long; corolla lobes lance-acuminate, tips inflexed.................*Cuscuta pentagona*, 354
   10. Flowers 2–5 mm. long; corolla lobes triangular-ovate.
      11. Ovary and capsule thickened at apex; calyx much shorter than corolla tube; flowers in loose inflorescences.
         12. Corolla lobes incurved at tips...*Cuscuta indecora*, 353
         12. Corolla lobes not incurved at tips.
               Cuscuta suaveolens, 357
      11. Ovary and capsule not thickened at apex; flowers in rather dense inflorescences.
               Cuscuta campestris, 347
1. Plants with chlorophyll, not parasitic.
   13. Fertile stamens 2 or 4.
      14. Cells of ovary 4, 1 ovule or seed in each cell; fruit of 4 nutlets; (leaves opposite).
         15. Ovary deeply 4-lobed; style attached at the base of the lobes; stems mostly 4-angled; herbage mostly with an aromatic scent.
            16. Stamen strongly upcurved, very long exserted.
               Trichostema dichotomum, 402
            16. Stamens not upcurved, included or slightly exserted.
               17. Calyx-teeth bristly, recurved; plant woolly.
               Marrubium vulgare, 391
               17. Calyx-teeth not recurved.
                  18. Anther-bearing stamens 2; flowers in axillary whorls.
                     19. Annuals; corolla 2-lipped...*Hedeoma pulegioides*, 384
                     19. Perennial with rootstocks; corolla nearly regular.
                        20. Leaves serrate; calyx-teeth shorter than the mature nutlets................*Lycopus uniflorus*, 390
                        20. Leaves (at least the lower) pinnately lobed; calyx-teeth longer than the mature nutlets.
                           Lycopus americanus, 389
                  18. Anther-bearing stamens 4; perennials.
                     21. Corolla nearly regular.
                        22. Flowers in terminal spikes.
                           23. Leaves sessile; spikes slender.
                              Mentha spicata, 395
                           23. Leaves petioled; spikes oblong.
                              Mentha piperita, 394
                        22. Flowers in axillary whorls.
                           24. Stems pubescent with recurved hairs.
                              Mentha arvensis, 392

24. Stems glabrous or nearly so .... *Mentha gentilis,* 393
21. Corolla 2-lipped.
  25. Plants prostrate or creeping.
    26. Stems hirsute; annuals............ *Stachys arvensis,* 400
    26. Stems not hirsute; perennials.
      27. Length of leaves less than 1 cm.
        Thymus serpyllum, 401
      27. Length of leaves more than 1 cm.
        28. Leaves orbicular-reniform.
          Glechoma hederacea, 383
        28. Leaves ovate-oblong.......... *Prunella vulgaris,* 398
  25. Plants erect or ascending.
    29. Flowers in corymbose clusters of heads subtended by purplish bracts............. *Origanum vulgare,* 397
    29. Flowers not in corymbose clusters.
      30. Leaves palmately lobed or cleft; calyx-teeth tipped with rigid spines............. *Leonurus cardiaca,* 388
      30. Leaves neither palmately lobed nor cleft.
        31. Upper pair of stamens longer than the lower pair.
          32. Leaves with crenate margin; bracts small.
            Nepeta cataria, 396
          32. Leaves with serrate margin; bracts large, bristly toothed........ *Dracocephalum parviflorum,* 381
        31. Upper pair of stamens shorter than lower pair.
          33. Calyx closed in the fruit; bracts large, reniform................... *Prunella vulgaris,* 398
          33. Calyx open in the fruit; bracts not as above.
            34. Calyx-teeth tipped with rigid spines.
              Galeopsis tetrahit, 382
            34. Calyx-teeth not tipped with spines.
              35. Calyx 10–13 nerved.... *Satureja vulgaris,* 399
              35. Calyx 4–6 nerved.
                36. Nutlets sharply 3-angled, truncate at apex.
                  37. Corolla 1.5–2.5 cm. long, broad (leaves often with a white spot).
                    Lamium maculatum, 386
                  37. Corolla 1–1.5 cm. long, slender.
                    38. Upper leaves sessile, clasping.
                      Lamium amplexicaule, 385
                    38. Upper leaves petioled.
                      Lamium purpureum, 387
                36. Nutlets obscurely 3-angled, rounded at apex............... *Stachys arvensis,* 400
15. Ovary not deeply lobed; style attached on top of the lobes; stems rarely somewhat 4-angled; herbage not aromatic scented.

Key to the Weeds

39. Spikes thick, with leafy bracts; procumbent annuals.
    Verbena bracteata, 377
39. Spikes slender, with bracts shorter than the calyx; erect perennials.
    40. Leaves sessile, downy with white hairs; flowers blue or purple .... *Verbena stricta*, 379
    40. Leaves petioled, not downy.
        41. Flowers white; fruits scattered on spike.
            Verbena urticifolia, 380
        41. Flowers blue or purple; fruits crowded.
            Verbena hastata, 378
14. Cells of ovary 2; fruit a capsule.
    42. Flowers scarious, small, greenish, in dense spikes, 4-merous; capsule circumscissile.
        43. Leaves cauline, opposite or whorled .... *Plantago indica*, 439
        43. Leaves radical, plant scapose.
            44. Shape of leaves broadly ovate, with long petioles; spikes long and slender; seeds plump.
                45. Capsule ovate, dehiscing near the middle.
                    Plantago major, 441
                45. Capsule cylindrical, dehiscing much below the middle.
                    Plantago rugellii, 444
            44. Shape of leaves linear, lanceolate or oblong-ovate.
                46. Leaf-blade lanceolate, oblong-ovate or obovate.
                    47. Blade lanceolate; sepals 3–3.5 mm. long.
                        Plantago lanceolata, 440
                    47. Blade not lanceolate; sepals 2–2.5 mm. long.
                        48. Plant perennial; leaves canescent; seeds flat on inner surface .... *Plantago media*, 442
                        48. Plant annual or biennial; leaves villous; seeds concave on inner surface .... *Plantago virginica*, 445
                46. Leaf-blade linear, silky-hairy or woolly.
                    49. Bracts longer than the flowers ...... *Plantago aristata*, 438
                    49. Bracts not longer than the flowers.
                        Plantago purshii, 443
    42. Flowers petaloid, 4–5-merous; ovary many-seeded; capsule not circumscissile.
        50. Fertile stamens 2.
            51. Flowers in axillary racemes; leaves hairy.
                52. Leaves alternate; annual or perennial.
                    53. Plant annual, prostrate or thinly ascending; sepals strongly ciliate; corolla 2–5 mm. wide.
                        Veronica hederaefolia, 430
                    53. Plant perennial, thickly matted; sepals with sparse and very short ciliate hairs; corolla up to 1 cm. wide.
                        Veronica filiformis, 429
                52. Leaves opposite; perennial.

92 Weeds

54. Racemes dense; pedicels shorter than the corolla.
   Veronica officinalis, 431
54. Racemes loose; pedicels longer than the corolla.
   Veronica chamaedrys, 428
51. Flowers not in axillary racemes.
   55. Flowers in a terminal spike or raceme.
      56. Leaves glabrous or nearly so, entire or shallow toothed; corolla pale blue or whitish.
         57. Creeping perennial............*Veronica serpyllifolia*, 435
         57. Erect annual...................*Veronica peregrina*, 432
      56. Leaves hairy, coarsely toothed; corolla bright blue; annual.
         Veronica arvensis, 427
   55. Flowers all solitary on long peduncles in axils of normal leaves.
      58. Mature pedicels 1.5–2.5 cm. long; corolla 10–12 mm. in diameter; capsule obcordate or broadly notched, reticulated.
         Veronica persica, 433
      58. Mature pedicels mostly less than 1 cm. long; corolla 4–6 mm. in diameter; capsule orbicular or slightly notched, not reticulated..............................*Veronica polita*, 434
50. Fertile stamens 4.
   59. Leaves opposite; corolla not spurred.
      60. Corolla 2-lipped, 1.5–3 cm. long.
         61. Corolla 2–3 cm. long, throat strongly inflated.
            Penstemon digitalis, 421
         61. Corolla 1.5–2 cm. long, throat flat......*Penstemon gracilis*, 422
      60. Corolla globular or cylindric, less than 1 cm. long.
         Scrophularia marilandica, 423
   59. Upper leaves alternate; corolla tubular, with a spur or sac at the base.
      62. Plants procumbent; leaves broad; flowers axillary; annuals.
         63. Leaves hastate, pubescent...............*Kickxia elatine*, 418
         63. Leaves reniform to orbicular, glabrous.
            Cymbalaria muralis, 416
      62. Plants erect; flowers in racemes.
         64. Leaves ovate or lanceolate, pubescent; corolla 4–6 cm. long; tall biennials.......................*Digitalis purpurea*, 417
         64. Leaves linear, spatulate-linear or oblong; corolla .5–3 cm. long.
            65. Flowers yellow; glabrous perennial.
               Linaria vulgaris, 420
            65. Flowers blue or purple; annual or biennial.
               66. Stems glabrous; seeds obpyramidal with sharp angles.
                  Linaria canadensis, 419
               66. Stems glandular-pubescent; seeds ovoid with several acute vertical ridges.
                  Chaenorrhinum minus, 415

Key to the Weeds 93

13. Fertile stamens 5 (rarely 6 or more).
 67. Ovaries 2, separate or nearly so; juice frequently milky.
  68. Stamen filaments separate; styles and stigmas united.
   69. Plant trailing; corolla blue; leaves persistent, opposite; seeds without hairy appendages.................*Vinca minor,* 337
   69. Plant erect; corolla white, yellow or pink; leaves not evergreen; seeds with hairy appendages.
    70. Corolla pinkish, 6–7 mm. long; flowers paniculate.
     Apocynum androsaemifolium, 335
    70. Corolla greenish-white, 2.5–4 mm. long; flowers corymbose.........................*Apocynum cannabinum,* 336
  68. Stamen filaments united; styles separate; stigmas united and fused with the anthers; seeds with hairy appendages.
   71. Stems twining.
    72. Leaves ovate to ovate-lanceolate; corolla rotate.
     73. Flowers greenish-white; corolla glabrous.
      Cynanchum vincetoxicum, 344
     73. Flowers purple; corolla pubescent within.
      Cynanchum nigrum, 343
    72. Leaves cordate-ovate with cordate base, petioles long; corolla campanulate; flowers white....*Ampelamus albidus,* 338
   71. Stems erect.
    74. Leaves alternate; flowers orange; juice not milky.
     Asclepias tuberosa, 341
    74. Leaves opposite or whorled; juice milky.
     75. Leaves linear, whorled; flowers greenish-white.
      Asclepias verticillata, 342
     75. Leaves broad, opposite; flowers pale purple or lavender.
      76. Hoods of crown short, obtuse......*Asclepias syriaca,* 340
      76. Hoods of crown long, lanceolate...*Asclepias speciosa,* 339
 67. Ovary 1; juice not milky; seeds without hairy appendages.
  77. The ovary deeply 4-lobed (at least in fruit).
   78. Leaves glabrous, narrowly lanceolate to obovate; stems prostrate or decumbent..........*Heliotropium curassavicum,* 371
   78. Leaves not glabrous.
    79. Nutlets armed with prickles, barbs or warts, attached laterally.
     80. Corolla orange or yellow.
      81. Plant erect; calyx-lobes narrow to linear.
       Amsinckia intermedia, 365
      81. Plant decumbent; calyx-lobes ovate.
       Amsinckia lycopsoides, 366
     80. Corolla blue or purple, occasionally white.
      82. Annual; nutlet 2–3 mm. long, margin and back with double row of barbed prickles.
       Lappula echinata, 372

82. Biennial or perennial; nutlet larger and without double row of prickles on back.
83. Biennial; nutlets covered with short barbed prickles on back.
84. Corolla 1–3 mm. wide.......... *Hackelia virginiana*, 370
84. Corolla 8–15 mm. wide...... *Cynoglossum officinale*, 367
83. Perennial; nutlets with a margin of flat awl-shaped prickles....................... *Hackelia floribunda*, 369
79. Nutlets unarmed, attached at or near the base.
85. Corolla regular, whitish; stamens equal.
86. Throat of corolla closed by 5 scales; perennial.
Symphytum officinale, 376
86. Throat of corolla open or with 5 crests.
87. Nutlets smooth, shiny, white; perennial.
Lithospermum officinale, 374
87. Nutlets roughened, gray and dull; annual.
Lithospermum arvense, 373
85. Corolla irregular, blue; stamens unequal.
88. Stamens exserted..................... *Echium vulgare*, 368
88. Stamens not exserted............... *Lycopsis arvensis*, 375
77. The ovary not 4-lobed.
89. Plants twining or climbing.
90. Stigmas capitate; annuals.
91. Corolla salverform or nearly so, red; stamens and style exserted............................... *Ipomoea coccinea*, 358
91. Corolla funnelform, purple, blue or white; stamens and style not exserted.
92. Leaves 3-lobed; flowers 2.5–4.5 cm. long.
Ipomoea hederacea, 359
92. Leaves entire; flowers 4.5–7 cm. long.. *Ipomoea purpurea*, 360
90. Stigmas linear or oblong; perennials.
93. Calyx with 2 large bracts at base; stigmas oblong.
Convolvulus sepium, 346
93. Calyx bractless at base; stigmas filiform.
Convolvulus arvensis, 345
89. Plants neither twining nor climbing.
94. Stamens 5; leaves not trifoliate.
95. Leaves opposite; ovary 1-celled; plants low, prostrate or creeping.
96. Plant creeping; corolla yellow.
Lysimachia nummularia, 334
96. Plant prostrate but not creeping; corolla red or white.
Anagallis arvensis, 333
95. Leaves alternate; ovary 2- to 4-celled.
97. Stamen filaments, at least some, bearded; fruit a 2-celled capsule; erect biennials.
98. Plant not woolly............... *Verbascum blattaria*, 424

Key to the Weeds

98. Plant woolly.
   99. Upper leaves decurrent; inflorescence a spike.
                                            Verbascum thapsus, 426
   99. Upper leaves not decurrent; inflorescence a panicle.
                                            Verbascum lychnitis, 425
97. Stamen filaments not bearded.
   100. Corolla rotate; anthers connivent; fruit a berry.
      101. Diameter of corolla 3–10 mm.; annuals.
         102. Stems, branches and leaves densely villous; berry yellow-red................................*Solanum villosum*, 414
         102. Stems, branches and leaves not densely villous, berry not yellow-red.
            103. Leaves wavy-toothed; berry purple to black.
                                        Solanum nigrum, 411
            103. Leaves pinnately lobed; berry greenish-yellow.
                                        Solanum triflorum, 413
      101. Diameter of corolla 1.5–3 cm.
         104. Plants silvery-hairy with close scurf-like branched hairs; perennial.....................*Solanum elaeagnifolium*, 410
         104. Plants not silvery-hairy.
            105. Plants prickly; fruit yellow; herbaceous.
               106. Flowers lavender or white; berry not spiny; perennial.
                                        Solanum carolinense, 408
               106. Flowers yellow; berry inclosed by spiny calyx; annual.
                                          Solanum rostratum, 412
            105. Plant not prickly, woody at base; flowers violet, purple or white; berry red...............*Solanum dulcamara*, 409
   100. Corolla tubular, funnelform or campanulate; anthers not connivent.
      107. Flowers crowded in bracteate clusters; calyx-lobes tipped with bristles; fruit a capsule; annuals.
         108. Plants glandular-viscid, with bad odor.
                                        Navarretia squarrosa, 362
         108. Plants not glandular-viscid, with no bad odor.
                                        Navarretia intertexta, 361
      107. Flowers not clustered; calyx-lobes not bristle-tipped.
         109. Calyx inflated in fruit.
            110. Fruit a 2-celled pendulous capsule.....*Ellisia nyctelea*, 363
            110. Fruit a berry.
               111. Corolla pale blue; berry 3- to 5-celled, dry, annual.
                                        Nicandra physalodes, 404
               111. Corolla white, yellow or greenish; berry 2-celled, fleshy; perennials.
                  112. Corolla white; fruiting calyx red.
                                        Physalis alkekengi, 405
                  112. Corolla yellow or greenish; fruiting calyx not bright red.

                113. Plants glandular-pubescent; mature berry yellow.
                                       Physalis heterophylla, 406
                113. Plants glabrous or nearly so; mature berry orange to purple.
                                       Physalis subglabrata, 407
           109. Calyx not inflated; fruit a capsule; annual.
                114. Corolla 6–15 cm. long; capsule 4-celled, spiny............*Datura stramonium,* 403
                114. Corolla 5–10 mm. long; capsule 2-celled, without spines.
                                       Phacelia purshii, 364
         94. Stamens 10; leaves trifoliate.
              115. Plants silky-hairy; corolla pink.
                                       Trifolium arvense, 269
              115. Plants not silky-hairy; corolla yellow.
                   116. Leaflets all sessile; plants erect.
                                       Trifolium agrarium, 268
                   116. Terminal leaflets stalked; plants spreading.
                        117. Head 20–40-flowered.
                                       Trifolium procumbens, 271
                        117. Head 3–20-flowered......*Trifolium dubium,* 270
DD. Ovary inferior.
  1. Flowers in involucrate heads.
    2. Stamens separate; flowers perfect.
      3. Stem and leaves prickly; chaff of receptacle with long stiff points............................*Dipsacus sylvestris,* 455
      3. Stem and leaves not prickly.
        4. Leaves simple, lanceolate; receptacle chaffy.
                                       Succisa australis, 457
        4. Leaves pinnatifid or compound; receptacle naked.
                                       Knautia arvensis, 456
    2. Stamens fused by the anthers, or, if separate, flowers imperfect and in two kinds of heads (Compositae).
      5. Flowers of the head all ligulate; juice milky (Liguliflorae).
        6. Pappus none (annuals)..............*Lapsana communis,* 542
        6. Pappus present.
          7. Pappus of scales; flowers blue........*Cichorium intybus,* 499
          7. Pappus of bristles or hairs.
            8. Bristles simple, often serrate.
              9. The stems unbranched, glabrous, naked (scapes) and bearing a single head.
                10. Achenes olive-green or brown; heads golden-yellow..................*Taraxacum officinale,* 565
                10. Achenes red or reddish-brown; heads sulfur-yellow............*Taraxacum erythrospermum,* 564

Key to the Weeds 97

9. The stems branched and leafy, or simple and hairy.
    11. Stems simple, usually bearing 1 or 2 bracts.
        12. Heads solitary; flowers yellow............*Hieracium pilosella*, 532
        12. Heads several on a stem.
            13. Flowers orange-red; plant stoloniferous.
                                                    Hieracium aurantiacum, 529
            13. Flowers yellow.
                14. Plant producing short, stout rootstock, and new shoots from creeping roots.
                    15. Basal leaves glabrous or with a few occasional bristles.
                                                    Hieracium florentinum, 530
                    15. Basal leaves finely stellate-pubescent on lower surfaces.
                                                    Hieracium praealtum, 533
                14. Plants producing slender rootstock and stolons.
                    16. Leaves glaucous, nearly glabrous above; stolons numerous...................*Hieracium floribundum*, 531
                    16. Leaves green, not glaucous, bristly hairy; stolons few.
                                                    Hieracium pratense, 534
    11. Stems branched, leafy.
        17. Stem-leaves reduced, linear or awl-shaped.
            18. Biennial; hairy below; flowers yellow.
                                                    Chondrilla juncea, 497
            18. Perennial; glabrous; flowers rose-purple.
                                                    Lygodesmia juncea, 544
        17. Stem-leaves broad, with toothed or irregular margin.
            19. Achenes flat.
                20. Body of achene beaked or with a narrow neck.
                    21. Flowers yellow; annual or biennial.
                        22. Leaf margins and midrib prickly, beak of achene 3-8 mm. long......................*Lactuca scariola*, 541
                        22. Leaves without prickles; beak of achene 1 mm. long.
                                                    Lactuca muralis, 539
                    21. Flowers blue; perennial...........*Lactuca pulchella*, 540
                20. Body of achene truncate at summit; heads 12-40 mm. in diameter; flowers yellow.
                    23. Perennial; heads large, 4 cm. in diameter.
                                                    Sonchus arvensis, 560
                    23. Annuals; heads small, 1.2-2.5 cm. in diameter.
                        24. Leaves spiny-toothed; achenes ribbed.
                                                    Sonchus asper, 561
                        24. Leaves soft-spiny or unarmed; achenes ribbed and cross-wrinkled..................*Sonchus oleraceus*, 562
            19. Achenes neither distinctly flattened nor beaked.
                25. Achenes tapering at both ends; ribs on achenes smooth.
                                                    Crepis capillaris, 505
                25. Achenes tapering upwards; ribs on achenes upwardly scabrous.............................*Crepis tectorum*, 506

8. Bristles plumose.
  26. Receptacle chaffy; leaves irregularly lobed.
                                               Hypochoeris radicata, 535
  26. Receptacle not chaffy.
    27. Leaves linear, grass-like, radical and cauline.
      28. Flowers purple............*Tragopogon porrifolius*, 567
      28. Flowers yellow.
        29. Peduncle slightly or not at all dilated below head; involucre 2–3 cm. high......*Tragopogon pratensis*, 568
        29. Peduncle noticeably dilated below head; involucre 4–7 cm. high................*Tragopogon major*, 566
    27. Leaves not grass-like.
      30. Plant leafy-stemmed.
        31. Outer bracts of involucre narrow; achenes beakless or nearly so; pappus slight plumose.
                                                 Picris hieracioides, 550
        31. Outer bracts of involucre ovate or subcordate; achenes beaked; pappus densely plumose.
                                                      Picris echioides, 549
      30. Plant scapose...............*Leontodon autumnalis*, 543
5. Flowers of the head tubular or only the marginal ligulate; juice not milky (Tubuliflorae).
  32. Involucre of the pistillate flowers closed and woody; heads unisexual.
    33. Involucre forming a spiny bur.
      34. Leaves with spines at the base........*Xanthium spinosum*, 571
      34. Leaves without spines...............*Xanthium orientale*, 570
    33. Involucre with a few sharp apical tubercles or spines.
      35. Tubercles in a double whorl; leaves white with hairs.
                                                  Fransoria discolor, 518
      35. Tubercles in a single whorl, forming a crown; leaves not white.
        36. Leaves 3-lobed, opposite..............*Ambrosia trifida*, 466
        36. Leaves pinnatifid, mostly alternate.
          37. Annual; leaves bipinnatifid....*Ambrosia artemisiifolia*, 464
          37. Perennial; leaves 1-pinnatifid...*Ambrosia psilostachya*, 465
  32. Involucre not closed and not woody.
    38. Heads without ray-flowers.
      39. Pappus capillary.
        40. Involucral bracts scarious; plants whitish.
          41. Basal leaves much larger than cauline leaves.
                                                  Antennaria plantaginifolia, 468
          41. Basal leaves absent or not different from the cauline leaves.
            42. Involucre papery-white; bracts striate, spreading.
                                               Anaphalis margaritacea, 467

Key to the Weeds 99

42. Involucre yellowish or brownish; bracts not striate, appressed.
   43. Plants low, much-branched; heads about 2 mm. high.
                                        Gnaphalium uliginosum, 522
   43. Plants erect, simple or sparingly branched; heads about 5 mm. high.
      44. Leaf-base clasping and decurrent.
                                         Gnaphalium macounii, 520
      44. Leaf-base tapering, not decurrent.
                                         Gnaphalium obtusifolium, 521
40. Involucral bracts not scarious, or if so, plants not white.
   45. Bracts in one series.
      46. Marginal flowers pistillate............*Erechtites hieracifolia,* 508
      46. Marginal flowers perfect..................*Senecio vulgaris,* 555
   45. Bracts in 2-many series.
      47. Leaves spiny; involucral bracts, at least the outer, tipped with a spine.
         48. Flowers yellow (Centaurea, see 68)
         48. Flowers purple, lavender or white.
            49. Pappus plumose.
               50. Plants perennial from creeping roots; heads small, clustered; flowers dioecious.............*Cirsium arvense,* 501
               50. Plants biennial, from a tap-root; heads large and few; flowers perfect.
                   51. Stem-leaves decurrent; bracts of the head all tipped with spreading spines.............*Cirsium vulgare,* 503
                 51. Stem-leaves not decurrent; bracts of the head appressed, the inner not spiny.
                      52. Leaves white-woolly beneath, nearly entire or with few lobes; head slender
                                          Cirsium altissimum, 500
                      52. Leaves green on both sides, with many lobes; heads 4-8 cm. broad...............*Cirsium pumilum,* 502
            49. Pappus not plumose.
               53. Heads solitary, 3-5 cm. broad, nodding, on long nearly naked peduncles...................*Carduus nutans,* 486
               53. Heads clustered at ends of winged branches, 1.5-2.5 cm. broad.......................*Carduus acanthoides,* 485
      47. Leaves not spiny.
         54. Corolla deeply lobed or lacerate; involucral bracts toothed; receptacle bristly. (Centaurea, see 68)
         54. Corolla toothed; involucral bracts entire; receptacle naked.
            55. Leaves finely pinnately divided, mostly alternate.
                                Eupatorium capillifolium, 514
            55. Leaves simple, whorled or opposite.
               56. Flowers purple or rose-colored; leaves whorled.
                                Eupatorium maculatum, 515

100   Weeds

    56. Flowers white; leaves opposite.
        57. Leaves petioled............*Eupatorium rugosum,* 517
        57. Leaves sessile, connate...*Eupatorium perfoliatum,* 516
39. Pappus not capillary.
    58. The pappus of awns or scales.
        59. Pappus of awns.
            60. Awns of pappus in two series, the outer much longer; achenes crowned with 10 teeth.........*Cnicus benedictus,* 504
            60. Awns of pappus not in two series.
                61. Achenes flat, with 2 awns.
                    62. Outer involucral bracts 4–8; corolla orange.
                        Bidens frondosa, 482
                    62. Outer involucral bracts 10–25.
                        63. Outer involucral bracts 12–25; leaves with 3–9 divisions; achenes 4.5–6.5 mm. long; awns 0.2–3 mm. long.
                            Bidens polylepis, 483
                        63. Outer involucral bracts 10–16; leaves usually with 3–5 divisions; achenes 6.5–17 mm. long; awns 3–9.5 mm. long........................*Bidens vulgata,* 484
                61. Achenes angular, with 3–4 awns.
                    64. Leaves pinnately divided...........*Bidens bipinnata,* 480
                    64. Leaves simple.......................*Bidens cernua,* 481
        59. Pappus of scales.
            65. Outer bracts of involucre about as long as the inner ones; petioles usually solid.....................*Arctium lappa,* 471
            65. Outer bracts of involucre shorter than the inner ones.
                          Arctium minus, 472
    58. The pappus of a crown or none.
        66. Receptacle conical when mature....*Matricaria matricarioides,* 547
        66. Receptacle flat.
            67. Heads over 1 cm. in diameter; bracts fimbriate or spiny.
                68. Outer involucral bracts terminating in spines; annuals or biennials.
                    69. Flowers purple, pink or rose-colored; stems not winged.
                        70. Color of flowers deep rose-purple; achenes without pappus....................*Centaurea calcitrapa,* 487
                        70. Color of flowers purplish-pink; achenes with a crown of pappus bristles...............*Centaurea iberica,* 490
                  69. Flowers yellow; stems winged or angled.
                      71. Spines stout, 1–2 cm. long, straw-colored.
                        Centaurea solstitialis, 495
                    71. Spines slender, .1–1 cm. long.
                      72. Spines 5–10 mm. long, purple; pappus often persistent.....................*Centaurea melitensis,* 493
                    72. Spines 1–4 mm. long; pappus deciduous.
                        Centaurea diffusa, 489

Key to the Weeds

68. Outer involucral bracts not terminating in spines, usually dilated or pectinate at the tip.
 73. Annual; marginal flowers mostly blue; stems and leaves with soft gray pubescence.
 Centaurea cyanus, 488
 73. Biennial or perennial; marginal flowers mostly purple or rose-colored.
  74. Involucral bracts with whitish, entire or nearly entire, appendages; achenes without a notch just above the base................*Centaurea repens,* 494
  74. Involucral bracts with dark pectinate tips.
   75. Leaves entire or occasionally toothed; pappus mostly wanting.
    76. Bracts with tan to brown tips. *Centaurea jacea,* 491
    76. Bracts conspicuously black on tips.
    Centaurea vochinensis, 496
   75. Leaves pinnatifid; pappus bristly and persistent.
   Centaurea maculosa, 492
67. Heads 1 cm. or less; bracts neither fimbriate nor spiny.
 77. Heads in corymbs, erect...........*Tanacetum vulgare,* 563
 77. Heads in racemes, spikes or panicles, drooping.
  78. Lower leaves opposite; receptacle chaffy.
   79. Leaves toothed or lobed, broad; annual.
   Iva xanthifolia, 538
   79. Leaves entire, narrowly tapering at base; perennial.
   Iva axillaris, 537
  78. Lower leaves alternate; receptacle naked or hairy.
   80. Leaves glabrous; annuals or biennials; stems solitary.
    81. Lobes of leaves rounded; heads in a loose spreading panicle; herbage sweet-scented.
    Artemisia annua, 474
    81. Lobes of leaves acute; heads in short axillary clusters; herbage not sweet-scented.
    Artemisia biennis, 475
   80. Leaves hairy, at least on the lower surface; perennials; stems usually in clumps.
    82. Leaves silky-gray on both surfaces; receptacle hairy..................*Artemisia absinthium,* 473
    82. Leaves white-woolly on the under surface; green above; receptacle smooth.
    Artemisia vulgaris, 477
38. Heads with ray-flowers and disk-flowers.
 83. Pappus capillary; receptacle naked.
  84. Ray-flowers yellow or cream-color.
   85. Involucral bracts in 1 series.

86. Heads solitary, on bracted scapose stems... *Tussilago farfara*, 569
86. Heads many, on leafy stem............... *Senecio jacobaea*, 554
85. Involucral bracts in 3 to many series.
  87. Heads large, 2–10 cm. in diameter.......... *Inula helenium*, 536
  87. Heads small, 0.5–1.5 cm. in diameter.
    88. Inflorescence flat-topped, corymbose.
                                    Solidago graminifolia, 558
    88. Inflorescence paniculate.
      89. Leaves rugose, with 1 main vein; stem villous.
                                      Solidago rugosa, 559
      89. Leaves not rugose, with 3 main veins.
        90. Stem below the panicle glabrous and glaucous.
                                      Solidago gigantea, 557
        90. Stem pubescent, at least above... *Solidago canadensis*, 556
84. Ray-flowers white, purple, violet or blue.
  91. Bracts in 3–5 series........................ *Aster simplex*, 478
  91. Bracts in 1–2 series.
    92. Ray-flowers about as long as the disk-flowers; heads greenish, small and numerous................ *Erigeron canadensis*, 510
    92. Ray-flowers much larger than the disk-flowers; heads medium or large.
      93. Heads 2.5–3.5 cm. in diameter; ray-flowers 1 mm. wide, blue........................... *Erigeron pulchellus*, 512
      93. Heads 1.5–2 cm. in diameter; ray-flowers less than 1 mm. wide.
        94. Ray-flowers about 100–150 in a head, pink; leaves clasping....................... *Erigeron philadelphicus*, 511
        94. Ray-flowers about 50 in a head, whitish; leaves not clasping.
          95. Leaves entire or toothed; pubescence sulgose.
                                      Erigeron strigosus, 513
          95. Leaves coarsely toothed; pubescence absent or scattered and hispid................. *Erigeron annuus*, 509
83. Pappus not capillary.
  96. Pappus of awns or scales.
    97. The pappus of awns.
      98. Involucre glutinous; receptacle naked.
                                    Grindelia squarrosa, 523
      98. Involucre not glutinous; receptacle chaffy (Bidens, see 61).
    97. The pappus of scales.
      99. Receptacle naked; ray-flowers yellow.
        100. Leaves lanceolate, decurrent on an angular stem.
          101. Cauline leaves toothed; ray-flowers fertile.
                                      Helenium autumnale, 524
          101. Cauline leaves entire; ray-flowers sterile.
                                      Helenium nudiflorum, 525

Key to the Weeds

100. Leaves linear, not decurrent.......*Helenium tenuifolium,* 526
99. Receptacle chaffy.
  102. Leaves pinnatifid or pinnately parted.
    103. Leaves opposite; pappus scales dissected into bristles.
Dyssodia papposa, 507
    103. Leaves alternate; pappus scales rounded.
Parthenium hysterophorus, 548
  102. Leaves toothed or entire.
    104. Ray-flowers white; heads less than 1 cm. in diameter.
Galinsoga ciliata, 519
    104. Ray-flowers yellow; heads more than 1 cm. in diameter.
      105. Perennial; leaves mostly opposite; propagating by tubers......................*Helianthus tuberosus,* 528
      105. Annual; leaves mostly alternate; without tubers.
Helianthus annuus, 527
96. Pappus of a small crown or none.
  106. Receptacle naked.
    107. Heads on tall leafy stems.
Chrysanthemum leucanthemum var. *pinnatifidum,* 498
    107. Heads solitary on short scapes.............*Bellis perennis,* 479
  106. Receptacle chaffy.
    108. Chaff of a single row of bracts, between the small yellow ray-flowers and the disk-flowers.
      109. Involucre longer than broad; herbage glandular-viscid; leaves broadly lanceolate to linear.
Madia sativa var. *congesta,* 546
      109. Involucre at least as broad as long; herbage densely hairy; glandular only in the inflorescence; leaves linear.
Madia glomerata, 545
    108. Chaff all over the receptacle.
      110. Color of ray-flowers white.
        111. Heads small, 3–7 mm. in diameter, clustered.
Achillea millefolium, 463
        111. Heads large, 12–30 mm. in diameter, solitary.
          112. Ray-flowers pistillate; leaves 1–2-pinnately divided, grayish-green; plant not strongly scented, the lower branches frequently rooting at nodes.
Anthemis arvensis, 469
          112. Ray-flowers neutral; leaves 3-pinnately dissected, yellowish-green; plant strong-scented; branches not rooting..............................*Anthemis cotula,* 470
      110. Color of ray-flowers yellow.
        113. Leaves undivided................*Rudbeckia serotina,* 552
        113. Leaves, at least some, trilobed or pinnately divided.
          114. Disc purple; chaff concave; achenes quadrangular.
Rudbeckia triloba, 553

104 Weeds

  114. Disc not purple; chaff truncate; achenes flattened-angular......*Ratibida pinnata*, 551
1. Flowers not in involucrate heads.
 2. Arrangement of leaves opposite or whorled.
  3. Leaves opposite.
   4. Stipules present, with bristly hairs; flowers axillary.
Diodia teres, 446
   4. Stipules absent; flowers cymose........*Valerianella olitoria*, 454
  3. Leaves whorled.
   5. Corolla funnelform....................*Sherardia arvensis*, 452
   5. Corolla rotate.
    6. Annual; flowers axillary, ovary and fruit bristly.
Galium aparine, 447
    6. Perennials; flowers in terminal panicles.
     7. Leaves 3-nerved, those on the main stem in whorls of 4, blunt; stem without bristles; ovary and fruit hairy.
Galium boreale, 449
     7. Leaves 1-nerved, those on the main stem in whorls of 6–8, cuspidate; ovary and fruit glabrous.
      8. Leaves of the main stem in whorls of 6; stem very rough-bristly....................*Galium asprellum*, 448
      8. Leaves of the main stem in whorls of 8; stem smooth or somewhat hairy.
       9. Leaves linear; corolla yellow........*Galium verum*, 451
       9. Leaves oblanceolate or nearly so; corolla white.
Galium mollugo, 450
 2. Arrangement of the leaves alternate or basal.
  10. Plants with slender creeping stems rooting at the nodes; leaves simple; flowers in heads; perennial.
Hydrocotyle sibthorpioides, 926
  10. Plants erect or climbing.
   11. Flowers in umbels; leaves compound; habit erect.
    12. Involucral bracts cleft, pinnatifid or bipinnatifid; corolla white; fruits with barbed prickles on the ribs.
     13. Annual; involucral bracts bipinnatifid.
Daucus pusillus, 325
     13. Biennial; involucral bracts cleft or pinnatifid.
Daucus carota, 324
    12. Involucral bracts entire or serrate or wanting; corolla white or yellow; fruit not prickly.
     14. Fruit flattened parallel with the partition; leaf-segments broad.
      15. Flowers yellow; leaves 1-pinnately compound, hairy, at least beneath..................*Pastinaca sativa*, 327

Key to the Weeds

  15. Flowers white; leaves ternately decompound, glabrous..................*Angelica atropurpurea*, 320
 14. Fruit not flattened parallel with the partition; flowers white.
  16. Fruit with hooked prickles or warts; annual.
                 Torilis japonica, 329
  16. Fruits without prickles or warts; biennial or perennial.
   17. Leaves pinnately compound.
    18. Leaflets filiform; roots fusiform or tuberous.
                   Carum carvi, 321
    18. Leaflets lanceolate, serrate; roots fibrous.
                   Sium suave, 328
   17. Leaves ternately decompound.
    19. Leaflets lanceolate or ovate, serrate; perennials.
     20. Plants with creeping rhizomes; leaflets ovate; oil tubes none....................*Aegopodium podagraria*, 319
     20. Plants with fascicles of fusiform roots; leaflets lanceolate; oil tubes solitary between ridges.
                   Cicuta maculata, 322
    19. Leaflets much cut into slender segments; biennials with fleshy tap-roots........*Conium maculatum*, 323
11. Flowers not in umbels; leaves simple.
 21. Corolla gamopetalous.
  22. Habit of plants climbing, twining or creeping.
   23. Plants clammy-pubescent; fruit 1-celled, 1-seeded, in clusters.
                 Sicyos angulatus, 459
   23. Plants nearly glabrous; fruit 2-celled, 2–4-seeded, solitary.
                 Echinocystis lobata, 458
  22. Habit of plants erect, not climbing.
   24. Corolla irregular; anthers fused............*Lobelia inflata*, 462
   24. Corolla regular; anthers separate.
    25. Leaves clasping; flowers axillary.....*Specularia perfoliata*, 461
    25. Leaves not clasping; flowers in a terminal raceme.
              Campanula rapunculoides, 460
 21. Corolla polypetalous; stems erect (capsule 4-celled).
  26. Flowers yellow; seeds not tufted.........*Oenothera biennis*, 318
  26. Flowers not yellow.
   27. Seeds naked.............................*Gaura biennis*, 317
   27. Seeds with a tuft of hairs.
    28. Petals notched; plant hirsute........*Epilobium hirsutum*, 316
    28. Petals entire; plants glabrous or puberulent.
              Epilobium angustifolium, 315

Equisetaceae (Horsetail Family)

1. *Equisetum arvense* L. Field horsetail, Horsetail fern, Meadow-pine, Pine-grass, Foxtail-rush, Scouring-rush, Bottle-brush, Horse-pipes, Snake-grass. Fig. 4, a–e.

Perennial; reproducing by spores and creeping tuber-bearing rootstocks. Common in moist fields and meadows; especially on sandy or gravelly soils. Native throughout the United States and southern Canada. April–May.

Description.—Stems annual, hollow, jointed, of two kinds, coming from a perennial, jointed, tuber-bearing, deep creeping rootstock. Fertile stems erect, 10–25 cm. high, unbranched, without chlorophyll, terminated by a cone of sporangia; sterile stems green, erect or prostrate, 10–50 cm. high, branches in whorls, ascending, frequently branched again. Leaves small, scale-like, 8–12 in a whorl, and joined by their edges into a sheath with black tips, "teeth."

Control.—The deep-seated system of roots and rootstocks with tuberous storage organs make it very difficult to eradicate the horsetail. Improved drainage (21),* clean cultivation (18) and the application of fertilizer to stimulate the crop plants, will gradually drive out the horsetail.

The fertile shoots appear very early in the spring and soon disappear. The sterile shoots come later and persist until autumn. This species, and probably every species of Equisetum, is poisonous to animals. When horses or cattle are fed on hay containing large quantities of this weed they become sick with a disease known as "equisetosis."

2. *Equisetum hyemale* L. Scouring-rush, Tall scouring-rush. Fig. 4, j–k.

Perennial; reproducing by spores and rootstocks. Moist or springy grasslands; on alluvial soils, banks and roadsides. Widespread from Canada to Mexico, locally abundant in the northeastern states. Native. May–September.

Description.—Stems evergreen, hollow, very stiff and rough, jointed, from 3 to 15 dm. high, dark green, unbranched or sparingly branched with simple branches; cones terminal. Leaves reduced, scale-like, in whorls, and joined into a sheath.

Control.—Since this weed usually occurs in places where cultiva-

---

* Figures in parentheses refer to paragraphs in Chapter III.

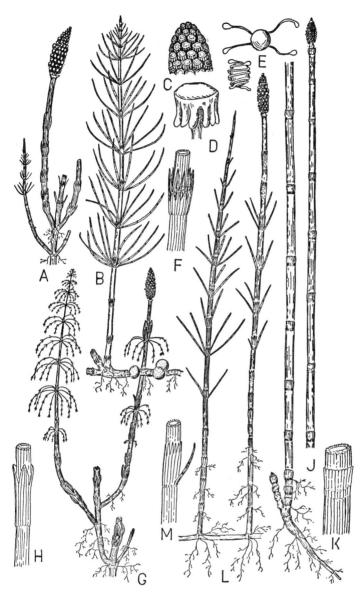

FIG. 4—**FIELD HORSETAIL**, *Equisetum arvense:* **A**, fertile shoot bearing cone (reduced to ⅓ natural size); **B**, sterile shoot (× ⅓); **C**, part of cone (× 3); **D**, sporangiophore (× 6); **E**, spores with elators (× 100).
WOOD HORSETAIL, *Equisetum sylvaticum:* **G**, sterile shoot with fertile shoots attached (× ⅓); **H**, section of stem with whorl of leaves at node (× 1).
SCOURING-RUSH, *Equisetum hyemale:* **J**, stem with cone (× ⅓); **K**, node (× 1).
HORSETAIL, *Equisetum palustre:* **L**, stem with cone (× ⅓); **M**, node (× 1).

tion is not practicable, improved drainage (21) and frequent close cutting (11) must be resorted to.

This species is represented by several varieties in different parts of the United States. It frequently forms very dense patches of shoots which persist all winter.

3. *Equisetum palustre* L. Horsetail, Shade horsetail. Fig. 4, l–m.

Perennial; reproducing by spores and creeping rootstocks. Locally common in low wet meadows and springy places. Newfoundland to Alaska, south to Oregon and New Jersey. Native. July.

Description.—Stems annual, hollow, jointed, erect, 5–10 dm. high, light green, simple or sparingly branched with simple ascending branches. Cones terminal on fertile stems which are very similar to the sterile stems or rarely on simple stems without chlorophyll. Leaves whorled, scale-like and joined by their edges into a loose sheath with 8–12 black teeth.

Control.— The same as for 1.

4. *Equisetum sylvaticum* L. Wood horsetail. Fig. 4, g–h.

Perennial; reproducing by spores and creeping rootstocks. Wet meadows, pastures and swampy woods. Widespread; Newfoundland to Alaska, south to Virginia, Ohio and Iowa. Native. May–June.

Description.—Stems annual, of two kinds; similar to *E. arvense* but the branches of the sterile stems drooping; fertile stems with 1 or 2 whorls of simple drooping branches. Teeth of the leaf-sheaths brown.

Control.—The same as for 1.

Equisetum telmateia Ehrh. Giant or Ivory horsetail. Similar to 1, but larger and with 20–30 teeth in a sheath, sometimes a weed on wet meadows and fields on alluvial soils of river valleys along the north Pacific Coast.

OSMUNDACEAE (Flowering Fern Family)

5. *Osmunda cinnamomea* L. Cinnamon fern.

Perennial; reproducing by spores. Low wet grasslands; about swamps and peat bogs. Common, Newfoundland to Minnesota, south to Florida and New Mexico. Native. May–June.

Description.—Fronds upright, in thick clumps from short thickened rootstocks, covered with rusty-brown wool when young. Fertile fronds appearing first, without chlorophyll, 2-pinnate, covered with

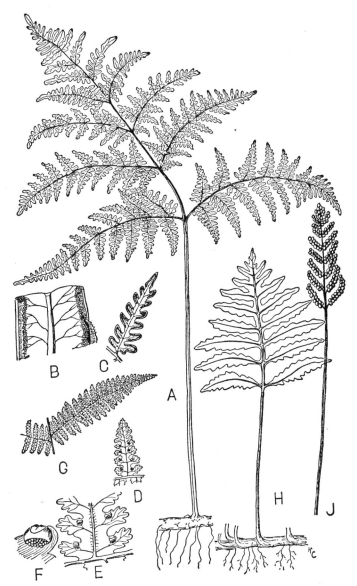

Fig. 5—**BRAKE FERN,** *Pteridium aquilinum* var. *latiusculum:* **A,** frond attached to rootstock (× ⅙); **B,** a small section of a pinnule showing marginal sporangia (× 4); **C,** a pinnule (× 1).
HAY-SCENTED FERN, *Dennstaedtia punctilobula:* **D,** a pinnule (× 1); **E,** portion of pinnule showing globular marginal sori (× 4); **F,** sorus (× 9); **G,** part of frond showing a pinna (× ⅓).
SENSITIVE FERN, *Onoclea sensibilis:* **H,** sterile frond (× ⅙); **J,** fertile frond (× ⅙).

cinnamon-colored, globular, naked sporangia, withering soon. Sterile fronds about 5–15 dm. high, 1-pinnate, with the pinnae lobed.

Control.—Improved drainage (21), followed by plowing and clean cultivation. Mow the fronds two or three times a year in pastures (11).

POLYPODIACEAE (Fern Family)

6. *Dennstaedtia punctilobula (Michx.)* Moore *(Dicksonia punctilobula* Gray). Hay-scented fern, Hairy dicksonia, Boulder fern. Fig. 5, d–g.

Perennial; reproducing by spores and creeping rootstocks. Hillside pastures, meadows and stony fields. This fern frequently forms extensive clumps or mats in old hillside pastures to the exclusion of all other vegetation. Eastern Canada to Minnesota, south to Georgia and Arkansas. Native. August–September.

Description.—Fronds alternate, from a long, slender, creeping rootstock, 4–10 dm. high, minutely glandular and hairy, lanceolate, mostly 2-pinnate, sweet-scented on drying. Sporangia in globular sori at or near the margin of the pinnae; sori surrounded by a cup-shaped indusium.

Control.—Mow close to the ground with a scythe (11) before spores are matured. Small patches can be grubbed out by hand (17). Cultivation will eradicate the fern, but land infested with it usually is so rocky or steep that plowing is impracticable.

7. *Onoclea sensibilis* L. Sensitive fern, Meadow brake, Polypod brake. Fig. 5, h–j.

Perennial; reproducing by spores and creeping rootstocks. Low wet meadows, swamps and marshes. Common, Newfoundland to Saskatchewan, southward to Florida, Oklahoma, and Texas. Most troublesome in the northeastern states. Native. August–September.

Description.—Rootstocks creeping, producing alternate scattered fronds of two kinds. Sterile fronds with erect stalks, 2–5 dm. long, bearing a nearly triangular blade about 1–3 dm. long, 1-pinnatifid; fertile fronds very dense and erect, 2-pinnate, the pinnules rolled up, pod-like, bearing many sporangia.

Control.—The same as for 5.

8. *Pteridium aquilinum* (L.) Kuhn var. *pubescens* Underw. Brake fern, Bracken, Hog brake.

Perennial; reproducing by spores and creeping rootstocks. Upland pastures, meadows and cultivated fields on recent clearings. Western

North America, most troublesome west of the Cascade Mountains. Native. August–September.

Description.—Plants with general habit and appearance similar to 9, but somewhat taller and the fronds pubescent or tomentose beneath.

Control.—Clean cultivation will destroy the rootstocks. Recently cleared land infested with brake fern should be plowed deep and the rootstocks harrowed out, piled and burned. In pastures mow or pull the fronds twice a year before spores are matured, usually July and August, but the exact dates vary somewhat with the locality. Fertilize infested areas and graze closely with sheep. Hogs are fond of the rootstocks of brake fern.

9. *Pteridium aquilinum* (L.) Kuhn var. *latiusculum* (Desv.) Underw. (*Pteris aquilina* L.). Brake fern, Bracken, Eagle fern, Hog brake, Upland fern. Fig. 5, a–c.

Perennial; reproducing by spores and creeping rootstocks. Upland pastures and abandoned fields, chiefly in sandy or sterile gravelly soils in the northeastern United States, and occasional south and west. Native. August–September.

Description.—Rootstocks creeping, often from 1–3 m. long, branched, dark brown or black, about 1–2 cm. in diameter. Fronds scattered, 5–20 dm. high, the stalk terminated by the broadly triangular blade which is divided into 3 main segments, each of which is usually 2-pinnate; the ultimate segments of the frond entire or lobed. Sporangia forming a continuous sorus along the margin of the pinnae, covered by the narrow recurved edge of the leaf which forms a marginal indusium. Spores light brown in mass.

Control.—Mow or pull the fronds in June before spores are produced and again in August. Fertilize infested areas and graze closely with sheep.

The brake fern is poisonous when eaten by stock. In central New York cows have been poisoned by eating the fronds during dry seasons or in late summer when other green vegetation is scarce.[1]

GRAMINEAE (Grass Family)

10. *Agropyron repens* (L.) Beauv. Quack-grass, Couch-grass, Witch-grass, Wheat-grass, Quitch-grass, Shelly-grass, Knot-grass, Devils-grass, Scutch-grass. Fig. 6.

[1] Hagan and Zeissig, 1927.

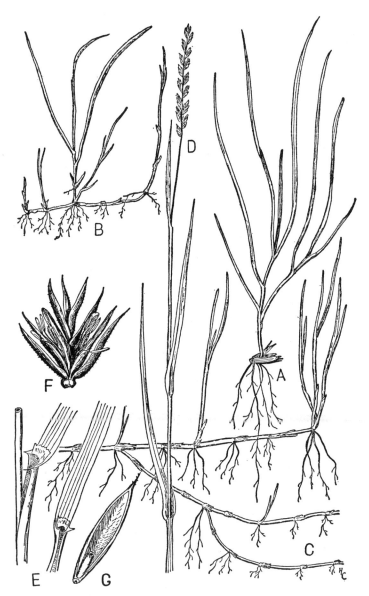

Fig. 6—**QUACK-GRASS**, *Agropyron repens:* **A**, seedling (× ⅓); **B**, a young plant showing the beginning of an extensive system of rootstocks (× ⅓); **C**, portion of rootstock of an older plant, showing nodes and scales (× ⅓); **D**, spike of spikelets (× ⅓); **E**, leaf-sheath showing ligule (× 1); **F**, spikelet showing flowers (× 3); **G**, seed (grain), (× 3).

Perennial; reproducing by seeds and rootstocks. Cultivated fields, grasslands and waste ground; especially on fertile soil. Abundant in the northeastern and north central states and adjacent Canada; local elsewhere. Introduced from Europe. June–July.

Description.—Stems erect, simple, 5–12 dm. high, from creeping jointed rootstocks with fibrous roots at the nodes. Leaves with glabrous sheaths which are shorter than the internodes; blades flat, narrow, rough on the upper surface. Spikelets 3–8-flowered, in 2 rows on opposite sides of the axis of a terminal spike which is 8–20 cm. long; spikelets with their broad sides against the axis of the spike; empty glumes stiff, 5–8 nerved, and with a sharp apex or awned; flowering glumes with 5–7 nerves, rounded on the back with a short terminal awn or pointed; palet 2-keeled. Grain 4–5 mm. long, narrowly oblong, yellow-brown, inclosed in the flowering glume.

Control.—Small areas of quack-grass can be destroyed by digging the rootstocks by hand (17) or covering them with mulch paper (24). Small patches of quack-grass should be treated as separate fields, otherwise the rootstocks will be dragged to other parts of the field by the harrow and other cultivating implements. A four-year rotation consisting of grain, hay and two years of a cultivated crop will usually control quack-grass without the loss of the use of the land for crops. (20).

The best time to start the control of quack-grass is after the meadow or pasture has been kept closely grazed for a year or two. This tends to bring the rootstocks closer to the surface of the soil. If the land is in cultivation, it should first be seeded to meadow or pasture for a year or two if possible, to bring the rootstocks near the soil surface. The following control work should be started in the dry hot weather about the first of July or as soon as the hay crop is removed. Plow shallowly, turning the sod over in order to expose the rootstocks to the drying action of the sun. As soon as the green shoots begin to appear, harrow with a spike-tooth or spring harrow, turning the sod and thus drying the roots. (The dried material can be raked into piles and burned.) Continue the harrowing until fall, harrowing whenever the quack-grass shoots appear above the soil surface. Never let them get higher than one inch. Leave the land rough over the winter. In the early spring as soon as the land can be worked, harrow as before or plow if necessary. Follow by a or b. (a) If it is an early and dry spring so that most of the quack-grass has been killed by corn planting time, plant a tilled crop, such as corn in check rows. This crop should be cultivated thoroughly, cutting the quack-grass out of the hills or rows with a hoe. (b) If the quack-grass is still abundant, keep up the harrowing until late June or early July and sow a heavy smother crop of buckwheat, fodder corn or sunflowers. This will tend to smother out the remaining quack-grass. After the harvest of the cultivated or smother crop the land should be plowed in autumn and left rough over winter. Next spring plant a good smother crop after the cultivated crop. If a good stand is secured, very little quack-grass will survive. If a smother crop was used the preceding summer, then follow with a cultivated crop. Slight modifications of this general method have been effective in a number of regions where quack-grass is troublesome.[2, 3, 4]

Quack-grass has some value as a pasture or hay crop. Its rootstocks are efficient soil-binders on slopes, embankments and sandy soils where a sod is necessary.

Agropyron smithii Rydb. Western wheat-grass, Colorado bluestem, Western quack-grass, Blue joint-grass, a perennial, native to the prairies from the north central states westward, often encroaches on cultivated fields.[5, 6] It can be distinguished from *Agropyron repens* by its bluish-green or grayish-green stem and leaves which tend to roll up lengthwise, and by its more slender and less extensive rootstocks. Its spikelets are less compact and have 6-12 flowers. The control is the same as for quack-grass.

11. *Aira caryophylleu* L. (*Aspris caryophyllea* Nash). Silver hair grass.

Annual; reproducing by seeds. Dry fields, pastures and waste ground; mostly on sandy or gravelly soil. Widespread across North America; but most abundant on the Pacific Coast. Introduced from Europe. May-June.

Description.—Stems single or a few in a cluster, slender, erect, 1-3 dm. high. Spikelets clustered toward the ends of thread-like branches of an open panicle; spikelets with 2 perfect flowers, about 3 mm. long; empty glumes thin, somewhat papery, nearly equal; flowering glumes about 2 mm. long, 2-toothed, and with a jointed awn on the back near the middle. Grain 1 mm. long, narrowly oval, yellow-brown, included within the glume and palet.

[2] Arny, 1915, 1928.
[3] Kephart, 1923.
[4] Runnels and Schaffner, 1931.
[5] Hume and Sloane, 1916.
[6] Norris, 1929.

Control.—Plow and sow a smother crop which should be turned under. Follow by a cultivated crop for a year before reseeding.

12. *Alopecurus geniculatus* L. Marsh foxtail, Bent foxtail, Water timothy, Water foxtail.

Perennial; reproducing by seeds and by stems rooting at the nodes. Moist meadows, pastures, along ditches, waste places. Widespread throughout the northern United States and southern Canada. Apparently native; possibly also introduced from Eurasia. June–August.

Description.—Stems slender, the lower part often bent and taking root at the nodes, the upper part erect, simple or sparingly branched, about 2–6 dm. high, smooth. Leaves with narrow rough blades and sheaths shorter than the internodes. Spikelets in dense cylindrical spikes (appearing like timothy), up to 1 dm. long; spikelets 1-flowered, about 3 mm. long; empty glumes unequal, with long ciliated keel; flowering glume shorter than the empty glumes, sac-like, bearing a slender bent awn near the middle, palet absent. Grain about 1 mm. long, flattened on one side, half-ovate, greenish-yellow.

Control.—Improve drainage (21). Clean cultivation (18).

Alopecurus pratensis L. Meadow foxtail-grass. Introduced from Europe into meadows and pastures in the eastern states. It has spikelets 5 mm. long with flowering glume as long as the empty glumes.

13. *Andropogon gerardii* Vitman (*A. furcatus* Muhl.). Beard-grass.

Perennial; reproducing by seeds. Dry meadows, pastures, fields and open waste places; especially on sandy soils. Widespread in eastern and central North America. Native. July–September.

Description.—Stems tufted, coarse, about 1–1.5 m. high, branched above. Leaves with glabrous sheaths and long narrow blades with scabrous margins. Racemes spike-like, in clusters of 2–6. Spikelets in pairs, 1 fertile and 1 staminate, on the jointed hairy rachis of the raceme; fertile spikelet 7–10 mm. long, sessile; empty glumes indurated, the first flattened, the second keeled; first flowering glume empty; fertile flowering glume with a twisted awn; staminate spikelet pedicelled and somewhat longer. Grain 3–4 mm. long, one side flattened, the other convex, dull, yellow-brown to reddish-brown.

Control.—Mow before seeds are matured. Plow and follow with a cultivated crop.

14. *Andropogon scoparius* Michx. Broom beard-grass, Wolf-grass, Poverty-grass.

Perennial; reproducing by seeds. Dry fields, pastures and waste ground; especially on sandy soil. Widespread in eastern and central North America. Native. July–September.

Description.—Stems tufted, 4–12 dm. high, branched above. Leaves with glabrous or hairy sheaths, blades usually hairy on the upper surface at base. Spikelets in pairs, 1 fertile and 1 staminate, on a solitary, spike-like, jointed, slender raceme; fertile spikelet of each pair 6–8 mm. long, sessile; empty glumes indurated, the first flattened, the second keeled; fertile flowering glume with a twisted awn; sterile spikelet pedicelled, reduced to a solitary awned glume, 2–4 mm. long. Grain 3–4 mm. long, dark brown, somewhat glossy.

Control.—The same as for 15.

15. *Andropogon virginicus* L. Beard-grass, Broom sedge, Sedgegrass.

Perennial; reproducing by seed. Dry meadows, pastures, fields and waste places. Widespread in the eastern and southern states; most troublesome in the South. Native. July–September.

Description.—Stems tufted, slender, smooth, 5–12 dm. high, sparingly branched above, becoming hard with age. Leaves 1–3 dm. long; blades hirsute near the base; sheaths glabrous or hirsute on the margin. Racemes slender, 2–3 in a cluster, about 3 cm. long, exceeded by the bract-like spathes; joints of raceme and the pedicel covered with long silky hairs. Spikelets in pairs; the fertile spikelet sessile, with 2 nearly equal hardened empty glumes and a colorless long-awned fertile flowering glume; the sterile spikelet pedicelled, reduced to a minute scale. Grain 1.5–2 mm. long, linear, one side flat, the other rounded, glossy, yellow-green.

Control.—Close grazing early in the season while the stems are still succulent and nutritious (10). Meadows should be cut early before the beard-grass has produced seeds. Plowing followed by a cultivated crop and a legume will eradicate this weed. Burning infested fields has been resorted to in the south; too frequent repetition of this practice depletes the humus content of the soil.

16. *Anthoxanthum odoratum* L. Sweet vernal-grass.

Perennial; reproducing by seeds. Dry fields, meadows and waste places; on somewhat poor soils. Widespread throughout the eastern and north central states; Washington to California; also eastern Canada. Introduced from Europe. May–June.

Description.—Stems slender, erect, 2–6 dm. high. Leaves flat, rough above, aromatic. Spikelets 1-flowered, 8–10 mm. long, in narrow erect spike-like panicles 4–8 cm. long, greenish-brown; empty glumes very unequal, somewhat pilose; fertile flowering glume without awn, palet 1-nerved; sterile flowering glumes 2-lobed, hairy, awned on the back. Grain inclosed in flowering glume, 1.5–2 mm. long, half-ovate, smooth, glossy, straw-colored.

Control.—Mow early before seeds are matured. Badly infested fields should be plowed and followed with a cultivated crop or cover-crop.

17. *Aristida dichotoma* Michx. Poverty-grass, Wire-grass. Fig. 15, d–e.

Annual; reproducing by seeds. Dry grasslands, sandy or gravelly pastures and neglected fields. Widespread in eastern North America, westward to Ontario and Texas. Native. August–October.

Description.—Stems tufted, wiry, much branched at base and forking at the nodes, 1–6 dm. high. Leaves with narrow inrolled blades and loose sheaths and hairy ligules. Spikelets in slender panicles, the lateral ones often enclosed in sheaths; spikelets 1-flowered; empty glumes 7–9 mm. long, unequal, slightly keeled, narrow, acuminate; flowering glume hard, closely folding over the grain and palet, terminated with 3 awns of which the lateral ones are short, 3–6 mm. long, but the central one is as long as the flowering glume. Grain 3–5 mm. long, linear-subulate, dull, greenish-yellow.

Control.—Plow, fertilize and follow with a cultivated crop or plow under a green-manure crop before reseeding.

18. *Aristida oligantha* Michx. Few-flowered aristida, Wire-grass, Triple-awn. Fig. 15, f–g.

Annual; reproducing by seeds. Dry grasslands, fields, roadsides and waste places. Widespread throughout the United States; most common in the southeastern states. Native. July–October.

Description.—Stems tufted, wiry, branched at base and forking at the nodes, 3–6 dm. high, often woolly at base. Leaves similar to 17. Spikelets in slender flexuous few-flowered panicles; spikelets similar to 17, but the 3 awns on each flowering glume 3–8 cm. long, nearly equal. Grain 10–14 mm. long, linear-subulate, dull, greenish-yellow.

Control.—The same as for 17.

19. *Avena fatua* L. Wild oats, Wheat oats, Oat-grass, Flax-grass. Fig. 7, a–c.

Annual; reproducing by seeds. Grain fields and waste places. Widespread; but most troublesome in the Northwest, the Pacific Coast and

FIG. 7—**WILD OATS,** *Avena fatua:* **A,** part of a plant showing habit (× ⅓); **B,** spikelet (× 1); **C,** grain (× 1).
WILD OAT-GRASS, *Danthonia spicata:* **D,** plant showing habit (× ⅓); **E,** spikelet (× 3); **F,** grain (× 3).
VELVET-GRASS, *Holcus lanatus:* **G,** plant showing habit (× ⅓); **H,** spikelet (× 3); **J,** grain (× 9).

the prairie provinces in western Canada. Introduced from Europe. June–July.

Description.—Stems erect, 4–12 dm. high, solitary or a few in a tuft. Leaves alternate; blades 5–8 mm. wide, glabrous; sheaths glabrous or slightly hairy. Spikelets 2–5 flowered, in open terminal panicles, pendulous, 2–2.5 cm. long, not including the awns, empty glumes many-nerved, smooth, exceeding the flowering glumes; flowering glumes 1–2 cm. long, pubescent with appressed brown hairs, also with a ring of hairs at base, 2-toothed at apex, provided with a twisted dorsal awn 3 cm. long or more. Grain 6–8 mm. long, oblong, grooved, hairy, straw-colored, inclosed in the flowering glume and palet.

Control.—Sow pure seed. Use a rotation in which the grain is followed by a cultivated crop, clover or alfalfa. Badly infested oat fields or meadows should be cut for hay in the dough stage to prevent seeds from being scattered.

20. *Bromus commutatus* Schrad. Smooth brome-grass.

Annual or biennial; reproducing by seeds. Fields, roadsides and waste places, usually on gravelly soils. Widespread throughout North America; most common in the northeastern states and on the Pacific Coast. Introduced from Europe. May–July.

Description.—Stems about 3–6 dm. high. Leaves with hairy sheaths. Panicles terminal, drooping, 1.5 dm. long. Spikelets 5–8-flowered, glabrous; flowering glumes about 7–10 mm. long, 5–7-nerved, with an awn 4–6 mm. long; palet with 2 barbed keels. Grain 6–8 mm. long, linear-obovate, grooved, minutely granular, yellowish-brown, adnate to the palet.

Control.—Mow before seeds are formed (11). Clean cultivation (6). Harrow infested meadows and pastures while the brome-grass seedlings are small.

21. *Bromus mollis* L. (*B. hordeaceus* L.) Soft chess, Soft brome. Fig. 8, g–j.

Annual; reproducing by seeds. Dry fields, pastures and waste places. Abundant on the Pacific Coast; infrequent in the eastern states. Introduced from Europe. June–July.

Description.—Stems 2–5 dm. high, pubescent. Sheath very hairy. Panicles terminal, erect, dense. Spikelets 6–10-flowered; flowering glumes about 1 cm. long, hairy or scabrous, with awn about 1 cm. long. Grain similar to 20.

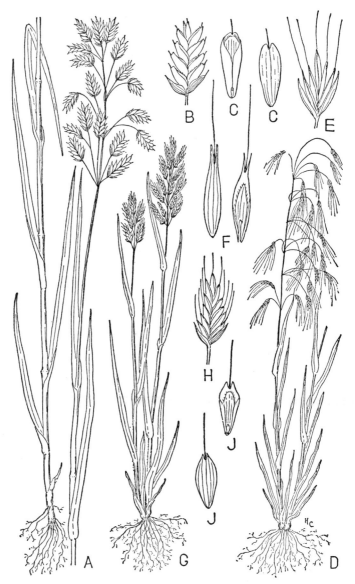

Fig. 8—CHESS, *Bromus secalinus:* **A**, plant showing general habit (× ⅓); **B**, spikelet (× 1); **C**, grains (× 3).
DOWNY BROME-GRASS, *Bromus tectorum:* **D**, plant (× ⅓); **E**, spikelet (× 1); **F**, grains (× 3).
SOFT CHESS, *Bromus mollis:* **G**, plant (× ⅓); **H**, spikelet (× 1); **J**, grain (× 3).

Control.—The same as for 20.

22. *Bromus racemosus* L. Upright chess.

Annual; reproducing by seeds. Fields, pastures and waste places. Widespread across the United States; frequent on the Pacific Coast. Introduced from Europe. June–July.

Description.—Stems erect, simple, often tufted, 3–6 dm. high. Leaves with hairy blades and sheaths. Panicles short, 4–7 cm. long, with erect or ascending branches. Spikelets 5–8-flowered, glabrous; empty glumes 6–8 mm. long, the first 3-nerved, the second 5–9-nerved; flowering glume rounded, about 7 mm. long, glabrous or scabrous, with a straight awn 6 mm. long; palet with 2 barbed keels. Grain linear, 6–7 mm. long, grooved, glossy, yellow-brown.

Control.—The same as for 24.

23. *Bromus rubens* L. Red brome, Red foxtail.

Annual; reproducing by seeds. Grasslands, waste places, new seedings of clover, alfalfa and lawns. Infrequent in the eastern states; common in the drier regions of the Pacific Coast. Introduced from Europe. May–July.

Description.—Stems erect, hairy, 1–4 dm. high. Leaves with pubescent blades and sheaths. Panicle erect, compact, about 3–7 cm. long, usually purplish. Spikelets 7–11-flowered, about 2.5 cm. long; empty glumes narrow acuminate, pubescent or smooth, the first 1-nerved, the second 3-nerved; flowering glumes 12–16 mm. long, lanceolate, keeled, with hyaline margins, 5-nerved, with an awn about 2 cm. long; palet with 2 hairy keels. Grain similar to 22 but slightly longer.

Control.—Cut the first crop of clover or alfalfa early before the red brome has matured seeds. Use a short rotation of crops (20). Clean cultivation (6).

24. *Bromus secalinus* L. Chess, Cheat, Wheat-thief, Cock-grass. Fig. 8, a–c.

Winter annual; reproducing by seeds. Grain fields, especially winter wheat and rye; waste places. Widespread throughout the United States and Canada where winter grain is grown. Introduced from Eurasia. June–July.

Description.—Stems 3–10 dm. high, solitary or a few in a tuft. Leaves with glabrous or somewhat hairy prominently nerved sheaths; blades somewhat hairy above. Panicles 7–12 cm. long, open, drooping at maturity. Spikelets 5–15-flowered, on long pedicels, 1–2 cm. long;

empty glumes unequal, 4–7 mm. long, the first 3–5-nerved, the second 7-nerved; flowering glumes elliptical, 7-nerved, about 6–9 mm. long, with or without a short awn, becoming thick and inrolled at margin when mature; palet with 2 barbed keels. Grain 6–7 mm. long, linear-obovate, deeply grooved, glossy, dark orange-brown.

In certain sections the erroneous belief is still held by some that chess is a degenerate form of wheat.

Control.—Use a short rotation of crops and do not include winter wheat more than once in four years (20). Use pure seed grain (1). In cultivated crops pull or hoe out scattered chess plants.

25. *Bromus sterilis* L. Barren brome-grass.

Annual or winter annual; reproducing by seeds. Waste places, roadsides, ballast grounds, stream banks. Widespread in the eastern United States; also Colorado and on the Pacific Coast. Introduced from Europe. May–July.

Description.—Stems 5–10 dm. high, often bent at base. Leaf-sheaths and blades somewhat hairy. Panicle open, drooping, with long slender branches, each usually with a solitary spikelet. Spikelets 4–10-flowered, 2.5–3.5 cm. long; empty glumes lanceolate, subulate, unequal, glabrous or scabrous; flowering glume rounded, linear, scabrous, with an awn about 2–3 cm. long from between 2 long teeth; palet with 2 barbed keels. Grain 10–12 mm. long, linear, grooved, ridged on the back, orange-brown.

Control.—The same as for 20.

26. *Bromus tectorum* L. Downy brome grass, Slender chess, Early chess. Fig. 8, d–f.

Annual or winter annual; reproducing by seeds. Fields, pastures, roadsides and waste places; on dry sandy or gravelly soils. Widespread throughout North America; very common in the western states. Introduced from Europe. May–July.

Description.—Stems slender, 3–6 dm. high, tufted. Leaf-blades and sheaths very pubescent. Panicle rather dense, drooping, 1-sided, 5–15 cm. long, with slender branches. Spikelets 3–6-flowered, 1–2 cm. long; empty glumes narrow, villous, unequal, the first 1-nerved, the second 3-nerved; flowering glume lanceolate, pubescent, 3–5-nerved, with hyaline margin, with an awn up to 1–1.5 cm. long; palet with 2 barbed keels. Grain 6–8 mm. long, linear, grooved, minutely striate, yellow to reddish-brown.

Control.—The same as for 20.

Bromus japonicus Thunb., Japanese chess, is listed as a weed (1954) in the north central states.

27. *Cenchrus longispinus* (Hack.) Fern. (*C. pauciflorus* of authors, not Benth., *C. carolianus* Walt.). Sandbur, Bur-grass, Sandbur-grass, Bear-grass, Hedgehog-grass. Fig. 11, d–f.

Annual; reproducing by seeds in the spiny burs. Fields, pastures, banks, orchards and waste places; mostly on sandy soil. Widespread from the northeastern states to Central America; most troublesome in the southern states; rare on the Pacific Coast. Native southward to Central America; introduced northward. July–September.

Description.—Stems flattened, branched, prostrate or ascending, 2–8 dm. long, often forming mats. Leaves flat or coiled, sheath ciliate. Spikelets 1-flowered, flattened, with thin glumes, 2–3 fused and surrounded by a hard spiny bur-like involucre; involucres about 6–8 mm. in diameter, about 6–15 in a raceme.

Control.—In grasslands and waste areas prevent seed production by hand hoeing or pulling the plants before seeds are matured. Burn plants with mature burs. Plow and follow with a clean cultivated crop (6).

28. *Cynodon dactylon* (L.) Pers. (*Capriola dactylon* Kuntze). Bermuda-grass, Scutch-grass, Dogs-tooth-grass, Wire-grass.

Perennial or annual in the North; reproducing by creeping rootstocks and seeds. Fields, meadows, lawns and waste places; mostly on sandy soil. Escaped rather freely and becoming a bad weed locally in the South where it is used extensively as a lawn and meadow grass; extending northward and on the Pacific Coast; not hardy north of Maryland. Introduced from Eurasia. July–August.

Description.—Stems prostrate and creeping, branching, with short wiry flattened culms. Leaves flattened, ligule with a fringe of whitish hairs. Spikes one-sided, about 4 cm. long, 3–5 in a cluster, radiating from the end of the culm. Spikelets 1-flowered, laterally compressed, sessile, arranged in 2 rows on the spikes; empty glumes unequal; flowering glume broad, keeled, ciliate, exceeding the empty glumes. Grain about 1.5 mm. long, oval, orange-red, free within the flowering glume.

Control.—Shallow plowing in late autumn so as to expose the rootstocks to the air over winter will kill the Bermuda-grass in the northern part of its range where the ground freezes. Farther south, where frosts are less severe, exposing the rootstocks over winter will not destroy them. Here the most successful method of controlling the Bermuda-grass [7] consists of plowing the field in September and sowing a winter smother crop such as rye, barley or oats. As soon as this crop is removed the field is again plowed or thoroughly disked and then planted with a crop of cowpeas or velvet beans. The dense shade of two successive cover-crops will destroy most of the Bermuda-grass and leave the soil fit for a cultivated crop.

29. *Dactyloctenium aegyptium* (L.) Richter. Egyptian-grass, Crow-foot-grass.

Annual; reproducing by seeds and creeping stems. Lawns, yards, gardens and waste places. Locally common in the eastern and north central states, also in the south, North Carolina to Texas. Introduced from Asia or northern Africa. July–September.

Description.—Stems prostrate or creeping, rooting at the lower nodes, branching. Leaves mostly glabrous. Spikes 2–6, spreading from the tip of the culm; with rows of sessile crowded spikelets on one side of a continuous rachis, the upper end of which is naked. Spikelets with several flowers, 2 of which are perfect, the upper ones imperfect; empty glumes unequal, keeled, the second with awn about 1 mm. long; flowering glume cuspidate, ciliate on the keel. Grain about 0.8 mm. long, obovate, with the loose pericarp irregular, transversely ridged, orange-brown.

Control.—In lawns, weed the seedlings by hand as soon as they can be recognized, the earlier the better. If this weed is already large, raise it with a rake and mow it close to the ground to prevent seed production.

30. *Danthonia spicata* (L.) Beauv. Wild oat-grass, Poverty-grass, Bonnet-grass, White horse, Wire-grass, Wild-cat-grass, Old fog. Fig. 7, d–f.

Perennial; reproducing by seeds. Old pastures, meadows, and neglected fields on dry, rather sterile, mostly acid soils. Widespread and locally abundant throughout the hilly sections of the eastern United States. Native. June–July.

Description.—Stems 2–7 dm. high, slender, tufted. Leaves numerous, mostly basal, tufted and often curled, those of the stem erect; blades and sheaths glabrous or somewhat hairy. Spikelets about 1 cm. long, several-flowered, the upper flowers imperfect, on short erect branches of a raceme-like panicle; empty glumes unequal, acuminate,

[7] Tracy, 1917.

exceeding the flowering glumes; flowering glumes 5 mm. long, sparsely hairy, with a flattened twisted awn from between 2 triangular teeth. Grain about 2 mm. long, obovate, finely striate lengthwise, reddish-brown, free from the flowering glume.

Control.—The presence of *Danthonia spicata* in agricultural soils is an indication that the soil fertility has been depleted. The soil must be fertilized and limed before crops or other grasses can compete. Extensive areas infested with this weed should be among the first to be converted from agricultural land into forests. Such land is better adapted for forest trees than agricultural crops and Danthonia cannot thrive in the forest shade.

31. *Digitaria filiformis* (L.) Koel. Slender crab-grass.

Annual; reproducing mainly by seeds. Sterile or sandy soil, fields, and open grounds; often a troublesome weed southward. New Hampshire, Michigan, and Iowa, south to Florida and Texas. Native. August–October.

Description.—Stem 0.5–14 dm. high, filiform or coarser, simple or branching at the leafy base. Leaf-blades 0.5–2 dm. long, 1–5 mm. wide, hirsute, or glabrous on the lower and scabrous on the upper surface; lower sheaths villous to hirsute. Racemes 1–6, unequal, 1–25 cm. long, slender, erect or ascending, often distinctly separated at base; rachis narrowly winged. Spikelets 1.5–2.5 mm. long, mostly in threes, appressed, well separated, scarcely overlapping, on slender, flexuous pedicels. Glume and sterile lemma sparsely or densely villous; the narrow glume shorter than the dark brown or dark purple fertile lemma.

Control.—The same as for 32.

32. *Digitaria ischaemum* (Schreb.) Muhl. (*D. humifusa* Pers.; *Syntherisma ischaemum* Nash). Small crab-grass, Finger-grass. Fig. 9, f–h.

Annual; reproducing by seeds and by stems rooting at lower nodes. Lawns, pastures, cultivated fields, roadsides and waste ground; mostly in light or sandy soil. Widespread in eastern North America; locally common elsewhere. Introduced from Europe. August–September.

Description.—Stems 1–4 dm. long, glabrous, much branched, spreading or forming prostrate mats, sometimes rooting at the lower nodes. Leaves alternate, 2–12 cm. long, 3–6 mm. wide, sheath and blade glabrous. Spikelets on short, terete, nearly glabrous pedicels, 1-flowered, in 2 rows on one side of a winged rachis of spike-like racemes; racemes 3–5, spreading from the top of the stem. Spikelets

Fig. 9—**LARGE CRAB-GRASS,** *Digitaria sanguinalis:* **A,** plant showing general habit (× ⅓); **B,** leaf (× ⅓); **C,** spike of spikelets (× ⅓); **D,** spikelet showing two empty glumes (× 4); **E,** grain (× 4).
SMALL CRAB-GRASS, *Digitaria ischaemum:* **F,** plant showing habit (× ⅓); **G,** spikelet showing two empty glumes (× 4); **H,** grain (× 4).

about 2 mm. long, with the first empty glume minute or lacking, second empty glume with thin flat hyaline margins, about as long as the spikelet; flowering glumes and palet thin, dark brown, with hyaline margins. Grain about 1.5 mm. long, oval, dull, pale yellow. This weed, and to a less extent 33, is one of the most troublesome to lawns during dry seasons in late summer. Its prostrate stems form unsightly brown patches in thin grass and bare places in lawns.

Control.—In cultivated fields practice clean cultivation followed by hoeing or hand weeding to destroy the plants that appear late in the season. In lawns weed the crab-grass seedlings by hand as soon as they can be recognized, the earlier the better. If the crab-grass is already large raise it with a rake and mow it close to the ground to prevent seed production. Apply fertilizer late in autumn or very early in the spring to stimulate the lawn to grow before the crab-grass seeds have germinated.

33. *Digitaria sanguinalis* (L.) Scop. (*Syntherisma sanguinale* Dulac). Large crab-grass, Purple crab-grass, Finger-grass, Polish millet, Crowfoot-grass, Pigeon-grass. Fig. 9, a–e.

Annual; reproducing by seeds and by stems rooting at lower nodes. Cultivated fields, waste places, bare patches in lawns and along ditches; especially on light soils. Widespread throughout North America; very troublesome in the eastern and southern states. Introduced from Europe. July–September.

Description.—Stems 2–10 dm. long, erect or ascending from a decumbent, often rooting, base. Leaf-blades 5–12 cm. long, 4–10 mm. wide, scabrous or hairy; sheaths and nodes somewhat hairy. Spikelets on angular scabrous pedicels, 1-flowered, in pairs on one side of the rachis of spike-like racemes; racemes 3–10, spreading from near the top of the stem. Spikelets about 3 mm. long, with the first empty glume minute; second empty glume one-half the length of the spikelet, with thin, flat, hyaline margins; flowering glumes and palet grayish-brown to purple, granular striate. Grain 1.5–2 mm. long, oval, finely granular, light yellow.

Control.—The same as for 32.

34. *Echinochloa crusgalli* (L.) Beauv. Barnyard-grass, Panic-grass, Cockspur-grass, Cocksfoot panicum, Barn-grass, Water-grass. Fig. 10, a–c.

Annual; reproducing by seeds. Spring grain fields, cultivated fields, farmyards, waste places and along ditches; especially on moist rich

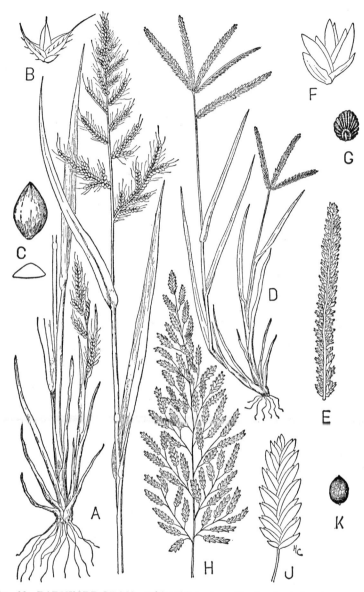

FIG. 10—**BARNYARD-GRASS**, *Echinochloa crusgalli*: **A,** plant showing general habit (× ⅓); **B,** spikelet (× 4); **C,** surface and cross-section views of seed (× 6).
GOOSE-GRASS, *Eleusine indica*: **D,** plant showing general habit (× ⅓); **E,** one-sided spike (× 1); **F,** spikelet (× 4); **G,** seed (× 6).
STINK-GRASS, *Eragrostis megastachya*: **H,** panicle (× ⅓); **J,** spikelet (× 4); **K,** seed (× 6).

soil; also a troublesome weed in rice fields. Widespread throughout the United States and Mexico; most troublesome in the South. Introduced from Europe. July–September.

Description.—Stems erect or ascending, stout and somewhat succulent, 5–15 dm. high. Leaves glabrous; the blade 1–2 dm. long. Panicle open, composed of nearly sessile one-sided racemes, with long hairs at the nodes. Spikelets with 1 perfect flower and often also 1 staminate flower below, with short awns; second empty glume with short spines on the midrib.

Control.—Clean cultivation (6); mow infested fields early before seeds are matured (11).

35. *Echinochloa pungens* (Poir.) Rydb. (*E. muricata* Fern.). Barnyard-grass.

Annual; reproducing by seeds. Low fields and grasslands on alluvial soil and near banks of streams. Widespread and locally common in the eastern United States, extending westward. Probably native. August–September.

Description.—Similar to 34. Panicle compact with short hairs at the nodes. Spikelets prominently bristly; second empty glume with spines on the midrib not shorter than those on the flowering glume; flowering glume rounded, hardened, glossy. Grain 1.5 mm. long, broadly oval, finely granular, yellow-brown.

Control.—The same as for 34.

36. *Eleusine indica* Gaertn. Goose-grass, Wire-grass, Yard-grass, Crowfoot-grass. Fig. 10, d–g.

Annual; reproducing by seeds. Grasslands, lawns, yards and waste places. Widespread in the eastern United States; west to Minnesota and South Dakota; most common in the South. Introduced from the warmer parts of Asia. June–September.

Description.—Stems flattened, decumbent or prostrate, branched and often forming mats. Leaves with loose, over-lapping, flattened sheaths and flat pale green blades. Spikes 2–10, radiating from the top of the stem, each with 2 rows of spikelets on one side extending to the tip of the rachis. Spikelets 3–5-flowered, about 5 mm. long, appressed, sessile; empty glumes unequal, with a rough keel; flowering glume keeled, glabrous; palet with 2 minutely barbed keels. Grain 1–1.5 mm. long, granular and ridged, loosely inclosed within the pericarp, reddish-brown.

Control.—Cut with a hoe or pull by hand before the seeds are formed. In lawns, pull the seedlings as soon as they can be recognized; if the weeds are already large, raise them with a rake and mow close several times.

37. *Eragrostis frankii* C. A. Mey. Love-grass.

Annual; reproducing by seeds and creeping stems rooting at the nodes. Cultivated fields and waste places; often in moist sandy situations. Eastern, north central, and southern states. Native; apparently introduced in some localities. August–September.

Description.—Similar to 38 but smaller and more slender. Leaf-sheaths glabrous; blades scabrous on upper surface; ligule pilose. Flowers 2–4 in a flattened spikelet; spikelet 2–3 mm. long. Grain as in 38.

Control.—The same as for 38.

38. *Eragrostis megastachya* (Koel.) Link (*E. cilianensis* Link; *E. major* Host.). Stink-grass, Meadow snake-grass, Meadow-grass, Strong-scented love-grass, Candy-grass. Fig. 10, h–k.

Annual; reproducing by seeds. Cultivated fields, roadsides and waste places. In much of the United States, especially in the South. Introduced, probably from Europe. June–September.

Description.—Stems 1–8 dm. long, erect or ascending from a decumbent base, branching freely below. Leaf-blades 5–15 cm. long, 3–6 mm. wide. Flowers 10–many, densely crowded in a flattened spikelet; spikelets 1–2 cm. long and about 2–3 mm. wide, in dense panicles; empty glumes keeled, much shorter than the spikelets; flowering glumes 3 nerved, broad, keeled, closely overlapping, the uppermost sterile. Grain 0.5–0.6 mm. long, oval, slightly flattened, orange-red.

Control.—In waste places cut with a hoe as soon as the seedlings can be recognized (5). Plow badly infested fields and follow with a clean cultivated crop; the last cultivation should be as late as possible so as to destroy late seedlings (7, 6).

39. *Eragrostis pectinacea* (Michx.) Nees (*E. pilosa* Beauv.). Tufted love-grass, Tufted spear-grass.

Annual; reproducing by seeds and by stems rooting at lower nodes. Dry fields, roadsides, paths and waste ground. Quebec to British Columbia, south through much of the United States to Mexico. Native; apparently also introduced in some localities. July–September.

Description.—Similar to 38 but smaller and more slender. Leaf-sheath pilose at the summit; blades 3–12 cm. long and 2–3 mm. wide. Flowers 5–many, in a spikelet; spikelets flattened, linear, 3–12 mm. long, 0.8–1.5 mm. wide; panicle open. Grain as in 38, 0.7–0.9 mm. long.

Control.—The same as for 38.

40. *Eragrostis poaeoides* Beauv. (*E. minor* Host.). Low love-grass.

Annual; reproducing by seeds. Cultivated soils, roadsides and waste places; especially near cities and towns. Throughout much of the United States. Introduced from Europe. July–September.

Description.—Similar to 38 but smaller and more slender. Flowers 8–many in a spikelet; spikelet 5–10 mm. long, 2 mm. wide; panicles open; flowering glumes not closely overlapping. Grain as in 38.

Control.—The same as for 38.

41. *Holcus lanatus* L. (*Ginannia lanata* Hub.; *Notholcus lanatus* Nash). Velvet-grass, Mesquite-grass. Fig. 7, g–j.

Perennial; reproducing by seeds. Meadows, grasslands and waste places; on rather damp rich soil, especially on peat lands. Widespread in eastern Canada and the northeastern and north central states; very troublesome in the Pacific Northwest. Introduced from Europe. June–August.

Description.—Entire plant grayish, velvety-hairy. Stems erect, 3–6 dm. high, solitary or in tufts. Leaves about 10–20 cm. long and 5–10 mm. wide. Spikelets 2-flowered, the lower perfect, the upper mostly staminate, in dense purple or grayish terminal panicles; empty glumes hairy, unequal, compressed, about 4 mm. long, exceeding the boat-shaped, somewhat hardened, flowering glumes; flowering glumes ciliate at apex, the upper with a hooked awn; palet thin, nearly as long as the flowering glume. Grain about 1.5 mm. long, oval, grooved, dull, yellowish-brown.

Control.—Pasture closely or mow infested meadows early before seeds are produced. Plow and follow with a smother crop or a clean cultivated crop for one or two years before reseeding.

42. *Hordeum jubatum* L. Squirrel-tail-grass, Wild barley, Skunk-tail-grass, Flicker-tail-grass, Tickle-grass, Foxtail-grass. Fig. 11, a–c.

Biennial or perennial; reproducing by seeds. Meadows, pastures, ranges, along ditches and on alkaline flats, waste places. Widespread in North America but most troublesome in the northern and Rocky Mountain states. Native. June–August.

Description.—Stems 3–6 dm. high, erect or bent at base, usually

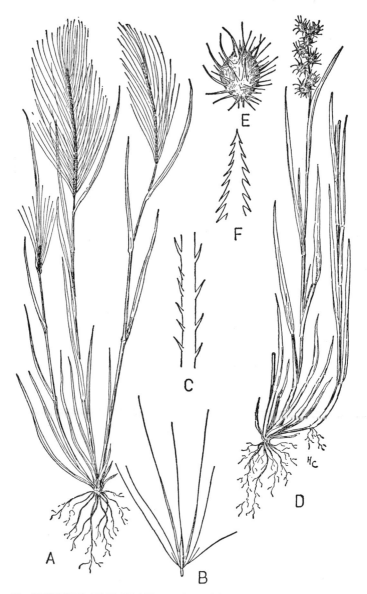

FIG. 11—**SQUIRREL-TAIL-GRASS,** *Hordeum jubatum:* **A,** plant showing general habit (× ⅓); **B,** cluster of three spikelets (× 1); **C,** a section of an awn or "barb" (× 50).
SANDBUR, *Cenchrus longispinus:* **D,** part of a plant showing habit (× ⅓); **E,** a bur-like spikelet (× 3); **F,** enlarged end of a spine from bur (× 50).

tufted. Leaves 3–5 mm. wide, scabrous. Flowers in nodding, jointed spikes, 5–12 cm. long; spikelets in groups of 3 at each joint, in 2 rows on opposite sides of the rachis of the spike; the middle spikelet 1-flowered, perfect, sessile; the 2 lateral spikelets of each group pedicelled, reduced to 1–3 spreading awns; empty glumes of perfect spikelet 3–6 cm. long, awn-like, spreading; flowering glume 6–8 mm. long and with an awn about 3–6 cm. long; all awns slender and barbed. Grain 2–3 mm. long, hairy at the tip, yellow. When the spikes are mature, the joints of the rachis with their 3 spikelets attached fall apart.

Control.—Plow deep and follow with an annual smother crop. Replow and seed to alfalfa if the land is adapted for it or plant a clean cultivated crop for a year before reseeding to grass. If sheep or cattle are allowed to graze on infested meadows early in the spring, the squirrel-tail-grass will be given a severe setback.

When animals feed on this grass, or hay containing it, the awns often work into the tissues of the mouth or into their eyes, causing wounds and ultimately infections of the gums, tongue tips and about the eyes.[8]

43. *Hordeum murinum* L. Wall barley, Barley-grass, Waybent.

Annual; reproducing by seeds. Dry fields, pastures and waste places. Western United States; infrequent near cities along the Atlantic Coast. Introduced from Europe. May–July.

Description.—Stems erect or decumbent at the base, usually in tufts, simple, smooth, 1–3 dm. high. Leaf-blades 4–10 cm. long, somewhat erect and rough above; sheaths loose, shorter than the internodes. Spike 5–10 cm. long, thick; spikelets in groups of 3 at a joint on the spike, all stalked; flowering glume of the fertile middle spikelet pedicelled; both empty glumes of the middle spikelet and the second empty glume of the sterile lateral spikelets ciliate, flattened, bearing awns 2–3 cm. long; lateral spikelets with outer empty glumes neither flattened nor ciliate and flowering glumes scabrous at the apex, long-awned. Grain 5–7 mm. long, hairy at apex, yellow.

Control.—Plow under and follow with a clean cultivated crop. Close grazing early in the season, or cutting to prevent seed formation are effective in pastures.

44. *Hordeum nodosum* L. Wild barley.

Annual or perennial; reproducing by seeds. Dry fields, pastures and

[8] Fleming and Peterson, 1919.

waste places. Widespread in the middle western states and to the Pacific Coast. Native to North America and Eurasia. May–July.

Description.—Similar to 43 but larger, up to 1 m. high. Spike 2–8 cm. long, slender, usually flexuous, at maturity separating into joints with 3 spikelets; flowering glume of middle spikelet sessile; flowering glumes of lateral spikelets sterile, awnless or short awned; empty glumes scabrous, all awn-like, about 2 cm. long or less. Grain 2.5–4 mm. long, narrowly obovate, hairy at apex, yellow.

Control.—The same as for 43.

45. *Hordeum pusillum* Nutt. Little barley.

Annual; reproducing by seeds. Pastures and waste places, especially on saline or alkaline soils. Western North America; also introduced in the southeastern states. Native to western North America. May–July.

Description.—Stems erect or decumbent at base, smooth, 1–4 dm. high. Leaf-blades 2–10 cm. long, erect, scabrous; sheath loose, shorter than the internodes. Spikes 2–8 cm. long, erect; spikelets in groups of 3 at a joint, the lateral sterile; middle spikelet with sessile flowering glume; empty glumes scabrous; first empty glume of lateral spikelets and both empty glumes of central spikelets much broader than the others, ending in slender awns 8–15 mm. long. Grain similar to 44.

Control.—The same as for 43.

46. *Lolium temulentum* L. Darnel, Poison rye-grass, Ivray. Fig. 19, c–d.

Annual; reproducing by seeds. Grain fields and waste places. Quebec and New England, west to Minnesota and Kansas and south to the Gulf; abundant in the Northwest and on the Pacific Coast. Introduced from Europe. July–August.

Description.—Stems erect, simple, often in clumps, smooth, 5–12 dm. high. Leaves with overlapping sheaths, blades smooth below, slightly roughened above. Spikes 2-ranked, 1–2 dm. long; spikelets 4–7-flowered, sessile, attached edgewise in alternate notches of the rachis; first empty glume wanting, the other like a bract, exceeding the lowest flowering glume; flowering glumes rounded, glabrous, bifid at apex, with or without awns. Grain 4–6 mm. long, narrowly obovate, dull, yellow-brown to orange-brown. The grains of darnel sometimes contain a fungus which is thought to be the cause of its poisonous properties.[9]

[9] Pammel, 1911.

Grass Family

Control.—Darnel does not persist in cultivated fields. Include a cultivated crop in a short rotation; avoid several grain crops in succession.

47. **Muhlenbergia frondosa** (Poir.) Fern. (*M. mexicana* Trin.). Mexican drop-seed, Satin-grass, Wood-grass, Knot-root-grass.

Perennial; reproducing by seeds and creeping rootstocks. Edges of fields, grasslands, cultivated ditches, roadsides and waste places; in rich gravelly soils. Common in eastern North America, to Minnesota and Texas. Native. August–October.

Description.—Stems decumbent, 6–9 dm. long, usually branching and rooting at the lower nodes. Rootstocks several or in clusters, very scaly and knotted. Leaves flat, scabrous, more or less ascending. Spikelets 1-flowered, in numerous dense panicles terminating the branches; empty glumes scabrous on the keel, acuminate or bristle-tipped, about as long as the acute or awned narrow, 3-nerved, flowering glume. Grain 1.5 mm. long, fusiform, chestnut-brown, usually surrounded by the flowering glume.

Control.—Plow the sod under and follow with a clean cultivated crop for a year before reseeding. Close grazing early in the season.

48. **Muhlenbergia schreberi** J. F. Gmel. Drop-seed, Nimblewill, Wire-grass.

Perennial; reproducing by seeds and creeping rootstocks. Pastures, meadows, roadsides, old orchards; mostly in rich gravelly soil. Widespread from the Atlantic Coast to Minnesota and Texas. Native. August–October.

Description.—Stems 3–8 dm. long, erect or ascending from a decumbent base, much branched and frequently rooting at the lower nodes; without clusters of scaly rootstocks. Spikelets similar to 47; empty glumes not more than one-fourth as long as the flowering glumes, mostly very minute or the first one lacking. Grain 1–1.4 mm. long, fusiform, chestnut-brown, usually free from the flowering glume.

Control.—The same as for 47.

49. **Panicum capillare** L. Old witch-grass, Tickle-grass, Witcheshair, Tumble weed-grass, Fool-hay. Fig. 12, a–b.

Annual; reproducing by seeds. Cultivated fields, pastures, roadsides, and waste places. Common in the eastern and north central United States; infrequent westward. Native. July–September.

Description.—Stems erect or decumbent and spreading, branching,

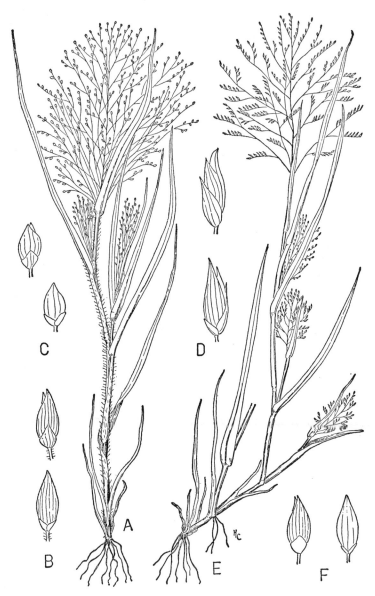

FIG. 12—**OLD WITCH-GRASS,** *Panicum capillare:* **A,** plant showing general habit (× ⅓); **B,** spikelets (× 6).
GATTINGERS WITCH-GRASS, *Panicum gattingeri:* **C,** spikelets (× 6).
SWITCH-GRASS, *Panicum virgatum:* **D,** spikelets (× 6).
SPREADING WITCH-GRASS, *Panicum dichotomiflorum:* **E,** plant (× ⅓); **F,** spikelet (× 6).

2–10 dm. high, hairy, especially at the nodes. Leaf-sheaths densely hairy; blades appressed-pubescent on both surfaces, 1–2.5 dm. long, 5–15 mm. wide. Spikelets 2–4 mm. long, 1-flowered or rarely with an additional staminate flower below the perfect one, in large spreading terminal panicles which frequently break from the plant when mature; empty glumes very unequal, the first acute or acuminate, about one-half as long as the second; flowering glume elliptical, very glossy, with 5 yellow nerves; palet glossy, with 2 nerves. Grain 1–1.2 mm. long, half-ovate, dull, yellow, free within the closed flowering glume and palet.

Control.—Clean cultivation until late in the season to prevent seed formation (6). Scattered plants should be pulled or hoed while small. Mow waste places before seeds are formed (11).

50. *Panicum dichotomiflorum* Michx. Spreading witch-grass, Sprouting crab-grass. Fig. 12, e–f.

Annual; reproducing by seeds, and rooting at the lower nodes of the stem. Low fields, cultivated ground, waste places, and roadsides. Common in the southern states; infrequent northward and westward. Native. July–October.

Description.—Stems glabrous, 3–10 dm. long, mostly spreading or decumbent, branched and rooting at the lower nodes, somewhat compressed and the lower nodes swollen. Leaf-sheaths loose, frequently purplish; blades 1–5 dm. long and 5–20 mm. broad. Spikelets similar to 49, in large spreading panicles; first empty glume not more than one-half as long as the flowering glume; flowering glume elliptical, 2 mm. long, gray-brown, with 5 yellow nerves, inclosing the grain and palet. Grain 1.5 mm. long, half-ovate, dull, yellow.

Control.—The same as for 49.

51. *Panicum gattingeri* Nash. Gattingers witch-grass. Fig. 12, c.

Annual; reproducing by seeds. Cultivated fields, pastures, roadsides, and waste places; often on moist soil. Widespread in the eastern and north central states. Native. August–September.

Description.—Similar to 49 but with a more spreading and branching habit, numerous smaller axillary panicles; spikelets smaller, 1.3–2.2 mm. long, and the first empty glume more obtuse; grain like 49, 1.2–1.3 mm. long.

Control.—The same as for 49.

52. *Panicum virgatum* L. Switch-grass. Fig. 12, d.

Perennial; reproducing by seeds and rootstocks. Low meadows, fields, waste places, and along streams; dry or moist, sandy or gravelly soils. Widespread from Manitoba to Mexico and eastward. Native. August–September.

Description.—Stems erect, tufted, glabrous or glaucous, 1–2 m. high, from creeping perennial rootstocks. Leaves narrow, flat, glabrous. Spikelets 4–5 mm. long, 1-flowered, in large open spreading panicles; the second empty glume and sterile flowering glume spreading, exceeding the glossy ovate-lanceolate flowering glume and grain. Grain about 2 mm. long, ovate, finely uneven, yellow with gray markings.

Control.—Plow the sod under and plant a clean cultivated crop for one or two years. In waste places mow several times, beginning early in the season.

53. *Paspalum ciliatifolium* Michx. Paspalum.

Perennial; reproducing by seeds. Grasslands, fields, dry open woods; mostly on low sandy soils. Common in the south Atlantic and Gulf states; infrequent northward. Native. August–October.

Description.—Stems tufted, erect or reclining, 4–8 dm. long. Leaves glabrous; blades 1–2.5 dm. long, about 1 cm. broad, often ciliate on the margin; sheaths compressed. Spikelets 1-flowered, plano-convex, nearly sessile, about 2 mm. long, glabrous, in pairs, on solitary or rarely clustered, slender, spike-like racemes; racemes 5–10 cm. long, with a narrow rachis with spikelets arranged in 2 rows on one side with the back of the fertile flowering glume toward the rachis, empty glumes about 2 mm. long; flowering glume broadly elliptical, strongly arched, hardened, glossy, straw-colored. Grain 1 mm. long, elliptical to circular, wrinkled, dull, plano-convex, pale yellow to white.

Control.—Hand pulling, hoeing (4, 5). Clean cultivation (6).

54. *Paspalum dilatatum* Poir. Dallis-grass.

Perennial; reproducing by seeds and rhizomes. Meadows, roadsides, ditches, and waste ground. New Jersey to Oregon, south to Florida and California; widely distributed throughout the southern states. Introduced from South America. May–October.

Description.—Stems few to many, tufted, glabrous, stout, erect, 4.5–17 dm. long, from a short rhizome. Leaves elongated, 3–12 mm. wide, up to 3 dm. long; sheaths loose, glabrous, or the lower ones softly pubescent. Spikelets ovoid, acuminate, 2.8–4 mm. long, 2–2.5

mm. broad; racemes 2–10, spreading or loosely ascending, 5–10 cm. long, much surpassing the reduced upper leaf. Glume and sterile lemma silky-villous, longer than the fruit.

Control.—The same as 53.

55. *Paspalum laeve* Michx. (*P. angustifolium* Le Conte).

Reproducing by seeds. Moist sandy fields, pinelands, thickets, and shores. New York to Indiana, south to Florida and the Gulf States. Native. July–October.

Description.—Stems erect or ascending, non-stoloniferous, 0.3–1.3 m. long, slightly tufted, usually several from one base. Leaves elongate, often crowded at base; sheaths keeled, usually loose, villous to glabrous; blades firm, flat, or folded, 5–25 cm. long, 3–10 mm. wide, pubescent to glabrous. Inflorescence terminal, usually exserted, with 2–8 distant ascending or spreading racemes 3–12 cm. long, with sometimes a second inflorescence included in or exserted from a lower sheath; rachis about 1 mm. wide, with a tuft of hairs at base; spikelets usually not paired, obtuse, glabrous, 2–3.2 mm. broad and 2.5–3.2 mm. long.

Control.—The same as 53.

Paspalum distichum L. Wire-grass. A serious weed in the southern states; also in Washington and Utah. It has creeping stems which root at the nodes, and racemes in pairs at the summit of the stem. The control is the same as for 53.

56. *Phalaris canariensis* L. Canary-grass, Bird-seed-grass.

Annual; reproducing by seeds. Gardens, roadsides and waste places. Widespread in eastern North America, but not abundant. Introduced from Europe. July–October.

Description.—Stems erect, tufted, 3–8 dm. high, glabrous or glaucous. Leaves flat. Spikelets 1-flowered, flattened, overlapping, in a crowded spike-like panicle which is from 2–3 cm. long; empty glumes keeled, much longer than the flowering glumes, green with white veins; sterile flowering glume small, like a hairy scale, attached to the fertile flowering glume; fertile flowering glume 4–5 mm. long, brownish-yellow, hard, and glossy in fruit, with the palet covering the grain which is oval and about 4 mm. long. The grains are a common ingredient of "bird seed."

Control.—The same as for 53.

57. *Poa annua* L. Annual spear-grass, Dwarf spear-grass.

Annual; reproducing by seeds. Lawns, cultivated ground and waste places; on moist rich soil. Widespread and common throughout the United States and even beyond. Introduced from Europe. April–October.

Description.—Stems 1–2 dm. high, tufted, flattened, decumbent at the base and frequently rooting at the lower nodes. Leaves very soft; sheaths loose. Spikelets 2–6-flowered, about 4 mm. long, in panicles 3–10 cm. long; empty glumes 1–3-nerved, keeled; flowering glumes awnless, distinctly 5-nerved, the nerves hairy at base; palea with 2 hairy keels; grain 1–1.5 mm. long, oval, dorsally keeled, dull, brown.

Control.—The spear-grass thrives best during cool weather in spring; in warm weather it soon turns brown and makes unsightly spots in lawns. Small patches in lawns should be weeded or hoed out early in the spring; more extensive areas should be mowed close and frequently so as to prevent seed formation. Clean cultivation will destroy this weed.

58. *Setaria glauca* (L.) Beauv. (*S. lutescens* Hub.; *Chaetochloa lutescens* Stuntz). Yellow foxtail, Summer-grass, Golden foxtail, Wild millet. Fig. 13, c–d.

Annual; reproducing by seeds. Cultivated ground, stubble fields, bare places in meadows and pastures, and waste places; on rich soils. Common throughout North America. Introduced from Europe. July–September.

Description.—Stems erect or ascending, 3–12 dm. high, branching and compressed at the base. Leaves flat, linear-lanceolate, glaucous. Spikelets 1-flowered, or rarely with a staminate flower below the terminal perfect one, surrounded by 5 or more tawny upwardly barbed bristles; spikelets 3 mm. long; in narrow spike-like panicles, 2–10 cm. long; first empty glume 3-nerved, the second 5-nerved, about one-half as long as the flowering glume; flowering glume oval, about 3 mm. long, with 3 small teeth at apex, indurated, yellow to brown, roughened by fine granular ridges. Grain about 2 mm. long, broadly ovate, plano-convex, slightly glossy, yellow with small dark markings.

Control.—Continued clean cultivation until late in the season to prevent seed formation. Scattered plants missed by the cultivator should be pulled by hand before seeds are matured. Surface cultivation of stubble fields in autumn does not induce foxtail seeds to

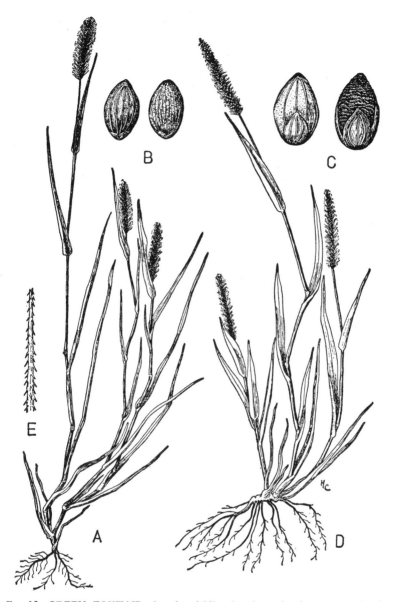

FIG. 13—**GREEN FOXTAIL,** *Setaria viridis:* **A,** plant showing general habit (× ¼); **B,** seed (× 6).
YELLOW FOXTAIL, *Setaria glauca:* **C,** seed (× 6); **D,** plant showing habit (× ¼).
BRISTLY FOXTAIL, *Setaria verticillata:* **E,** section of an awn showing recurved barbs (× 30).

germinate. Close grazing of the foxtail on grain stubble by sheep will prevent a new crop of seeds from infesting the field (10).

59. **Setaria verticillata** (L.) Beauv. (*Chaetochloa verticillata* Scribn.). Bristly foxtail, Pigeon-grass. Fig. 13, e.

Annual; reproducing by seeds. Waste places, stubble fields and cultivated ground. Infrequent in the eastern states and becoming more prevalent in certain sections of the Middle West.[10] Introduced from Europe. July–September.

Description.—Similar to 60 but with more spreading and branching habit, rougher stems and leaves; bristles downwardly barbed; spikelet 2–2.5 mm. long.

Control.—The same as for 58.

60. *Setaria viridis* (L.) Beauv. (*Chaetochloa viridis* Scribn.). Green foxtail, Bottle-grass, Pigeon-grass, Wild millet. Fig. 13, a–b.

Annual; reproducing by seeds. Cultivated ground, grain stubble, waste places; on rich soils. Common throughout North America. Introduced from Europe. July–September.

Description.—Similar to 58 but spikelets 2 mm. long, surrounded by 1–3 green or purple upwardly barbed bristles; panicle broader; flowering glume similar but smaller, light to dark brown, mottled, granular, striate. Grain 1.5 mm. long, oval, slightly glossy, yellow with dark markings.

Control.—The same as for 58.

Setaria faberi Herrm., Nodding foxtail, larger than *S. viridis*, with spikes nodding and leaves finely hairy on the upper surface, is an annual weed locally very abundant in the midwestern states.

61. *Sorgum halepense* (L.) Pers. (*Holcus halepensis* L.) Johnson-grass, Means-grass, Egyptian-grass, False guinea-grass, Millet-grass, Morocco millet. Fig. 14.

Perennial, reproducing by seeds and extensive rootstocks. Cultivated fields, meadows and waste places; especially on rich river bottom soils, also on irrigated lands in the Southwest. Widespread throughout the southern states; local extending northward. Introduced from the Mediterranean region and grown extensively as a forage and hay crop in the South.[11] June–July.

Description.—Stems from rootstocks, stout, erect, 1–2 m. high, 1–2 cm. in diameter at base, pith with a sugary juice. Leaves with glabrous sheaths and blades 3–5 dm. long, broad, flat, smooth. Panicles

[10] Keim and Frolik, 1934.   [11] Vinall, 1926.

Fig. 14—**JOHNSON-GRASS**, *Sorgum halepense:* **A**, basal part of plant ($\times \frac{1}{3}$); **B**, flower panicle ($\times \frac{1}{3}$); **C**, portion of fruiting branch ($\times 2$); **D**, grains ($\times 3$).

large, loose, with whorls of spreading branches. Spikelets in groups of 3, the central one sessile and fertile, sometimes awned, with a purplish flowering glume with fine appressed hairs; the 2 lateral spikelets pedicelled, staminate or empty; flowering glume about 5 mm. long, hardened, hairy or barbed near apex. Grain about 2.5 mm. long, oval, finely striate, reddish-brown.

Control.—The methods of controlling Johnson-grass are largely based on the nature and habits of its extensive system of rootstocks. The studies of Cates and Spillman [12] revealed that Johnson-grass produces three different kinds of rootstocks, primary, secondary and tertiary. The primary rootstocks include all those which are alive in the ground at the beginning of the growing season in the spring; they die at the close of the growing season. The secondary rootstocks are produced by the primary rootstocks, grow to the surface and form crowns or new plants; their length is determined by the depth of the primary rootstocks from which they start and usually they do not exceed the latter in diameter. The tertiary rootstocks are those that are sent out from the base of the crown of the Johnson-grass about the time when it blossoms. The size and depth of the tertiary rootstocks increase with the size of the tops and the length of time that they are allowed to grow. If Johnson-grass is left undisturbed in loose rich soil, it may produce thick tertiary rootstocks which penetrate the soil for a meter. On the other hand, if the tops are mowed or grazed constantly, the tertiary rootstocks are much smaller and remain very near to the surface of the soil. The control of Johnson-grass should be begun when the rootstocks are in this condition. If the rootstocks are not near the soil surface, turn the infested field into a meadow or pasture and cut it for hay before blossoming every time a new growth is produced, or pasture closely for two years. This will bring the rootstocks into the upper layers of soil where they can be destroyed by shallow plowing followed by frequent harrowing to prepare the ground for the next crop. Plant an intertilled crop and cultivate frequently. Frequent mowing before blossoming not only prevents the spread of Johnson-grass by seeds but it also weakens the rootstocks. Summer fallowing has been effective in the more arid regions.

These general control methods are based on the work of Cates and Spillman [13] and results reported more recently from several states

[12] Cates and Spillman, 1907.   [13] Cates and Spillman, 1907.

in the South and Southwest that have been summarized by Talbot.[14]

Sorgum vulgare Pers. var. Wild corn, Chicken corn. Reported as a serious weed in corn fields in the overflow lands of the Ohio River in Indiana and Kentucky [15] and also in the lower Mississippi Valley. Delaying the date of planting of the corn, followed by clean cultivation and hand weeding within the rows is effective in controlling the wild corn.

62. *Sporobolus neglectus* Nash. Drop-seed, Rush-grass.

Annual; reproducing by seeds. Dry sandy pastures, fields and waste places. From eastern Canada to North Dakota, southward to Virginia and Texas. Native. August–September.

Description.—Stems erect or the base decumbent, simple or branched, usually in tufts, very slender, 1–3 dm. high, smooth. Leaves very slender, long tapering, smooth below and rough at base above; sheaths much inflated, more than half the length of the internode. Spikelets in slender panicles which are at first more or less hidden in the sheaths; spikelet 1-flowered, 2.5–3 mm. long, awnless; empty glumes, flowering glume and palet all nearly equal, acute, glabrous, glossy, nearly white. Mature grain separating from the spikelet, its pericarp thin, free from the seed, about 1 mm. long, oval, orange-red.

Control.—Plow, fertilize and plant a cultivated crop; or plow under a green-manure crop before reseeding.

63. *Sporobolus poiretii* (R. and S.) Hitchc. Smut-grass, Black-seed grass. (The common name is due to a black fungus which frequently infects it.)

Perennial; reproducing mainly by seeds. Sandy soil, roadsides, and ditch banks. Virginia to Kentucky and Arkansas, south to the Gulf; adventive in New Jersey. Naturalized from tropical America. May–October.

Description.—Stems 3–10 dm. long, erect, wiry, tufted, solitary or a few together. Leaves 10–30 cm. long, 2–5 mm. wide, narrowing to a fine point. Panicle 1–3 dm. long, slender, stiff; spikelets 1.5–2 mm. long, shining, crowded on the slender, erect glumes; glumes obtuse, first glume 0.5–0.75 mm. long, second glume 0.7–1.1 mm. long. Lemma 1.6–2 mm. long, about equal to the palea.

Control.—The same as for 62.

Sporobolus vaginiflorus Wood. Drop-seed. Very similar to *S.*

[14] Talbot, 1928.   [15] Hansen, 1923.

neglectus and sometimes confused with it, occurs in similar situations. Its spikelets are 4 mm. long and have pubescent flowering glumes.

64. *Stipa comata* Trin. and Rupr. Needle-and-thread-grass, Western stipa.

Perennial; reproducing by seeds. Dry native grasslands, ranges, and pastures. From the Great Plains westward to the Pacific Coast. Native. June–July.

Description.—Stems in tufts, erect, smooth, glabrous, simple, 3–12 dm. high. Leaf-blades often rolled lengthwise and with thread-like tips; sheaths overlapping, mostly crowded at the base, the upper long, loose and often covering the base of panicle. Spikelets 1-flowered, in a loose, erect, or spreading panicle; empty glumes 2–2.8 cm. long, narrow, acute, tapering into a slender awn; flowering glume 9–11 mm. long, hairy, hardened and covering the grain, ending in a 1–2.5 dm. long, jointed awn which is hairy below and scabrous and twisted above the joint. Grain about 5 mm. long, cylindrical, dull, yellow-brown.

Control.—Close mowing or close grazing to prevent seed formation. Plow, and sow a green-manure crop and plow it under before reseeding.

65. *Stipa spartea* Trin. Porcupine-grass, Needle-grass, Weathergrass, Auger-seed. Fig. 15, a–c.

Perennial; reproducing by seeds. Dry native grasslands, meadows and pastures. North central states to the Rocky Mountains. Native. June–July.

Description.—Stems tufted, erect, stout, simple, smooth, 6–12 dm. high. Leaves mostly with flat blades; sheaths mostly overlapping, somewhat rough. Spikelets 1-flowered, in long slender panicles with erect branches; empty glumes 2.8–3.5 cm. long, attenuate, awns 11–20 cm. long; flowering glumes as in 64 but with awns 11–20 cm. long. Grain 11–14 mm. long, cylindrical, yellow-brown.

Control.—The same as for 64.

Aegilops cylindrica Host. Goat-grass. A winter annual grass introduced from Russia, with the general growth habits of wheat; local about New York, Indiana, Missouri, and Oklahoma, Colorado, and New Mexico. It was first reported as troublesome in wheat fields of south central Kansas and adjacent Oklahoma.[16] Control of goat-

[16] Johnston, 1931.

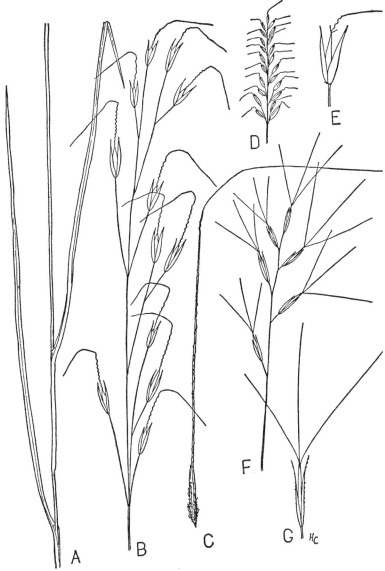

FIG. 15—**PORCUPINE-GRASS,** *Stipa spartea:* **A,** part of stem (× ⅓); **B,** panicle (× ⅓); **C,** grain, with twisted awn (× ⅔).
POVERTY-GRASS, *Aristida dichotoma:* **D,** part of panicle (× ⅓); **E,** spikelet (× 1½).
FEW-FLOWERED ARISTIDA, *Aristida oligantha:* **F,** part of panicle (× ⅓); **G,** spikelet (× ⅔).

grass is not difficult when rotation with cultivated crops is practicable.
Arrhenatherum elatius (L.) Mert. and Koch var. *bulbosum* (Willd.) Spenner. Tuber oat-grass, is listed (1954) as a weed in the northeastern states.

Oryza sativa L. Red rice is a noxious plant (1954) in several southern states.

CYPERACEAE (Sedge Family)

66. *Cyperus diandrus* Torr. Sedge, Galingale.

Annual; reproducing by seeds. Locally common in meadows and pastures; on sandy or mucky soil. Widespread in eastern North America, west to Minnesota and North Dakota. Native. August–October.

Description.—Stems erect, simple, solid, triangular, 1–4 dm. high. Neither tubers nor corms formed; roots fibrous. Leaves mostly basal, 3-ranked. Spikelets of several to many flowers, flattened, 2-ranked, in terminal clusters on 2–5 very unequal branches, subtended by an involucre of 3 unequal leaf-like bracts; spikelets lance-oblong, 5–10 mm. long; their scales marked with dark brown or purple; flowers with 2 stamens; styles 2-cleft. Achenes oblong-obovate, about 1.2 mm. long, base truncate, apex with a small projection to which the style remnant often persists, dull, olive-gray to brown, with a network of light gray lines.

Control.—Improve the drainage (21). Plow and follow by clean cultivation (18).

67. *Cyperus esculentus* L. Yellow nut-grass, Chufa, Northern nut-grass, Coco, Coco sedge, Rush nut, Nut sedge, Edible galingale, Earth almond. Fig. 16, a–b.

Perennial; reproducing by seeds and tubers. Cultivated fields and gardens, and sometimes in grain fields; on rich or sandy soil; often limited to low poorly drained areas of fields. Widespread in eastern North America; also along the Pacific Coast; occasional elsewhere. Native to North America and also to Eurasia. July–September.

Description.—Stems erect, simple, triangular, yellow-green, 3–9 dm. high, from perennial tuber-bearing rootstocks; tubers about 1–2 cm. long. Roots fibrous. Leaves in 3 ranks, narrow, grass-like, about as long as the stem, with closed sheaths, mostly basal. Spikelets straw-color or chestnut-brown, acute, 1–3 cm. long, of several flowers, flattened, 2-ranked, in more or less compound terminal umbels;

FIG. 16—**YELLOW NUT-GRASS**, *Cyperus esculentus:* **A,** plant showing habit (× ⅓); **B,** spikelet (× 3).
SEDGE, *Cyperus strigosus:* **C,** plant showing habit (× ⅓); **D,** spikelet (× 3).

umbel subtended by long bracts; flowers with 3 stamens; style 3-cleft. Achenes 3-angled, linear to oblong-cylindric, about 1.5 mm. long, base and apex blunt, yellowish-brown, granular-striate.

Control.—Improve the drainage if the soil is too wet (21). Clean cultivation followed by hand weeding in the hills (6 and 5). The most practicable method of cleaning large areas badly infested by nut-grass is to fallow them for an entire season (19).

68. *Cyperus iria* L.

Annual; reproducing by seeds. Swamps, clearings, ditches, and roadsides, especially about rice fields. Virginia south to Florida and Texas, in the Gulf States and adventive in New York. Introduced from Eurasia. July–October.

Description.—Stems one to several, erect, up to 6 dm. long, without stolons or hardened rhizomes or tubers; roots tufted, fibrous. Umbel simple or with several ascending rays; rays up to 1 dm. long. Involucral leaves 3–5, some of them longer than the rays. Spikes loose and open, 1–2 cm. long; spikelets loosely fasciculate and ascending, yellowish or brownish, in elongated clusters with short tips. Scales thin, broadly obovate, retuse, 1–1.7 mm. long, sides golden brown, midvein green. Rachilla slender and flexuous, wingless. Achene ellipsoid, 3-angled, 1–1.5 mm. long, brown to blackish, densely puncticulate, stipitate.

Control.—The same as for 66.

69. *Cyperus rotundus* L. Nut-grass, Nut sedge, Coco sedge, Cocograss.

Perennial; reproducing by seeds and small tubers. Cultivated fields and gardens, especially in the cotton belt. Southeastern United States from New Jersey southward and westward to Texas; occasional northward. Introduced from Asia. July–September.

Description.—Stems erect, simple, triangular, 1–6 dm. high, from perennial tuber-bearing rootstocks. Leaves similar to 67, but shorter than the stem. Spikelets chestnut-brown, similar to 67. Seeds linear-oblong, 3-angled, 1.5 mm. long, base and apex obtuse, granular, dull, olive-gray to brown, covered with a network of gray lines.

Control.—Frequent stirring of the soil early in the season to induce seeds and tubers to germinate; followed by continued late cultivation to prevent the formation of new seeds or tubers.

70. *Cyperus strigosus* L. Sedge. Fig. 16, c–d.

Perennial; reproducing by seeds and corm-like tubers. Damp

Sedge Family

meadows, low fields and swamps. Widespread, especially in the eastern and central United States; local on the Pacific Coast. Native. July–September.

Description.—Stems erect, simple, solid, triangular, 2–10 dm. high, from a hard corm-like base. Roots fibrous. Leaves mostly basal, 3-ranked, flat, soft. Spikelets flattened, several-flowered, in terminal, simple or compound umbels. Achenes similar to 67 but larger, 1.7 mm. long.

Control.—Improve the drainage (21). Plow in autumn and the next spring follow by clean cultivation (18).

71. *Scirpus atrovirens* Willd. Meadow rush, Dark green bulrush, Club rush.

Perennial; reproducing by seeds and creeping rootstocks. Low and wet places in meadows and pastures, most troublesome on bottom lands subject to overflow. Common; Maine to Georgia, westward to Saskatchewan and Missouri. Native. June–July.

Description.—Stems simple, stout, erect, solid, 1–1.5 m. high. Leaves in 3 ranks on the stem, pale green, with scabrous margins. Spikelets several-flowered, dull greenish-brown, several-ranked, ovoid or cylindric, 4–10 mm. long, in dense clusters of 10–30; umbels of spikelets subtended by 2 or more leaf-like involucral bracts; flowers perfect, subtended by barbed bristles about as long as the achene. Achene obovoid-oblong, about 1 mm. long, irregularly 3-angled, finely granular, dull, yellow.

Control.—Improve the drainage (21); dig out scattered clumps in pastures (17); in extensive areas mow early to prevent seed production (11); plow in autumn and follow by a clean cultivated crop (18).

72. *Scirpus cyperinus* (L.) Kunth Wool-grass.

Perennial; reproducing by seeds and rootstocks. Wet meadows, bogs, and swamps. Newfoundland to Saskatchewan, south to Florida and Louisiana. Native. August–October.

Description.—Stems smooth, obtusely 3-angled or nearly terete, 1–1.8 m. high, from dense tussocks. Leaves linear, long and rigid, basal leaves many and curving, principle blades 3–10 mm. wide, bright green. Bracts 3–6, leaf-like, unequal, spreading, usually drooping at the tip, pigmented at base. Rays of the inflorescence several; ultimate branchlets bearing 1–several sessile spikelets and often 1 or 2 spikelets with pedicels; spikelets numerous, ovoid to cylindric, woolly at maturity, 3–10 mm. long, in dense heads or clusters. Scales elliptic to oval, 1.5–2 mm. long, obtuse to acute. Achenes pale or nearly white, obovate, sharply beaked, 0.7–1 mm. long; bristles smooth, long, flexuous, much longer than the achene.

Control.—The same as for 71.

Carex lupulina Muhl., Hop sedge, is listed (1954) as a weed in the north central states.

The Sedge family contains a number of other native plants, especially in the genera Carex, Cyperus and Scirpus, which in certain regions persist as weeds in grasslands on recently drained areas. These forms do not persist under cultivation.

COMMELINACEAE (Spiderwort Family)

73. *Commelina communis* L. Dayflower.

Annual; reproducing by seeds and creeping stems. Gardens, dooryards, neglected fields, and waste places; on rich soils. Massachusetts to Florida, westward to Kansas and Texas. Introduced from Asia. July–September.

Description.—Stems creeping, slender, rooting at the swollen nodes. Leaves alternate, simple, entire, parallel-veined, lanceolate, 2–5 cm. long, with sheathing petioles. Cymes inclosed by a cordate clasping bract. Flowers perfect, irregular; sepals 3, unequal, the 2 lateral partly fused; petals 3, the 2 lateral enlarged and rounded, blue, lasting but a day; stamens 6, unequal, only 3 fertile, the others sterile; pistil 3-celled, the dorsal with 1 ovule and the others each with 2 ovules usually forming 2 seeds. Seeds 3–4 mm. long, flat on one side and very convex on the other, hilum circular, pitted, granular-netted, dull grayish-brown.

Control.—Clean cultivation (6); hand hoeing while the plants are small (5).

JUNCACEAE (Rush Family)

74. *Juncus bufonius* L. Toad rush.

Annual; reproducing by seeds. Low pastures, wet places in fields and meadows and along ditches. Widespread throughout North America. Native to North America and also to Europe. June–July.

Description.—Stems slender, low, 1–3 dm. high, leafy, often branched and usually clustered. Roots fibrous. Leaves alternate, flat or terete, wiry. Flowers in terminal cymes, small, greenish, perfect; sepals and petals linear-lanceolate, awl-pointed; stamens 6; carpels 3, united into a compound pistil; ovary superior. Fruit a capsule with many seeds; seeds 0.5 mm. long, narrowly ovoid or ellipsoidal, marked with irregular brown lines, yellowish-brown.

Control.—Improved drainage (21); clean cultivation (18).

75. *Juncus effusus* L. Bog rush, Rush, Soft rush. Fig. 19, a–b.

Perennial; reproducing by seeds and rootstocks. Low pastures, meadows, borders of marshes and along ditches; especially on mucklands and alluvial soils subject to overflow. Represented by several varieties throughout North America. Troublesome in lowland pastures in the Puget Sound region. Native to North America; also to Europe. June–August.

Description.—Stems leafless, 3–12 dm. high, from creeping rootstocks with fibrous roots, often forming dense hummocks or tussocks. Leaves basal, mostly reduced to sheaths. Flowers appearing in lateral cymes, perfect; sepals 3; petals 3, very acute, greenish; stamens 3; carpels 3, united into a compound pistil; ovary superior. Fruit a greenish-brown capsule with many seeds; seeds 0.4 mm. long, narrowly ovoid or half-oval, base oblique, finely ribbed and cross-lined.

Control.—The appearance of the bog rush is an indication of the need of improved drainage. The tussocks should be grubbed out, piled in holes where they can be buried, or dried and burned. Land cleaned of bog rush should be cultivated for one or two years before seeding to grass.

76. *Juncus gerardii* Loisel. Black-grass, Hog-rush.

Perennial; reproducing by seeds and rhizomes. Salt marshes and saline places; occasional inland as a railroad weed. Quebec and Newfoundland to New York, Virginia and Minnesota, occasionally south to Florida and local inland. Native. June–September.

Description.—Stems rigid, 1.5–8 dm. high, in small tufts, scarcely flattened, from slender, dark, horizontally spreading rhizomes. Leaf sheath extending about half way up the culm; blades soft, slender, tapering, flat, involute, up to 2 dm. long, green; cauline leaves 1 or 2, the upper divergent at or near the middle of the stem; sheaths entire at the summit. Inflorescence 2–8 cm. long, narrow, many-flowered, with ascending or erect branches, usually longer than the involucral leaf. Flowers brown and green, 2.3–5 mm. long, mostly sessile or with short pedicals. Capsule 2.4–3.3 mm. long, ovoid, obtuse, mucronate, dark brown, shining, 3-celled. Seeds obovoid, 0.4–0.5 mm. long, lightly ribbed, and cross-lined, brown.

Control.—The same as for 75.

77. *Juncus tenuis* Willd. Path rush, Field rush, Slender yard rush, Wire-grass, Slender rush, Poverty rush.

Perennial; reproducing by seeds. Bare places and along paths in pastures, meadows and in waste places; both on dry and wet soil. Widespread in the United States and in Canada. Native. June–August.

Description.—Stems wiry, tufted, 1–6 dm. high. Roots fibrous. Leaves linear, flat, with sheath covering about one-half of stem. Flowers clustered in cymes near the tips of the branches, perfect; sepals 3; petals 3, green, lanceolate, very acute; ovary similar to 75. Fruit a falsely 1-celled capsule with numerous seeds; seeds similar to 75, but somewhat smaller, orange-brown, translucent, sometimes winged along one side.

Control.—Plow and follow by a cultivated crop. Grasslands that cannot be plowed should be fertilized, harrowed and seeded with a mixture containing white clover.

LILIACEAE (Lily Family)

78. *Allium canadense* L. Wild onion, Meadow garlic. Fig. 17, e–f.

Perennial; reproducing by bulbs, bulblets, and seeds. Low meadows and pastures; on rich alluvial soils, also in open woods. From the Atlantic Coast westward to Ontario, Minnesota, and Texas. Native. May–June.

Description.—Stems simple, leafless, about 3 dm. high, glabrous, bulbous at base. Roots fibrous. Leaves basal, linear, from a fibrous-coated bulb, about 1–2 cm. in diameter. Flowers in umbels which also contain many small bulblets; perianth of 6 distinct pinkish segments; stamens 6, distinct; carpels 3, fused into a compound pistil; ovary superior. Fruits a 3-valved capsule with 3–6 seeds; seeds ovoid, glossy, black, about 3 mm. in diameter; a small wing-like ridge extends from the hilum around the seed.

Control.—Plow in autumn and follow with a clean cultivated crop for one or two years before reseeding. In small areas dig or hoe out the bulbs before new bulblets are formed (5).

The entire plant has a strong scent of onion or garlic. In the early

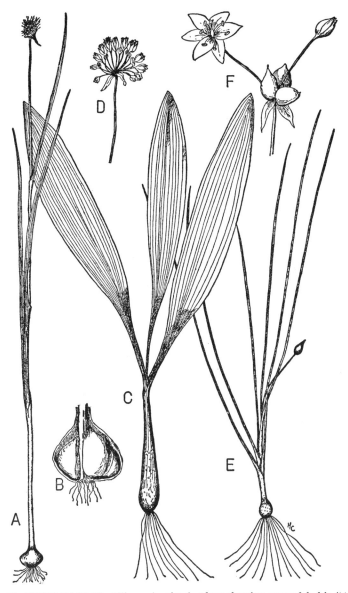

Fig. 17—**WILD GARLIC,** *Allium vineale:* **A,** plant showing general habit (× ⅕); **B,** sectioned bulb showing hard bulbils (outside) and soft bulbils (middle).
WILD LEEK, *Allium tricoccum:* **C,** plant showing bulb and leaves (× ⅓); **D,** flower-cluster (× ⅓).
WILD ONION, *Allium canadense:* **E,** plant showing bulb, leaves and unopened bud (× ⅓); **F,** opened bud bearing bulblets and flowers (× 2).

spring, cows feeding on this plant, or on 79, produce milk with a strong "garlic taint."

79. *Allium tricoccum* Ait. Wild leek. Fig. 17, c–d.

Perennial; reproducing by seeds and bulbs. In rich humus in woods or woodland pastures; on gravelly or sandy soil. Locally common from eastern Canada to Minnesota, southward to Georgia, Tennessee, and Iowa. Native. July.

Description.—Stems simple, leafless, 2–4 dm. high, from a pointed bulb. Roots fibrous. Leaves 2 or 3 basal, 3–6 cm. wide, appearing in early spring and disappearing before the flowers develop. Flowers in many-flowered umbels without bulblets; perianth of 6 greenish-white segments; stamens 6, distinct; carpels 3, fused into a compound pistil; ovary superior. Fruit a capsule, 3-valved and strongly 3-lobed with 3 glossy black seeds similar to 78.

Control.—Dig or hoe out the bulbs (5) as soon as the leaves appear.

80. *Allium vineale* L. Wild garlic, Wild onion, Crow garlic, Field garlic. Fig. 17, a–b.

Perennial; reproducing by bulbs and bulblets; rarely by seed in the northern states. Grain fields, meadows, pastures, lawns and waste places; on sandy or gravelly soils. Common along the Atlantic Coast from Massachusetts to Georgia, westward to Michigan and Arkansas. Introduced from Europe. May–June.

Description.—Stems slender, the lower half covered with sheathing leaf-bases. Bulbs of two kinds, soft bulbs which germinate during the first autumn and hard bulbs which remain dormant over winter and germinate the next year, or even later. Leaves slender, hollow, grooved above, glabrous, sheathing at base. Umbels with densely crowded bulblets. Flowers similar to 79, but seldom producing seed in the northern part of its range. Seeds about 2.8 mm. long, flattened, convex on one side, dull, black, wrinkled, and covered with minute tubercles.

Control.—On large areas disk to destroy the tops and plow in late autumn, follow by early spring plowing and plant a clean cultivated crop (6); repeat for two or three years.[1, 2] Dig scattered clumps or plants by hand (17).

Herbage with strong scent of garlic. The bulblets are about the size of wheat grains, and wheat grown in regions where wild garlic

[1] Talbot, 1929.   [2] Runnels and Schaffner, 1931.

is common is frequently contaminated with them. Flour made from wheat containing garlic bulblets becomes tainted with garlic flavor. Cows feeding on this plant produce garlic-tainted milk.

A number of other species of Allium, native to the western United States, are sometimes common on ranges, in meadows, pastures, and waste places, especially on the Pacific Coast. Most of these reproduce freely by seeds and bulbs, but do not produce aerial bulblets.

81. *Convallaria majalis* L. Lily-of-the-valley. Fig. 18, e–f.

Perennial; reproducing by rootstocks; seldom by seed when introduced. Roadsides and abandoned yards and adjacent meadows. Widely planted in yards; occasionally persisting and spreading, in the northeastern United States and on the Pacific Coast. Introduced in the northern United States; native in Europe and in the southern Appalachian Mountains. May–June.

Description.—Stems leafless, from slender running rootstocks with fibrous roots. Leaves 2 or 3, basal, oblong, glabrous. Flowers nodding, in one-sided racemes; perianth bell-shaped with 6 fused segments; stamens 6, included, inserted on the base of perianth; carpels 3, fused into a 3-celled pistil with a stout style. Fruit a few-seeded berry, seldom produced when introduced; seeds about 4 mm. in diameter, wrinkled, somewhat glossy, black.

Control.—Dig the rootstocks by hand and follow by clean cultivation.

All parts of the plant are poisonous when eaten.

82. *Hemerocallis fulva* L. Day-lily, Tawny orange lily.

Perennial; reproducing by tuberous roots. Roadsides, neglected meadows, banks, and waste places; on rich, damp, gravelly soils. Locally common as an escape where it was formerly planted extensively as an ornamental; most troublesome in the northeastern states. Introduced from Eurasia. June–July.

Description.—Stems simple, leafless, from fleshy tuber-like roots. Leaves long, linear, keeled, 2-ranked, basal, glabrous. Flowers several, ephemeral, with a funnelform lily-like perianth of 6 fused orange segments with wavy margins; stamens 6, inserted on the throat of the perianth; carpels 3, fused into a 3-celled pistil with a long style. Fruit a capsule, 3-valved; does not produce seeds.

Control.—Plow and rake out the roots in autumn; follow by clean cultivation (6). In small areas where plowing is not practicable dig the roots out by hand (17).

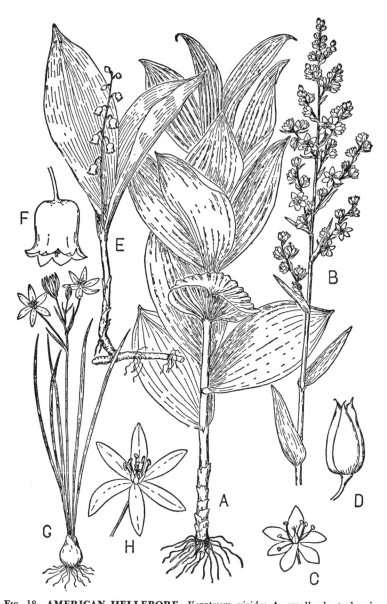

Fig. 18—**AMERICAN HELLEBORE,** *Veratrum viride:* **A,** small plant showing habit (× ⅓); **B,** flowering shoot (× ⅓); **C,** flower (× 1); **D,** capsule (× 1).
LILY-OF-THE-VALLEY, *Convallaria majalis:* **E,** plant showing habit (× ⅓); **F,** flower (× 2).
STAR-OF-BETHLEHEM, *Ornithogalum umbellatum:* **G,** plant showing habit (× ⅓); **H,** flower (× 1).

83. *Ornithogalum umbellatum* L. Star-of-Bethlehem. Fig. 18, g–h.

Perennial; reproducing mostly by bulbs; rarely by seeds in America. Escaped in lawns, meadows, roadsides, and waste places. Locally troublesome from New York and North Carolina to Ohio, Tennessee, and Kansas, infrequent elsewhere. Introduced from Europe. May–June.

Description.—Stems from ovate bulbs, in clumps, erect, 1–3 dm. high, slender, scapose, terminated by a scarious-bracted corymb of white star-shaped flowers. Leaves basal, linear, somewhat fleshy, with a light green midrib, about as long as the stem. Flowers perfect; perianth-segments 6, white, with a green vein on the back; stamens 6; pistil 3-celled, stigma and style 3-angled. Capsule 3-lobed, several-seeded; seeds about 1.5 mm. long, ovoid to irregular, coarsely granular, black.

Control.—Scattered plants should be dug out by hand, at flowering time when they can be easily seen, and the bulbs should be destroyed by drying or burning. The bulbs are destroyed by plowing the sod and following with a clean cultivated crop (18).

Formerly planted more extensively as an ornamental. The flowers open only in bright sunshine. All parts of the plant are poisonous when eaten.

84. *Veratrum viride* Ait. American hellebore, Indian poke, White hellebore, False hellebore, Green hellebore, Bear corn. Fig. 18, a–d.

Perennial; reproducing by seeds and short rootstocks. Low woods, pastures and meadows; especially on rich alluvial soils, in the northeastern states. Widespread throughout southern Canada and the northern United States; southward in the mountains to Georgia and Oregon. Native. June–July.

Description.—Stems stout, simple, 6–20 dm. high, very leafy, from a short rootstock with many coarse fibrous roots. Leaves broadly oval, plaited, pointed, clasping, and somewhat pubescent. Flowers polygamo-monoecious, in a dense panicle of racemes; perianth of 6 yellowish-green more or less persistent segments; carpels 3, united into a 3-celled pistil with 3 styles. Capsule many-seeded; seeds about 9 mm. long, flattened or half obovate, seed-coat with a broad wing around the seed, surface finely granular, light buff.

Control.—Improved drainage (21), followed by cultivation (18). Grub out scattered plants or patches in pastures (17).

The entire plant is poisonous. Because of its very acrid taste it is seldom eaten by animals.

85. *Zigadenus venenosus* S. Wats. Meadow death camas, Soap plant, Alkali-grass, "Lobelia." Fig. 19, e–g.

Perennial; reproducing by seeds and bulbs. Wet meadows and grasslands, also on stony slopes. Pacific Coast to Idaho, Nevada and Utah. Native. May–June.

Description.—Stem from a bulb, erect, unbranched, glabrous, 2–6 dm. high. Leaves alternate, mostly basal, grass-like and keeled, often folded. Flowers perfect, regular, usually in simple racemes; perianth-segments 6, white or greenish-yellow, all clawed; stamens 6; pistil of 3 carpels fused, maturing into a capsule, 3-lobed to cylindric, 3-celled, several-seeded; seeds about 5 mm. long, oval to obovate, flattened, wrinkled, light brown.

Control.—Dig the bulbs when the soil is wet, and destroy them.

Zigadenus venenosus, as well as several other species in this genus, is poisonous when eaten by animals. Most losses occur among sheep, although cattle, horses and human beings may also be affected. The more common poisonous species are the following.[3]

Zigadenus gramineus Rydb. Grassy death camas. Common on hillsides, especially in the northern Rocky Mountains, and less so on the Great Plains; leaves very narrow.

Zigadenus paniculatus S. Wats. Foothill death camas. Common in grassy places on gravelly or stony areas in the Great Basin region, especially in Utah and Nevada; leaves broader, flowers in panicles.

Zigadenus elegans Pursh. Mountain death camas. Common in moist meadows and prairies, Iowa to Missouri, west to Alaska and Arizona; ovary inferior.

Zigadenus nuttallii Gray. Nuttalls death camas.[4] From the upland prairies of Arkansas, Kansas, Oklahoma and eastern Texas; stamens exceeding the sepals and petals.

Smilax glauca Walt. Saw brier, and several other species of *Smilax*, Green brier, common woody vines in woods and thickets along the Atlantic Coast and Gulf Coast, sometimes appear in meadows and fields. They are usually prickly and have tendrils. To eradicate these briers, the tops should be mowed and burned and the roots grubbed out or plowed when possible.

[3] Marsh and Clawson, 1922.
[4] Marsh, Clawson and Roe, 1926.

FIG. 19—**BOG RUSH,** *Juncus effusus:* **A,** basal and upper part of plant (× ⅓); **B,** flower (× 3).
DARNEL, *Lolium temulentum:* **C,** parts of plant (× ⅓); **D,** spikelet (× 2).
MEADOW DEATH CAMAS, *Zigadenus venenosus:* **E,** plant showing habit (× ¼); **F,** capsule (× 1); **G,** flower (× 1).

MYRICACEAE (Sweet Gale Family)

86. *Comptonia peregrina* (L.) Coult. (*Myrica asplenifolia* L.) Sweet-fern, Fern gale, Meadow fern, Shrubby fern, Sweet bush, Fernwort, Spleenwort bush. Fig. 59, a–c.

A small shrub; reproducing by seeds and creeping rootstocks. Hillside pastures, old fields and open woods; on sterile, sandy, gravelly or stony acid soils. Eastern Canada to North Carolina, rare westward to Minnesota and Illinois and south to Georgia. Native. April–May.

Description.—A low much branched shrub, 3–8 dm. high. Leaves alternate, lanceolate, pinnately lobed with rounded lobes, sweet-scented. Flowers mostly monoecious, in catkins; staminate catkins short-cylindrical, scaly; pistillate catkins globular; flowers much reduced. Fruit a small nut surrounded by 8 awl-shaped persistent scales; nut about 5 mm. long, ovoid or nearly cylindric, olive-brown, glossy, somewhat ridged near the base.

Control.—Mow close to the ground for two or three seasons (11); grub out with a mattock and follow with an application of fertilizer and lime and reseed.

URTICACEAE (Nettle Family)

87. *Cannabis sativa* L. Hemp, Red-root, Gallow-grass, Neckweed, Marijuana.

Annual; reproducing by seeds. Escaped on rich soils, neglected fields, farmyards, waste places and roadsides. Locally established from Quebec to British Columbia and southward. Introduced from Asia. July–September.

Description.—Stems 1–2 m. high, stout, simple or sparingly branched, rough. Leaves opposite, alternate above, compound with 5–7 lanceolate coarsely toothed leaflets. Flowers dioecious, small and green; the pistillate flowers in spike-like clusters, the staminate with 5 perianth-segments and 5 stamens, in axillary compound racemes or panicles. Fruit an achene about 4 mm. long, ovoid to nearly round, with obtuse edges, yellow to olive-brown.

Control.—Mow plants growing in waste places before seeds are formed (11). Hemp does not persist under clean cultivation.

88. *Urtica dioica* L. Stinging nettle, Tall nettle, Slender nettle.

Perennial; reproducing by seeds and creeping rootstocks. Neglected yards, waste places and roadsides; on rich soil. Widespread

Nettle Family

but infrequent, throughout the eastern states. Introduced from Eurasia. July–September.

Description.—Stems from a rootstock, erect, 1–2 m. high, ridged, bristly-hairy with stinging hairs. Leaves opposite, simple, ovate to heart-shaped, coarsely serrate, hairy, usually more than twice as wide as the length of the petiole, stipules distinct, green to pale brown. Flowers mostly dioecious, small, greenish, in branching panicled spikes; staminate flowers with 4 perianth-segments and 4 stamens; pistillate flowers with 4 perianth-parts and a 1-celled ovary. Fruit an achene about 1 mm. long, flattened, ovate, minutely granular, yellow to grayish-tan, calyx and remnant of style often attached.

Control.—Mow close to the ground to prevent seed formation (11); grub the rootstocks out and kill by drying.

89. *Urtica gracilis* Ait. Stinging nettle, Slender nettle, Tall nettle.

Perennial; reproducing by seeds and creeping rootstocks. Barnyards, fence-rows, waste places, and neglected fields; in damp rich loamy soils. Widespread across the northern United States and southern Canada. Native. July–August.

Description.—Similar to 88, but more slender and less bristly-hairy. Leaves narrow, not as wide as the length of the petiole, margin finely serrate, with 3–5 main veins from the base, nearly glabrous; stipules distinct, green to pale brown. Flowers monoecious, in slender spikes grouped into a loose panicle, at least the upper flowers pistillate; staminate flowers with 4 perianth-segments and 4 stamens; pistillate flowers with 4 perianth-parts and a 1-celled ovary. Fruit a small achene similar to 88.

Control.—The same as for 88.

90. *Urtica urens* L. Stinging nettle, Small nettle.

Annual; reproducing by seeds. Waste places and neglected gardens, fields and orchards. Widely distributed but rare in southern Canada and in the eastern states; locally common on the Pacific Coast. Introduced from Europe. May–August.

Description.—Stems erect, branched from the base, slightly hairy, about 1–6 dm. high. Leaves opposite, ovate, coarsely dentate, petioles 1–3 cm. long. Flower-clusters axillary, shorter than the petioles; flowers mostly dioecious, similar to 88. Achene 1.6 mm. long, similar to 88.

Control.—Clean cultivation (6).

Urtica procera Muhl., until recently included with *Urtica gracilis* by eastern American authors, is the common native nettle in eastern Canada and from the eastern United States to North Carolina and Louisiana.[1]

In western North America two native nettles, *Urtica lyallii* Wats., from Alaska to Oregon, and *Urtica holosericea* Nutt., from the Great Basin to California, frequently occur as weeds about farmyards, along fences, and in waste places on rich moist soils.

KEY TO THE STINGING NETTLES

1. Annuals..*U. urens*
1. Perennials.
   2. Plants dioecious..*U. dioica*
   2. Plants not dioecious
      3. Mature stipules scarious to herbaceous in texture, green to pale-brown.
         4. Stem glabrous above; leaves glabrous on both surfaces or sparingly pilose above...............................*U. gracilis*
         4. Stem cinereous, pilose or puberulent above; leaves cinereous-puberulent beneath..............................*U. procera*
      3. Mature stipules coriaceous, deep-brown to chestnut-brown.
         5. Leaves nearly glabrous above, somewhat pubescent beneath.
                                                                                  U. lyallii
         5. Leaves soft-pubescent on both sides..............*U. holosericea*

POLYGONACEAE (Buckwheat Family)

91. *Polygonum argyrocoleon* Steud. Silver-sheathed knotweed. Annual; reproducing by seeds. Alfalfa fields, the seeds occurring as impurities in alfalfa seed. Southern California and adjacent states. Introduced from the Caspian Sea region.

Description.—Stems erect, branched, 3–6 dm. high. Leaves linear or elliptic to lanceolate, 1–7 cm. long, alternate, entire; petiole short, from a conspicuous silvery sheath. Flowers in a leafless interrupted spike, 2–5 in a cluster on a pedicel which usually exceeds the pinkish perianth. Fruit, "seed," a 3-angled, glossy, minutely-dotted, brown achene, slightly more than 2 mm. in length.

Control.—Use clean seed (1).

92. *Polygonum aviculare* L. Knotweed, Knot-grass, Door-weed, Mat-grass, Pink-weed, Bird-grass, Stone-grass, Way-grass, Goose-grass. Fig. 21, h–k.

Annual; reproducing by seeds. Chiefly in hard trampled ground

[1] Fernald, 1926.

Fig. 20—**BLACK BINDWEED,** *Polygonum convolvulus:* **A,** plant showing general habit (× ⅓); **B,** fruit (× 2); **C,** seed (achene), (× 4).
HEDGE BINDWEED, *Polygonum scandens:* **D,** portion of a plant showing general habit (× ⅓); **E,** flower (× 4); **F,** fruit (× 2); **H,** seed (achene), (× 4).

about yards, roadsides, paths and waste places. Common throughout the northern United States and southern Canada. Native to North America but also introduced from Eurasia. June–October.

Description.—Stems slender, striate, much branched, mostly forming prostrate mats from thin tap-roots. Leaves lanceolate, 5–20 mm. long, alternate, entire, acute, bluish-green; petiole very short, adnate to the short sheath formed by the stipules. Flowers in axillary clusters, perfect, small; perianth less than 2 mm. long, with 5–6 green segments with pinkish margins; stamens mostly 8; pistil solitary, ovary 1-celled, styles 3. Fruit, "seed," an achene, about 2 mm. long, 3-angled, dark reddish-brown to black, dull, finely granular-striate; remains of perianth usually attached.

Control.—Hand hoeing (5), hand pulling (4).

93. *Polygonum coccineum* Muhl. Water smartweed, Swamp knotweed, Tansy mustard.

Perennial; reproducing by seeds or rootstocks. Variable, found in shallow or deep water, wet ground or dry soil, meadows and fields. Quebec and Nova Scotia, west to Washington and south to North Carolina and California. Native. July–October.

Description.—Stems from creeping rootstocks. Terrestrial form with ascending stems, aquatic with glabrous floating or submersed leaves and branches. Terrestrial with ovate-lanceolate leaves, long-attenuate, 1–2 dm. long; aquatic forms smoother, leaves thinner with rounded to cordate bases; petioles longer. Racemes 1 or 2, cylindric, often narrow at summit, 4–15 cm. long, about 1 cm. thick; peduncle pubescent to glabrous. Calyx 4.5–5 mm. long, sepals scarlet to pink, rarely white. Flowers of two forms. Fruiting calyx veiny, about 7 mm. long. Achene dark brown to black, shining, 2.5–3 mm. long and about the same width.

Control.—The same as for 97.

94. *Polygonum convolvulus* L. Black bindweed, Wild buckwheat, Knot bindweed, Bear-bind, Ivy bindweed, Climbing bindweed, Cornbind. Fig. 20, a–c.

Annual; reproducing by seeds. Cultivated fields, gardens, grain fields and waste places. Common throughout the northern United States and Canada. Introduced from Europe. July–October.

Description.—Stems twining or creeping, branched at base; roots fibrous. Leaves alternate, simple, entire, heart-shaped or somewhat

Buckwheat Family

halberd-shaped, taper-pointed. Flowers in panicled racemes; perianth of 5 segments, the 3 outer strongly keeled; stamens 3–9; pistil with 1-celled ovary. Fruit a minutely roughened, dull black, triangular achene about 3 mm. long.

Control.—Cultivation (6), hand weeding (4), harrowing (8).

95. *Polygonum cuspidatum* Sieb. and Zucc. Japanese knotweed, Mexican bamboo. Fig. 23.

Perennial; reproducing by long stout rhizomes, also by seeds. An escaped ornamental; waste places and neglected gardens; spreading rapidly and becoming obnoxious. Newfoundland to Ontario, and in many parts of the northeastern United States, west to Minnesota and Iowa south to Maryland. Native of Japan. August–September.

Description.—Stems stout, erect, 1–3 m. high, glaucous, often mottled, shrubby. Leaves broadly ovate, petiolate, truncate to somewhat cuneate at base, 5–15 cm. long, 5–12 cm. broad, basal angle prominent. Dioecious species with greenish-white flowers in axillary panicles; outer sepals narrowly winged along the midrib; styles 3; stigmas minute. Fruiting calyx wing-angled, 8–9 mm. long. Achene triangular, shiny, 3–4 mm. long.

Control.—Frequent cultivation to grub out the rhizomes; should not be allowed to form seeds.

Polygonum sachalinense F. Schmidt, another Polygonum introduced from Asia, is similar to *P. cuspidatum* but coarser and taller, with angular-striate stem, cordate leaves and greenish flowers. It is spreading in a similar manner in neglected gardens and waste ground, Massachusetts and New York to Maryland.

96. *Polygonum erectum* L. Erect knotweed. Fig. 21, a–c.

Annual; reproducing by seeds. Yards, roadsides and waste places; on rich soil. Locally common throughout the northern United States and southern Canada, south to Georgia and Kansas. Native. July–October.

Description.—Similar to 92, but with more erect stems. Leaves elliptical, obtuse, yellowish-green. Calyx yellowish-green, 3 mm. long. Achenes a little larger than in 92.

Control.—The same as for 92.

97. *Polygonum hydropiper* L. Smartweed, Water pepper, Water smartweed, Biting knotweed, Pepper plant, Red shanks. Fig. 22, e–g.

Annual; reproducing by seeds. Low meadows, pastures, cultivated

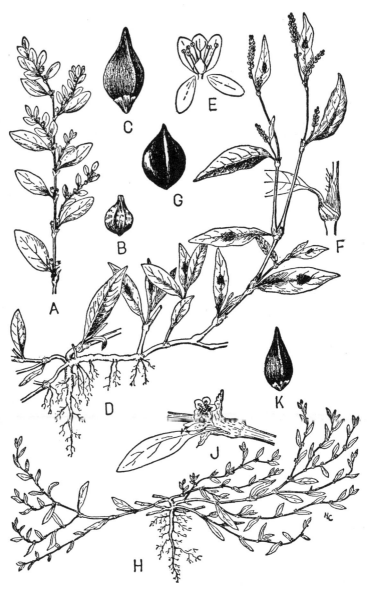

FIG. 21—**ERECT KNOTWEED**, *Polygonum erectum:* **A**, branch showing habit (× ⅓); **B**, fruit, surrounded by calyx (× 3); **C**, "seed" (× 9).
LADYS-THUMB, *Polygonum persicaria:* **D**, part of plant showing habit (× ¼); **E**, flower (× 3); **F**, node with sheath (× 1); **G**, "seed" (× 9).
KNOTWEED, *Polygonum aviculare:* **H**, plant showing habit (× ⅓); **J**, node with flowers and leaf (× 3); **K**, "seed" (× 9).

ground and waste places. Widespread and locally common. Native to North America and Europe. Probably introduced in many parts of the United States. August–September.

Description.—Stems jointed, 3–6 dm. high, glabrous, branched below. Leaves alternate, simple, narrowly lanceolate, entire, with very peppery taste; stipules fused into sheaths (ocreae) with a bristly ciliate margin. Flowers in close, terminal, nodding spikes; perianth with 5 greenish segments, dark glandular-dotted; stamens 6; pistil with 1-celled ovary, style 2–3 parted. Fruit a dull purplish-black achene, about 2 mm. long, somewhat flattened to irregularly 3-angled, minutely granular-striate, with remains of the perianth usually attached.

Control.—Clean cultivation (6), hand hoeing (5), harrowing (8), frequent mowing in pastures and meadows (11), improved drainage (21).

The herbage of this weed contains a pungent juice which causes smarting when it comes in contact with the eyes.

98. *Polygonum hydropiperoides* Michx. Mild water pepper, Mild smartweed. Fig. 22, h–k.

Perennial; reproducing by seeds and rooting stems. Wet meadows and lands subject to inundation. Locally common throughout the United States. Native. August–September.

Description.—Similar to 97, but the stems often decumbent and rooting at the lower nodes; without peppery taste, and with more slender erect spikes. Perianth flesh-colored or nearly white; the segments not dark glandular-dotted; stamens 8. Achene similar to 97 but smooth and glossy, more regularly 3-angled, and perianth separating freely.

Control.—The same as for 97.

99. *Polygonum lapathifolium* L. Pale smartweed, Pale persicaria, Willow-weed, Knotweed.

Annual; reproducing by seeds. Fields, meadows and waste places, especially on rich moist soils. Locally common throughout the northern United States and along the Pacific Coast. Introduced from Europe. July–September.

Description.—Stems erect or decumbent, jointed, branched, often rooting at the lower nodes. Leaves similar to 101. Flowers similar to 101, but in slender drooping spikes on glabrous or slightly glandular peduncles; stamens 6. Fruit an achene, about 2 mm. long, flattened, dark brown, finely granular to glossy surface; with perianth remains often attached.

Control.—The same as for 97.

100. *Polygonum orientale* L. Princes-feather.

Annual; reproducing by seeds. Gardens and waste places. Escaped from cultivation about many cities and towns, especially in the eastern states and southward. Introduced from Asia. August–October.

Description.—Stems stout, about 1 m. high, branching above, hairy. Leaves alternate, simple, ovate to oblong, pointed, sheaths ciliate and usually with a spreading border. Flowers in close, cylindrical, drooping spikes; perianth rose-colored, 5-parted; stamens 7. Achene broadly circular, flattened, about 2.5–3 mm. long, reddish-brown to black, minutely granular, dull or slightly glossy; perianth usually separating.

Control.—Hand pulling (4); hand hoeing (5); cultivation (6).

101. *Polygonum pensylvanicum* L. Pennsylvania smartweed, Purple-head, Glandular persicary, Hearts-ease, Swamp persicary. Fig. 22, a–d.

Annual; reproducing by seeds. Damp grasslands, cultivated ground, waste places and along ditches. Locally common throughout the United States except in the higher mountainous regions. Native. July–October.

Description.—Stems jointed, swollen at the nodes, erect, 3–6 dm. high, branched. Leaves alternate, simple, lanceolate, entire, glabrous on lower surface; stipules fused into a sheath which is free from ciliate bristles. Flowers in close erect spikes on glandular hairy peduncles; perianth bright pink or rose-color, with 5 segments which are appressed in fruit; stamens 8; pistil 1, ovary 1-celled. Fruit an achene, glossy, nearly orbicular, at least one surface concave, sometimes 3-angled, 2.5–3 mm. long, reddish-brown to black, minutely granular; remains of perianth usually attached.

Control.—Clean cultivation (6), hand hoeing (5), harrowing (8), frequent mowing in pastures and meadows (11), improved drainage (21).

102. *Polygonum persicaria* L. Ladys-thumb, Persicary, Spotted smartweed, Heartweed, Spotted knotweed, Red shanks, Willow-weed, Lovers-pride. Fig. 21, d–g.

Annual; reproducing by seeds. Cultivated ground, roadsides, waste

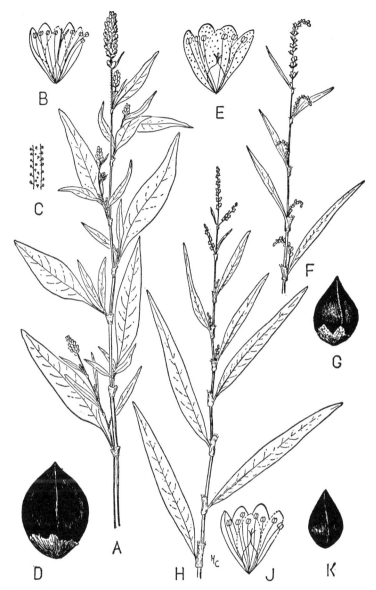

FIG. 22—**PENNSYLVANIA SMARTWEED,** *Polygonum pensylvanicum:* **A,** branch (× ⅓); **B,** flower (× 3); **C,** glandular hairs on peduncle (× 4); **D,** achene (× 6).
SMARTWEED, *Polygonum hydropiper:* **E,** flower (× 6); **F,** branch showing flower-cluster (× ⅓); **G,** seed (achene), (× 6).
MILD WATER PEPPER, *Polygonum hydropiperoides:* **H,** branch showing flower-cluster (× ⅓); **J,** flower (× 6); **K,** seed (achene), (× 6).

places. Common throughout the northern United States and southern Canada. Introduced from Europe. July–October.

Description.—Similar to 101. Leaves usually marked with a triangular dark spot near the middle; sheaths bristly ciliate; peduncles not glandular-hairy. Stamens mostly 6. Achenes 2 mm. long, broadly ovate, flattened or triangular, purplish-black, glossy.

Control.—Clean cultivation (6), hand hoeing (5), harrowing (8), frequent mowing in pastures and meadows (11).

103. *Polygonum scandens* L. Hedge bindweed, Hedge buckwheat, Climbing false buckwheat. Fig. 20, d–h.

Annual; reproducing by seeds. Moist alluvial, sandy or gravelly soil, along fence-rows and edges of fields. Locally common in the northern United States and southern Canada, south to Florida and Texas. Native. July–September.

Description.—Similar to 94, but somewhat coarser. Fruit a glossy black triangular achene about 4.5 mm. long. Described as a perennial in most manuals but all specimens studied were strictly annual.

Control.—Hand pulling (4), mowing (11), burning (13).

104. *Polygonum setaceum* Baldw.

Perennial; reproducing by seeds and by rhizomes. Swamps, low woods, shores, wet clearings. New York west to Michigan, south to Florida and Texas. Native. July–October.

Description.—Stems firm, erect, from long horizontal rhizomes, 2.5–6 mm. thick toward base, 5–10 dm. high, often branching from median nodes. Sheaths with marginal bristles 5–15 mm. long. Leaves oblong-lanceolate, strigose or glabrous on both sides, dark green; primary ones 0.4–1.8 dm. long and 1.2–3.5 cm. broad. Racemes 1— several, loosely flowered, mostly with peduncles, erect, very slender, 1–8 cm. long; pedicels exserted; sepals at maturity 3–3.5 mm. long, white or pink. Achene 2–3 mm. long, oblong, angled, black, shining.

Control.—The same as for 96.

A number of other species of Polygonum native to various parts of the United States may become weeds, especially in low wet fields and along ditches.

105. *Rumex acetosa* L. Sour dock, Garden sorrel, Meadow sorrel, Tall sorrel, Green sorrel. Fig. 24, j–m.

Perennial; reproducing by seeds and creeping roots. Old pastures and permanent meadows on moist alluvial soil; waste places. South-

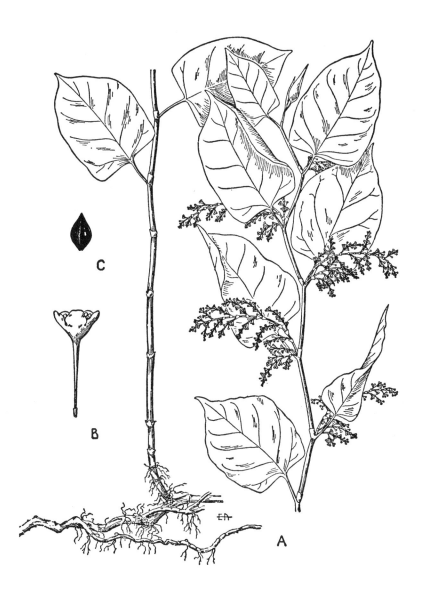

Fig. 23—**JAPANESE KNOTWEED,** *Polygonum cuspidatum:* **A,** habit showing rootstocks (× ⅓); **B,** flower (× 5); **C,** seed (× 3).

ern Canada; locally abundant in north Atlantic states; infrequent elsewhere. Introduced from Eurasia. May–July.

Description.—Stems erect, slender, nearly 1 m. high, frequently tufted, or scattered from creeping roots. Leaves alternate, simple, entire, oblong to broadly lanceolate, arrow-shaped, light green, glabrous; stipules forming a sheath. Flowers dioecious, very small, in open terminal panicles; calyx of 6 sepals, the 3 inner much enlarged, about 4 mm. wide and longer than the mature achenes; stamens 6; pistil 1, triangular, with 3 styles, forming an achene about 2 mm. long, irregularly 3-angled, reddish-brown.

Control.—In gardens, clean cultivation (18); scattered plants in pastures, hand pulling (4) or hand hoeing (5); on old meadows, crop rotation (20).

The herbage has a sour taste. The plant has been introduced as a pot-herb and is even now occasionally grown in some gardens for that purpose.

106. *Rumex acetosella* L. Sheep sorrel, Field sorrel, Horse sorrel, Sour-weed, Sour-grass, Red-top sorrel, Cow sorrel, Red-weed, Mountain sorrel. Fig. 24, e–h.

Perennial; reproducing by seeds and creeping roots. Dry, sandy or gravelly sterile fields and meadows. Common throughout the United States and southern Canada. Introduced from Eurasia. May–September.

Description.—Stems low, 3 dm. high, usually scattered, produced from creeping roots. Leaves alternate, simple, entire, narrowly lanceolate or the upper almost linear, the lower halberd-shaped. Flowers dioecious, very small, in terminal panicles; panicles appearing reddish when mature; sepals 6, very small, not as long as the mature achene. Achenes triangular, about 1 mm. long, reddish-brown, glossy, but usually surrounded by the persistent roughened perianth.

Control.—In cultivated crops, clean cultivation (18); old meadows, use a short crop rotation (20). Apply fertilizer, especially nitrogen.

Sheep sorrel is frequently considered as an indicator of acid soil. It also thrives on neutral or slightly alkaline soils, especially if these are low in nitrate nitrogen.

107. *Rumex altissimus* Wood. Smooth dock, Water dock, Pale dock.

Perennial; reproducing by seeds. Swamps and wet rich soil. New Hampshire to Washington, south to Georgia, Texas and New Mexico. Native. April–August.

FIG. 24—**PATIENCE DOCK**, *Rumex patientia*: **A**, branch and basal leaf showing habit (× ⅓); **B**, nutlet surrounded by calyx (× 2); **C**, cross-section of calyx (× 2); **D**, seed (nutlet), (× 6).

SHEEP SORREL, *Rumex acetosella*: **E**, part of plant showing habit (× ⅓); **F**, nutlet surrounded by calyx (× 3); **G**, cross-section of calyx (× 3); **H**, seed (× 6).

SOUR DOCK, *Rumex acetosa*: **J**, flowering branch and lower leaves (× ⅓); **K**, nutlet surrounded by calyx (× 3); **L**, cross-section of calyx (× 2); **M**, seed (× 6).

Description.—Stems erect, grooved, up to 2 m. high, with short axillary branches. Leaves flat, thickish, lanceolate to ovate-lanceolate, acute or acuminate, obtuse or rarely rounded at base, obscurely veiny, lower stem leaves 1–3 dm. long. Inflorescence loose racemes in panicles, 1–3 dm. long, nearly leafless, with few short ascending branches. Flowers perfect; perianth light green, 2 mm. long; pedicels 3–5 mm. long, jointed just above the base. Usually only one conspicuous grain on one of the valves; when more than one, often unequal. Achene about 3 mm. long.

Control.—The same as for 108.

108.—*Rumex crispus* L. Curly dock, Sour dock, Yellow dock, Narrow-leaved dock. Fig. 25, a–d.

Perennial; reproducing by seeds. Meadows, pastures, lawns and waste places. Common in the United States and southern Canada. Introduced from Eurasia. June–September.

Description.—Stems about 1 m. high, branched above, glabrous, ridged, with enlarged nodes; during the first year the plant forms a dense rosette of leaves on a stout tap-root. Leaves alternate, simple, lanceolate, prominently curly and wavy along the margins; stipules fused to form a sheath just above each node. Flowers in several whorls crowded into a cluster of ascending racemes; calyx of 6 greenish sepals more or less persistent, the 3 inner enlarged (in the fruit called valves), heart-shaped and nearly entire, 4–6 mm. wide, and each bearing a round plump grain (tubercle); pistil 1, with 3 styles. Achenes triangular, glossy, reddish-brown, about 2 mm. long.

Control.—In pastures and lawns remove scattered plants, including all roots, with a spud (12). If fields are overrun, plow (7), follow by clean cultivation (18).

109. *Rumex hastatulus* Baldw. Wild sorrel. Fig. 26, a–c.

Perennial; reproducing by seed. Sandy soil along coast, Massachusetts to Florida, west to Texas, local on dry soil north to Kansas and southern Illinois. Native. May–August.

Description.—Stems one to several from stout woody taproot, 1.5–13 dm. high. Numerous long-petioled basal leaves with blades blunt, lanceolate to oblong, usually with 2 widely divergent basal lobes (or merely undulate), 5–15 cm. long; cauline leaves linear to hastately 3-lobed. Panicle 1–3 dm. long, leafless, upright, with many ascending racemes; species dioecious, with flowers on slender recurved pedicels about 2.5 mm. long; outer sepals of the pistillate

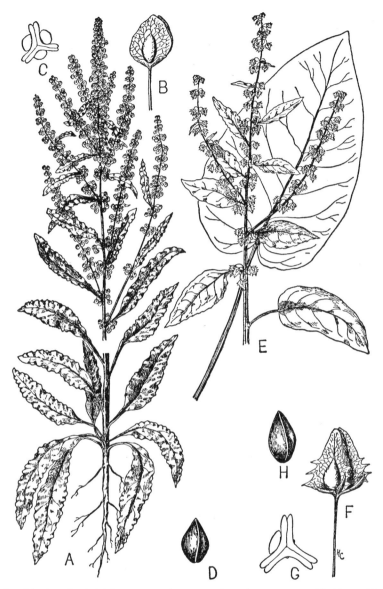

FIG. 25—**CURLY DOCK**, *Rumex crispus:* **A,** plant showing general habit (× ⅓); **B,** a fruit surrounded by persistent calyx (× 3); **C,** diagram of cross-section of calyx showing three tubercles (× 3); **D,** seed (achene), (× 6).
BROAD-LEAVED DOCK, *Rumex obtusifolius:* **E,** basal leaf and fruiting shoot (× ⅓); **F,** fruit surrounded by calyx (× 3); **G,** diagram of cross-section of calyx showing one tubercle (× 3); **H,** seed (achene), (× 6).

Fig. 26—**WILD SORREL,** *Rumex hastatulus:* **A,** habit (× ⅓); **B,** flower (× 3); **C,** fruit with winged perianth (× 3).
BARILLA, *Halogeton glomeratus:* **D,** habit (× ⅓); **E,** flower (× 3); **F,** fruit (× 3).

flower reflexed in fruit, scarcely 1 mm. long; inner sepals becoming broadly round-cordate, 2.5–4.5 mm. long. Achene 1.5 mm. long, brown.

Control.—The same as for 105.

110. *Rumex mexicanus* Meissn. Pale dock, Willow-leaved dock, White dock.

Perennial, reproducing by seeds. Meadows and waste places; on moist rich soil. Locally common in the northeastern states and the Rocky Mountain and Pacific Coast states, also in southern Canada. Native. June–September.

Description.—Similar to 108. Leaves flat, with entire margins, pale green. Calyx dull, greenish-brown; sepals triangular, nearly entire, the 3 inner (valves) each with a narrow tubercle (grain). Achenes triangular, about 2 mm. long, glossy, reddish-brown to light brown.

Control.—The same as for 108.

111. *Rumex obtusifolius* L. Broad-leaved dock, Bitter dock, Celery seed. Fig. 25, e–h.

Perennial; reproducing by seeds and new shoots from roots. Fields, pastures, lawns; in rich moist or shaded soil. Widespread, but local, throughout the United States and southern Canada. Introduced from Eurasia. July–September.

Description.—Similar to 108. Leaves broad, with nearly entire slightly wavy margin; the lower ones ovate to heart-shaped, the upper oblong-lanceolate. The 3 inner sepals (valves) ovate, reticulated, with prominent long teeth, at least near the base; only 1 sepal of calyx bearing a grain. Achenes 2 mm. long, 3-angled, glossy, reddish-brown.

Control.—In pastures and lawns remove scattered plants, including all roots, with a spud (12). If fields are overrun, plow (7), follow by clean cultivation (18). Also improve drainage (21).

112. *Rumex patientia* L. Patience dock, Blunt-leaved dock. Fig. 24, a–d.

Perennial; reproducing by seeds. Roadsides and waste places; on rich loamy soil. Locally common in the northeastern states and eastern Canada; rare elsewhere. Introduced from Eurasia. June–July.

Description.—Similar to 108. Three inner sepals (valves) nearly entire; grains absent or only 1 valve producing a small grain. Achene 3 mm. long, 3-angled, glossy, pale yellowish-brown.

Control.—Pull out or cut individual plants with a spud (11, 12). In certain localities other native species of Rumex sometimes become weeds in meadows and pastures, especially on low wet land.

CHENOPODIACEAE (Goosefoot Family)

113. *Atriplex patula* L. Orache, Lambs-quarters, Fat-hen, Saltbush. Fig. 27, a–c.

Annual; reproducing by seeds. Waste places, cultivated fields and gardens; on sandy or alkaline soils. Widespread, but only locally common. Native, but introduced in some sections. August–October.

Description.—Stems much branched, spreading, ascending or erect, ridged, glabrous, or scurfy. Leaves alternate, simple, the lower opposite, entire or distantly toothed, lanceolate or hastate, the upper often linear and sessile. Flowers monoecious, the two kinds together or in separate spikes; staminate flowers with 3–5 sepals, 5 stamens; pistillate flowers consisting of a pistil surrounded by a pair of bracts. Pistil forming a utricle inclosed by the pair of enlarged, triangular, ovate, often toothed, valvate bracts which may fuse at the base; seed lens-shaped, about 1.5 mm. in diameter, black, glossy when the papery pericarp is removed.

Control.—The young seedlings are easily destroyed. Clean cultivation early in the season (6); followed by hand pulling of scattered plants to prevent seed formation (4).

114. *Atriplex patula* L. var. *hastata* (L.) Gray. Halberd-leaved orache. Fig. 27, d.

Annual; reproducing by seeds. Waste ground and neglected fields, especially on alkaline or saline soils. Locally common in the northeastern states and westward, including the Rocky Mountains. Native, but probably introduced in many sections. August–October.

Description.—Similar to 113, but with the lower leaves broadly triangular and hastate at the base. Seed somewhat larger, brownish, dull, and often somewhat wrinkled.

Control.—The young seedlings are easily destroyed. Clean cultivation early in the season (6); follow up with hand pulling of scattered plants to prevent seed formation (4).

115. *Atriplex rosea* L. Red scale, Red orache, Tumbling atriplex.

Annual; reproducing by seeds. Waste places and fields. Common in waste places about the irrigated sections from the northern Rocky

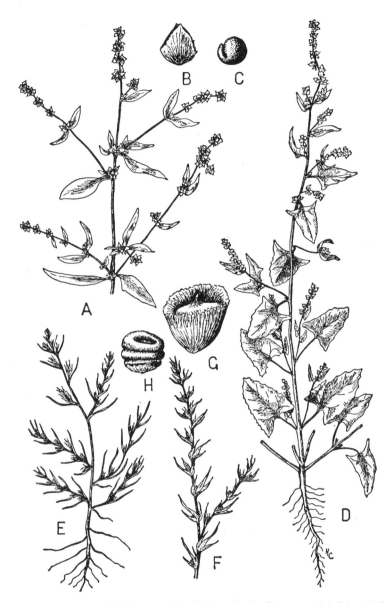

Fig. 27—ORACHE, *Atriplex patula*: **A,** branch showing general habit (× ⅓); **B,** fruit inclosed in two bracts (× 2); **C,** seed (× 2).
HALBERD-LEAVED ORACHE, *Atriplex patula* var. *hastata*: **D,** plant (× ⅓).
RUSSIAN THISTLE, *Salsola kali* var. *tenuifolia*: **E,** young plant (× ⅓); **F,** branch with spine-like leaves (× ⅓); **G,** fruit with husk (× 6); **H,** seed (× 6).

Mountain states to California;[1,2] very local, New York to Wisconsin and south. Introduced from Europe. July–August.

Description.—Stems much branched and spreading, covered with a silvery scurf, 3–10 cm. high. Leaves alternate, short-petioled or the upper sessile, oblong to rhombic-ovate, with a coarsely dentate or wavy margin, silvery-scurfy. Flowers similar to 113; bracts of pistillate flowers in fruit coarsely dentate and warty. Seed about 1.7 mm. in diameter, flattened, with convex sides, yellowish-brown, dull, unevenly wrinkled.

Control.—Clean cultivation (6) and hand hoeing (5) in cultivated fields; close mowing to prevent seed formation (11).

116. *Axyris amaranthoides* L. Russian pigweed.

Annual; reproducing by seeds. Grain fields, roadsides and along railroads and waste places. Western Ontario to the dry belt of British Columbia, southward to Minnesota, North Dakota and Montana. Introduced from Siberia. June–August.

Description.—Stems erect, much branched, 5–12 dm. high, very leafy, pale green, grooved, often breaking off at the ground and forming a "tumble weed." Leaves alternate, simple, pale green; in young leaves the lower surface whitish with stellate hairs. Flowers small, greenish, imperfect, the staminate in the upper part and the pistillate at the base of the racemose clusters; pistillate flowers with a pair of separate bracts and 3–4 sepals which turn white in fruit. Seeds oval, flattened, about 2 mm. long, gray or yellowish-brown, with a silky luster, surface minutely wrinkled or striate, with a short thin groove across the base.

Control.—Stray plants should be pulled by hand (4) or hoed (5) before seeds are produced. Harrow grain fields early in the spring (8). Summer fallowing (19), followed by a clean cultivated crop or a grain crop which is harrowed early in the spring has been recommended for large areas infested with Russian pigweed.[3]

117. *Bassia hyssopifolia* (Pall.) Ktze. Thorn orache.

Annual; reproducing by seeds. Seashores and waste places. Massachusetts to New York, southwestern United States, also in Nevada and the Pacific Coast states. Introduced from Eurasia. July–October.

Description.—Stems stiff, whitish, up to 4 dm. or more high. Leaves flat, oblong-linear, acute; the floral ones shorter and oblong. Habit much like *Chenopodium album*. Flower-clusters in the axils of small

[1] Garrett, 1921. [2] Piemeisel, 1932. [3] Tice, 1932.

Goosefoot Family 183

bracts, borne in short or elongate, slender, paniculately arranged, woolly spikes, at first dense but later interrupted; spikes about 4 mm. in diameter. Each of the 5 sepals at maturity bearing a slender incurved spine.

Control.—The same as for 118.

118. *Chenopodium album* L. Lambs-quarters, Pigweed, White goosefoot, Fat-hen, Mealweed, Frost-blite, Bacon-weed. Fig. 28, a–d.

Annual; reproducing by seeds. Gardens, cultivated fields, grain fields and waste ground. Common throughout the United States. Introduced from Eurasia. June–September.

Description.—Stems erect, branched above, angular or ridged, glabrous, 5–20 dm. high. Leaves alternate, simple, ovate to lanceolate, without stipules, margin with a few low broad teeth, the upper sometimes linear and sessile, grayish-green and mealy below. Flowers perfect, small, greenish, sessile, in irregular spikes clustered in panicles; calyx of 5 sepals; sepals somewhat keeled and nearly covering the mature fruit; stamens 6; pistil 1, with 2 or 3 styles; ovary 1-celled, attached at right angles to the flower axis (horizontal). Fruit a utricle (a seed covered by the thin papery pericarp which often persists); seed lens-shaped, with a marginal notch, black and somewhat glossy, about 1.3 mm. in diameter.

Control.—Clean cultivation (6) and hand weeding (4) in cultivated crops; harrowing of cultivated fields and grain fields while the crop plants are small (8).

119. *Chenopodium ambrosioides* L. Wormseed, Mexican-tea, Jerusalem-tea, Spanish-tea, Strong-scented pigweed. Fig. 29, f–j.

Annual; reproducing by seeds. Neglected yards, waste places and fields. Common in the middle Atlantic and southern states; occasional northward and westward. Introduced from tropical America. July–September.

Description.—Stems erect or decumbent, ridged, much branched and spreading, about 1 m. long. Leaves alternate, simple, with a petiole or sessile, lanceolate, coarsely toothed or nearly entire, glandular-hairy and strongly aromatic. Flowers small, greenish, clustered in a broad panicle; with 3–5 sepals; otherwise similar to 118. Seed lens-shaped, about 0.6 mm. in diameter, glossy, reddish-brown to nearly black.

Control.—This is difficult to pull by hand on account of its strong root, and when mowed off it sprouts up again. The surest way to

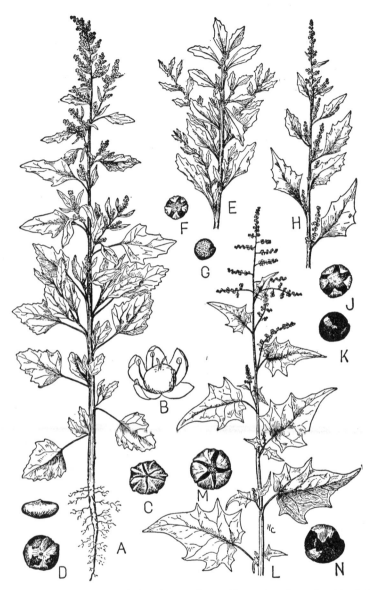

Fig. 28—**LAMBS-QUARTERS**, *Chenopodium album:* **A**, plant showing habit (× ¼); **B**, flower (× 18); **C**, fruit surrounded by calyx (× 9); **D**, "seed" (× 9).
OAK-LEAVED GOOSEFOOT, *Chenopodium glaucum:* **E**, branch showing habit (× ¼); **F**, fruit surrounded by calyx (× 9); **G**, "seed" (× 9).
GOOSEFOOT, *Chenopodium urbicum:* **H**, branch with flowers (× ¼); **J**, fruit surrounded by calyx (× 9); **K**, "seed" (× 9).
MAPLE-LEAVED GOOSEFOOT, *Chenopodium hybridum:* **L**, branch with flowers (× ¼); **M**, fruit surrounded by calyx (× 9); **N**, "seed" (× 9).

destroy it is to hoe off the crowns below the surface of the ground before flowers are produced.

C. ambrosioides var. *anthelminticum* is grown in Maryland and South Dakota for the volatile oil which is distilled from the tops of the plant.

120. *Chenopodium bonus-henricus* L. Good King Henry, Perennial goosefoot. Fig. 29, k–n.

Perennial; reproducing by seeds. Gardens, roadsides and waste places; on rich soil. Infrequent, chiefly about cities and towns in the eastern states. Introduced from Europe. June–August.

Description.—Stems stout, from a perennial root, erect, mostly simple, but often clustered, somewhat mealy, 0.5–2 m. high. Leaves broadly triangular, with hastate base, margin wavy or nearly entire, mealy. Flowers in panicles of spike-like clusters, small, green, similar to 118. Seeds vertical (parallel to the flower axis), about 2 mm. in diameter, with convex sides, one edge straight, often notched, dull, finely roughened, purplish-black.

Control.—Hand hoeing (5); cultivation (6).

The plant was originally introduced as a pot-herb, for which purpose it is still grown in a few gardens.

121. *Chenopodium botrys* L. Jerusalem-oak, Feather-geranium, Turnpike-geranium. Fig. 29, a–e.

Annual; reproducing by seeds. Limited mostly to sterile waste ground and roadsides, sometimes appearing in sandy or gravelly fields. Locally common throughout the United States. Introduced from Europe. July–August.

Description.—Stems erect, forked above, 2–3 dm. high. Leaves alternate, simple, glandular-hairy and strongly aromatic-scented, pinnately lobed, the lobes broad and obtuse. Flowers sessile, pubescent, in open cymes which are grouped into loose clusters; sepals 3–5, pointed, hairy, almost covering the utricle. Seed lens-shaped, about 0.7 mm. in diameter, purplish-black, dull, minutely roughened.

Control.—Clean cultivation (6) and hand weeding in cultivated crops (4); harrowing of cultivated fields and grain fields while the crop plants are still small (8).

This plant is frequently black with dust clinging to its glandular-viscid hairs. Animals seldom touch it on account of its strong odor and taste.

122. *Chenopodium capitatum* (L.) Aschers.(*Blitum capitatum* L.). Strawberry blite, Blite mulberry, Strawberry spinach, Strawberry pigweed.

Annual; reproducing by seeds. Cultivated ground, especially in new clearings in woodland regions; waste places. Infrequent or locally common, mostly in the northeastern states and the Rocky Mountain states, also in Canada and Alaska. Native to North America and also native to Europe. July–September.

Description.—Stems ascending, much branched. Leaves triangular or halberd-shaped, margin wavy or toothed, somewhat mealy. Flowers small, greenish, in small clusters which become fleshy, red, and berry-like when mature; sepals 5; stamens 5; pistil 1, ovary 1-celled, developing into a utricle surrounded by the fleshy red calyx when mature. Seeds vertical, lens-shaped, about 1 mm. in diameter, dull, black.

Control.—Clean cultivation (6) or hand hoeing (5) early in the season to prevent seed production.

123. *Chenopodium glaucum* L. Oak-leaved goosefoot. Fig. 28, e–g.

Annual; reproducing by seeds. Cultivated ground, waste places and roadsides; mostly in sandy or gravelly soils. Locally common throughout the northern United States and southern Canada; south to Virginia, Missouri, and Nebraska. Introduced from Europe. June–August.

Description.—Stems low, much branched, and spreading, glaucous or mealy, without glands. Leaves alternate, simple, coarsely toothed, or pinnately lobed, pale green on the upper surface and white on the lower surface. Flowers small, greenish, clustered in axillary spikes; sepals 5, greenish, not keeled, spreading; stamens 5, pistil 1, ovary 1-celled, forming a utricle. Seeds vertical, the upper ones sometimes horizontal, lens-shaped, about 1 mm. in diameter, brown to reddish-brown, finely roughened.

Control.—Clean cultivation (6) and hand weeding (5) in cultivated crops; harrowing of cultivated fields and grain fields while the crop plants are small (8).

124. *Chenopodium hybridum* L. Maple-leaved goosefoot. Fig. 28, l–n.

Annual; reproducing by seeds. Fields, roadsides, waste places and open woods; mostly in rich gravelly soils or shady places. Widespread and locally common in the northeastern states and southern Canada; rare elsewhere. Native. June–September.

Description.—Stems erect, ridged, much branched, spreading, glabrous, 1–2 m. high. Leaves alternate, simple, triangular or heart-shaped, thin, the lower coarsely toothed or palmately lobed, not mealy. Flowers in large spreading panicles, similar to 118; sepals thin, slightly or not at all keeled, spreading in fruit. Seed lens-shaped, about 1.5 mm. in diameter, finely uneven, dull, black.

Control.—Clean cultivation (6), hoeing (5) or hand pulling (4) before the flowers appear.

125. *Chenopodium murale* L. Nettle-leaved goosefoot, Sow-bane, Swine-bane.

Annual; reproducing by seeds. Gardens, cultivated fields and waste places. Widespread and locally common throughout the United States and southern Canada. Introduced from Europe. June–September.

Description.—Stems erect or decumbent, glabrous, 3–10 dm. high, much branched. Leaves alternate, on slender petioles, ovate to rhombic-ovate, coarsely dentate or wavy-margined. Flowers perfect, small, greenish, in short, branched, axillary clusters; calyx of 5 sepals without keels, not completely covering the fruit; stamens 6; pistil and fruit similar to 118. Seed lens-shaped, 1 mm. in diameter, finely granular, dull, black.

Control.—Clean cultivation (6) and hand weeding (4) in cultivated crops; harrowing of cultivated fields and grain fields while the crop plants are small (8).

126. *Chenopodium paganum* Reichenb. Pigweed. Fig. 29, o–r.

Annual; reproducing by seeds. Rich cultivated fields and waste places. Widely distributed, especially in the northeastern United States. Introduced from Eurasia. July–September.

Description.—Similar to, but somewhat larger than, 118, with which it is often confused. Leaves yellowish-green or dark green, not so mealy. Sepals prominently keeled. Seeds lens-shaped, 1.4–1.6 mm. in diameter, black, glossy, finely striate; pericarp usually persisting.

Control.—Clean cultivation (6) and hand weeding (4) in cultivated crops; harrowing of cultivated fields and grain fields while the crop plants are small (8).

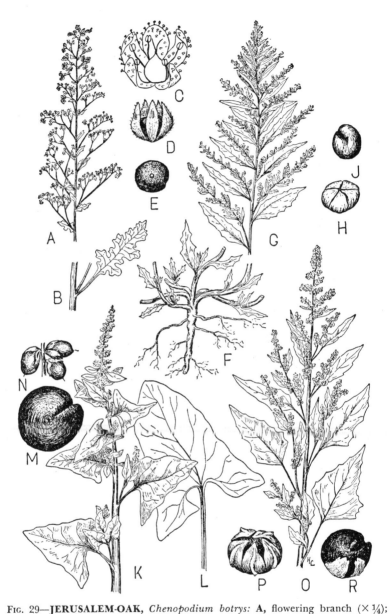

FIG. 29—**JERUSALEM-OAK**, *Chenopodium botrys:* **A**, flowering branch ($\times \frac{1}{4}$); **B**, leaf ($\times \frac{1}{3}$); **C**, flower ($\times 18$); **D**, fruit surrounded by calyx ($\times 9$); **E**, seed ($\times 9$).
WORMSEED, *Chenopodium ambrosioides:* **F**, basal part of plant ($\times \frac{1}{4}$); **G**, flowering branch ($\times \frac{1}{4}$); **H**, fruit surrounded by calyx ($\times 9$); **J**, seed ($\times 9$).
GOOD KING HENRY, *Chenopodium bonus-henricus:* **K**, flowering branch ($\times \frac{1}{4}$); **L**, leaf ($\times \frac{1}{4}$); **M**, seed ($\times 9$); **N**, cluster of fruits ($\times 3$).
PIGWEED, *Chenopodium paganum:* **O**, flowering branch ($\times \frac{1}{4}$); **P**, fruit surrounded by calyx ($\times 9$); **R**, seed ($\times 9$).

127. *Chenopodium polyspermum* L. Many-seeded goosefoot.

Annual; reproducing by seeds. Local on waste ground or cultivated land; naturalized in scattered localities. Quebec to Ontario, south to Tennessee and Iowa; adventive in Oregon. Adventive from Eurasia. July–September.

Description.—Stems erect or decumbent, up to 1 m. high; the lower branches prostrate or divergent. Plants green and glabrous throughout. Leaves thin but opaque, neither glandular nor farinose, entire or with an obscure tooth at widest part, ovate to elliptic, 2–8 cm. long, obtuse or sub-acute, rounded or cuneate to a short slender petiole. Inflorescence axillary, borne along entire length of plant in dense spikes or loose cymes, about the length of leaves or shorter, many-flowered. Calyx-lobes thin, spreading at maturity, smooth, rounded on the back. Seeds shiny, black, about 1 mm. wide; pericarp firmly adhering to the seed.

Control.—The same as for 121.

128. *Chenopodium rubrum* L. Coast-blite, Red goosefoot.

Annual; reproducing by seeds. Salt marshes and moist saline or alkaline soil. Newfoundland to New Jersey, more abundant inland from Minnesota and Iowa westward to British Columbia and California; occasionally adventive in waste places elsewhere. Indigenous in North America and Europe. July–November.

Description.—Stems erect, angled, simple to much branched from the base, 1–9 dm. high, rather fleshy, glabrous, green to dull yellow. Larger leaves rhombic-ovate, -deltoid, or -lanceolate, hastate, with a conspicuous lateral tooth on each side, up to 1.5 dm. long, tapering both at base and at tip, green on both sides, becoming reddish. Flowers in dense axillary leafy spikes, forming in vigorous plants a terminal leafy panicle; calyx-lobes fleshy, obtuse, becoming reddish. Seeds mostly vertical, dark brown, shiny, 0.7–1 mm. wide, margins rounded; pericarp adhering firmly.

Control.—The same as for 126.

129. *Chenopodium urbicum* L. Goosefoot. Fig. 28, h–k.

Annual; reproducing by seeds. Sandy or gravelly gardens and waste places. Infrequent or locally common in the northeastern and central states and eastern Canada; also on the Pacific Coast. Introduced from Europe. July–September.

Description.—Plants similar to 118, but smaller and less mealy. Flowers in narrow raceme-like panicles; sepals thin, not keeled, more or less spreading in fruit. Seeds lens-shaped, about 1 mm. in diameter, smooth, glossy, black; pericarp often persisting.

Control.—Clean cultivation (6) and hand weeding (4) in cultivated crops; harrowing of cultivated fields and grain fields while the crop plants are small (8).

130. *Corispermum hyssopifolium* L. Bugseed.

Annual; reproducing by seeds. Grain fields, cultivated fields and waste places, especially on sandy soils. From the Great Lakes westward to the Rocky Mountains and southward to Mexico. Native. July–September.

Description.—Stems erect or with irregularly spreading branches, pale green, 3–10 dm. high, often breaking off at the ground and forming a "tumble weed"; succulent when young but becoming hard and dry when mature. Leaves alternate, sessile, linear to awl-shaped, 1-nerved, somewhat hairy when young, often dilated at the base. Flowers mostly perfect, small, greenish, solitary in the axils of the upper awl-shaped leaves, with a dilated base; calyx of a single sepal, or absent; stamens 1 or 2; pistil 1, with 2 styles; ovary developing into a flat oval utricle with a concave upper surface. Fruit about 4 mm. long, with a winged margin, brownish-black, finely granular; the pericarp and the 2 styles adhering to the erect seed.

Control.—Clean cultivation until late in the season (6). Early cutting of grasslands (11) before flowers are produced and while the stems are still succulent.

131. *Cycloloma atriplicifolium* (Spreng.) Coult. Winged pigweed, Tumble weed.

Annual; reproducing by seeds. Fields and waste places, chiefly in sandy or alkaline soils. From the Great Lakes westward and southward through the Great Plains; also locally common in Quebec, the northeastern states, and on the Pacific Coast. Native to the western United States and introduced eastward. July–September.

Description.—Stems erect, 2–8 dm. high, with many spreading branches, pale green, often ridged or grooved, becoming purplish with age, often breaking off at the ground to form "tumble weeds." Leaves alternate, simple, with short petioles or the upper sessile, margin wavy-toothed. Flowers perfect or pistillate; calyx of 5 keeled sepals nearly covering the fruit; stamens 5; pistil 1, with 2 or 3 styles; ovary developing into a utricle surrounded by a broad, circular, winglike outgrowth from the base of the sepals. Seeds lens-shaped, about

2 mm. in diameter, nearly circular with a marginal notch, finely granular, glossy, black; pericarp often persisting.

Control.—Clean cultivation to prevent seed formation (6); hand pulling (4).

132. *Halogeton glomeratus* (C. A. Meyer) Coult. Barilla. Fig. 26, d–f.

Annual; reproducing by seeds. Arid and desert land. Rocky Mountain states where it is a rapidly spreading poisonous weed. Introduced from Europe.

Description.—Herb with stiff, erect to spreading, succulent branches. Leaves alternate, linear, nearly cylindric or somewhat spatulate, nearly glabrous and somewhat glaucous and fleshy; the upper leaves reduced and bracteose. Flowers minute, sessile, axillary in bracts, crowded, 3–5 in a glomerate cluster; calyx of 5 sepals; stamens 3–5; pistil with 1-celled ovary and 2-lobed stigma, forming a compressed utricle, surrounded by the 5-winged calyx.

Control.—The same as for 134.

133. *Kochia scoparia* (L.) Roth. Summer-cypress, Kochia, Burning-bush, Fireball, Mexican fireweed.

Annual; reproducing by seeds. Waste places, ballast grounds and occasionally in fields; mostly on dry soils. Widespread, but most common in the middle western states. Introduced from Eurasia. July–September.

Description.—Stems erect, slender, much branched, forming a "cypress-like" crown, 3–10 dm. high. Leaves alternate, simple, lanceolate to linear, nearly glabrous, with ciliate margins. Flowers perfect, sessile, small, in axillary clusters; calyx 5-lobed, each lobe developing into a wing-like appendage. Seed about 1.8 mm. long, ovate, flattened, with a groove on each side from the narrow end; surface dull, finely granular, brown with yellow markings.

Control.—Mow early before seeds are matured (11). In cultivated ground hand weed or hoe as early as possible (4, 5).

Frequently grown, under the name of "summer-cypress," as an ornamental for its bright autumnal color.

134. *Salsola kali* L. var. *tenuifolia* Tausch (*S. pestifer* Nelson). Russian thistle, Saltwort, Prickly glasswort, Russian cactus, Russian tumble weed, Tumbling thistle, Wind witch. Fig. 27, e–h.

Annual; reproducing by seeds. Cultivated ground and waste places, chiefly in dry regions of the western United States and western

Canada, and the north central states. Occasional along railroad tracks or in waste places in sandy or saline soils, in eastern states. Introduced from Eurasia. July–October.

Description.—Stems with many ascending or spreading branches, 2–6 dm. high, often reddish or reddish striped. Leaves alternate, simple, the first linear, fleshy, and succulent, the later ones awl-shaped, stiff and ending in spines. Late in the season the whole plant becomes hard and woody and breaks loose from the soil and is blown about as a "tumble weed," scattering its seeds. Flowers small, greenish, axillary, with 2 small bracts, perfect; calyx of 5 lobes, becoming horizontally winged with nearly orbicular spreading wings on the back which nearly cover the fruit; stamens 5; pistil 1, with 2 styles; ovary 1-celled and 1-seeded. Fruit conical, snail-shaped, about 1.5 mm. long; seed gray to yellowish-brown, dull, finely granular, with a spirally coiled embryo and no endosperm.

Control.—Clean cultivation, followed by hand hoeing to kill stray plants (6). Harrow the growing crop while the plants are small (8). Clean out fence corners and mow scattered plants; pile and burn them to prevent seeds from being scattered.

AMARANTHACEAE (Amaranth Family)

135. *Amaranthus albus* L. (*A. graecizans* L.) Tumble weed, Tumbling pigweed, White pigweed. Fig. 30, e.

Annual; reproducing by seeds. Fields and waste places. Most common in the dry plains and prairie regions; locally common on trucklands in the eastern states. Native to western North America. July–August.

Description.—Stems whitish, much branched, erect or ascending, 2–5 dm. high; when mature breaking from the soil, and carried about by the wind. Leaves alternate, simple, small, obovate or spatulate, apex obtuse. Flowers each with 3 bracts, polygamous, greenish, in small, crowded, axillary clusters; sepals 3, acuminate; stamens 2–3; pistil 1. Fruit a rugose circumscissile utricle; seed lens-shaped, 9 mm. in diameter, glossy, black; embryo coiled about the endosperm.

Control.—The same as for 136.

136. *Amaranthus graecizans* L. (*A. blitoidea* Wats.) Prostrate pigweed, Spreading pigweed, Mat amaranth. Fig. 30, g.

Annual; reproducing by seeds. Neglected fields and waste places; mostly on dry soils. Common from Minnesota westward; locally in-

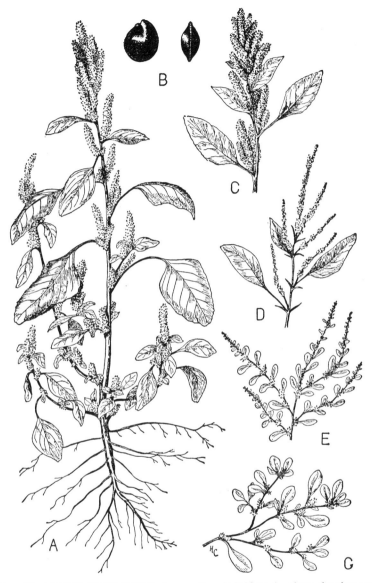

FIG. 30—**GREEN AMARANTH,** *Amaranthus hybridus:* **A,** plant showing general habit ($\times \frac{1}{6}$); **B,** seeds ($\times 9$).
AMARANTH PIGWEED, *Amaranthus retroflexus:* **C,** branch ($\times \frac{1}{6}$).
SPINY AMARANTH, *Amaranthus spinosus:* **D,** branch ($\times \frac{1}{6}$).
TUMBLEWEED, *Amaranthus albus:* **E,** branch ($\times \frac{1}{4}$).
PROSTRATE PIGWEED, *Amaranthus graecizans:* **G,** branch ($\times \frac{1}{4}$).

troduced in the eastern states. Native to the drier regions of western North America. June–August.

Description.—The general appearance of this species is much like 135, but its habit is prostrate or decumbent and its leaves are larger. Sepals obtuse or acute. Fruit not rugose; seed lens-shaped, about 1.5 mm. in diameter, glossy, black.

Control.—Early surface tillage (8) to induce the seeds to germinate, followed by a clean cultivated crop (6) with hand weeding or hoeing (4) to prevent individual plants from maturing seeds.

137. *Amaranthus hybridus* L. Green amaranth, Spleen amaranth, Rough pigweed, Amaranth pigweed, Red amaranth. Fig. 30, a–b.

Annual; reproducing by seeds. Cultivated land, and waste places on rich soil. Widespread and locally common throughout the United States and southern Canada. Introduced from tropical America. July–September.

Description.—Similar to 141, but less rough and deeper green, with more slender and spreading spikes and shorter bracts. Seed 1–1.2 mm. long, broadly ovate, slightly notched, sides convex, margin prominent, glossy, black or with reddish tint.

Control.—The same as for 141.

138. *Amaranthus lividus* L.

Annual; reproducing by seeds. Local on waste ground. Massachusetts and New York to Maryland; also in scattered stations southward and westward. Introduced from the Tropics. August–October.

Description. Stems rather coarse, prostrate or ascending to erect, up to 1 m. high, frequently red. Leaves firm, mostly rhombic-ovate, 1–8 cm. long, with retuse tip; petioles about equaling the blade. Flowers usually in dense clusters and thick terminal panicles; bracts oblong to ovate, less than half the length of the calyx; 2 sepals of the pistillate flowers oblong, 1.5–2 mm. long, a third, shorter and narrower, often lateral. Utricle broadly conic-ovoid, 1.7–2.2 mm. long, thin-walled, indehiscent, smooth. Seed nearly circular, 1–1.2 mm. wide.

Control.—The same as for 136.

139. *Amaranthus palmeri* S. Wats. Palmers amaranth.

Annual; reproducing by seeds. Local in waste places and dry soil. A common weed in Texas; Massachusetts to Pennsylvania west to Texas and California. Native in southwestern United States. September–October.

Description.—Stems erect, stout, simple or with short branches, glabrous throughout to somewhat pubescent above, 0.3–1.5 m. high. Leaves with long petioles, rhombic-ovate to lanceolate or oblong, blades 0.3–1.7 cm. long, tapering to an obtuse or rounded mucronate tip, prominently veined beneath. Flowers in erect, terminal spikes, 1–5 dm. long, sometimes with smaller axillary spikes at base; dioecious; subulate, pungent bracts 2 or 3 times as long as flowers; sepals of pistillate flowers obovate-spatulate, 2–3 mm. long. Utricle circumscissile at the middle, subglobose, rugose above, 1.5 mm. long. Seeds oval, 1–1.3 mm. long, dark reddish-brown.

Control.—The same as for 136.

140. *Amaranthus powellii* S. Wats.

Annual; reproducing by seeds. Waste places, loose or sandy soil. Maine to Pennsylvania; Wyoming and New Mexico to Oregon and California. Native in western America. August–September.

Description.—Differing from the closely related *A. retroflexus* in having stamens mostly 3, bracts 2 or 3 times as long as acute sepals, and terminal panicle a prolonged stiff central spike and few lateral spikes.

Control.—The same as for 136.

141. *Amaranthus retroflexus* L. Amaranth pigweed, Green amaranth, Red-root, Rough pigweed, Chinamans greens, Careless weed. Fig. 30, c.

Annual; reproducing by seeds. One of the commonest weeds of gardens, cultivated fields and waste places. General throughout the United States and southern Canada. Introduced from tropical America. July–September.

Description.—Stems erect, strict, branched above, rough and somewhat hairy. Root reddish or pink above. Leaves alternate, simple, with long petioles, ovate or rhombic, margin wavy. Flowers in thick spikes crowded in a close stiff terminal panicle; flowers polygamous, small, green, with 3 stiff spine-like persistent bracts, 4–6 mm. long; calyx with 5 sepals; stamens 5; pistil 1, 1-celled. Fruit a utricle, circumscissile, opening with a lid; seed lens-shaped, ovate, 0.9–1.3 mm. long, notched at the narrow end, margin inconspicuous, glossy, black.

Control.—Land infested with pigweed should be harrowed several times in the spring to induce seeds to germinate (8); follow by a clean cultivated crop (6) with hand weeding (4) or hoeing (5) to prevent stray plants from producing seeds. Use a crop rotation (20) including alfalfa or clover.

142. *Amaranthus spinosus* L. Spiny amaranth, Thorny amaranth, Soldier weed, Prickly careless weed. Fig. 30, d.

Annual; reproducing by seeds. Cultivated ground, neglected yards, roadsides and waste places. New England to Minnesota and southward. Introduced from tropical America. June–September.

Description.—Similar to 141 in general appearance. Leaves with a pair of axillary spines. Flowers monoecious, yellowish-green; the staminate above in long slender spikes; the pistillate in globular, mostly axillary clusters. Utricle not circumscissile; seed 0.8 mm. long, similar to 141.

Control.—The same as for 141.

Acnida altissima Riddell (*A. tuberculata* Moq.). Water hemp. A native annual locally common as a weed along ditches and wet meadows in the middle Atlantic states, and rarely farther north and west. It is dioecious, has naked pistillate flowers, and the general habit of *Amaranthus retroflexus*.

Iresine celosia L. (*I. paniculata* Ktze.). Jubas bush or Blood-leaf. An annual introduced from tropical America, escaped as a weed in waste places and dry sandy or gravelly fields in the southern states. It has opposite leaves and large showy panicles of small silvery-white flowers, each with a 5-parted calyx. The calyx of the pistillate flower exceeds the bracts and is silky at the base.

NYCTAGINACEAE (Four-o'clock Family)

143. *Mirabilis nyctaginea* (Michx.) MacM. (*Oxybaphus nyctagineus* Sweet, *Allionia nyctaginea* Michx.). Wild four-o'clock, Umbrella-wort.

Perennial; reproducing by seeds and roots. Prairies, dry meadows and roadsides. Native to the middle western states; locally naturalized in the northeastern states. May–September.

Description.—Stems several, from a thick fleshy tap-root, erect, 3–10 dm. high, with broadly spreading forked branches, glabrous or somewhat hairy, ridged and often purplish. Leaves opposite, entire, glabrous or nearly so, ovate to cordate, petioled, or the upper sessile. Flowers perfect, small, 3–5 in an involucre; these involucres borne on hairy peduncles which are grouped in umbels in forked terminal clusters; involucres 5-lobed, persistent, becoming enlarged, veiny and

Pokeweed Family

colored in fruit; each flower with a 5-lobed reddish calyx, 3-5 stamens, a pistil with 1 style and a 1-celled ovary. Fruit a hard nutlet, cylindric-obovoid, 5-ribbed and angled, warty, pubescent, grayish-brown, about 4 mm. long.

Control.—Plow up infested meadows (7) and plant a cultivated crop for one or two years. Hoe out scattered plants in pastures and waste places where plowing is not practicable (5).

Mirabilis hirsuta (Pursh) MacM. (*Oxybaphus hirsutus* Sweet) with sessile lanceolate leaves, and *Mirabilis linearis* (Pursh) Heimerl (*Oxybaphus linearis* Robinson) with linear leaves, perennials with habits similar to *M. nyctaginea*, sometimes become weeds in the middle western states, where they are native, and in the eastern states, where they have been sparingly introduced.

PHYTOLACCACEAE (Pokeweed Family)

144. *Phytolacca americana* L. (*P. decandra* L.). Pokeweed, Pokeberry, Virginia poke, Scoke, Pigeon berry, Garget, Inkberry, Red-ink plant, Coakum, American cancer, Cancer jalap. Fig. 31.

Perennial; reproducing by seeds. Rich pasture lands, waste places and open places in woodlands; mostly on deep, rich, gravelly soils. Locally common from Maine to Minnesota and southward. Native. July–August.

Description.—Stems stout, glabrous, 2–3 m. high, branched above, single or several from a large, fleshy, white root. Leaves alternate, simple, entire, with long petioles. Flowers in long, narrow, terminal racemes which become lateral and opposite to the leaves; flowers small; sepals 5, white, petal-like and rounded; stamens 10; pistil 1, of about 10 united carpels each with 1 vertical seed. Fruit a dark purple, depressed, 10-seeded berry with crimson juice; seeds lens-shaped, about 3 mm. in diameter, glossy, black.

Control.—Plow up badly infested grasslands (7) and plant a cultivated crop for one or two years. In pastures or waste places where plowing is not practicable, scattered plants are easily destroyed by cutting them below the crown with a spud or spade (12).

The roots and berries of the pokeweed are poisonous. Both are used in the preparation of certain medicines. The young shoots are often seen in the eastern markets. When thoroughly cooked, they make excellent greens or potherbs.

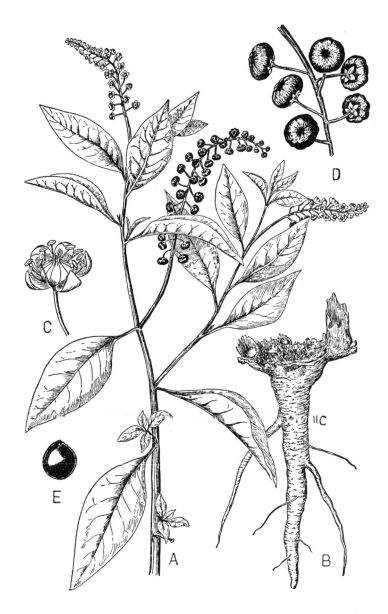

Fig. 31—**POKEWEED**, *Phytolacca americana*: **A**, branch showing general habit (× ⅓); **B**, root (× ¼); **C**, flower (× 2); **D**, berries (× 2); **E**, seed (× 3).

Carpet-weed and Purslane Families

AIZOACEAE (Carpet-weed Family)

145. *Mollugo verticillata* L. Carpet-weed, Indian chickweed, Whorled chickweed, Devils-grip. Fig. 32, f–j.

Annual; reproducing by seeds. Gardens, lawns, waste places; on dry, gravelly, or sandy soils. Locally common in the eastern and middle western states; south to Florida and Texas. Introduced from tropical America. Also native to Africa. July–September.

Description.—Stems much branched, forming prostrate mats, 1–2 dm. in diameter. Leaves whorled, simple, spatulate, entire, without stipules. Flowers in umbel-like clusters at the nodes; sepals 5, white inside; stamens 3, rarely 5; pistil 1, 3-celled, forming a 3-valved capsule with many seeds. Seeds about 0.5 mm. long, flattened, kidney-shaped with a protuberance on the concave side, with 1 or more ridges, orange-red.

Control.—Clean cultivation continued until late in the season (6); hand hoe scattered plants (5).

PORTULACACEAE (Purslane Family)

146. *Portulaca oleracea* L. Purslane, Pusley, Pursley, Wild portulaca. Fig. 32, a–e.

Annual; reproducing by seeds. Gardens, cultivated fields and waste places; mostly on rich soil. Widespread and locally common throughout the United States and southern Canada. Introduced from Europe. July–September.

Description.—Stems prostrate, much branched and forming mats, succulent, glabrous, often reddish in color. Leaves alternate or clustered, simple, obovate or wedge-shaped, entire, thick and fleshy. Flowers small, axillary, sessile, from flattened pointed buds; calyx 2-cleft, the tube fused with the ovary below; petals 5, small, yellowish, inserted on the calyx, opening only in the sunshine, fugacious; stamens 7–12, inserted on the calyx (perigynous); pistil 1-celled, with a 5–6-lobed style. Capsule globular, many-seeded, opening by a lid with the upper part of the calyx attached; seeds 0.7 mm. in diameter, flattened, broadly ovate, with a yellowish scar and small concave area at the smaller end, edges rounded, roughened by curved rows of minute rounded tubercles, slightly glossy, black.

Control.—Kill purslane while it is in the seedling stage by frequent and continued shallow cultivation or hoeing (5, 6) from the

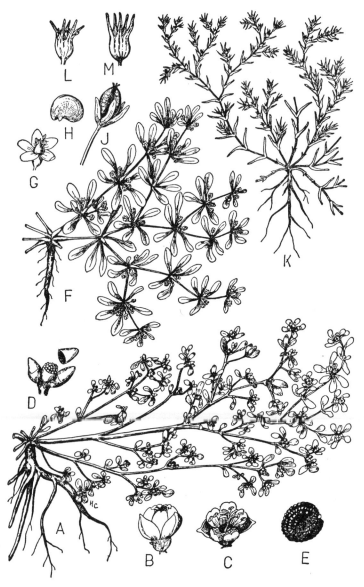

Fig. 32—PURSLANE, *Portulaca oleracea:* **A,** part of plant showing habit (× ⅓); **B,** flower before opening (× 3); **C,** flower after opening (× 3); **D,** capsule showing dehiscence (× 1); **E,** seed (× 18).
CARPET-WEED, *Mollugo verticillata:* **F,** part of plant showing habit (× ½); **G,** flower (× 3); **H,** seed (× 12); **J,** capsule (× 3).
KNAWEL, *Scleranthus annuus:* **K,** part of plant (× ⅓); **L,** flower (× 4); **M,** fruit surrounded by calyx (× 5).

Pink Family

time the first seedlings appear until frost. Plants that have blossomed should be removed from the soil and burned or composted, otherwise the fleshy stems will take root again and mature seeds. Most purslane seedlings do not appear until the weather is very warm, or after most other weeds have been destroyed by cultivation. The continued late cultivation is essential for the destruction of purslane.

CARYOPHYLLACEAE (Pink Family)

147. *Agrostemma githago* L. Corn cockle, Purple cockle, Corn rose, Corn campion, Crown-of-the-field, Corn mullein, Old maids pink. Fig. 33.

Winter annual; reproducing by seeds. Grain fields, especially in winter wheat and rye. Widely distributed throughout the United States. Introduced from Eurasia. May–July.

Description.—Stems erect, branched above, jointed, silky, hairy. Leaves opposite, entire, linear, silky, about 1 dm. long. Flowers in cymes or solitary on long peduncles; sepals 5, fused below into a tube, with 5 long lobes and 10 prominent ribs; petals 5, purple; stamens 10; pistil with a 1-celled superior ovary with free-central placenta; style-branches 5. Capsule with several seeds; seeds 2–3 mm. in diameter, triangular, rounded on the back, with rows of pointed tubercles, black.

Control.—Sow clean seed (1); practice a crop rotation in which wheat is not planted more than once in four years (20); hand pull scattered plants in grain fields before seeds are produced (4).

Corn cockle seeds rarely retain their viability for more than one year when plowed under in the soil. The seeds contain a poisonous substance. Flour made from wheat contaminated by them produces a poorly rising bread with an unpalatable flavor. Feeds made from screenings containing a large percentage of corn cockle are dangerous to stock and poultry.

148. *Arenaria serpyllifolia* L. Thyme-leaved sandwort, Sandwort chickweed. Fig. 38, h–k.

Annual; reproducing by seeds. Sandy or gravelly cultivated fields and waste places. Common throughout the United States and southern Canada. Introduced from Eurasia. All summer.

Description.—Stems much branched, rough and somewhat hairy, about 1 dm. high. Leaves opposite, simple, sessile, ovate, less than

FIG. 33—**CORN COCKLE**, *Agrostemma githago*: **A**, plant showing general habit (× ⅓); **B**, autumn rosette stage (× ⅓); **C**, flower with calyx and corolla removed (× 1); **D**, capsule with calyx (× 1); **E**, vertical section of capsule (× 1); **F**, seed (× 4).

Pink Family

1 cm. long, stipules none. Flowers in leafy cymes; sepals 5, separate; petals 5, entire, white; stamens 10; pistil with 3 styles. Capsule flask-shaped, splitting into 3 or 6 valves, with numerous seeds; seeds short, kidney-shaped, 0.5 mm. long, dark reddish-brown, covered with curved rows of tubercles.

Control.—The application of fertilizer to stimulate the crop plants will drive out the sandwort; clean cultivation (6).

149. *Cerastium arvense* L. Field chickweed, Meadow chickweed. Fig. 39, n–p.

Perennial; reproducing by seeds and creeping stems. Lawns, grassy fields and rocky banks. Locally common throughout the United States, but most troublesome in lawns and grasslands in the northeastern states. Native to Europe and North America. Probably mostly introduced where it appears in lawns. April–June.

Description.—Stems slender, 1–2 dm. long, ascending or erect, usually forming mats or tufts, hairy or nearly glabrous rooting at the lower nodes. Leaves opposite, simple, sessile, linear to narrowly lanceolate. Flowers few to several in cymose clusters at the tip of nearly naked stems; sepals 5 (rarely 4); petals 5 (rarely 4), 2-cleft, white, about 1 cm. long; stamens usually 10; pistil with superior ovary and free-central placenta; styles mostly 5. Capsule 1-celled, about 1 cm. long, many-seeded; seeds about 0.7 mm. long, flat on the sides, the back curved and roughened by irregular rows of elongated tubercles, chestnut-brown.

Control.—This weed does not persist under cultivation. In lawns raise the creeping stems with a rake and mow close to the ground; small patches can be dug or hoed out and reseeded or planted with new sod.

150. *Cerastium viscosum* L. Mouse-ear chickweed.

Annual; reproducing by seeds. Lawns, meadows, fields and waste places. Most common in the middle Atlantic states, the South and the Pacific Coast states. Introduced from Europe. April–July.

Description.—Stems ascending or in clumps, viscid-hairy, 5–15 cm. high. Leaves opposite, sessile, entire, ovate to obovate, hairy. Flowers small, in dense, terminal, cymose clusters which become more open later; pedicels about as long as sepals; flowers similar to 149, but smaller. Fruit a cylindrical capsule with many seeds; seeds similar to 149, 0.5 mm. long, chestnut-brown.

Control.—The same as for 149.

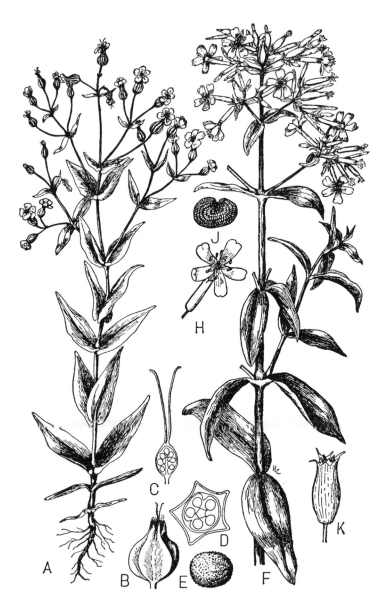

Fig. 34—**COW COCKLE**, *Saponaria vaccaria*: **A**, plant showing habit (× ⅓); **B**, pistil surrounded by angular calyx (× 1); **C**, vertical section of pistil (× 2); **D**, cross-section of capsule and calyx (× 1); **E**, seed (× 5).

BOUNCING BET, *Saponaria officinalis*: **F**, portion of plant with flowers (× ⅓); **H**, flower (× ⅓); **J**, seed (× 5); **K**, capsule surrounded by calyx (× 1).

Pink Family

151. *Cerastium vulgatum* L. Mouse-ear chickweed. Fig. 39, h–m.
Perennial; reproducing by seeds and creeping stems. Chiefly in lawns, pastures and meadows, also in cultivated land. Widespread throughout the United States and southern Canada. Introduced from Europe. May–October.
Description.—Plant similar to 150 in general appearance but the stems more prostrate and often rooting at the lower nodes and forming mats. Leaves mostly oblong to spatulate. Flowers similar to 149 but smaller; sepals 4–6 mm. long; petals not longer than the sepals, sometimes absent. Fruit a small cylindrical capsule with many seeds; seeds similar to 149, tubercles more distant, chestnut-brown.
Control.—The same as for 149.

152. *Dianthus armeria* L. Deptford pink, Grass pink. Fig. 38, d–f.
Annual or winter annual; reproducing by seeds. Sandy or stony meadows, pastures and waste places. Locally common in the eastern states and the Pacific Northwest. Introduced from Europe. June–September.
Description.—Stems erect, branched, 2–4 dm. high. Leaves opposite, simple, linear, hairy. Flowers in dense cymose clusters; calyx subtended by 2 or more awl-shaped bracts, of 5 sepals fused into a tube with 5 teeth; petals 5, rose-colored, with white dots, margin crenate; pistil with 1-celled ovary and free-central placenta; styles 2. Capsule 4-valved, many-seeded; seeds about 1.3 mm. long, curved, roughened by minute tubercles, dull, black.
Control.—Clean cultivation (6). In grasslands, hand pulling (4) or early close mowing to prevent seed production (11) will control this weed.

153. *Lychnis alba* Mill. White cockle, White campion, Evening lychnis, Snake cuckoo, Thunder flower, Bull rattle, White robin. Fig. 35, a–e.
Biennial or perennial; reproducing by seeds and short rootstocks. Grasslands and new clover and alfalfa seedings. Widespread and locally common; especially on rich well-drained soils, in the eastern states, north central states and the Pacific Northwest, also in southern Canada. Introduced from Europe. June–August.
Description.—Stems spreading or nearly erect, hairy, jointed. Leaves opposite, simple, ovate to lanceolate, hairy. Flowers dioecious, solitary on long peduncles or in cymose clusters; sepals 5, fused into a tube with 5 long tapering teeth; petals 5, white or pink, much ex-

ceeding the calyx, opening in the evening; stamens 10; pistil 1, with free-central placenta; styles 5. Capsule ovoid-conical, many-seeded; seeds short kidney-shaped, about 1.4 mm. long; hilum surrounded by a raised ridge; surface roughened by blunt tubercles forming irregular gray lines where their bases meet, gray to light tan.

Control.—Sow pure clover and alfalfa seed (1). Plow badly infested fields and follow with a clean cultivated crop in a short rotation (7). Scattered plants should be pulled or hoed off early to prevent seed production (4, 5).

154. *Lychnis dioica* L. Red campion, White soapwort, Red birdseye, Red robin.

Biennial; reproducing by seeds. Meadows, fields and waste places. Infrequent in eastern Canada and the eastern United States. Introduced from Eurasia. June–September.

Description.—Similar to 153. Flowers dioecious or polygamous; calyx-tube with 5 triangular-lanceolate acute teeth; petals red or rarely white, opening in the daytime. Capsule globose, many-seeded; seeds similar to 153, dark violet-gray, with pointed tubercles.

Control.—The same as for 153.

155. *Lychnis flos-cuculi* L. Ragged robin, Meadow pink, Meadow campion, Cuckoo flower, Crow flower, Indian pink, Ragged jade. Fig. 35, f–j.

Perennial; reproducing by seeds and short rootstocks. Moist rich meadows and pastures. Locally common in the middle Atlantic states, also northward to Canada and westward to Iowa. Introduced from Europe. June–July.

Description.—Stems from short rootstocks, erect, 3–6 dm. high, downy below, somewhat sticky above. Leaves opposite, simple, narrowly lanceolate, entire. Flowers perfect, in open panicles; sepals fused into a short tube with 4–5 teeth; petals 5, red, each with 4 linear lobes; stamens 10; pistil with 5 styles and a 1-celled ovary. Capsule conical, many-seeded; seeds 0.8 mm. long, short kidney-shaped, roughened by curved rows of small pointed tubercles, violet-gray.

Control.—The same as for 153.

156. *Saponaria officinalis* L. Bouncing bet, Soapwort, Fullers herb, Scourwort, Old maids pink, London pride, Hedge pink, Wild sweet william, Sweet betty. Fig. 34, f–k.

Perennial; reproducing by seeds and short rootstocks. Roadsides, railroad embankments, waste places and neglected fields. Common

FIG. 35—**WHITE COCKLE,** *Lychnis alba:* **A,** plant showing habit (× ⅓); **B,** staminate flower (× 1); **C,** pistillate flower (× 1); **D,** pistil (× 1); **E,** seed (× 10).
RAGGED ROBIN, *Lychnis flos-cuculi:* **F,** plant showing habit (× ⅓); **H,** flower (× 1); **J,** pistil (× 2).

throughout eastern North America, local on the Pacific Coast. Introduced from Europe. July–September.

Description.—Stems stout, erect, simple, jointed, from a short rootstock. Leaves opposite, simple, sessile, ovate-lanceolate, entire. Flowers perfect, in dense, terminal, cymose clusters; sepals fused into a long tube with 5 very small unequal teeth; petals 5 (sometimes double), pink to rose-colored, rarely white, with appendages on the claws; stamens 10; pistil with 2 styles. Capsule opening with 4 teeth, 1-celled or 2–4-celled at the base, with numerous seeds; seeds about 1.8 mm. long, short kidney-shaped, surface roughened by tubercles, black.

Control.—Small areas are best disposed of by hand digging (17). Larger areas should be mowed close to the ground every time the first flowers open (11).

Bouncing bet is somewhat poisonous. If its crushed leaves are stirred in water, they form a soapy emulsion due to the saponin glucoside contained in them.

157. *Saponaria vaccaria* L. Cow cockle, Cow-herb, Spring cockle, Pink cockle. Fig. 34, a–e.

Annual; reproducing by seeds. Chiefly in grain fields, occasionally in new alfalfa and clover seedings, waste places. Common in the grain sections of the northwestern states and western Canada. Elsewhere often appearing wherever western feed is used but usually it does not spread or persist. Introduced from Eurasia. June–July.

Description. Stems erect, stout, much branched above, jointed, glaucous, from a narrow tap-root. Leaves opposite, simple, sessile, ovate to lanceolate, glaucous. Flowers perfect, in open cymose clusters; calyx 5-toothed and 5-angled, enlarging as it matures; petals 5, pale red; stamens 10; pistil with 2 styles. Capsule opening with 4 teeth, 1-celled, many-seeded; seeds globular, 2 mm. in diameter, covered with very minute tubercles, except at the whitish, nearly circular, depressed hilum, dull, black.

Control.—Sow clean wheat seed (1); harrow grain fields to destroy seedlings (8); pull scattered plants by hand (4). Practice a short rotation with cultivated crops in which wheat is not grown oftener than once in three years.

Cow cockle seeds have poisonous properties similar to corn cockle.

158. *Scleranthus annuus* L. Knawel, German knot-grass. Fig. 32, k–m.

Annual or winter annual; reproducing by seeds. Gardens, lawns, fields and waste places; on dry, gravelly or sandy soils. Eastern United States and Canada; west to Minnesota and south to Florida and Mississippi; infrequent on the Pacific Coast; most common along the Atlantic Coast. Introduced from Europe. May–October.

Description.—Stems much branched, low, spreading, 5–10 cm. high. Leaves opposite, simple, awl-shaped, without stipules. Flowers small, green, sessile, in axillary clusters; sepals 5, fused below into a cup which surrounds the 1-seeded fruit; stamens 5 or 10, on the calyx; pistil 1, with 2 styles. Fruit a utricle, inclosed in the 5-toothed, 10-angled, straw-colored calyx, about 3–4 mm. long.

Control.—Early and clean surface cultivation to prevent seed formation (6).

159. *Silene antirrhina* L. Sleepy catchfly, Tarry cockle. Fig. 36, a–c.

Annual; reproducing by seeds. Meadows, fields, open woods and waste places; mostly on dry gravelly or sandy soils. Widespread, but only locally common, especially in the northeastern states. Native. May–July.

Description.—Stems very slender, erect, 2–8 dm. high, jointed, glabrous but with a sticky dark band on each internode. Leaves opposite, simple, linear or lanceolate. Flowers perfect, in panicles; sepals 5, fused into a 10-nerved tube; petals 5, obcordate, ephemeral; stamens 10; pistil with 3 styles. Capsule 1-celled, many-seeded; seeds 0.6 mm. long, short kidney-shaped, surface roughened by curved rows of minute blunt tubercles, violet-gray.

Control.—Badly infested meadows should be mowed early before seeds are formed (11), follow by fall plowing and plant a clean cultivated crop the following year (6). Scattered plants can be pulled or hoed by hand (4, 5); harrow grain fields (8).

160. *Silene cserei* Baumg. Campion. Fig. 37, a–g.

Biennial; reproducing by seeds. Railroads, roadsides, and waste places; spreading rapidly. New York to Montana, south to New Jersey and Missouri. Introduced from Europe. May–September.

Description.—Stems 1–few, erect. Leaves, like those of *S. cucubalus,* thick and glaucous, 2–4 cm. wide; the cauline leaves clasping, oblong or oblong-ovate. Flowers in simple elongate narrow panicle or in panicle with 2 slender, erect branches; calyx in fruit cylindric-ellipsoid, little inflated, veinlets not conspicuously reticulate; capsule subsessile and exserted. Seeds 0.6–1 mm. long, papillate.

Control.—The same as for 163.

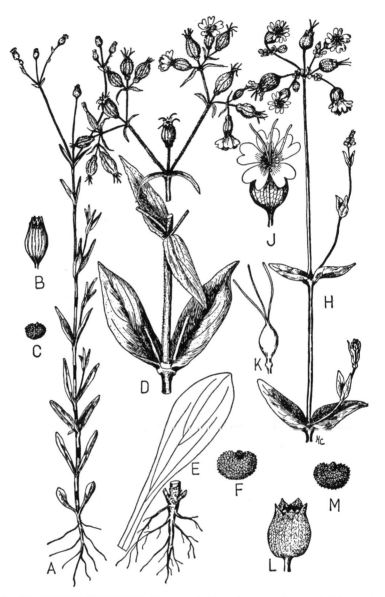

Fig. 36—**SLEEPY CATCHFLY,** *Silene antirrhina:* **A,** plant showing habit (× ⅓); **B,** capsule (× 1); **C,** seed (× 2).
NIGHT-FLOWERING CATCHFLY, *Silene noctiflora:* **D,** portion of plant showing habit (× ⅓); **E,** basal leaf (× ⅓); **F,** seed (× 6).
BLADDER CAMPION, *Silene cucubalus:* **H,** portion of flowering shoot (× ⅓); **J,** flower (× 1); **K,** pistil (× 3); **L,** capsule (× 1); **M,** seed (× 6).

Pink Family

161. *Silene cucubalus* Wibel (*S. latifolia* Britten and Rend., *S. inflata* Sm.). Bladder campion, White bottle, Cow-bell, Bubble poppy. Fig. 36, h–m.

Perennial; reproducing by seeds and by shoots developing from roots. Meadows and clover fields, waste places; mostly on sandy or gravelly soils. Common in the northeastern states, local in the Pacific Northwest; infrequent elsewhere. Introduced from Eurasia. June–September.

Description.—Stems spreading or erect, 3–10 dm. high, branched, jointed, glaucous, from a short rootstock. Leaves opposite, simple, ovate-lanceolate, entire or the rosette leaves denticulate, glaucous. Flowers perfect, in panicle-like cymes; calyx of 5 fused sepals, nearly globular, pale green or whitish, much inflated; petals 5, white, 2-cleft; stamens 10; pistil with 3 styles. Capsule 1-celled, many-seeded; seed similar to 153, mouse-gray, 1.3 mm. long.

Control.—Sow pure clover and alfalfa seed (1). Plow badly infested fields and follow with a clean cultivated crop in a short rotation (7). Scattered plants may be carefully dug out to destroy all roots (17).

162. *Silene dichotoma* Ehrh. Forked catchfly, Hairy catchfly.

Annual or winter annual; reproducing by seeds. Meadows, pastures and new seedings; on gravelly soils. Widespread and locally common in the eastern United States; also on the Pacific Coast. Introduced from Europe. June–July.

Description.—Stems erect or spreading, branched, jointed, hirsute, sticky. Leaves opposite, simple, lanceolate or oblanceolate, entire. Racemes compound; flowers perfect, on short pedicels; calyx-tube with 10 ribs, hirsute; petals 5, white or pink; pistil with 3 styles. Capsule 1-celled, many-seeded; seeds about 1.5 mm. long, short kidney-shaped, sides flattened, roughened by rows of blunt tubercles, dull gray.

Control.—The same as for 159.

163. *Silene gallica* L. (*S. anglica* L.). English catchfly, Windmill pink.

Annual; reproducing by seeds. Grain fields, gardens and waste places. Locally common on the Pacific Coast; infrequent in the middle western and eastern states. Introduced from Europe. May–June.

Description.—Stems erect, slender, sparingly branched, jointed,

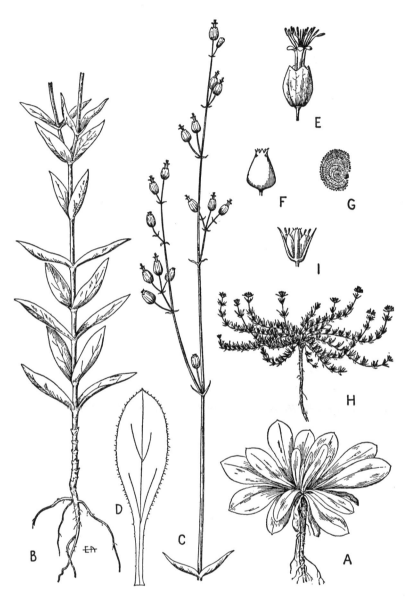

Fig. 37—**CAMPION,** *Silene cserei:* **A,** rosette stage (× ⅓); **B-C,** habit of flowering plant (× 1); **D,** rosette leaf (× 1); **E,** flower (× 1); **F,** capsule (× 1); **G,** seed (× 8). **RED SAND-SPURRY,** *Spergularia rubra:* **H,** habit (× ⅓); **I,** flower (× 2).

pubescent, somewhat viscid, 3–6 dm. high. Leaves opposite, spatulate, mucronate. Racemes simple, one-sided; flowers similar to 162 with ovoid calyx and smaller corolla. Seeds 1 mm. long, short kidney-shaped, sides flattened and ridged, surface roughened by curved rows of small tubercles, purplish-gray.

Control.—Badly infested meadows should be mowed early before seeds are formed (11), follow by fall plowing and plant a clean cultivated crop the following year (6). Scattered plants can be pulled or hoed by hand (4, 5); harrow grain fields (8).

164. *Silene noctiflora* L. Night-flowering catchfly, Clammy cockle, Sticky cockle. Fig. 36, d–f.

Annual or winter annual; reproducing by seeds. Meadows and new clover and alfalfa seedings; mostly on rich gravelly soils. Widespread and common in eastern North America; recently becoming locally common in the Pacific Northwest. Introduced from Europe. June–September.

Description.—Stems erect, stiff, branched above, 3–10 dm. high, jointed, covered with sticky hairs. Leaves opposite, simple, spatulate to lanceolate, entire. Flowers perfect, in open cymes or solitary on long peduncles; calyx-tube hairy, 5–10 nerved, with sharp teeth; petals 5, white or yellowish, opening at night; pistil with 3 styles. Capsule 1-celled, many-seeded; seeds about 1.2 mm. long, short kidney-shaped, hilum between 2 knob-like raised projections; surface roughened by curved rows of small blunt tubercles, their bases forming wavy dark brown lines, gray.

Control.—Sow pure clover and alfalfa seed (1). Badly infested meadows should be mowed early before seeds are formed (11), follow by fall plowing and plant a clean cultivated crop the following year (6). Scattered plants can be pulled or hoed by hand (4, 5); harrow grain fields (8).

165. *Spergula arvensis* L. Spurry, Devils-gut, Sandweed, Pick-purse, Yarr. Fig. 38, a–c.

Annual; reproducing by seeds. Grain fields, cultivated fields and gardens; especially on sandy or gravelly soils. Very abundant in the Pacific Northwest, widespread and locally common throughout the eastern states and Canada. Introduced from Europe. June–August.

Description.—Stems very slender, jointed, cord-like, much branched and spreading, glabrous or slightly sticky. Leaves whorled,

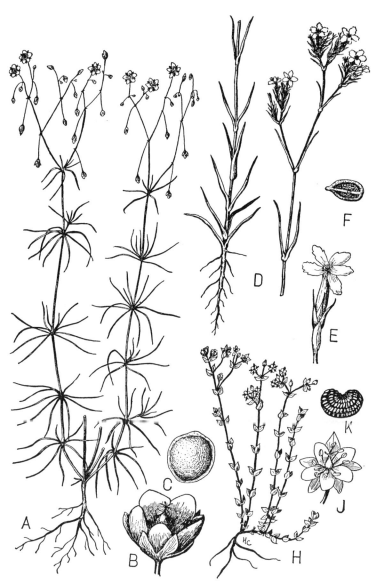

Fig. 38—**SPURRY,** *Spergula arvensis:* **A,** plant showing habit (× ⅓); **B,** flower (× 2); **C,** seed (× 10).
DEPTFORD PINK, *Dianthus armeria:* **D,** plant showing habit (× ⅓); **E,** flower subtended by bracts (× 1); **F,** seed (× 10).
THYME-LEAVED SANDWORT, *Arenaria serpyllifolia:* **H,** plant showing habit (× ½); **J,** flower (× 3); **K,** seed (× 20).

Pink Family 215

thread-like, bright green, with minute stipules. Flowers small, perfect, in open cymes; sepals 5, nearly separate; petals 5, white; stamens 5–10; pistil with 5 styles, 5-valved. Capsule 1-celled, many-seeded; seeds 1.3 mm. in diameter, lens-shaped, dull, black with a light brown winged margin, surface roughened by minute tubercles and flecked with light brown prickles.

Control.—Disk or harrow grain fields after the crop is removed, to induce seeds to germinate (8). Clean cultivation (6), followed by hand hoeing and hand pulling (4, 5).

166. *Spergularia rubra* (L.) J. and C. Presl. Red sand-spurry. Fig. 37, h–i.

Annual or short-lived perennial; reproducing by seed. Wasteland weed on sandy or gravelly soil. Newfoundland to British Columbia, south to Alabama and California. Introduced from Europe. April–October.

Description.—Stems simple or much branched, up to 3.5 dm. high, often diffusely matted or prostrate, glabrous or somewhat glandular-pubescent. Leaves in fascicles, mucronate, scarcely fleshy, linear-filiform, up to 2.5 cm. long; conspicuous, scarious, lanceolate stipules longer than broad. Flowers in small, leafy, many-flowered cymes; petals pink or reddish, shorter than the sepals; sepals 3.5–5 mm. long, oblong-lanceolate, usually pubescent; stamens 10 or as few as 6. Capsule 3.5–5 mm. long. Seeds 0.4–0.6 mm. long, deeply sculptured, dark brown, not winged.

Control.—The same as for 165.

167. *Stellaria graminea* L. Grass-leaved stitchwort, Grassy starwort. Fig. 39, e–g.

Perennial; reproducing by seeds and creeping stems. Damp grassy fields; in sandy or gravelly soils. Locally common in eastern Canada and the northeastern United States. Introduced from Eurasia. May–July.

Description.—Stems slender, ascending or nearly prostrate, angular, glabrous. Leaves opposite, simple, narrowly lanceolate, broadest near the base, entire. Flowers perfect, in long-stalked terminal, spreading, many-flowered cymes; sepals 4–5, separate, spreading; petals 4–5, 2-cleft, white, about 5 mm. long; stamens usually 10, sometimes 8 or less; pistil with 3 (rarely more) styles. Capsule ovoid, 1-celled, opening by about 6 valves, many-seeded; seeds about 1 mm. in diameter, nearly circular to broadly ovate with a con-

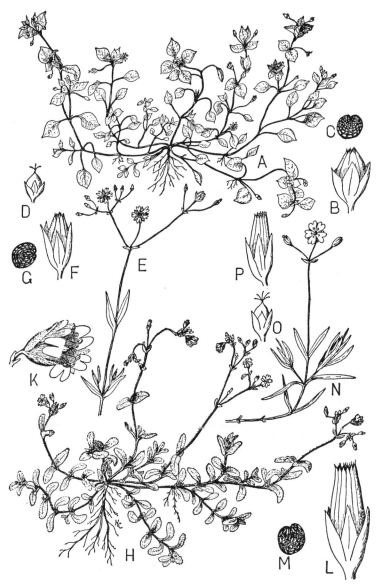

Fig. 39—**CHICKWEED**, *Stellaria media:* **A,** plant showing general habit (× ¼); **B,** capsule (× 2½); **C,** seed (× 6); **D,** pistil (× 1).
GRASS-LEAVED STITCHWORT, *Stellaria graminea:* **E,** branch showing leaves and flowers (× ¼); **F,** capsule (× 2½); **G,** seed (× 6).
MOUSE-EAR CHICKWEED, *Cerastium vulgatum:* **H,** plant showing general habit (× ¼); **K,** flower (× 2); **L,** capsule (× 2½); **M,** seed (× 12).
FIELD CHICKWEED, *Cerastium arvense:* **N,** branch showing leaves and flowers (× ¼); **O,** pistil (× 1); **P,** capsule (× 2½).

spicuous notch, sides flat or convex, roughened by small irregular ridges, dull, gray-brown.

Control.—The same as for 168.

168. *Stellaria media* Cyrill (*Alsine media* L.). Chickweed, Starwort, Starweed, Bindweed, Winter-weed, Satin flower, Tongue-grass. Fig. 39, a–d.

Annual or winter annual; reproducing by seeds and creeping stems rooting at the nodes. Gardens, cultivated fields and new lawns; on rich soils. Common throughout North America wherever gardens or cultivated fields are planted. Introduced from Europe. March–December.

Description.—Stems low, slender, much branched, creeping or ascending, with rows of hairs, rooting at the nodes. Leaves opposite, simple, ovate to oblong, entire, the lower on hairy petioles, the upper mostly sessile. Flowers perfect, axillary or in cymose clusters; sepals 5, separate; petals 5, 2-parted, shorter than the sepals; stamens 5 (3–10); pistil with 3–4 styles. Capsule 1-celled, many-seeded; seeds 1 mm. in diameter, nearly circular to ovate, with a conspicuous notch, sides slightly convex, roughened by curved rows of small tubercles, dull, reddish-brown.

Control.—Clean cultivation, especially while the seedlings are small (6); hand weeding (4). In new lawns rake and mow close (11).

The common chickweed makes its most vigorous growth during cool weather and when considerable moisture is available. In mild regions and in mild seasons in the North it blossoms and produces seeds throughout the winter.

RANUNCULACEAE (Crowfoot Family)

169. *Adonis annua* L. Pheasants eye, Birds-eye.

Annual; reproducing by seeds. Fields, along banks and in waste places. Introduced from Europe as an ornamental. Occasionally escaped in the northeast and the south, chiefly in the lower Mississippi Valley; local elsewhere. Late summer.

Description.—Stems 2–7 dm. high. Leaves alternate, sessile or subsessile, finely dissected; leaf-blades 2–5 cm. long. Flowers solitary at the end of the stem or its branches; petals 6–10, orange or red, usually with a dark spot at base; sepals 5, green, 6–9 mm. long. Head of achenes cylindric, 1–2 cm. long; achenes rough, 3–5 mm. long.

Control.—The same as for 172.

170. *Delphinium menziesii* DC. Low larkspur, Poisonweed, Cowpoison, Staggerweed, Peco, Stavesacre. Fig. 40, a–d.

Perennial; reproducing by seeds and fleshy tuberous roots. Open hillsides and ranges, especially at higher altitudes; also in sandy or gravelly fields and meadows on lowlands. Rocky Mountains westward to the Pacific Coast. Native. May–June.

Description.—Stems erect, simple, or sparingly branched above, hairy or nearly glabrous, 2–5 dm. high, from a cluster of tuberous roots. Leaves few, alternate, palmately lobed or dissected into linear segments, long-petioled or sessile above, slightly pubescent or glabrous. Flowers perfect, irregular, in few-flowered open racemes, mostly deep blue or rarely pale blue or yellowish-white; pedicels slender, spreading, with 1–2 small bracts; calyx of 5 sepals, the upper spurred; corolla of 2–4 petals, the upper 2 spurred; stamens several; carpels 3, each developing into a several-seeded, spreading, pubescent follicle. Seeds 2 mm. long, somewhat angled, truncate at base, winged around the hilum; glossy, dark with light brown wings.

Control.—Grub out the plants and tuberous roots when they begin to blossom. Graze with sheep.

171. *Delphinium strictum* A. Nels. (*D. simplex* Dougl.). Larkspur. Fig. 40, e.

Perennial; reproducing by seeds. Meadows and ranges in Idaho and eastern Washington and Oregon. Native. June–July.

Description.—Stems erect, simple or branched above, more leafy than 170, 5–10 dm. high. Leaves alternate, the basal petioled and with broader lobes, the upper sessile and with linear lobes. Racemes close, often branched, pedicels shorter than the spurs; flowers similar to 170. Seeds similar to 170, but smaller.

Control.—The same as for 170.

Delphinium menziesii, D. strictum and several other native species common on the higher ranges and grazing lands in the western United States, are poisonous to cattle and horses. Sheep seem to be able to eat Delphinium without much harm.[1, 2]

172. *Ranunculus abortivus* L. Small-flowered buttercup, Smooth-leaved crowfoot. Fig. 41, a–c.

Biennial; reproducing by seeds. Lowland meadows, pastures and fields; on gravelly soils. Common in eastern North America; also found in north central states. Native. April–July.

[1] Marsh, Clawson and Marsh, 1916. [2] Clawson, 1933.

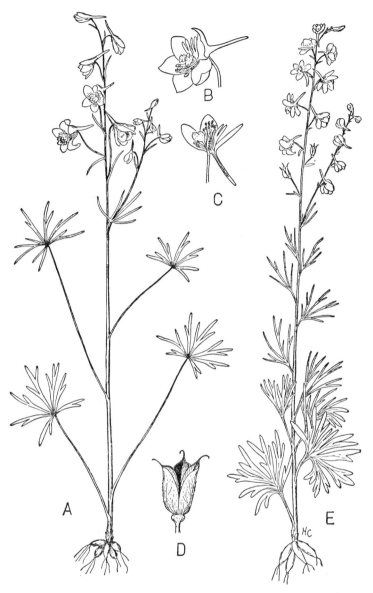

Fig. 40—**LOW LARKSPUR,** *Delphinium menziesii:* **A,** plant showing habit (× ⅓); **B,** flower (× ½); **C,** vertical section of flower (× ½); **D,** follicles (× 1). **LARKSPUR,** *Delphinium strictum:* **E,** plant showing habit (× ⅓).

Description.—Stems erect, sparingly branched, 2–6 dm. high, somewhat fleshy, glabrous or slightly hairy. Leaves on the stem alternate, 3-lobed or parted, nearly sessile, the basal leaves petioled, mostly heart-shaped or kidney-shaped, with crenate margins. Flowers perfect, regular, solitary or in corymbose clusters, 1 cm. in diameter or less; sepals 3–6, green, reflexed; petals usually 5, pale yellow, shorter than the sepals; stamens numerous; carpels numerous, separate, with sessile stigmas. Achenes numerous, in a globular cluster or head, about 1.5 mm. long, flattened, with a short curved beak, dull, wrinkled, yellowish-brown.

Control.—Plow badly infested fields and follow by a clean cultivated crop (6). Meadows and pastures should be mowed close as soon as the first flowers appear (11).

173. *Ranunculus acris* L. Tall field buttercup, Meadow buttercup, Tall crowfoot, Butter flower, Blister plant, Gold cup, Butter-rose. Fig. 41, d–f.

Perennial; reproducing by seeds. Pastures and meadows; chiefly on heavy moist soils. Widespread; most abundant in the northeastern states, becoming common in southwestern British Columbia and northwestern Washington. Introduced from Eurasia. June–July.

Description.—Stems erect, often in clusters, branched, hairy, about 5–10 dm. high, from thick fibrous roots. Leaves alternate, palmately 3-divided, the divisions sessile and parted into narrow segments. Flowers perfect, regular, solitary or in cymose clusters; calyx mostly of 5 separate spreading sepals; petals 5–7, obovate, bright yellow, much exceeding the sepals; stamens numerous; carpels numerous, separate, with short, recurved, persistent styles. Achenes numerous, in globular heads, 3 mm. long, flattened, with a prominent nearly straight beak, dull, minutely pitted, dark brown, margin often lighter.

Control.—Since field buttercups thrive best in wet soil, first of all improve the drainage (21). Field buttercups will not thrive under cultivation. Badly infested meadows should be plowed and followed by clean cultivation for a year (6). Fertilize and harrow pastures in the spring, mow the clumps of buttercups as soon as the first flowers bloom and again when the second crop appears.

The field buttercup has a very acrid juice which acts as an irritant when eaten by stock, often blistering the mouth or intestinal tract. Animals usually do not touch it if there is plenty of other vege-

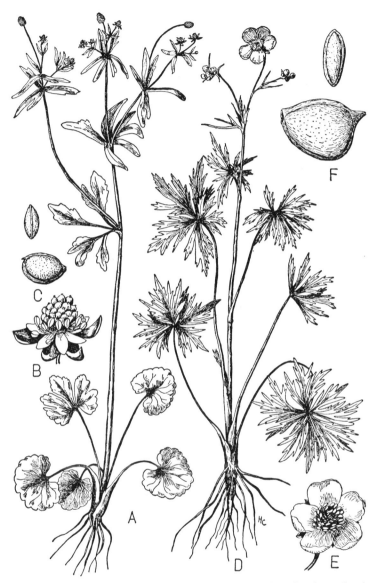

Fig. 41—**SMALL-FLOWERED BUTTERCUP,** *Ranunculus abortivus:* **A,** plant showing habit (× ⅕); **B,** flower (× 3); **C,** achenes (× 10).
TALL FIELD BUTTERCUP, *Ranunculus acris:* **D,** plant showing habit (× ⅙); **E,** flower (× 1); **F,** achenes (× 10).

tation. For this reason it produces seed freely and often overruns pastures.

174. *Ranunculus arvensis* L. Field buttercup, Hunger weed. Fig. 43, a–c.

Annual; reproducing by seeds. Waste places. New Jersey, North Carolina, Ohio, Utah, Oregon and California; occasional along the Atlantic Coast. Introduced from Europe. May–September.

Description.—Stems erect, mostly solitary, branched above, 1.5–5 dm. high, sparsely pubescent. Leaves divided deeply into many linear and entire or narrowly cuneate and incised lobes; lower leaves petioled. Flowers on long peduncles; petals 5, obovate, yellow, 5–8 mm. long; sepals lanceolate, 5–7 mm. long, pubescent, yellowish; stamens 10–15. Achenes short-stalked, obovate, 4.5–10 mm. long, bearing numerous spines up to 2.5 mm. long; beak subulate, 2–4 mm. long, flat, slightly curved.

Control.—The same as for 172.

175. *Ranunculus bulbosus* L. Bulbous crowfoot, Bulbous buttercup, Yellow weed, Blister flower. Fig. 42, d–e.

Perennial; reproducing by seeds and bulb-like corms. Meadows and pastures; mostly on limy or gravelly soils. Locally common in the northeastern United States; infrequent elsewhere. Introduced from Eurasia. June–July.

Description.—Similar to 173. Stems erect, 2–4 dm. high, from a short thick bulb-like base. Basal leaves 3-divided, the middle division stalked, the lateral ones sessile. Sepals 5, reflexed; petals 5–7, round, deep glossy yellow, wedge-shaped at the base; carpels numerous, with short recurved persistent styles. Achenes numerous, occurring in globular heads, similar to 173 but with a hooked beak.

Control.—The same as for 173.

176. *Ranunculus repens* L. Creeping buttercup. Fig. 42, f–h.

Perennial; reproducing by seeds and runners. Moist meadows and pastures, especially on mucklands; lawns and along ditches. Widespread in the northeastern states and very abundant on the north Pacific Coast. Introduced from Europe. June–August.

Description.—Stems low, hairy, creeping and rooting at the nodes. Leaves alternate, long-petioled, 3-divided or 3-lobed, hairy, dark green or sometimes with light spots. Flowers perfect, regular, solitary or in corymbose clusters; sepals 5, separate, green, not reflexed; petals 5–7, bright glossy yellow, about 1 cm. long; stamens numerous;

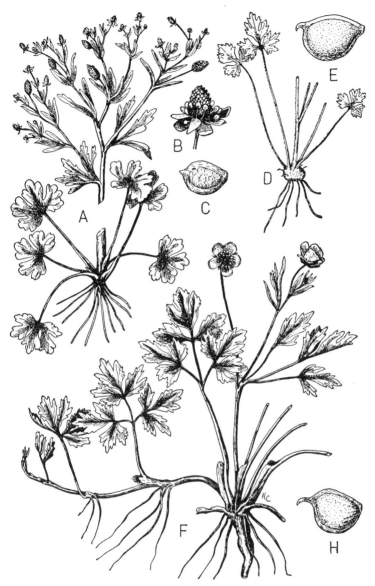

Fig. 42—**CURSED CROWFOOT**, *Ranunculus sceleratus:* **A**, plant showing general habit (× ⅓); **B**, flower (× 3); **C**, achene (× 8).
BULBOUS CROWFOOT, *Ranunculus bulbosus:* **D**, basal part of plant showing bulbous stem (× ⅓); **E**, achene (× 6).
CREEPING BUTTERCUP, *Ranunculus repens:* **F**, plant showing creeping habit (× ⅓); **H**, achene (× 6).

carpels numerous, separate, with short recurved styles. Achenes numerous, in globular heads, 2.5 mm. long, with a hooked beak, blackish-brown.

Control.—Plow infested fields and follow with a clean cultivated crop (6) for one or two years.

177. *Ranunculus sceleratus* L. Cursed crowfoot, Celery-leaved crowfoot, Bog buttercup. Fig. 42, a–c.

Annual or winter annual; reproducing by seeds. Wet pastures, meadows and fields, along ditches, and near shores of streams and ponds. Widespread and locally common throughout the United States and southern Canada. Native to North America and Eurasia. May–July.

Description.—Stems erect, branched above, glabrous, pale green, stout, somewhat fleshy, often hollow. Leaves alternate, glabrous, the upper 3-parted or compound with 3–5 narrow sessile leaflets; basal leaves palmately parted or lobed with blunt toothed lobes. Flowers similar to 172. Achenes very numerous, on a cylindrical or oblong head, about 1 mm. long, with a short straight beak, granular or ridged on the sides, yellow.

Control.—Since the cursed crowfoot usually causes most trouble in situations where cultivation is not practicable, the plants should be destroyed by pulling or mowing as soon as the first flowers appear. The plants must be piled in a heap and composted or placed where they will dry out, otherwise they will grow again.

178. *Ranunculus testiculatus* Crantz. Buttercup. Fig. 43, d–e.

Perennial; reproducing by seeds. Along dry roadsides. Sagebrush regions in the Northwest. Introduced from Europe. Summer.

Description.—Stem pilose, less than 1 dm. high. Leaves all basal; blade extending as wing the length of petiole, 3-parted, divisions cleft. Sepals persistent, tomentose, elliptical, 3–5 mm. long; petals 5, oblanceolate, yellow, 4–6 mm. long. Achenes many in long, narrow spike, each achene beak lanceolate, tomentose, 4 mm. long, 1 mm. broad.

Control.—The same as for 172.

The buttercups are seldom eaten by animals because they contain an acrid juice which is a disagreeable irritant and somewhat poisonous. Dried buttercup plants in hay are harmless because the poisonous principle is volatile and is dissipated when the hay is cured. The juice of some of the common buttercups will raise

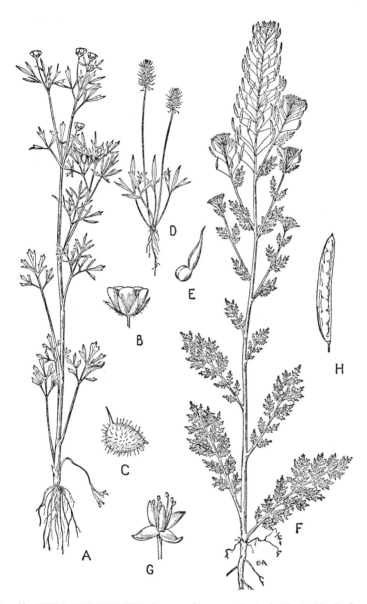

Fig. 43—**FIELD BUTTERCUP,** *Ranunculus arvensis:* **A,** habit (× ⅓); **B,** flower (× 1); **C,** fruit (× 3).
BUTTERCUP, *Ranunculus testiculatus:* **D,** habit (× ⅓); **E,** fruit (× 3).
TANSY MUSTARD, *Descurainia pinnata:* **F,** habit (× ⅓); **G,** flower (× 4); **H,** fruit (× 2).

blisters on the skin of many individuals when they come in contact with it.

BERBERIDACEAE (Barberry Family)

179. *Berberis vulgaris* L. Barberry, European barberry. Fig. 59, g–h.

Shrubby perennial; reproducing by seeds and new shoots from spreading roots. Gravelly and stony hillside pastures, fence-rows, woodlands and waste places. Locally common in the northeastern and north central states. Introduced from Europe. May–June.

Description.—Stems woody, branched, forming shrubs 1–3 m. high; wood and inner bark yellow. Leaves alternate on the long twigs of the season, reduced to 3-branched spines near the end of the season's growth; leaves on the short spur-like shoots in false whorls or clusters; leaves simple, with toothed margins. Flowers in drooping axillary racemes, perfect, regular; sepals 6, separate; petals 6, separate, yellow; stamens 6; pistil with a very short stigma; ovary superior. Fruit an ellipsoid red berry with 1 or a few seeds; seeds 5–6 mm. long, obovate, one side flattened, the other rounded, reticulated or nearly smooth, chocolate-brown.

Control.—Grub out scattered bushes, being sure to get all the roots, otherwise they will send up new shoots.

The common barberry is an alternate host for the black stem-rust of wheat and other cereals, harboring the cluster cup stage of the rust. Other species of Berberis and the related Mahonia serve as rust hosts in the northwest. The United States Department of Agriculture, in its war on the barberry in the wheat sections of the north central states, has destroyed many millions of bushes.[1, 2]

PAPAVERACEAE (Poppy Family)

180. *Argemone mexicana* L. Mexican poppy, Prickly poppy.

Annual or winter annual; reproducing by seeds. Fields, meadows and waste places. Locally established in the eastern states, more common in the southern and southwestern states. Introduced from Mexico. June–September.

Description.—Stems erect, stout, sparingly branched, usually prickly, with yellow juice, 3–8 dm. high. Leaves alternate, coarsely lobed and spiny along the margin and midrib, often glaucous and

[1] Schulz and Thompson, 1925.   [2] Kempton and Thompson, 1925.

Poppy Family

blotched with white, mostly clasping at the base. Flowers solitary, showy; calyx of 2–3 prickly sepals, falling as the flower opens; corolla of 4–6 separate lemon-yellow or cream-colored petals, 1–3 cm. long; stamens many, distinct; pistil solitary, with sessile 3–6 radiate stigma on a very short style; ovary maturing into an ellipsoid prickly capsule with 3–6 valves and many seeds. Seeds nearly 2 mm. long, ovoid-spherical, surface with angular depressions and a crest along one side, blackish-brown.

Control.—Hand pulling or hoeing (4, 5). Clean cultivation (6).

Argemone intermedia Sweet. Prickly poppy, native from southern Illinois southward and westward, sometimes becomes a weed. It has a larger white corolla and more glaucous foliage than 180. Control same as for 180.

181. *Chelidonium majus* L. Celandine, Great celandine, Swallow-wort, Wartweed.

Biennial; reproducing by seeds. Fields, neglected yards, roadsides and waste places; on damp, rich, gravelly soil. Northeastern United States southward to Georgia, Tennessee and Missouri. Introduced from Eurasia. May–July.

Description.—Stems erect, branched, spreading, hairy, fleshy, and very brittle, with an orange-colored acrid juice. Leaves alternate, 1- or 2-pinnately divided or irregularly cut, hairy. Flowers perfect, in umbels; sepals 2, falling very soon; petals 4, separate, yellow; stamens about 20; pistil 1, with an almost sessile 2-lobed stigma. Capsule long, cylindrical, opening from the base into 2 narrow valves, each with several seeds; seeds about 1.4 mm. long, ovoid, surface marked with rows of square pits, glossy, gray-brown to black, with a yellow crest on one side.

Control.—Hand pulling or hoeing (4, 5). Clean cultivation (6).

182. *Papaver rhoeas* L. Corn poppy, Field poppy, Red poppy.

Annual; reproducing by seeds. Waste places, ballast grounds, rarely in grain fields and new seedings of grass or clover. Occasional, in the eastern and southern United States, but it does not seem to spread much. Introduced from Europe. June–August.

Description.—Stems erect, sparingly branched, bristly-hairy, with milky juice. Leaves alternate, pinnately lobed, hairy. Flowers perfect, regular, solitary on long peduncles; flower-buds nodding; sepals mostly 2, falling as the buds open; petals 4, separate, bright red, often darker at the base; stamens numerous; pistil with a flat 4–20-rayed stigma. Capsule obovoid, with 4–20 parietal, many-seeded placentae which appear like imperfect partitions, opening with several pores under the edge of the stigma; seeds 0.7 mm. long, kidney-shaped, slightly flattened, marked with rows of depressions, purplish-gray.

Control.—Hand pulling or hoeing (4, 5).

Papaver dubium L. Field poppy. Introduced from Europe, similar to 182 but taller and with more finely divided leaves and with a smooth club-shaped capsule; occurs locally in fields, grasslands and waste places in the eastern and southern states and along the Pacific Coast.

Roemeria refracta, Roemeria poppy, is a noxious weed (1954) in Utah.

Fumaria officinalis L. Fumitory, of the family Fumariaceae. An introduced annual in fields and waste places, mostly in the southern states and locally from western Washington to California. Its flowers are borne in racemes, have a purplish 1-spurred corolla, and the pistil develops into a small globular 1-seeded nutlet.

CAPPARIDACEAE (Caper Family)

183. *Cleome serrulata* Pursh. Rocky Mountain bee-plant, Stinking clover.

Annual; reproducing by seeds. Dry fields, pastures, ranges and waste places. North central states westward and northward. Native. July–September.

Description.—Stems erect, 5–20 dm. high, nearly glabrous, branched above. Leaves alternate, with 3 lanceolate nearly entire leaflets, glabrous, with a strong fetid odor. Flowers in terminal bracteate racemes, perfect; sepals 4, separate; petals 4, separate, clawed and entire, rose-purple to white; stamens 6, separate; pistil solitary, 1-celled. Capsule borne on an elongated stipe, linear to oblong, recurved, many-seeded; seeds 3–4 mm. long, obovate, with a projection at base, with a groove on each side from hilum to apex, surface dull, roughened, gray-brown.

Control.—Hoe or pull scattered plants before seeds are formed (5, 4). In large areas a rotation including a cultivated crop every three or four years will control this weed (20).

184. *Polanisia graveolens* Raf. Clammy-weed, Stinkweed, Stinking clover.

Annual; reproducing by seeds. Sandy or gravelly fields, along banks of streams and lakes, waste places. Local, Quebec to Maryland, westward to the Great Plains. Native. July–August.

Description.—Stems erect, 2–7 dm. high, with many ascending branches, with glandular-sticky hairs. Leaves alternate, with 3 oblong leaflets. The whole plant with fetid odor and glandular-clammy hairs. Flowers in leafy racemes, perfect; sepals 4, separate, purplish; petals 4, separate, 4–6 mm. long, clawed and notched, yellowish-white; stamens 6–12, separate, about as long as the petals; pistil solitary, on a very short stipe, 1-celled, maturing into an erect, cylindrical or curved, glandular, many-seeded capsule. Seeds about 2 mm. in diameter, flattened, notched near hilum, granular, reddish-brown.

Control.—Hoe or pull by hand before seeds are formed (5, 4).

CRUCIFERAE (Mustard Family)

The flowers of the weeds belonging to the mustard family are remarkably uniform. Their general characteristics are recorded here and are not repeated under the descriptions of the individual species. They consist of 4 separate sepals which usually fall very early; 4 separate petals arranged like a cross; 6 separate stamens, 4 long and 2 short ones; pistil solitary with a 2-celled superior ovary which develops into a capsule or pod which usually splits open by 2 valves, each valve containing 1 or several seeds. The herbage of most members of the mustard family has a very pungent or peppery flavor.

185. *Alliaria officinalis* Andrz. Garlic-mustard.

Biennial; reproducing by seeds. Moist shaded soil; roadsides, open woods, waste places and near buildings. Quebec and Ontario to Virginia, Kentucky and Kansas. Introduced from Europe. April–June.

Description.—Stems up to 1 m. high, simple or slightly branched, glabrous or nearly so. Lower leaves reniform, petioled; upper ones sessile; cauline leaves deltoid, 3–6 cm. long, acute, coarsely toothed. Flowers white; petals 5–6 mm. long; siliques 2.5–6 cm. long, widely divergent, on short, stout pedicels, 4-angled when dry. Seeds black, about 3 mm. long.

Control.—In meadows hand pull scattered plants as soon as they begin to blossom (4). Badly infested fields should be plowed early and followed by a clean cultivated crop (6). Cultivation should be continued late into the summer to destroy the last seedlings to appear. Surface cultivation after the crops are removed will destroy the autumn seedlings.

186. *Arabidopsis thaliana* (L.) Heynh. Mouse-ear-cress.

Annual or winter annual; reproducing by seeds. Fields, roadsides and waste places, especially in sandy soil. Widespread, Massachusetts to Georgia westward to Wisconsin and Arkansas; waste places on dry ground in Oregon and Washington. Introduced from Europe. March–July.

Description.—Stems slender, erect, simple or branched from the hairy base, sparingly branched above, 5–50 cm. high. Leaves mostly in a basal rosette, stellate-pubescent, oblong to narrowly obovate, entire or shallowly toothed, usually petiolate, 1–5 cm. long; upper leaves smaller, linear to narrowly oblong, sessile. Flowers in open racemes; petals white, 2–4 mm. long, spatulate, twice the length of the sepals; style very short. Capsules glabrous, 1–2 cm. long, about 1 mm. wide on divergent or ascending capillary pedicels, 0.4–1.5 cm. long. Seeds 40–68, 1-seriate, barely 0.5 mm. long.

Control.—The same as for 188.

187. *Barbarea verna* (Mill.) Aschers. Scurvy-grass.

Perennial or biennial: reproducing by seeds. Fields, meadows and waste places. Widespread across the United States, especially in the eastern and north central states. Introduced from Europe. May–June.

Description.—Similar to 188, but of more slender habit. Lower leaves with 4–10 pairs of lateral lobes. Pods longer, with pedicels as thick as pods; seeds similar to 188 but larger, 1.6–2 mm. long.

Control. The same as for 188.

188. *Barbarea vulgaris* R. Br. Winter cress, Yellow rocket, St. Barbaras cress, Bitter cress, Rocket cress, Yellow-weed, Water mustard, Pot-herb. Fig. 44.

Perennial or biennial; reproducing by seeds and sometimes by new shoots from old crowns. New meadows, clover fields and cultivated land; especially on rich alluvial soils. Abundant in the northeastern and north central states, occasional westward to the Pacific Northwest. Introduced from Eurasia; also native to northern North America. May–June.

Description.—Stems erect, simple or branched, ridged, glabrous, often in clusters, from perennial or biennial roots. Lower leaves with 1–4 pairs of lateral lobes and a large rounded terminal lobe, dark green, glabrous, glossy, thick and often persisting all winter; stem-leaves alternate, cut, toothed or lobed at the base. Flowers in

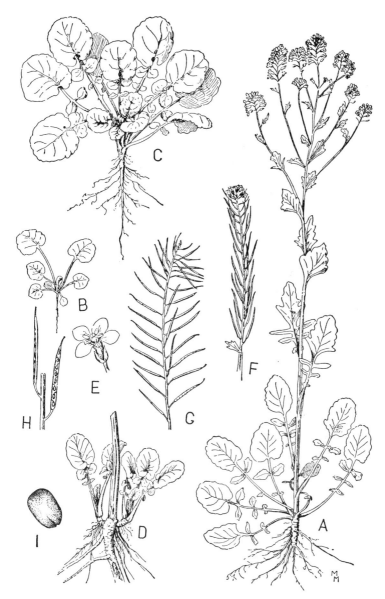

FIG. 44—**WINTER CRESS**, *Barbarea vulgaris:* **A,** plant showing general habit early in second season ($\times \frac{1}{5}$); **B,** seedling ($\times \frac{1}{5}$); **C,** rosette of leaves of first season ($\times \frac{1}{4}$); **D,** base of plant, early spring of third year, showing part of dead flowering stem of second season and rosettes which will produce flowering stems in third season ($\times \frac{1}{5}$); **E,** flower; **F,** seed-pods (common form), ($\times \frac{1}{3}$); **G,** seed-pods (shade form), ($\times \frac{1}{3}$); **H,** mature fruits (slightly reduced); **I,** seed ($\times 8$).

naked racemes; petals bright yellow. Seed-pods erect or ascending, on spreading very slender pedicels, linear, slightly 4-angled, about 2–3 cm. long, splitting into 2 valves which extend to within about 2 mm. of the apex, each valve with several seeds arranged in 1 row; seeds 1–1.5 mm. long, broadly oval to oblong, notched at one end, roughened, dull, gray-brown, cotyledons accumbent.

Control.—Pull or hoe out scattered plants just as they begin to blossom (4, 5). Badly infested fields should be disked or harrowed several times in late summer and early autumn to destroy the rosettes and to induce the seeds to germinate (26); plow under next spring and follow with a cultivated crop. In new meadows mow the tops as soon as the flowers appear, and low enough to catch the lowest flower-stalks (11).

189. *Berteroa incana* (L.) DC. Hoary alyssum. Fig. 55, a–e.

Perennial or biennial; reproducing by seeds. Meadows, pastures and waste places; mostly on dry, sandy or gravelly soils. Locally common in the northeastern states and infrequent in the central states; noxious (1954) in Michigan and Minnesota. Introduced from Europe. June–September.

Description.—Stems erect or spreading, branched, about 3–7 dm. high, covered with branched hairs. Leaves alternate, simple, lanceolate, entire, pale green and covered with branched hairs. Flowers in racemes; petals white, 2-parted. Seed-pods elliptic, about 1 cm. long, somewhat flattened parallel to the broad partition, each valve containing a few seeds, seeds about 1.5 mm. long, flattened, broadly ovate to circular, minutely granular, purplish-brown, cotyledons accumbent.

Control.—Pull or hoe out scattered plants (4, 5); clean cultivation (6).

190. *Brassica hirta* Moench (*B. alba* Boiss., *Sinapis alba* L.). White mustard, Charlock, Senvre, Kedlock. Fig. 45, d–f.

Annual; reproducing by seeds. Cultivated ground and waste places. Widespread, but usually appearing locally as an escape from cultivation. Introduced from Eurasia. June–August.

Description.—Stems erect, 3–6 dm. high, with spreading branches, hairy. Leaves alternate, hairy, pinnately lobed or divided, with a large terminal lobe and several smaller lateral lobes or irregular segments. Flowers in racemes; petals yellow; fruiting pedicel spreading, about 1 cm. long. Seed-pod bristly hairy; beak flattened, often

Fig. 45—**INDIAN MUSTARD**, *Brassica juncea:* **A,** part of plant showing habit (× ⅓); **B,** fruit (× 1); **C,** seed (× 5).
WHITE MUSTARD, *Brassica hirta:* **D,** part of plant showing habit (× ⅓); **E,** fruit (× 1); **F,** seed (× 5).

containing 1 seed, sword-shaped, usually longer than the valves; seeds globular, 2–2.5 mm. in diameter, yellow, minutely pitted; cotyledons conduplicate.

Control.—The same as for 192.

The white mustard is grown as a crop in Europe and also in California.[1]

191. *Brassica juncea* (L.) Coss. Indian mustard, Leaf mustard. Fig. 45, a–c.

Annual or winter annual; reproducing by seeds. Grain fields, cultivated fields and waste places. Widespread in the United States and in western Canada. Introduced from Asia. June–August.

Description.—Stems erect, 5–10 dm. high, branched, nearly glabrous, somewhat glaucous. Leaves alternate, the lower lyrate, the upper oblong, nearly entire, with tapering base, glabrous or nearly so. Flowers in racemes with slender spreading pedicels 7–10 mm. long. Seed-pods 2.5–7 cm. long, spreading; beak without a seed, very narrow, conical, one-fourth or less the total length of the pod; seeds globular, 1.5–2 mm. in diameter, dark brown, roughened by a minute network of about 10 meshes to a square mm., cotyledons conduplicate.

Control.—The same as for 192.

192. *Brassica kaber* (DC) L. C. Wheeler var. *pinnatifida* (Stokes) L. C. Wheeler (*B. arvensis* Ktze., *Sinapis arvensis* L.). Wild mustard, Field mustard, Charlock, Field kale, Kedlock. Fig. 46.

Annual; reproducing by seeds. Spring grain fields, especially oats; also cultivated land and waste places. Throughout the grain sections of North America; the commonest mustard of grain fields in the northeastern states. Introduced from Europe. June–August.

Description.—Stems, leaves and flowers similar to 190, but the upper leaves not pinnately lobed. Fruiting pedicel spreading, about 5–7 mm. long. Seed-pod glabrous or nearly so; beak angular, shorter than the valves, often containing 1 seed; seeds globular, about 1.5 mm. in diameter, black or dark purplish-brown, smooth or minutely reticulated.

Control.—Use only mustard-free seed. Clean cultivation followed by hand weeding in cultivated crops (6, 4). Spring weeding with a harrow on a warm dry day when the wheat is about 4 inches high will destroy most of the mustard seedlings (8). Frequent disking or

[1] Sievers, 1930.

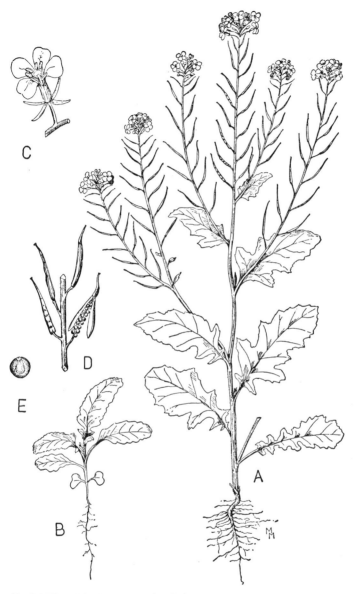

FIG. 46—**WILD MUSTARD,** *Brassica kaber* var. *pinnatifida*: **A,** plant showing general habit (× ⅕); **B,** seedling in about the stage when spraying is most effective (× ¼); **C,** flower (× 1); **D,** mature fruits (seed-pods), (× ⅓); **E,** seed (× 3).

harrowing of stubble fields will induce seed germination (26). Hand pull scattered plants in grain fields and meadows as soon as the plants begin to blossom.

193. *Brassica nigra* (L.) Koch. Black mustard, Cadlock, Warlock, Scurvy, Senvil, Brown mustard. Fig. 47, e–h.

Annual or winter annual; reproducing by seeds. Neglected fields, pastures, waste places, banks of streams and ditches; usually in rich soils. Not as troublesome in grain fields as 192. Widespread and locally common throughout the United States and southern Canada. Introduced from Eurasia. July–October.

Description.—Stems erect, with spreading branches above, 1–2 m. high, hairy. Leaves alternate, the lower lyrate with a large terminal lobe, the upper lanceolate, nearly entire, with tapering base. Flowers in long slender racemes with closely appressed erect pedicels 3–7 mm. long. Seed-pods 1–2 cm. long, appressed; beak without a seed, narrow, conical, less than one-fourth the length of the pod; seeds ellipsoidal, about 1.5 mm. long, dark brown or light reddish-brown, surface reticulated, cotyledons conduplicate.

Control.—The same as for 192.

194. *Brassica rapa* L. (*B. campestris* L.) Rutabaga. Fig. 47, a–d.

Winter annual or biennial; reproducing by seeds. Fields and waste places. Widespread; very abundant in the valley lands of the Pacific Coast, infrequent or locally common in the north central and eastern states. Introduced from Europe. June–August.

Description.—Stems erect, 6–10 dm. high, branched, glaucous, slightly hairy when young. Leaves alternate, irregularly pinnately lobed or dissected, glaucous, the upper leaves ovate to cordate with clasping base. Flowers in long racemes, pale yellow; fruiting pedicel spreading. Seed-pods glabrous; beak conical, about one-half the length of the valves; seeds globular, about 2 mm. in diameter, dark brown or black, cotyledons conduplicate.

Control.—The same as for 192.

195. *Camelina microcarpa* Andrz. Small-seeded false flax, False flax, Dutch flax, Western flax, Siberian oil seed. Fig. 52, e–h.

Annual; reproducing by seeds. Grain fields, flax fields and waste places. Western Canada and the Northwest, westward to British Columbia and Washington; occasional in the eastern states. Introduced from Europe. At present frequently brought into the eastern states with feeds made from grain grown in the northwestern states. May–June.

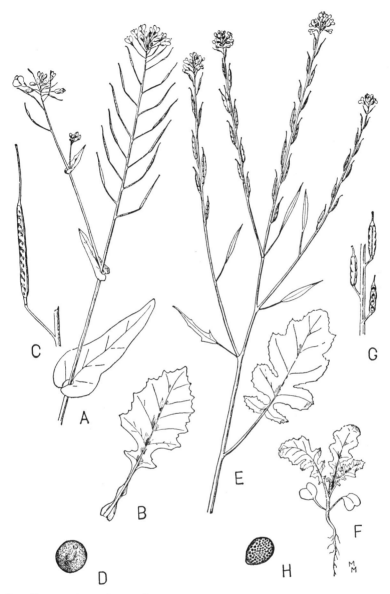

Fig. 47—**RUTABAGA**, *Brassica rapa*: **A**, upper part of flowering stem (× ⅓); **B**, leaf from lower part of stem (× ¼); **C**, mature fruit (seed-pod), (× ½); **D**, seed (× 6).
BLACK MUSTARD, *Brassica nigra*: **E**, upper part of flower stem (× ⅓); **F**, seedling in about the stage when spraying is most effective (× ¼); **G**, mature fruit (seed-pod), (× 1); **H**, seed (× 6).

Description.—Stems erect, slender, with a few erect branches, slightly hairy. Leaves alternate, simple, sessile, lanceolate to arrow-shaped, slightly toothed, hairy. Flowers in long slender racemes; pedicels 1–3 cm. long; petals pale yellow. Seed-pods obovoid, not flattened, 4–6 mm. broad, splitting by 2 valves; seeds several in each valve, oblong, about 1 mm. long, ridged, with a groove on each side, surface granular, reddish-brown, cotyledons incumbent.

Control.—Hand pull or hoe (4, 5) stray plants before seeds are formed. Harrow or disk grain fields several times after the crop is removed, repeat in the spring, plow two weeks later. Harrow badly infested grain fields two or three times when the grain is 3–6 inches high. Clean cultivation (6) will destroy this weed.

Camelina dentata Pers., Flat-seeded false flax, has been introduced with flax into Manitoba. It has flattened yellow to pale brown seeds 2 mm. long. *Camelina sativa* Crantz, Large-seeded false flax, occurs in grain fields and with flax in western Ontario, Manitoba, the Dakotas and Minnesota; also in grain fields westward. The seed-pods are larger than *C. microcarpa,* and its seeds are about 2 mm. long. The methods of control are the same as for 195.

196. *Capsella bursa-pastoris* (L.) Medic. Shepherds-purse, Shepherds-bag, Pepper plant, Case weed, Pick-purse. Fig 48, m–p.

Annual or winter annual; reproducing by seeds. Gardens, cultivated fields, grain fields and waste places. Very common throughout North America. Introduced from Europe. March–December.

Description.—Stems erect, branched, slender, from a thin tap root. Leaves alternate, simple, variously toothed or lobed; the stem-leaves sessile, arrow-shaped. Flowers in slender racemes, on slender pedicels; petals small, white. Seed-pods triangular, of 2 valves, flattened at right angles to the partition; valves boat-shaped, each with several to many seeds; seeds about 1 mm. long, oblong, grooved on each side, orange-brown, cotyledons incumbent.

Control.—Clean cultivation begun early and continued late into the summer (6). Hand hoeing and weeding in gardens and small areas (4, 5). Badly infested fields should be harrowed several times after the crop is removed. Infested grain fields should be harrowed while the grain is from 3–6 inches high (8).

197. *Cardamine pratensis* L. Cuckoo flower.

Perennial; reproducing by seeds and short rootstocks. Low meadows and pastures, and sometimes common in lawns. Apparently native from the northeastern states northward; also introduced from Europe. May–June.

Description.—Stems from a short rootstock, erect or ascending, simple, glabrous, 3–6 dm. high. Leaves alternate, pinnately compound; leaflets linear to oblong, entire or nearly so. Flowers in racemes, 2–3 cm. in diameter; petals white or rose-colored. Seed-pods 2–3 cm. long, nearly cylindrical, with short beak, several-seeded; seeds oblong to nearly circular, about 1.5 mm. long, flattened and somewhat winged at base, cinnamon-brown, cotyledons accumbent.

Control.—Improved drainage in meadows (21). Plow and cultivate for a year (6, 7). Hoe out from lawns and reseed bare spots.

198. **Cardaria draba** (L.) Desv. (*Lepidium draba* L.). Hoary cress, Hoary pepperwort, White-top, Devils cabbage, Perennial peppergrass, White-weed. Fig. 48, g–j.

Perennial; reproducing by seeds, rootstocks and creeping roots. Cultivated fields, meadows, grain fields, and waste places. North Atlantic states, Middle West to Pacific Coast; most troublesome in the Rocky Mountain region. Introduced from Eurasia. June–August.

Description.—Stems erect or spreading, 3–8 dm. high, branched, hairy. Leaves alternate, simple, oval or oblong, dentate, the upper with broad clasping base. Flowers in showy corymb-like racemes; petals white. Seed-pods heart-shaped, 2–3 mm. long, often oblique and inflated, with a prominent persisting style, not notched, dehiscent, each valve with 1 seed; seeds about 2 mm. long, obovate, slightly flattened, granular, reddish-brown; cotyledons incumbent.

Control.—Small areas where cultivation is not practicable can be eradicated by hand digging (17); covering with mulch paper (24). Large infestations can be controlled by summer fallow (19); [2] clean cultivation of intertilled crops (18), followed by a smother crop like alfalfa, sweet clover or sudan-grass (22).

Lepidium repens Boiss., also called Hoary cress, a native of Afghanistan, has been reported from several places in California and Oregon.[3] It is similar to *Cardaria draba* but has lens-shaped pods.

199. **Cardaria pubescens** (C. A. Mey.) Rollins (*Hymenophysa pubescens* C. A. Mey.). Whitetop, Siberian mustard, Globe-podded hoary cress.

Perennial; reproducing by seeds. Roadsides, waste land, and fields, especially in alfalfa. Washington to California; local in the east. Adventive, probably from Asia. May–September.

[2] Hulbert, Spence and Benjamin, 1934.   [3] Bellue, 1933a.

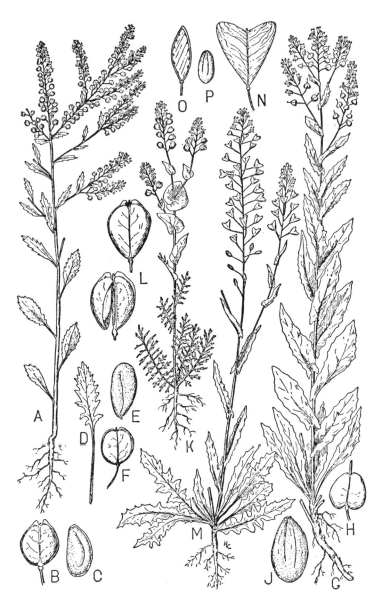

FIG. 48—**PEPPER-GRASS**, *Lepidium virginicum:* **A**, part of plant showing general habit (× ⅓); **B**, fruit (× 3); **C**, seed (× 9).
GREEN-FLOWERED PEPPER-GRASS, *Lepidium densiflorum:* **D**, leaf (× ⅓); **E**, seed (× 3); **F**, fruit (× 9).
HOARY CRESS, *Cardaria draba:* **G**, plant (× ⅓); **H**, fruit (× 3); **J**, seed (× 9).
CLASPING-LEAVED PEPPER-GRASS, *Lepidium perfoliatum:* **K**, plant showing general habit (× ¼); **L**, fruits (× 3).
SHEPHERDS-PURSE, *Capsella bursa-pastoris:* **M**, part of plant showing general habit (× ⅕); **N**, fruit (× 3); **O**, fruit with valves fallen away (× 3); **P**, seed (× 9).

Description.—Stem 1–4 dm. high, puberulent, branching above, leafy. Leaves auriculate-clasping, sessile, 1–3.5 cm. long. Flowers in corymbose racemes; racemes 4–10 cm. long; petals white. Fruits on slender, ascending pedicels, subglobose, inflated, 2.5–4 mm. wide, terminating in the slender, persistent style. Seeds 2–4, brown, wingless, slightly reticulate.

Control.—The same as for 198.

200. *Conringia orientalis* (L.) Dumort. Hares-ear mustard, Treacle mustard, Klinkweed, Rabbit-ears. Fig. 52, a–d.

Annual; reproducing by seeds. Cultivated fields, gardens and waste places. Common in grain fields in western Ontario, Manitoba and the northwestern states; infrequent elsewhere. Introduced from Europe. May–July.

Description.—Stems erect, up to 1 m. high, slender, simple or sparingly branched, glabrous, stiff and wiry when mature. Leaves alternate, simple, elliptical, entire, with clasping base, glaucous, somewhat fleshy. Flowers in slender racemes; petals small, pale yellow. Seed-pods erect, linear, 4–10 cm. long, angled, 2-celled, each cell with 1 row of seeds; seeds 2–2.5 mm. long, oblong, with 2 vertical grooves separated by a ridge, surface coarsely granular, dark grayish-brown; cotyledons incumbent.

Control.—The same as for 195.

201. *Coronopus didymus* (L.) Smith (*Carara didyma* Britt.). Swine cress, Wart cress, Carpet cress. Fig. 56, d–e.

Annual or biennial; reproducing by seeds. Gardens, fields and waste places. The Atlantic Coast and southern states, westward to the Pacific Coast. Introduced from Eurasia. March–June.

Description.—Stems prostrate, much branched, hairy, forming mats from 2–8 dm. in diameter. Leaves alternate, deeply lobed and the lobes sometimes also lobed again, the lower petioled, the upper sessile. Flowers minute, in slender axillary racemes; corolla white. Capsule 2-seeded, deeply notched at apex, splitting into 2 nutlets; nutlets about 1.5 mm. long, half-ovate, with an elongated opening on the flat side, roughened by an irregular network of ridges, dull, yellow; cotyledons incumbent.

Control.—Clean cultivation followed by hand hoeing (6, 5).

This weed has a strong disagreeable odor which is imparted to dairy products from animals feeding upon it.

Coronopus procumbens Gilibert has been found in waste places along the Atlantic Coast and also in California. Its pods are not notched at the apex.

202. *Descurainia pinnata* (Walt.) Britt. Tansy mustard. Fig. 43, f–h.

Annual; reproducing by seeds. Dry or sandy soil, open woods, prairies, waste places; especially abundant in arid and semiarid regions. Throughout the United States from Canada to Mexico. Native. March–August.

Description.—Stem 1–8 dm. high, erect, simple or much branched, green, densely canescent, glabrous to glandular. Leaves alternate, pinnately dissected, often with grayish stellate pubescence, the lower ones the larger, oblong to oblanceolate. Racemes up to 3 dm. long, glandular to glabrous. Petals yellow or yellowish green to nearly white. Capsule clavate, 0.4–2 cm. long, on widely divergent to erect pedicels; seeds in 2 rows, less than 1 mm. long.

Control.—The same as for 198.

203. *Descurainia richardsonii* (Sweet) O. E. Schulz (*Sisymbrium incisum* Engelm.). Tansy mustard.

Biennial; reproducing by seeds. Grain fields, roadsides, and waste places. Common in the western United States and Canada; occasional as a wayside weed eastward. Native to western North America. July–September.

Description.—Stems erect, 6–12 dm. high, much branched above. Leaves alternate, pinnatifid, with narrow segments, covered with appressed gray hairs. Flowers in numerous elongated racemes; petals yellow, about 2 mm. long. Seed pods cylindric, appressed ascending, with seeds 1-ranked and 4–8 in each locule; seeds similar to 224.

Control.—Clean cultivation will kill this weed (6). Hand pull or hoe stray plants before seeds are formed (4, 5). Harrow or disk grain fields several times after the crop is removed, repeat in the spring, and plow two weeks later. Harrow badly infested grain fields two or three times when the grain is 3–6 inches high.

204. *Descurainia sophia* (L.) Webb (*Sisymbrium sophia* L.). Flixweed.

Annual or biennial; reproducing by seeds. Fields, gardens, and waste places. Widely established Quebec to Washington, south to Delaware, Missouri, and California; more common in Canada, eastern Oregon, Washington, and the Great Basin region. Introduced from Europe. June–August.

Description.—Stems erect, mostly simple, 2–6 dm. high, hoary-

Fig. 49—**WORM-SEED MUSTARD,** *Erysimum cheiranthoides:* **A,** plant showing general habit (× ⅓); **B,** flower (× 2); **C,** mature fruit (seed-pods), (× ½); **D,** seed (× 10).
TREACLE MUSTARD, *Erysimum repandum:* **E,** leaf (× ⅓); **F,** fruits (× ⅓); **G,** mature fruit (× 1); **H,** seed (× 10).

pubescent with branched hairs. Leaves alternate, 2- or 3-times pinnately compound, the segments very narrow or linear, hairy. Flowers in slender racemes; petals very small, yellow or greenish-yellow. Seedpods cylindrical, slender, 1–2 cm. long, ascending, 2-valved; seeds in 1 row in each valve, about 1 mm. long, similar to 224.

Control.—The same as for 203.

205. *Diplotaxis tenuifolia* (L.) DC. Large sand rocket, Wall rocket. Fig. 50, k–n.

Perennial; reproducing by seeds. Gardens, waste places and on ballast ground; mostly on dry, gravelly or sandy soils. Local in the eastern states and adjacent Canada; rare in Idaho and California.[4] Introduced from Europe. June–September.

Description.—Stems slender, erect, spreading or decumbent, branched, somewhat hairy. Leaves alternate, oblong, pinnately lobed or toothed, hairy, with a disagreeable odor. Flowers in slender naked racemes; pedicels spreading, fruiting pedicels 1–3 cm. long. Seedpods 2–4 cm. long, 2-celled, dehiscent; seeds in 2 rows in each cell, about 1 mm. long, somewhat flattened, grooved, orange-yellow; cotyledons conduplicate.

Control.—The same as for 196.

Diplotaxis muralis (L.) DC., introduced from Europe, is locally common in the northeastern states, adjacent Canada, and west to Iowa. It is an annual or biennial with less leafy stems, smaller flowers and shorter pedicels.

206. *Draba verna* L. Whitlow-grass.

Annual or biennial; reproducing by seeds. Dry fields, lawns and waste places. Locally common in the eastern and north central states. Introduced from Eurasia. April–May.

Description.—Stems erect, 2–10 cm. high, naked. Leaves all radical, in a rosette, oblong or lanceolate, about 1 cm. long, covered with branched hairs. Flowers in small racemes, petals white. Seedpods oval to oblong-lanceolate, flattened parallel to a broad partition, with many-seeded valves; seeds about 0.7 mm. long, flattened, obovate, notched at base, dull, orange-brown; cotyledons accumbent.

Control.—Raking, followed by close mowing of lawns (11). Close grazing by sheep in pastures (10). Clean cultivation and hand weeding to prevent seed formation (6).

207. *Eruca sativa* Mill. Garden rocket, Rocket salad. Fig. 50, f–j.

[4] Tidestrom, 1934.

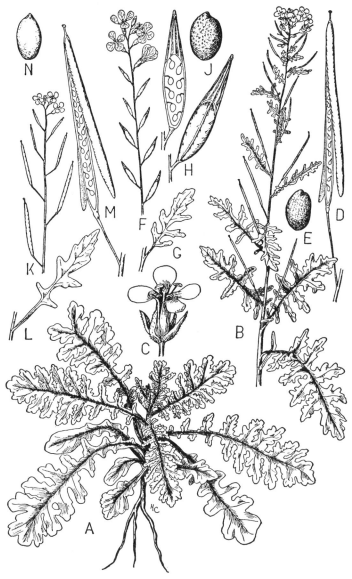

FIG. 50—**DOG MUSTARD**, *Erucastrum gallicum:* **A**, winter rosette (× ⅓); **B**, flowering shoot (× ⅓); **C**, flower (× 2); **D**, capsule (× ⅓); **E**, seed (× 9).
GARDEN ROCKET, *Eruca sativa:* **F**, flowering shoot (× ⅓); **G**, leaf (× ⅙); **H**, capsule (× ⅓); **J**, seed (× 8).
LARGE SAND ROCKET, *Diplotaxis tenuifolia:* **K**, flowering shoot (× ⅓); **L**, leaf (× ⅓); **M**, capsule opened to show two rows of seeds (× 1); **N**, seed (× 9).

Annual or winter annual; reproducing by seeds. Fields and waste places; on sandy soils. Infrequent as an escape in the eastern and north central states and eastern Canada where it has been cultivated as a pot-herb or salad plant. Introduced from Eurasia. June–September.

Description.—Stems erect, with stiff, spreading, rough branches, about 3–8 dm. high. Leaves alternate, pinnately lobed or divided, fleshy, somewhat hairy, strong-scented. Flowers in racemes; petals pale yellow with purple veins. Seed-pods cylindrical, about 2 cm. long; beak about as long as the valves, strongly keeled; seeds several in each valve, about 1.8 mm. long, oblong, with a projection on one side of hilum, dull, yellow-brown; cotyledons conduplicate.

Control.—Hoeing and hand weeding (5, 4).

208. *Erucastrum gallicum* (Willd.) O. E. Schulz (*E. pollichii* Schimp. and Spenn.). Dog mustard. Fig. 50, a–e.

Annual or winter annual; reproducing by seeds. Grain fields, waste places, fields, roadsides and railroad tracks. Common in the Dakotas, western Minnesota, and parts of Manitoba and Saskatchewan.[5] Frequently introduced in the northeastern states and eastern Canada, also west to British Columbia and south to Missouri. First reported in the United States from Milwaukee, Wisconsin, in 1903.[6,7] Introduced from Europe. July–October.

Description.—Stems slender, 3–8 dm. high, with spreading or erect branches. Leaves alternate, pinnatifid with obtuse lobes, the upper sessile and toothed, somewhat hairy, gray-green. Flowers in slender leafy-bracted racemes; petals pale yellow, about 1 cm. long. Seed-pods 3–5 cm. long, slender, cylindrical or slightly angled, many-seeded; seeds about 1 mm. long, oblong, orange-brown, finely reticulated, dull; cotyledons incumbent.

Control.—The same as for 195.

The dog mustard grows vigorously after the grain is cut, sometimes blossoming long after severe frosts have killed most other vegetation.

209. *Erysimum cheiranthoides* L. Worm-seed mustard, Treacle mustard. Fig. 49, a–d.

Annual or winter annual; reproducing by seeds. Grain fields, new meadows, cultivated fields and gardens; on gravelly or sandy soils. Widespread across the northern United States and southern Canada.

[5] Groh, 1933.   [6] Robinson, 1911.   [7] Blake, 1924.

Mustard Family

Native to some parts of North America and also to Europe. Probably introduced in many localities. June–September.

Description.—Stems slender, erect, branching, somewhat rough, appressed, pubescent. Leaves alternate, lanceolate, margin wavy or somewhat toothed, with branched or forked hairs. Flowers in racemes; petals pale yellow. Seed-pods 1–2 cm. long, linear, 4-sided, erect on spreading pedicels, each valve with a prominent keel, dehiscent; seeds several, in one row in each valve, about 1 mm. long, ovoid to triangular in outline, a vertical ridge with groove on each side, dull, orange-brown, minutely granular; cotyledons incumbent.

Control.— Clean cultivation followed by hand hoeing (6, 5). Grain fields should be disked or harrowed several times after the the crop is removed to destroy autumn seedlings.

210. *Erysimum repandum* L. Treacle mustard, Spreading mustard. Fig. 49, e–h.

Annual or winter annual; reproducing by seeds. New seedings and lawns, cultivated fields and waste places. Eastern Washington and Oregon to central California, eastward to the Atlantic states. Introduced from Europe. July–September.

Description.—Stems erect or decumbent and spreading, branched. Leaves alternate, simple, lanceolate, margin with a few low teeth. Flowers in racemes; petals yellow. Seed-pods 4–8 cm. long, linear, 4-sided, spreading, on short thick pedicels, dehiscent, each valve with a prominent keel; seeds several in one row in each valve, about 1 mm. long, oblong, 3-angled, grooved, orange-brown; cotyledons incumbent.

Control.—Clean cultivation (6); hand pulling (4); hoeing (5).

211. *Isatis tinctoria* L. Woad. Fig. 56, a–c.

Biennial; reproducing by seed. Grain fields, gardens, and waste places. Locally common and spreading in the Great Basin; infrequent in Newfoundland, Virginia, and California; probably elsewhere. Introduced from Eurasia. May–June.

Description.—Stems erect, 5–15 dm. high, simple or branched, hairy below, glabrous and glaucous above. Leaves alternate, entire or nearly so, obtuse, bluish-green, with a broad whitish nerve on the upper surface of the blade, the upper leaves sessile, clasping. Flowers in branched corymbose racemes; sepals about 2 mm. long; petals about 3–4 mm. long, bright yellow. Seed-pods 1–2.5 cm. long, ovate to obovate, 1-celled, usually 1-seeded, surrounded by a broad

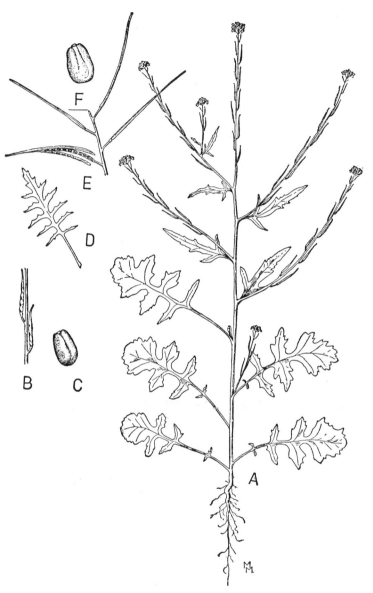

Fig. 51—**HEDGE MUSTARD,** *Sisymbrium officinale:* **A,** plant showing general habit ($\times$ 1/6); **B,** mature fruit ($\times$ 1/3); **C,** seed ($\times$ 12).
TUMBLE MUSTARD, *Sisymbrium altissimum:* **D,** leaf from near base of stem ($\times$ 1/4); **E,** mature fruit ($\times$ 1/3); **F,** seed ($\times$ 12).

Mustard Family

wing, indehiscent, the short style often persisting as a point; seeds about 3 mm. long, narrowly ovoid, with 2 deep grooves, granular, orange; cotyledons incumbent.

Control.—The same as for 195.

212. *Lepidium campestre* (L.) R. Br. Field pepper-grass, Downy pepper-grass, Field cress, Cow cress. Fig. 53, g–m.

Biennial or winter annual; reproducing by seeds. Grain fields, new seedings of clover or alfalfa, waste places. Widespread and locally abundant throughout the northeastern and north central states and Ontario; infrequent elsewhere. Introduced from Europe. June–July.

Description.—Stems erect, 2–6 dm. high, often clustered, with stiff ascending branches above, downy-hairy. Leaves alternate, crowded or the lower in rosettes; the upper leaves arrow-shaped with clasping base, the lower spatulate to lanceolate, often pinnately lobed or toothed, downy. Flowers in racemes; petals whitish. Seed-pods ovate, about 6 mm. long and 4 mm. wide, winged, on short spreading pedicels, flattened at right angles to the partition, curved upward and the upper surface concave, dehiscent, each valve with 1 seed; seeds 2–2.5 mm. long, obovoid, one side flattened, coarsely granular, dark brown; cotyledons incumbent.

Control.—In meadows hand pull scattered plants as soon as they begin to blossom (4). New seedings of clover and alfalfa should be cut early before the weed seeds are mature (11). Badly infested fields should be plowed early and followed by a clean cultivated crop (6). Cultivation must be continued late into the summer to destroy the last seedlings to appear. Surface cultivation after the crops are removed will destroy the autumn seedlings.

213. *Lepidium densiflorum* Schrad. (*L. apetalum* Willd.). Green-flowered pepper-grass, Wild tongue-grass. Fig. 48, d–f.

Annual or winter annual; reproducing by seeds. Grain fields, meadows, pastures, and neglected yards; especially on light sandy soils. Widespread throughout the United States and southern Canada; common in the West and Middle West. Introduced from Eurasia. June–September.

Description.—Stems erect, branched above, nearly glabrous except in the inflorescence. Leaves alternate, lanceolate, pinnately lobed or toothed. Flowers in racemes; petals mostly absent; stamens 2. Seed-pods nearly orbicular, 1.5–2.5 mm. wide, notched at the top, flattened at right angles to the partition, dehiscent, each valve with 1 seed; seeds about 1.5 mm. long, obovate, flattened on one side, slightly winged, finely granular, dull, orange; cotyledons incumbent.

Control.—The same as for 196.

214. *Lepidium perfoliatum* L. Clasping-leaved pepper-grass. Fig. 48, k–l.

Annual; reproducing by seeds. Grain fields, clover fields, pastures, waste ground and about railroads. Abundant and troublesome in the Great Basin, but also appearing locally from New York to the Pacific Coast. Introduced from Europe and spreading very rapidly. May–July.

Description.—Stems slender, erect, branched. Leaves alternate; the lower much dissected, with linear segments; the upper ovate, with clasping base. Flowers in racemes; petals white or yellowish-green. Seed-pods nearly orbicular, about 5 mm. in diameter, flattened at right angles to the partition, dehiscent, each valve with 1 seed; seeds about 2 mm. long, obovate, each side ridged, margin winged, granular, reddish-brown; cotyledons incumbent.

Control.—The same as for 195.

215. *Lepidium ruderale* L. Pepper-grass.

Annual or biennial; reproducing by seeds. Waste places and roadsides. Newfoundland to Michigan, Delaware and Texas. Introduced from Europe. May–September.

Description.—Stems 1–5 dm. tall, minutely pubescent, commonly branched at the base; plant fetid. Basal leaves bipinnatifid, early withering, the cauline leaves more persistent; the lower divided, the upper mostly linear and entire, rounded at tips. Racemes mostly simple, loosely flowering, often 1 dm. or more long, at maturity with 6–10 fruits per cm. of length; flowers numerous; sepals linear; petals lacking. Fruit usually oval or ovate, 2.2–3 mm. long, wingless, and marginless; cotyledons incumbent.

Control.—The same as for 198.

216. *Lepidium virginicum* L. Pepper-grass, Birds pepper, Poor mans pepper, Tongue-grass. Fig. 48, a–c.

Annual or biennial; reproducing by seeds. Grain fields, cultivated ground, new seedings and waste places; on dry soils. Widespread from the Atlantic Coast to the Rocky Mountains. Native. May–October.

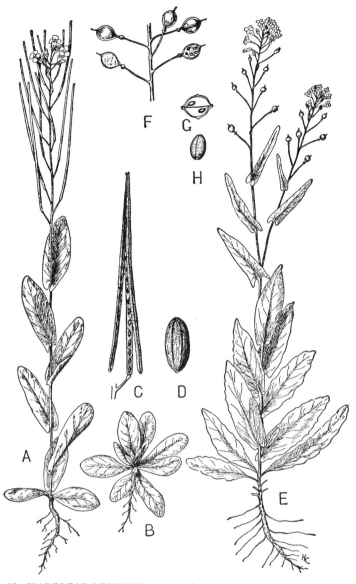

FIG. 52—**HARES-EAR MUSTARD,** *Conringia orientalis:* **A,** plant showing general habit (× ⅓); **B,** rosette stage (× ⅓); **C,** mature seed-pod splitting open (× ½); **D,** seed (× 6).
SMALL-SEEDED FALSE FLAX, *Camelina microcarpa:* **E,** plant showing general habit (× ⅓); **F,** mature seed-pods (× 1); **G,** cross-section of seed-pod (× 2); **H,** seed (× 6).

Description.—Very similar to 213. Flowers with small white or greenish petals. Seed-pods larger, 2.5–3 mm. wide; seeds 1.5–2 mm. long, obovate, with one edge straight, the other rounded, slightly winged at apex, granular, chestnut-brown; cotyledons accumbent.

Control.—The same as for 196.

Lepidium latifolium L., Perennial peppercress, a tall weed with subterranean rhizomes, has been locally naturalized from Europe along beaches, shores, and waste places. It was declared noxious (1954) in California and Washington.

217. *Neslia paniculata* Desv. Ball mustard. Fig. 55, f–h.

Annual or winter annual; reproducing by seeds. Grain fields and waste places. Common in the northwestern states and western Canada; infrequent eastward. Adventive from Europe. June–July.

Description.—Stems slender, erect, somewhat hairy with branched hairs. Leaves alternate, simple, oblong, entire or nearly so, with clasping base. Flowers in slender racemes, on slender spreading pedicels; petals yellow. Seed-pods globular or slightly flattened, 2–3 mm. in diameter, beaked, 1- or 2-celled, netted on the surface, grayish-brown, indehiscent, 1 or rarely 2 or more seeded; cotyledons incumbent.

Control.—The same as for 195.

218. *Raphanus raphanistrum* L. Wild radish, Jointed charlock, White charlock, Jointed radish, Wild kale, Wild turnip, Cadlock. Fig. 57, a–f.

Annual or winter annual; reproducing by seeds. Grain fields, cultivated ground and waste places. Common in northeastern and central states and Canada; Pacific Northwest. Introduced from Europe. July–September.

Description.—Stems erect or spreading, 3–10 dm. high, much branched, hairy, from a stout tap-root. Leaves alternate, lyrate, with more or less regular, rounded lobes. Petals pale yellow, or rarely white, with purple veins. Seed-pods linear, jointed or necklace-like, 2–10-seeded, indehiscent but breaking crosswise at the joints into fragments each with 1 or 2 seeds; seeds 2–3 mm. long, ovoid, finely reticulated; cotyledons conduplicate.

Control.—Clean cultivation followed by hand weeding in cultivated crops (6, 4). Spring weeding with a harrow on a warm dry day when the grain is about 4 inches high will destroy most of the seedlings (8). Frequent disking or harrowing of stubble fields will in-

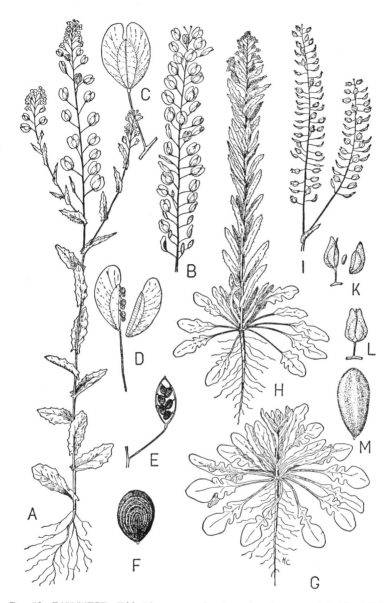

Fig. 53—**FAN-WEED**, *Thlaspi arvense:* **A,** plant showing general habit (× ⅓); **B,** ripe seed-pods (× ⅓); **C-E,** mature pods (× 2); **F,** seed (× 6).
FIELD PEPPER-GRASS, *Lepidium campestre:* **G,** winter rosette stage (× ⅓); **H,** plant showing general habit in spring (× ⅓); **I,** stalk with ripe seed-pods (× ⅓); **K-L,** mature pods splitting to discharge seeds (× 1½); **M,** seed (× 6).

duce seed germination (26). Hand pull scattered plants in grain fields and meadows as soon as the plants begin to blossom.

219. **Raphanus sativus** L. Radish, Wild radish. Fig. 57, h.

Annual; reproducing by seeds. Grain fields and cultivated ground. Abundant in the Pacific Coast states; locally common in the Lake Ontario region. Introduced from Europe. This is the escaped form of the garden radish. June–September.

Description.—Stems erect or spreading, much branched, 4–10 dm. high, hairy, from a fleshy tap-root. Leaves alternate, lyrate or irregularly lobed or with several lateral lobes or segments and a large terminal lobe. Flowers in long racemes; petals purplish, pink or white, 1–2 cm. long. Seedpods fleshy and spongy, 5–10 mm. in diameter, tapering, 3–8 cm. long, slightly constricted, 2–10-seeded, indehiscent and not jointed; seeds 3–4 mm. long, ovoid, finely reticulated, orange or reddish-brown; cotyledons conduplicate.

Control.—The same as for 218.

220. **Rorippa austriaca** (Crantz) Bess. (*Radicula austriaca* Crantz). Austrian field cress. Fig. 54, j–l.

Perennial; reproducing by seeds, creeping roots and rootstocks. Cultivated fields, pastures and waste places. Orange County, New York,[8] and adjacent fields in New Jersey; also limited areas in Minnesota,[9] Wisconsin,[10] California,[11] Saskatchewan,[12] and elsewhere. The first American record of this weed was in 1910 from Orange County, New York. Introduced from southeastern Europe. June–August.

Description.—Stems erect, 5–10 dm. high, branched, rough, glabrous, from creeping perennial roots. Leaves alternate, simple, toothed or nearly entire; lower petioled; upper sessile, cordate, clasping. Flowers in racemes; sepals yellow, 1–2 mm. long; petals yellow, 2–3 mm. long. Seed-pods globular, 1.5–3 mm. long, on a spreading pedicel 7–15 mm. long, dehiscent, with 2 convex nerveless valves; seeds 6–12 in each valve, about 1 mm. long, light brown, finely warted-reticulate; cotyledons accumbent.

Control.—The same as for 198.

221. **Rorippa islandica** (Oeder) Borbas. (*R. palustris* Bess., *Radicula palustris* Moench). Marsh cress, Yellow water cress. Fig. 54, e–h (including *R. islandica* var. *hispida*).

Annual, biennial or perennial; reproducing by seeds and new

[8] Hansen, 1922. [9] Army, 1932. [10] Small, 1923.
[11] Bellue, 1933b. [12] Groh, 1933.

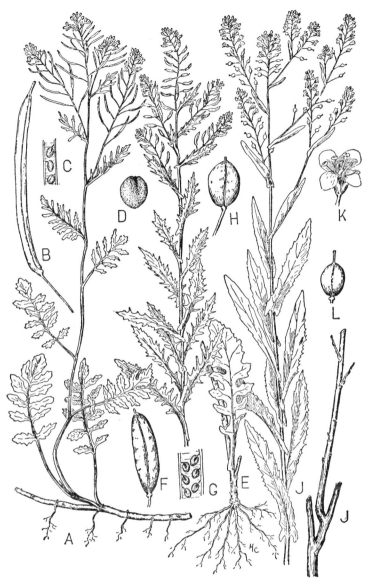

Fig. 54—**YELLOW CRESS,** *Rorippa sylvestris:* **A,** part of plant showing habit (× ⅓); **B,** seed-pod (× 3); **C,** section of pod (× 4); **D,** seed (× 18).
MARSH CRESS, *Rorippa islandica:* **E,** part of plant (× ⅓); **F,** seed-pod (× 3); **G,** section of pod (× 4).
MARSH CRESS, *Rorippa islandica* var. *hispida:* **H,** seed-pod (× 3).
AUSTRIAN FIELD CRESS, *Rorippa austriaca:* **J,** part of plant showing general habit (× ⅓); **K,** flower (× 3); **L,** seed-pod (× 3).

shoots from crown or roots. Poorly drained fields, meadows and pastures; in oat fields following corn stubble without plowing.[13] Widespread in the eastern and north central states; infrequent as a weed elsewhere. Native; also native to Eurasia. June–September.

Description.—Stems erect, 3–8 dm. high, glabrous or somewhat hairy. Leaves alternate, pinnately lobed, cleft or parted, the upper irregularly toothed, glabrous. Flowers in racemes; sepals and petals yellow. Seed-pods subglobose,[14] ovate or short-cylindrical, dehiscent, with 2 convex nerveless valves each with several seeds; seeds about 0.6 mm. long, ovate, slightly flattened, slightly notched, yellow-brown; cotyledons accumbent.

Control.—Improve drainage (21). Plow infested fields and include a cultivated crop in a short rotation before sowing grain (20). Early and late clean cultivation of intertilled crops (6).

222. **Rorippa sylvestris** Bess. (*Radicula sylvestris* Druce). Yellow cress, Creeping yellow cress. Fig. 54, a–d.

Perennial; reproducing by seeds, rootstocks and creeping roots. Wet meadows, fields, nurseries, rock-gardens, and lawns; especially on rich moist soils. Eastern Canada to Virginia, westward to the Mississippi Valley. Introduced from Europe. June–September.

Description.—Stems slender, ascending or spreading, irregularly branched, from creeping shoots, rootstocks or roots. Leaves alternate, pinnately parted or lobed, the narrow divisions toothed or cut, glabrous. Flowers in racemes; petals yellow. Seed-pods 10–25 mm. long, linear, on slender spreading pedicels; 2 valves each with several seeds arranged in 1 row; seeds about 0.8 mm. long, nearly circular with a notch, orange-brown; cotyledons accumbent.

Control.—Hand digging of all roots (17); early and late clean cultivation (6); cover with mulch paper (24).

223. **Sibara virginica** (L.) Rollins.

Annual or short-lived biennial; reproducing by seeds. Sandy or stony ground, open woods, waste places, clearings, and cultivated fields. Virginia, Illinois, Kansas, south to Florida, Texas, and California. Native. March–June.

Description.—Stems 1–4 dm. high, short-hirsute or glabrous

---

[13] Pammel, King and Hayden, 1929.

[14] The form with subglobose pods and hirsute stems has been described as var. *hispida* (Desv.) Butt. & Abbe.

FIG. 55—**HOARY ALYSSUM**, *Berteroa incana:* **A,** plant showing habit (× ⅓); **B,** seedling (× ⅓); **C,** fruit (× 2); **D,** seed (× 6); **E,** fruit split open (× 6).
BALL MUSTARD, *Neslia paniculata:* **F,** plant showing habit (× ⅓); **G,** seed (× 4); **H,** fruit (× 4).

above, with decumbent or divergent stiff branches. Leaves lyrate-pinnatifid with linear to spatulate segments. Flowers in close racemes, much elongated in fruit; petals white or pinkish, 1.5–3 mm. long, oblanceolate to linear. Capsules linear, 1.5–3 cm. long, 1–2 cm. wide, flattened on widely ascending stout pedicels. Seeds narrowly winged, suborbicular, 1.2–1.5 mm. long.

Control.—The same as for 226.

224. *Sisymbrium altissimum* L. (*Norta altissima* Britton). Tumble mustard, Jim Hill mustard, Tall hedge mustard. Fig. 51, d–f.

Annual or winter annual; reproducing by seeds. Grain fields, new seedings, cultivated ground and waste places. Widespread across the northern United States and southern Canada; most common and troublesome in the Northwest and on the Pacific Coast. Introduced from Europe. June–September.

Description.—Stems erect, 1–2 m. high, with spreading branches above, pubescent with simple hairs, pale green. Leaves alternate, pinnately compound or deeply cleft with narrow segments. Flowers in racemes; petals pale yellow. Seed-pods 5–10 cm. long, narrow, cylindrical, spreading, stiff, on short thick pedicels, dehiscent into 2 many-seeded valves with 1–3 nerves; seeds about 1 mm. long, oblong to 3-angled, flattened on one side, light brown; cotyledons incumbent.

Control.—Clean cultivation will kill this weed (6). Hand pull or hoe stray plants before seeds are formed (4, 5). Harrow or disk grain fields several times after the crop is removed, repeat in the spring, and plow two weeks later. Harrow badly infested grain fields two or three times when the grain is 3–6 inches high.

225. *Sisymbrium officinale* (L.) Scop. Hedge mustard. Fig. 51, a–c.

Annual or winter annual; reproducing by seeds. Fields, gardens, and waste places. Common throughout the United States and southern Canada. Introduced from Europe. June–September.

Description.—Stems erect, stiff, with a few spreading branches above, 4–10 dm. high. Leaves alternate, pinnately lobed or parted, with broad, irregular toothed, lateral segments and a large terminal lobe. Flowers small, in racemes; petals yellow. Seed-pods awl-shaped, 1–2 cm. long, somewhat hairy, on very short pedicels, lying close

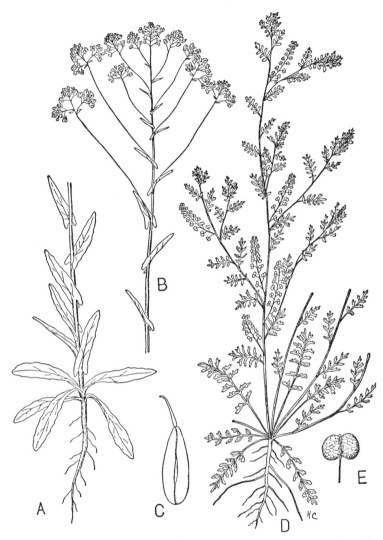

FIG. 56—**WOAD**, *Isatis tinctoria:* **A-B,** plant showing habit (× ⅓); **C,** fruit (× 1). **SWINE CRESS**, *Coronopus didymus:* **D,** plant showing general habit (× ⅓); **E,** fruit (× 3).

against the axis of the raceme, dehiscent, with 2 valves, each with several seeds; seeds about 1.5 mm. long, similar to 224.

Control.—The same as for 224.

Sisymbrium irio L., is a noxious weed (1954) in Arizona.

226. *Thlaspi arvense* L. Fan-weed, Penny-cress, French-weed, Stinkweed, Bastard cress. Fig. 53, a–f.

Annual or winter annual; reproducing by seeds. Grain fields, grasslands, gardens, and waste places. Widespread, but most abundant and troublesome in the prairie provinces of western Canada and the northwestern states. Frequently brought into the eastern states with western feed. Introduced from Europe. May–June.

Description.—Stems erect, branched above, 1–5 dm. high, glabrous. Leaves alternate, simple, toothed; lower petioled; upper sessile, somewhat clasping. Flowers in racemes; petals white. Seed-pods orbicular to obovate, about 1 cm. in diameter, broadly winged, notched at the top, flattened at right angles to the partition, dehiscent by 2 winged valves, each with 2–8 seeds; seeds 1.5–2 mm. long, obovate to ovate, flattened, with 10–14 curved, granular ridges on each side, dark reddish-brown; cotyledons accumbent.

Control.—Surface cultivation in late autumn and early spring (26). Harrow grain fields when the grain is 3–4 inches high (8). Disk or harrow stubble fields after the grain is removed (9, 8). Summer fallow (19).

227. *Thlaspi perfoliatum* L. Penny-cress.

Annual or winter annual; reproducing by seeds. Cultivated ground and waste places. Local in Ontario and New York; south to Virginia, Kentucky, and Missouri. Introduced from Europe. April–May.

Description.—Similar to 226 but smaller. Upper leaves perfoliate. Seed-pods about 5 mm. in diameter; seeds 1–1.4 mm. long, ovate to obovate, flattened, with a curved groove on each side, minutely granular, yellow-brown; cotyledons accumbent.

Control.—The same as for 226.

Chorispora tenella (Willd.) DC. Blue mustard, adventive from Asia, is a local weed along roadsides and in fields in northern United States.

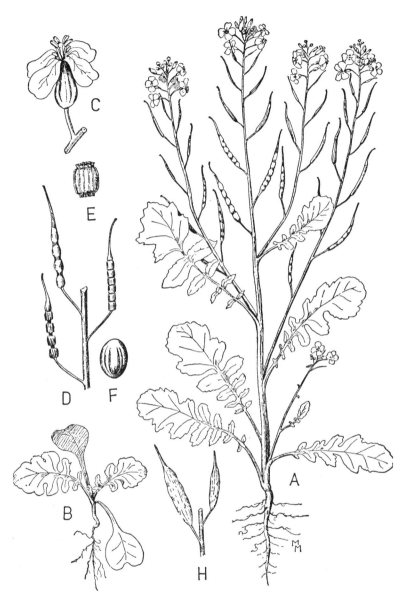

FIG. 57—**WILD RADISH,** *Raphanus raphanistrum:* **A,** plant (× ⅙); **B,** seedling in about the stage when spraying is most effective (× ⅓); **C,** flower (× 1); **D,** mature seed-pods (× ⅓); **E,** piece of pod containing one seed (× 2); **F,** seed (× 4).
RADISH, *Raphanus sativus:* **H,** mature seed-pods (× 1).

RESEDACEAE (Mignonette Family)

228. *Reseda lutea* L. Yellow mignonette, Cut-leaved mignonette, Dyers rocket. Fig. 58, e–g.

Perennial; reproducing by seeds. Meadows, old fields, and roadsides; on sandy soils. Local in the northeastern states; west to Michigan and south to Missouri. Introduced from Europe. June–October.

Description.—Stems erect or ascending, usually clustered, 3–10 dm. high. The roots and stems of this weed have a pungent or peppery taste. Leaves alternate, 1–2-pinnately compound or irregularly parted, with small gland-like stipules. Flowers in long slender racemes, perfect, irregular; sepals 6, forming a spreading calyx; petals 6, pale yellow; stamens 15–20, on a one-sided disk. Fruit a 3-lobed, 1-celled, open capsule with 3 parietal placentae; seeds about 1.5 mm. long, obovate, the coiled embryo visible through the seed-coat, surface smooth, glossy, dark green to nearly black.

Control.—Pull or hoe scattered plants before seeds are formed (4, 5). Clean cultivation will destroy this weed (6).

CRASSULACEAE (Orpine Family)

229. *Sedum acre* L. Mossy stonecrop.

Perennial; reproducing by stems rooting at the nodes; rarely by seeds. Lawns, pastures, cemeteries, roadsides and waste places; especially in dry stony soils and on rocky banks. Widely planted as an ornamental; escaped chiefly in the northeastern United States; also locally in the Pacific Northwest. Introduced from Europe. June–July.

Description.—Stems slender, prostrate, spreading on the ground with a moss-like habit; rooting very freely at the nodes. Leaves alternate, very small, ovate, very thick and fleshy, often overlapping, glabrous. Flowers and fruits like 230, but flowers with bright yellow petals.

Control.—The same as 230.

230. *Sedum purpureum* (L.) Link (*Sedum triphyllum* S. F. Gray). Live-for-ever, Garden orpine. Fig. 58, a–d.

Perennial; reproducing by fleshy tuber-like roots, stems rooting at the nodes, and rarely by seeds. Old stony meadows, pastures, roadsides, and waste places. Widely introduced throughout the northern United States, but apparently troublesome as a weed chiefly in the

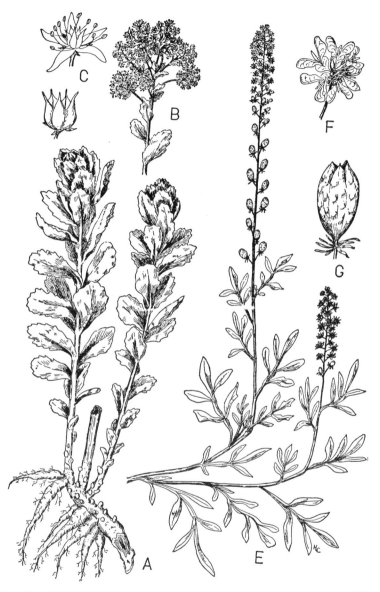

Fig. 58—**LIVE-FOREVER**, *Sedum purpureum:* **A,** plant showing habit (× ⅓); **B,** flower-cluster (× ⅓); **C,** flower (× 2); **D,** fruit capsules (× 2).
YELLOW MIGNONETTE, *Reseda lutea:* **E,** branch showing habit (× ⅓); **F,** flower (× 3); **G,** capsule (× 2).

northeastern states and adjacent Canada. Introduced from Eurasia. August–September.

Description.—Stems erect, usually in tufts, unbranched, stout, fleshy, 2–7 dm. high. Leaves alternate, simple, sessile, oval, with blunt teeth, very fleshy, glabrous. Flowers in crowded compound cymes, perfect, regular; sepals 5; petals 5, separate, purplish, oblong-lanceolate; stamens mostly 10; carpels 5, separate, maturing into abruptly pointed, several-seeded follicles.

Control.—Small patches or scattered clumps should be dug out, roots and all (17). Large areas should be cultivated for one or two years (6). When cultivation is not practicable, close grazing with sheep will control this weed (10).

ROSACEAE (Rose Family)

231. *Agrimonia gryposepala* Wallr. Agrimony, Feverfew, Beggar-ticks, Stickweed.

Perennial; reproducing by seeds. Old neglected fields, meadows and hedge-rows; on dry soil. Common in eastern North America, Tennessee, Missouri, and Kansas; infrequent on the Pacific Coast. Native. July–August.

Description.—Stems erect, 6–12 dm. high, hirsute, from a short rootstock with fibrous roots. Leaves alternate, irregularly pinnately compound; leaflets 5–9, ovate to elliptic-oblong, with rounded teeth, sometimes with small leaflets between the larger ones, slightly hairy on the lower surface. Flowers in long slender racemes, perfect, regular; sepals fused below into a top-shaped furrowed tube with hooked bristles on the throat and with 5 lobes above, becoming hard and inclosing 2 achenes when mature; petals 5, yellow; stamens 5–15, perigynous; pistils 2, with terminal styles. Achenes 1 or 2, inclosed in the dry hard calyx-tube which is 6–8 mm. long, top-shaped, and furrowed, with a margin of hooked bristles and 5 lobes at the apex.

Control.—Pull or hoe out the scattered plants in pastures as soon as the blossoms appear (4, 5). Badly infested fields should be plowed and planted to a cultivated crop for a season.

232. *Crataegus* spp. Thorn-apple, Hawthorn, White thorn.

Perennial shrubs or small trees; reproducing by seeds. Hillside pastures and neglected fields; especially in limestone regions of eastern North America. A number of species of Crataegus are common in old pastures in the northeastern states. They are the most troublesome of the shrubby weeds in New York state. Native. June.

Description.—Stems woody, much branched, forming shrubs or small trees, provided with prominent sharp thorns. Leaves alternate, simple, lobed or variously toothed, with narrow stipules which usually fall very early. Flowers in corymbs on short spur-like branches; calyx cup-shaped below, with 5 lobes; petals 5, white or pinkish, separate, inserted on the calyx disk; stamens 5–25, inserted in 1–3 rows on the calyx rim; carpels mostly 2–5, fused, and surrounded below by the calyx; styles 2–5, free. Fruit a fleshy pome with 1–5 bony nutlets, each with 1 seed; nutlets 5–8 mm. long, usually half-ovate, yellowish-brown.

Control.—Grub out small bushes (17). Pull larger bushes with a team and chain. Trees should be cut in midsummer, piled to dry and burned.

233. *Duchesnea indica* (Andr.) Focke. Indian strawberry, Mock strawberry.

Perennial; reproducing by seeds and runners. Lawns and grassy banks, waste places; especially about cities. Infrequent in the eastern states and the Mississippi Valley. Introduced from Eurasia. May–June.

Description.—Stems creeping or decumbent, rooting at the nodes. Leaves alternate, trifoliate, hairy; stipules fused with the base of the long petiole. Flowers solitary or a few in a cluster, perfect, regular; sepals 5, fused below, and alternating with 5 larger bract-like appendages; petals 5, yellow, separate; stamens numerous, separate; carpels numerous, separate. Fruit a spongy, but not juicy, red receptacle with numerous achenes scattered over its surface; achenes about 1 mm. long, dark red.

Control.—Rake the runners from lawns and mow close (11). Hoe out scattered plants and reseed the bare spots (5).

234. *Geum aleppicum* Jacq. var. *strictum* (Ait.) Fern. (*G. strictum* Ait.) Yellow avens.

Perennial; reproducing by seeds. Low meadows, pastures and thickets; usually on rich soils. Northeastern and central states, southern Canada, Pacific Coast. Native; also native to Asia. June–July.

Description.—Stems erect, stout, hairy, 8–15 dm. high. Lower leaves interruptedly pinnate, with wedge-shaped toothed leaflets; stem-leaves with 3–5 rhombic-ovate or oblong acute leaflets; stipules

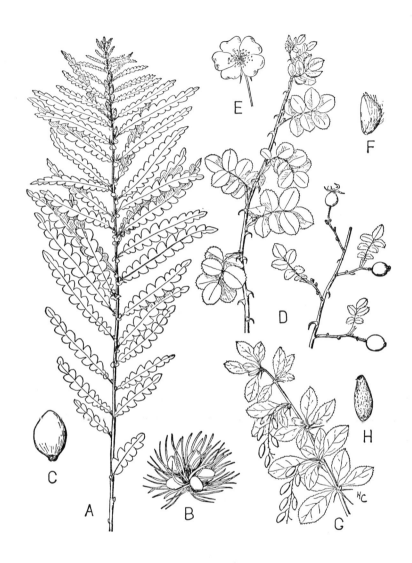

Fig. 59—**SWEET-FERN,** *Comptonia peregrina:* **A,** branch showing habit (× ⅓); **B,** cluster of nutlets (× 1); **C,** nutlet (× 3).
SWEET BRIER, *Rosa eglanteria:* **D,** branches showing leaves and fruits (× 1); **E,** flower (× ½); **F,** nutlet (× 3).
BARBERRY, *Berberis vulgaris:* **G,** branch showing racemes of berries (× ⅓); **H,** seed (× 3).

toothed, 15–40 mm. long. Flowers solitary or in corymbose clusters, perfect, regular; calyx bell-shaped, deeply 5-cleft, with 5 small bractlets in the notches; petals 5, bright yellow; stamens numerous; carpels numerous, separate, with jointed hairy styles, on a downy receptacle, forming an obovoid head of achenes in fruit. Achenes about 3 mm. long, elliptical, flattened, slightly ridged, hairy, with a hooked style exceeding the achene.

Control.—The same as for 231.

Geum macrophyllum Willd., a widespread species of the northern states, is a common weed in the Pacific Northwest. Other native species sometimes act as weeds in thickets or neglected meadows.

235. *Potentilla argentea* L. Silvery cinquefoil, Hoary cinquefoil. Fig. 61, d–f.

Perennial; reproducing by seeds. Dry, sandy, or gravelly fields, pastures, and lawns. Widespread and locally common in the limestone regions, mostly in the northeastern and north central states. Naturalized from Europe. June–August.

Description.—Stems ascending or nearly prostrate, 1–5 dm. long, branched, white-woolly. Leaves alternate, palmately compound, usually with 5 wedge-shaped toothed or lobed leaflets, green above and with white woolly hairs on the lower surface, margins revolute. Flowers in open leafy cymes, perfect, regular; calyx flat, white-hairy, deeply 5-cleft with a bractlet in each notch so that it appears 10-cleft; petals 5, rounded, yellow; stamens about 20; carpels numerous, separate, with slender terminal styles without glands. Achenes on a crowded dry hairy receptacle, about 0.7 mm. long, half-ovate, slightly flattened, smooth or reticulated, yellow-brown.

Control.—Plow, plant a smother crop (22), and turn under to increase the organic matter in the soil. Hand hoe or pull scattered plants (4, 5) in lawns and fields.

236. *Potentilla canadensis* L. (*P. pumila* Poir.) Cinquefoil. Fig. 60, c.

Perennial; reproducing by seeds and runners. Dry, sandy, or gravelly fields. Widespread in eastern North America. Native. May–June.

Description.—Very similar to 242 in general habit and appearance, but with more prostrate stems, and the lowest axillary flower produced at the node above the first long internode. Leaflets silky-

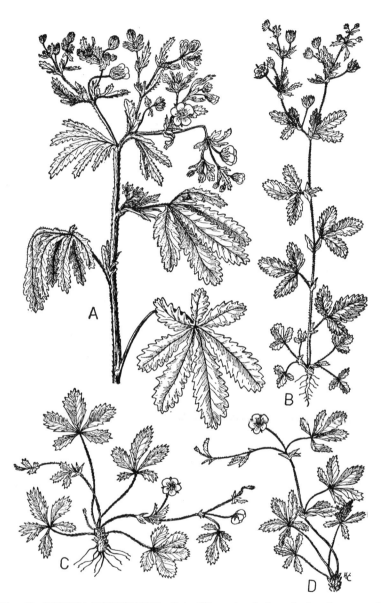

Fig. 60—**SULFUR CINQUEFOIL,** *Potentilla recta:* **A,** branch showing general habit (× ¼).
ROUGH CINQUEFOIL, *Potentilla norvegica:* **B,** plant showing habit (× ¼).
CINQUEFOIL, *Potentilla canadensis:* **C,** plant showing general habit (× ¼).
CINQUEFOIL, *Potentilla simplex:* **D,** plant showing general habit (× ¼).

hairy on the under surface, dentate. Achenes similar to 242 but more orange-brown.

Control.—The same as for 242.

237. *Potentilla fruticosa* L. Shrubby cinquefoil, Black brush, Prairie weed, Yellow hardhack. Fig. 61, a–c.

Shrubby perennial; reproducing by seeds and new shoots from roots. Old pastures and meadows; especially on wet limy soils. Common in the northeastern states and eastern Canada; local in the Great Lakes region, also in the mountainous regions of the western states. Native; also native of Eurasia. June–August.

Description.—Stems woody, with shreddy bark; much branched, low, spreading shrub, 2–8 dm. high. Leaves alternate, pinnately compound, with 5–7 closely crowded, narrow, entire leaflets about 1 cm. long, silky-hairy, with curled margins; stipules forming a sheath-like structure about the stem. Flowers perfect, regular, solitary or a few in a cluster; calyx 5-cleft, with 5 bractlets; corolla of 5 separate, round, yellow petals; stamens numerous; pistils numerous, with styles attached below the middle. Achenes on a hairy dry receptacle, similar to 242, but hairy.

Control.—Improve the drainage of pastures (21) and grub out the bushes (17), harrow or disk several times and reseed pastures. When plowing is practicable, plant a clean cultivated crop for a year before reseeding.

238. *Potentilla intermedia* L. Downy cinquefoil.

Perennial; reproducing by seeds. Dry stony pastures, meadows and waste places. Very local in the northeastern states. Introduced from Europe. June–July.

Description.—Stems erect, stout, grayish-hairy, rough, branched, 3–7 dm. high. Leaves alternate, palmately compound, with 3–5 oblanceolate to narrowly obovate dentate leaflets, green on the upper surface and grayish below. Flowers perfect, regular, in open somewhat leafy cymes; calyx hairy, 5-cleft; corolla of 5 large yellow petals; stamens about 20; pistils numerous, with filiform terminal styles developing into numerous achenes on a dry receptacle. Achenes about 1 mm. long, glabrous, half-ovate to kidney-shaped, marked with curved branched ridges, yellow-brown.

Control.—The same as for 231.

239. *Potentilla norvegica* L. (*P. monspeliensis* L.). Rough cinquefoil, Tall five-finger. Fig. 60, b.

Annual or biennial; reproducing by seeds. Fields, meadows, and old pastures. Widespread throughout the northern states and Canada. Introduced from Eurasia. Some of the forms or varieties included in this species appear to be native to North America. July–October.

Description.—Stems erect, rough, hairy, branched, about 2–9 dm. high. Leaves alternate, trifoliate; leaflets narrowly oblong, the upper often 3–5-toothed near the end. Flowers perfect, regular, in open leafy cymes; calyx 5-cleft, hairy; corolla of 5 small yellow petals; stamens about 10–20; pistils numerous, with thick terminal styles, developing into achenes on a dry receptacle. Achenes about 0.8 mm. long, similar to 238.

Control.—Clean cultivation will destroy this weed (18).

240. *Potentilla recta* L. Sulfur cinquefoil, Rough-fruited cinquefoil, Tormentil. Fig. 60, a.

Perennial; reproducing by seeds. In meadows and pastures; on dry, gravelly, or stony soils; most troublesome in limestone regions. Locally abundant in the northeastern states and Ontario. Introduced from Europe. June–August.

Description.—Stems erect or spreading, 3–7 dm. high, branched above, hairy. Leaves alternate, palmately compound, with 5–9 oblanceolate hairy leaflets with dentate margins. Flowers perfect, regular, in many-flowered, compact, almost leafless cymes; calyx hairy, 5-cleft, with 5 bractlets; corolla of 5 pale yellow petals, about 2 cm. in diameter; stamens numerous, pistils numerous, with filiform terminal styles, maturing into glabrous achenes on a dry receptacle. Achenes similar to 238 but with a winged margin.

Control.—Pull or hoe scattered plants from meadows when the first flowers appear (4, 5). Badly infested meadows should be mowed early to avoid spreading of seeds. Plowing followed by clean cultivation for a year will destroy this weed.

241. *Potentilla reptans* L.

Annual; reproducing by seeds and by rooting from the tips of stems. Lawns, grassy roadsides, and waste places. Local, Nova Scotia and Ontario, south to Massachusetts and Virginia. Introduced from Europe. May–August.

Description.—Stems prostrate and creeping, elongate, up to 1 m. long. Leaves palmately compound, principle leaves long-petioled, cauline ones short-petioled, all with 5 or 7 leaflets (occasionally 4 or 3), obovate, oblong-obovate, or oblanceolate, the median 2–5 cm. long, the lateral progressively smaller, all crenate-serrate. Flowers 1 or rarely 2 from the node, yellow, 1.5–2.5 cm. wide, on long peduncles, bractlets often longer than the sepals.

Control.—The same as for 240.

242. *Potentilla simplex* Michx. (*P. canadensis* L.) Cinquefoil, Five-finger, Barren strawberry. Fig. 60, d.

Perennial; reproducing by seeds and runners. Dry, gravelly, or sandy fields and old meadows. Common from New England to the Great Lakes and southward. Native. May–June.

Description.—Stems slender, spreading, procumbent, often rooting at the tip, hairy. Leaves alternate, palmately compound, with 3 leaflets, the 2 lateral ones parted so that there appear to be 5 leaflets, margin serrate; petioles slender, with stipules. Flowers perfect, regular, solitary in the leaf-axils; first flower at the node above the second long internode; calyx flat, 5-cleft, with 5 bractlets so that it appears 10-cleft; corolla of 5 separate yellow petals; stamens numerous; pistils numerous, with lateral styles, maturing into glabrous achenes on a dry hairy receptacle. Achenes about 1.2 mm. long, half-ovate, marked with wavy broken ridges and rounded tubercles, yellow-brown.

Control.—When practicable, increase the soil fertility and organic matter by adding fertilizer and plowing under a crop of rye. Land infested with this weed is often acid and otherwise poorly adapted for any purpose but reforestation.

243. *Prunus virginiana* L. Choke-cherry.

Shrubby perennial; reproducing by seeds and new shoots from roots. Fence-rows, borders of neglected fields and waste places. Common throughout eastern North America, Missouri and Kansas. Native. May.

Description.—Stems erect, woody, much branched, forming bushes about 1–3 m. high, rarely small trees; bark gray, the inner layers with an odor and taste of bitter almonds. Leaves alternate, simple, ovate to obovate, with finely serrate margin; petiole with 1 or more glands near the upper end. Flowers perfect, regular, in slender drooping racemes; calyx bell-shaped with 5 lobes, deciduous after flowering; corolla of 5 spreading separate white petals attached on the calyx rim; stamens about 20, attached on the calyx. Pistil solitary, maturing into a fleshy red to nearly black fruit with a hard

Fig. 61—**SHRUBBY CINQUEFOIL,** *Potentilla fruticosa:* **A,** branch showing leaves and flowers (×⅓); **B,** flower (× 1); **C,** seed (achene), (× 12).
SILVERY CINQUEFOIL, *Potentilla argentea:* **D,** plant showing habit (× ⅓); **E,** cluster of achenes (× 2); **F,** seed (achene), (× 12).

Rose Family

stone (drupe); stone about 5 mm. long, ovoid, granular, yellow-brown.

Control.—Grub out small bushes (17). Pull large bushes with a team and chain. Fence-rows and large areas should be cut in midsummer and burned.

244. *Rosa eglanteria* L. (*R. rubiginosa* L.). Sweet brier, Eglantine, Sweetleaf rose. Fig. 59, d–f.

Shrubby perennial; reproducing by seeds. Pastures, old fields, and waste places. Widespread in eastern North America; locally common in western Washington and Oregon. Introduced from Europe. June–July.

Description.—Stems erect or recurved, branched, armed with stout hooked prickles. Leaves alternate, pinnately compound; leaflets doubly serrate, resinous on the lower surface, sweet-scented; stipules fused with the petiole for about one-half its length. Flowers perfect, regular, solitary; calyx urn-shaped, contracted above, with 5 lobes; corolla of 5 separate pink petals inserted on the calyx rim; stamens numerous on the calyx rim; pistils numerous, hairy, inserted within the urn-shaped calyx-tube. Ovaries maturing into hard bony achenes surrounded by the fleshy calyx-tube with deciduous sepals, producing the "rose hip" or "rose apple"; achenes about 4 mm. long, ovate or irregularly flattened, often hairy, pale yellow or brown.

Control.—The same as for 243.

Rosa arkansana Porter, Prairie rose, is listed (1954) as a weed in the north central states.

Some of the native wild roses, *Rosa* spp., in the Great Plains states and the prairie provinces of Canada sometimes persist in fields, meadows and pastures.

245. *Rubus flagellaris* Willd. (*R. villosus* of Gray's Manual, ed. 7). Dewberry, Running blackberry, Trailing bramble.

Trailing shrubby perennial; reproducing by seeds and runners. Fields, meadows and pastures; on dry sandy or gravelly soils. Widespread in the northeastern United States; Minnesota, Quebec, and Ontario; less frequent southward. Native. June.

Description.—Stems somewhat woody, biennial, prostrate or trailing, slender, often 2–3 m. long, armed with stout recurved prickles, during the second year producing erect flowering branches, 1–3 dm. high. Leaves alternate, compound, mostly with 5 leaflets on the new stems and 3 leaflets on the flowering stems; leaflets hairy, with serrate margins. Flowers in short, open, leafy, corymb-like racemes; calyx 5-parted; corolla of 5 white petals, about 2–3 cm. in diameter; stamens numerous, inserted on the calyx rim; pistils numerous, maturing into small black drupes which cling to the receptacle, "a berry."

Control.—Grub out scattered clumps (17). Extensive areas should be mowed and burned (13). Plow in autumn, disk several times and plant a smother crop the following spring; follow with a cultivated crop before reseeding.

Several other kinds of brambles, Rubus spp., are sometimes troublesome on logged-off lands, new clearings or in old neglected fields.

246. *Sanguisorba minor* Scop. Garden burnet, Salad burnet, Pimpernelle.

Perennial; reproducing by seeds. Dry gravelly fields, meadows and pastures. Local in the northeastern states, south to Virginia and Tennessee. Introduced from Eurasia. June–July.

Description.—Stems ascending or spreading, mostly tufted, 3–5 dm. high. Leaves alternate, pinnately compound, leaflets 1–2 cm. long, ovate, deeply toothed. Flowers in a globular, greenish head, the lower perfect, the upper pistillate; calyx with 4-angled tube with 4 petaloid lobes; petals none; stamens 12 or more, with slender drooping filaments; pistils 1–3, maturing into achenes inclosed in the dry, 4-angled, winged calyx-tube, about 3 mm. long, brown.

Control.—Clean cultivation (6).

247. *Spiraea alba* DuRoi (*S. salicifolia* of Gray's Manual, ed. 7, not L.). Meadowsweet. Fig. 62, a–c.

Shrubby perennial; reproducing by seeds and rootstocks. Moist pastures and meadows; usually in limy soils. Widespread in eastern North America. Native. August–September.

Description.—Stems woody, erect, stiff, yellowish-brown, mostly in clumps; shrubs 1–2 m. high. Leaves alternate, simple, lanceolate-oblong, finely serrate, mostly entire toward the base, about 4–7 cm. long, green on both sides. Flowers in a stiff, erect, panicle-like cluster, perfect, regular; calyx-tube very short, 5-cleft, persistent; corolla of 5 separate white petals, inserted with the numerous stamens on the calyx rim; pistils 5, rarely more, maturing into few- to several-seeded persisting follicles. Seeds about 2 mm. long, linear, flattened or 3-angled, yellow-brown.

Control.—Improve drainage (21). Grub out scattered clumps with

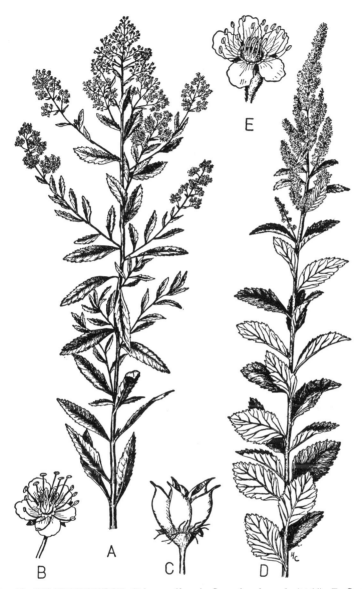

Fig. 62—**MEADOWSWEET**, *Spiraea alba:* **A,** flowering branch (× ⅓); **B,** flower (× 2); **C,** mature fruit (follicles), (× 2).
STEEPLEBUSH, *Spiraea tomentosa:* **D,** flowering branch (× ⅓); **E,** flower (× 2).

a mattock (17), harrow or disk several times and reseed. Large clumps can be pulled with a team and chain. Dense growths of brush should be cut with a brush hook, burned, and the land plowed when possible, and followed by a clean cultivated crop for a year before reseeding.

248. *Spiraea douglasii* Hook. Hardhack, Hackbrush.

Shrubby perennial; reproducing by seeds and rootstocks. Low wet pastures and meadows; especially on peat soil. British Columbia to Oregon, west of the Cascade Mountains. Native. June–July.

Description.—Stems woody, erect, stiff, mostly in clumps, 1–2 m. high. Plants with general habit like 247, but with leaves broader and woolly on the lower surface; flowers pink or rose-colored; seed-pods glabrous; seeds shorter than in 247.

Control.—The same as for 247.

249. *Spiraea latifolia* (Ait). Borkh. Meadowsweet, Hardhack.

Shrubby perennial; reproducing by seeds and rootstocks. Pastures and old fields; on rocky or sandy soils. Widespread in eastern North America. Native. July–August.

Description.—Very similar to 247, but mostly somewhat lower and with more spreading branches with red or purplish-brown bark. Leaves elliptical, 1.5–4 cm. broad and more coarsely serrate. Corolla pink or white. Seeds similar to 247.

Control.—The same as for 247.

250. *Spiraea tomentosa* L. Steeplebush, Hardhack, Woolly meadowsweet, Silver-weed, White cap. Fig. 62, d–e.

Shrubby perennial; reproducing by seeds and shoots from creeping roots. Moist fields and pastures; on low sandy or stony soils. Locally common in eastern North America. Native. July–August.

Description.—Similar to 247, but with less branched, stiff, woolly stems frequently forming dense clumps about 1 m. high. Leaves densely white-woolly on the lower surface. Flowers in short, dense, erect, spike-like panicles; petals rose-color or sometimes pale or nearly white. Follicles persisting from year to year; seeds similar to 247 but shorter and russet-brown.

Control.—The same as for 247.

LEGUMINOSAE (Pulse Family)

251. *Alhagi camelorum* Fisch. Camel thorn.

Shrubby perennial; reproducing by seeds and extensive deep-rooted

Pulse Family

rootstocks. In date plantings, alfalfa fields and other crops, waste places. California from Merced County to the Imperial Valley, in 1925, but now in many of the western states. Introduced from Eurasia, probably prior to 1915. Possibly also introduced from the date-producing sections of North Africa.[1] June.

Description.—Stems woody, but dying to the ground in winter, 3–10 dm. high, much branched and thorny. Leaves simple, alternate, petioled, entire, glabrous and thick. Flowers in panicles, perfect, irregular; sepals 5, fused at the base; petals 5, a standard, 2 wings, and the 2 lower fused into a keel, purple; stamens 10. Carpel 1, developing into a bead-like pod with 1 seed in each of the several lobes; seeds similar to alfalfa seeds.

Control.—Clean cultivation (18). Flooding the soil has proved effective in some localities.[2]

252. *Cassia fasciculata* Michx. (*C. chamaecrista* L.). Partridge pea, Large-flowered sensitive plant.

Annual; reproducing by seeds. Dry sandy fields, meadows and pastures. From the Atlantic Coast to Minnesota and South Dakota, southward to Mexico. Native. July–September.

Description.—Stems erect or ascending, branched below, somewhat hairy. Leaves alternate, pinnately compound, with 10–15 pairs of linear-oblong oblique leaflets; stipules striate, persistent. Flowers in small clusters above the leaf-axils, on slender pedicels; sepals 5, separate, except at base; petals 5, unequal, spreading, yellow or some of them with a purple spot near the base; stamens 10, all with anthers, unequal. Pistil 1, with slender style, maturing into a several-seeded flat legume; seeds 3–4 mm. long, mostly oblong-rectangular with convex sides, with rows of shallow depressions, dull, dark brown with lighter edges. The plants are cathartic and produce "scours" in cattle feeding on them.

Control.—Clean cultivation (6). Plants in pastures should be mowed or pulled as soon as the blossoms appear (11).

253. *Cassia marilandica* L. Wild senna.

Perennial; reproducing by seeds. Pastures and old fields; especially on alluvial soils. Widespread in the eastern United States, west to Iowa, Kansas, and Texas. Native; sometimes introduced northward. July–August.

Description.—Stems erect, often in clumps, coarse, glabrous, about

[1] Bottel, 1933.   [2] Ball and Robbins, 1933.

1 m. high. Leaves alternate, pinnately compound with 5–9 pairs of lance-oblong obtuse leaflets, petiole with a slender club-shaped basal gland; stipules linear, deciduous. Flowers in short axillary racemes; sepals 5; petals 5, yellow, spreading; stamens 10, 7 with anthers. Legume linear, flat, curved, 6–12 cm. long, 8–11 mm. wide, segmented and many-seeded; seeds 4–5 mm. long, obovate, flattened, dull, grayish-brown with darker area in center of each side.

Control.—Clean cultivation (18). Mow pastures (11). Grub out scattered plants in pastures (17).

254. *Cassia nictitans* L. Partridge pea.

Annual; reproducing by seeds. Sandy fields, meadows and waste places. Widespread throughout the eastern and southern states. Native. July–September.

Description.—Similar to 252. Leaves with 10–20 pairs of leaflets. Flowers smaller, 4–8 mm. broad, on short pedicels; anther-bearing stamens 5, nearly equal. Seeds similar to 252, 2–3 mm. long, glossy.

Control.—The same as for 252.

255. *Cassia tora* L. Coffee weed, Sicklepod.

Annual; reproducing by seeds. Waste places, fields, moist woods, barnyards, rich land. Pennsylvania to Florida, west to Missouri, Kansas and Mexico. Probably introduced from tropical America. July–September.

Description.—Stems 4–15 dm. high, nearly glabrous. Leaves alternate, pinnately-compound with 4–6 elliptic to obovate leaflets, the terminal pair larger, 9–7 cm. long, petiolar gland between lower pair of leaflets. Flowers large in axillary, loose, few-flowered racemes; petals yellow, 10–15 mm. long. Stipules prominent. Legume falcate, 4-angled, 3–4 mm. wide, dehiscent. Seeds thick, shiny, truncate at ends, with diagonal stripe on sides.

Control.—The same as for 252.

Cassia occidentalis L. Coffee senna, an annual introduced from the tropics, occurs in waste places and fields in the southern states.

256. *Crotalaria sagittalis* L. Rattle-box, Rattleweed, Wild pea. Fig. 64, f–g.

Annual; reproducing by seeds. Fields and waste places; on sandy soil. Along the Atlantic Coast, Massachusetts to Florida; north central states and lower Mississippi Valley. Native. June–July.

Description.—Stems erect or spreading, branched, 6–30 cm. high, hairy. Leaves alternate, simple, oval or oblong-lanceolate, entire, with very short petioles; stipules prominent, decurrent on the stem. Flowers solitary or a few on a peduncle; calyx 5-cleft, somewhat 2-lipped; corolla irregular, yellow, not longer than the calyx, with a large heart-shaped petal (standard) and a scythe-shaped keel formed by the fusion of the 2 lower petals, and 2 lateral petals; stamens 10, fused by the filaments into a cleft sheath, 5 stamens with smaller anthers. Pistil 1, maturing into a many-seeded, blackish, inflated pod; when the pod is dry, the loose seeds rattle in it; seeds 2–3 mm. long, short kidney-shaped, glossy, brown.

Control.—Clean cultivation (6).

257. *Cytisus scoparius* L. Scotch broom.

Shrubby perennial; reproducing by seeds. Atlantic Coast, mostly from Virginia southward; a pest on the Pacific Coast from British Columbia to Oregon. Introduced from Europe as an ornamental shrub. May–June.

Description.—Woody, erect, 1–3 m. high, glabrous or nearly so, much branched with many slender, stiff, angled, green branches. Leaves alternate, mostly with 3 small obovate entire leaflets, the upper often reduced to a single leaflet. Flowers solitary or in pairs on slender peduncles in the axils of upper leaves; sepals 5, fused into a 2-lipped calyx; petals 5, a standard, 2 wings, and the 2 lower forming a broad keel; bright yellow; stamens 10, monadelphous; pistil with a long spirally curved style. Legume flat, several-seeded; seeds 2.5–4 mm. long, oblong to ovate, base truncate or notched, sides flat or convex, with caruncle attached, glossy, olive-brown to reddish-brown.

Control.—Prevent seed formation (3). Grub out scattered bushes (17). Cut off the large clumps and burn them when dry.

258. *Desmodium canadense* (L.) DC. Tick trefoil, Canada tick trefoil, Sainfoil.

Perennial; reproducing by seeds. Old fields, pastures, thickets and borders of woodlands; mostly on gravelly soils. Eastern North America, southward to North Carolina and Kansas. Native. August–September.

Description.—Stems erect, coarse, branched above, 5–15 dm. high, hairy. Leaves alternate, trifoliate; leaflets oblong-lanceolate, obtuse, entire, 3–8 cm. long, exceeding the petiole; stipules small, mostly deciduous. Flowers in dense racemes, perfect; calyx somewhat 2-lipped; corolla irregular, about 1 cm. long, with 5 rose-colored to

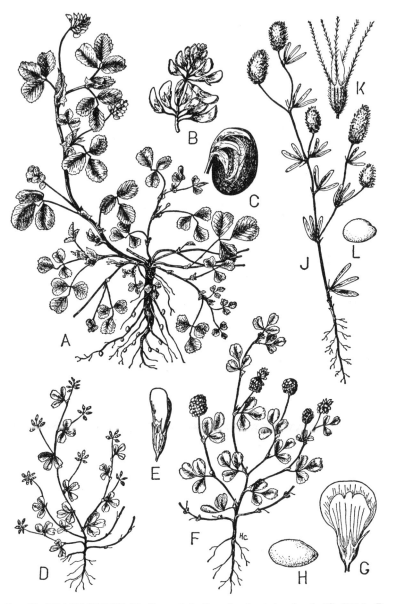

Fig. 63—**BLACK MEDIC**, *Medicago lupulina:* **A,** plant showing habit (× ½); **B,** flower-cluster (× 6); **C,** one-seeded legume (× 6).
LITTLE HOP CLOVER, *Trifolium dubium:* **D,** plant (× ⅓); **E,** flower (× 4).
LOW HOP CLOVER, *Trifolium procumbens:* **F,** plant showing habit (× ⅓); **G,** flower (× 4); **H,** seed (× 12).
RABBIT-FOOT CLOVER, *Trifolium arvense:* **J,** plant showing habit (× ⅓); **K,** flower (× 4); **L,** seed (× 12).

purple petals, the upper petal (standard) obovate, the 2 lower petals fused into a straight keel with the lateral petals (wings) adhering to it; stamens 10, 9 fused by their filaments and 1 separate. Pistil 1, maturing into a flat deeply jointed pod, separating into 3–6 joints each with 1 seed, and with hooked hairs on the surface; seeds about 3 mm. long, kidney-shaped, dull, reddish-brown.

Control.—Mow pastures (11); close grazing (10); clean cultivation (6).

Several of the native Desmodiums of the eastern United States may encroach upon pastures, especially those bordering on woodlands.

259. *Genista tinctoria* L. Dyers broom. Dyeweed, Woad waxen, Dyers greenweed.

Shrubby perennial; reproducing by seeds and creeping rootstocks. Dry, sterile, sandy fields, hillsides, and pastures. Northeastern states, local west to Michigan. Introduced from Europe. June–July.

Description.—Low, erect, woody shrubs, about 1 m. high, with slender, erect, ridged or angular branches. Leaves alternate, simple, lanceolate, entire. Flowers in spike-like racemes, perfect, irregular; calyx 2-lipped; corolla yellow, of 5 petals, oblong spreading standard, 2 wings, and the lower 2 forming a straight deflexed keel; stamens 10, fused by their filaments into a complete sheath (monadelphous), with long and short anthers alternating. Pistil 1, maturing into a flat several-seeded legume, seed 2–3 mm. long, nearly circular with a notch at hilum, glossy, yellow to reddish-brown.

Control.—The same as for 257.

260. *Glycyrrhiza lepidota* (Nutt.) Pursh. Wild licorice, Sweet root, American licorice.

Perennial; reproducing by seeds and creeping roots. Prairies, meadows, pastures, and waste places. From the prairie provinces of Canada and Minnesota southward and westward to the Pacific Coast, adventive eastward. Native. May–August.

Description.—Stems erect, branched, 3–9 dm. high, scurfy or glandular-viscid. Leaves alternate, pinnately compound with 11–19 entire mucronate leaflets, speckled with scales or dots when mature; stipules minute. Flowers in short axillary spikes on long peduncles; calyx 5-toothed; corolla with a narrow standard and blunt keel, whitish; stamens 9 fused by their filaments and 1 separate. Pistil maturing into a legume 1–2 cm. long, few-seeded, covered with stout hooked prickles, scarcely dehiscent, bur-like and widely distributed by animals; seeds about 3 mm. long, more or less kidney-shaped, smooth, dull, greenish to reddish-brown.

Control.—Clean cultivation (6); close and frequent mowing (11).

261. *Lathyrus palustris* L. Wild pea, Marsh vetchling.

Perennial; reproducing by seeds. Rich moist fields and thickets. Locally common from the St. Lawrence Valley to the Great Lakes region; also in the Pacific Northwest. Native; also native to Eurasia. July–August.

Description.—Stems slender, mostly winged, glabrous, about 1 m. high, more or less climbing by tendrils. Leaves alternate, pinnately compound, mostly with 4–10 lanceolate or elliptic firm leaflets, 3–7 cm. long; stipules broadly heart-shaped, oblique. Flowers perfect, irregular, 3–8 on an axillary peduncle; calyx 5-toothed, irregular; corolla purplish, 1.5–2.5 cm. long, of 5 irregular petals, a standard and 2 lateral wings somewhat fused to the keel; stamens 10, 9 fused by the filaments and 1 free (diadelphous). Pistil 1, with a flat style which is hairy on the inner side, maturing into a several-seeded legume; seed globular, 3 mm. in diameter, hilum elliptical, only slightly depressed, smooth, dull, reddish-brown with black markings.

Control.—Clean cultivation (18); close cutting of meadows before seeds are ripe (11).

262. *Lathyrus pratensis* L. Meadow pea, Yellow vetchling, Yellow tare, Craw pea, Tar-fitch.

Perennial; reproducing by seeds. Meadows, fields and waste places. Newfoundland to Ontario, south to New England and west to Illinois. Introduced from Europe. June–August.

Description.—Stem low, straggly, or climbing. Leaves alternate, compound, with 2 bright green, lanceolate or linear, acute leaflets and a tendril. Flowers 3–9 on a peduncle, similar to 261 but smaller and with yellow petals. Seeds similar to 261 but somewhat smaller, purplish-brown or yellowish-brown with purplish spots, glossy.

Control.—The same as for 261.

263. *Lespedeza violacea* (L.) Pers. Bush clover.

Perennial; reproducing by seeds. Dry sterile pastures, meadows and waste land. Widespread eastern, southern and north central United States. Native. July–September.

Description.—Stems erect or spreading, slender, branched, 3–8 dm. high, somewhat hairy. Leaves alternate, trifoliate; leaflets oval or oblong, mostly 2–5 cm. long, hairy on the lower surface; stipules slender, 5–8 mm. long. Flowers in raceme-like panicles, irregular, of two kinds; the larger flowers perfect, with violet-purple petals, 6–8 mm. long, 10 stamens, 9 fused and 1 separate, 1 pistil, but seldom maturing a fruit; the smaller flowers apetalous, pistillate, maturing into a single 1-seeded flat legume, 4–6 mm. long. Seeds about 2 mm. long, ovate, with slight protuberance at hilum, yellow to purplish-brown, smooth, dull.

Control.—Mow close to the ground as soon as the first blossoms appear. Close grazing (10). Plow badly infested fields, fertilize, and sow a smother crop (22) before reseeding.

Lespedeza striata (Thunb.) H. and A., introduced from Asia, occurs in old fields and waste places from New Jersey to Kansas and southward. Several of the native Lespedezas, common in dry open woodlands, sometimes grow in old pastures and fields in the eastern states.

264. *Lupinus perennis* L. Lupine, Quakers bonnets, Sun dial, Indian beans, Blue pea. Fig. 64, a–e.

Perennial; reproducing by seeds and rootstocks. Dry fields, pastures and woodlands; chiefly in sandy soils. Widespread throughout eastern North America. Native. May–June.

Description.—Stems erect, stout and succulent, hairy, 3–6 dm. high. Leaves alternate, on slender petioles, palmately compound with 7–11 oblanceolate entire leaflets, hairy. Flowers in long terminal racemes; calyx deeply 2-lipped; corolla very showy, purplish-blue, standard with turned-back sides, 2 wings closing over the scythe-shaped keel; stamens 10, fused by the filaments into a closed sheath. Pistil forming a broad hairy 4–7-seeded legume; seeds 4–5 mm. long, obovate, smooth, yellowish-white with brown or black markings.

Control.—Pull or hoe scattered plants by hand (4, 5). In fields plow deep (7) and follow by clean cultivated crop (6).

A number of native lupines occurring among the native vegetation on ranges and meadows in western North America are sometimes considered as weeds on account of their poisonous properties.[3] Most of these do not persist under cultivation.

265. *Medicago lupulina* L. Black medic, Trefoil, Black clover, None-such, Hop medic. Fig. 63, a–c.

[3] Marsh, Clawson and Marsh, 1916.

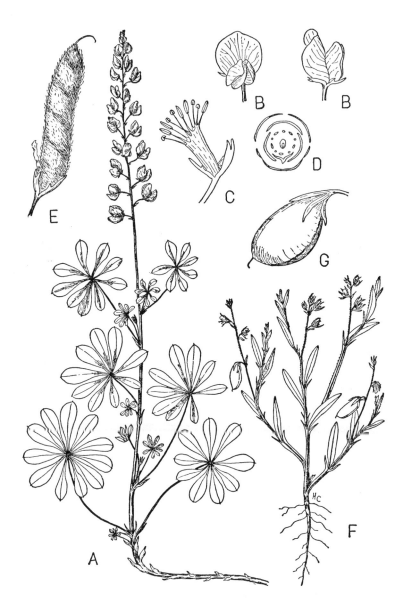

Fig. 64—**LUPINE**, *Lupinus perennis*: **A**, plant showing habit ($\times \frac{1}{4}$); **B**, flower ($\times 1$); **C**, flower with parts removed to show stamens ($\times 2$); **D**, diagram of cross-section of flower; **E**, mature seed-pod ($\times 1$).
RATTLE-BOX, *Crotalaria sagittalis*: **F**, plant showing habit ($\times \frac{1}{4}$); **G**, seed-pod ($\times 1$).

Pulse Family

Annual or winter annual; reproducing by seeds. Fields, gardens, lawns and waste places; on somewhat sterile soils. Widespread throughout North America. Introduced from Eurasia. May–September.

Description.—Stems procumbent or prostrate, branched below, rough and hairy. Leaves alternate, trifoliate; leaflets wedge-shaped, toothed at the apex, the terminal one on a jointed stalk; stipules prominent. Flowers in dense short spikes, perfect, irregular; calyx 5-cleft, persistent; corolla yellow, of 5 petals (standard, 2 lateral wings and the 2 lower petals fused into a keel); stamens 9 fused and 1 separate. Pistil solitary, maturing into a kidney-shaped or coiled, black, ridged, 1-seeded pod (legume); seeds 1.5–2 mm. long, oval to short kidney-shaped, with a protuberance at the hilum, sides convex, smooth, dull, greenish-brown to orange-brown; pericarp often persisting. Black medic seeds are often confused with those of *Medicago sativa*, Alfalfa, which are about 2–2.5 mm. long, mostly kidney-shaped, often angular or flattened near the ends with the hilum in a depression near the middle of the seed.

Control.—Clean cultivation (6).

266. *Melilotus alba* Desr. White sweet clover, Honey clover, Tree clover.

Biennial; reproducing by seeds. Gravelly or sandy fields, roadsides and waste places. Widespread across the United States and southern Canada. Introduced from Eurasia. June–October.

Description.—Stems erect or nearly so, branched, 1–2 m. high, glabrous. Leaves alternate, trifoliate; leaflets glabrous, narrowly obovate to oblong, serrate, blunt or indented at the apex, the terminal leaflet on a jointed stalk; stipules prominent. Flowers in small spike-like racemes, perfect, irregular, similar to 265 but corolla white; the standard longer than the 2 wings and keel. Pistil maturing into an ovoid, reticulate, 1- or rarely 2-seeded legume, 2–3.5 mm. long, with reticulate surface, legume often persisting over the seed; seed about 2 mm. long, oblong to oval, notched near one end, smooth, dull, yellowish-green to orange-brown.

Control.—Pull or hoe scattered plants (4, 5). In fields plow deep and follow with a clean cultivated crop.

The white and also the yellow sweet clover (267) are grown rather extensively for forage, pasture and green-manure crops in certain regions.

267. **Melilotus officinalis** (L.) Lam. Yellow sweet clover.

Annual or biennial; reproducing by seeds. Neglected fields, pastures, roadsides and waste places. Widespread, across the United States and southern Canada. Introduced from Eurasia. June–October.

Description.—Very similar to 266 in general appearance, but stems less erect and usually not so stout, much branched; petals yellow, nearly equal in length. Legume 2.5–3.5 mm. long, with prominent transverse ridges, very slightly netted, glabrous; seed the same as in 266.

Control.—The same as for 266.

This plant is sometimes confused with the tall yellow sweet clover, *Melilotus altissima* Thuill., which also occurs in the eastern states. The latter also has a yellow corolla, but its legumes are 4.5–5.5 mm. long, and netted and hairy on the surface.

268. **Trifolium agrarium** L. Yellow clover, Hop clover, Hop trefoil.

Annual; reproducing by seeds. Old fields, meadows and pastures; on dry rather poor soils. Northeastern United States and adjacent Canada. In many pastures and fields on the acid soils on the higher hills in southern New York this clover provides a considerable amount of forage. Introduced from Europe. June–August.

Description.—Stems erect or spreading, 1–3 dm. high, glabrous. Leaves alternate, palmately compound, with 3 obovate-oblong leaflets, nearly sessile and attached at the same point; stipules narrow, fused to the petiole over half its length. Flowers small, perfect, irregular, pedicelled, in loose heads; pedicels recurved with age; calyx 5-cleft, persistent; corolla yellow, withering and persistent, becoming striate, dry and brown with age; the petals more or less fused; stamens 10, 9 fused by the filaments and 1 separate. Pistil 1, maturing into a 1- or few-seeded legume; seed 1 mm. long, oblong, notched near one end, smooth, dull, yellow-green to orange-brown.

Control.—Close pasturing to prevent seed formation (10); clean cultivation (6).

269. **Trifolium arvense** L. Rabbit-foot clover, Stone clover. Old-field clover. Fig. 63, j–l.

Annual; reproducing by seeds. Fields, meadows and waste places; in dry, sandy, or stony soil. Quebec to Florida, west to the Pacific. Introduced from Europe. July–September.

Description.—Stems erect, slender, much branched, silky-hairy, 1–4 dm. high. Leaves alternate, compound, with 3 narrow oblanceolate leaflets. Flowers perfect, sessile, in ovoid-cylindrical heads, very grayish-hairy when mature; calyx 5-cleft, with 5 bristly silky-hairy teeth, exceeding the pale pink corolla; otherwise similar to the flowers in 268. Legume 1-seeded; seed similar to 268, but only slightly notched.

Control.—Improve the soil by applying fertilizer or plow under a green-manure crop, and follow with a clean cultivated crop.

270. *Trifolium dubium* Sibth. Little hop clover. Fig. 63, d–e.

Annual; reproducing by seeds. Dry, gravelly, or sandy fields, pastures, and lawns. Widespread; most common in the northeastern United States and on the Pacific Coast. Introduced from Europe. June–July.

Description.—Similar to 268 but much smaller, with weak, ascending or creeping stems; the terminal leaflet of each leaf on a short jointed stalk; heads with 5–12 flowers; corolla not striate; seeds similar to 268 but glossy and larger, 1–1.3 mm. long.

Control.—In lawns raise the stems with a rake and mow close (11). Hoe out scattered plants (5). In fields plow under and follow with a cultivated crop (7).

271. *Trifolium procumbens* L. Low hop clover, Hop trefoil. Fig. 63, f–h.

Annual; reproducing by seeds. Pastures, lawns and waste places; mostly on gravelly soil. Widespread throughout North America. Introduced from Europe. June–August.

Description.—Similar to 268 but smaller, with procumbent or creeping hairy stems, 1–2 dm. high; terminal leaflet of each leaf on a short jointed stalk; heads many-flowered; seeds similar to 268 but larger, 1–1.5 mm. long.

Control.—The same as for 270.

272. *Ulex europaeus* L. Gorse, Furze.

Shrubby perennial; reproducing by seeds and creeping roots. Common on the Pacific Coast; local along the Atlantic Coast from Massachusetts to Virginia. Introduced from Europe as an ornamental. May–June.

Description.—Stems woody, densely branched, 1–2 m. high, with many stiff, spreading, green and leafy thorns. Leaves alternate, mostly reduced to sharp pointed petioles. Flowers in axils of upper leaves; calyx deeply 2-lipped, large, yellow; corolla yellow, with ovate standard, 2 wings and keel of equal length; stamens 10, fused by the filaments into a sheath. Pistil developing into a short-oblong legume; seeds 2–3 mm. long, ovate, often somewhat angled, base truncate or notched, with a caruncle, smooth, glossy, greenish-yellow to reddish-brown.

Control.—Grub out bushes (17). Cut and burn extensive infestations (13).

273. *Vicia angustifolia* Reichard. Narrow-leaved vetch, Wild tare, Wild pea.

Annual or winter annual; reproducing by seeds. Meadows, fields, roadsides and waste places; on rich gravelly soils. Widespread in the eastern states and Canada; local westward to the Pacific Coast. Introduced from Europe. June–August.

Description.—Stems slender, more or less climbing by tendrils at the ends of leaves, glabrous or nearly so. Leaves alternate, pinnately compound, terminated by a tendril; leaflets 2–5 pairs, oblong-linear, 1.5–3 cm. long. Flowers perfect, irregular, on very short peduncles or sessile, a few in the leaf-axil; calyx with nearly equal teeth, about 1 cm. long; corolla of 5 unequal purplish petals, a standard, 2 wings fused to the keel; stamens 10, 9 fused by their filaments and 1 separate; pistil solitary with a filiform style which is hairy all around the apex. Fruit a several-seeded legume, 4–6 cm. long; seeds 2–3 mm. in diameter, globular, hilum elliptical, not depressed; surface dull, velvety, black or grayish-purple with irregular black spots.

Control.—Mow infested fields before seeds are ripened (11). Plow and follow with a cultivated crop for one season (7, 6).

274. *Vicia cracca* L. Wild vetch, Tufted vetch, Cow vetch, Bird vetch, Cat peas.

Perennial; reproducing by seeds and rootstocks. Fields, meadows and thickets; on gravelly or sandy soil. Northeastern United States, Michigan, Illinois, and eastern Canada; also on the Pacific Coast. Probably introduced from Eurasia; possibly also native. June–July.

Description.—Stems slender, climbing or spreading, with appressed hairs. Leaves alternate, pinnately compound, terminated by a tendril; leaflets 4–12 pairs, oblong-lanceolate, mucronate. Flowers perfect, irregular, in many-flowered, dense, one-sided, axillary ra-

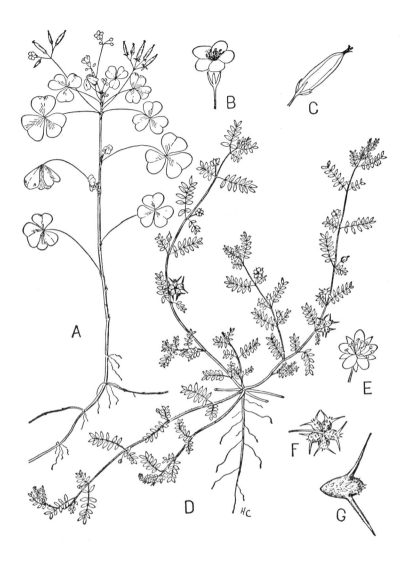

Fig. 65—**YELLOW WOOD SORREL**, *Oxalis europaea:* **A,** part of a plant showing habit (× 1/3); **B,** flower (× 1); **C,** fruit (× 1).
PUNCTURE VINE, *Tribulus terrestris:* **D,** part of plant showing habit (× 1/3); **E,** flower (× 1); **F,** fruit-cluster of five spiny carpels (× 1); **G,** a mature fruit with spines (× 2).

cemes; corolla purple, turning blue (rarely white); legume with 6–12 seeds; otherwise the flowers and fruits are very similar to 273. Seeds similar to 273 but slightly smaller.

Control.—The same as for 273.

275. *Vicia tetrasperma* (L.) Moench. Slender vetch, Smooth tare, Lentil tare, Four-seeded vetch.

Annual; reproducing by seeds. Sandy or gravelly fields, meadows, and waste places. Locally common in eastern North America. Introduced from Europe. June.

Description.—Stems very slender, climbing by tendrils on the leaves, or spreading. Leaves alternate, pinnately compound, terminating in a tendril, with 4–6 pairs of linear, oblong, obtuse leaflets. Flowers perfect, 2–4 mm. long, 1–6 on a distinct axillary peduncle; calyx with 5 unequal teeth; corolla bluish, irregular; stamens 10, 9 fused by their filaments and 1 separate; pistil solitary with a filiform style which is hairy around the apex. Fruit a small, narrow, smooth, 4-seeded legume; seeds similar to 273, about 1.6 mm. in diameter, gray-green with irregular dark purple spots.

Control.—The same as for 273.

Vicia sativa L., Spring vetch, and *Vicia villosa* Roth., Hairy or Winter vetch, both of which are grown extensively in certain regions for cover-crops and green-manuring, sometimes escape or persist; however, they do not survive long under cultivation.

Several native perennial plants belonging to the legume family and known as loco weeds occur in the arid and semi-arid regions of the West and Southwest. When eaten by horses, and to a less extent by cattle or sheep, these plants cause the loco disease with characteristic symptoms. The commonest kinds of loco weeds are *Astragalus mollissimus* Torr., Purple loco, *Astragalus diphysus* Gray, Blue loco, and *Oxytropis lambertii* Pursh, White loco.[4]

Two other members of the legume family, *Aeschynomene virginica* (L.) BSP, curly indigo, and *Sesbania exaltata* (Raf.) Cory, tall indigo, weeds of rice fields, are considered noxious weeds (1954) in Arkansas.

Swainsonia salsula (Pall.) Taub., Austrian field pea, is noxious (1954) in Colorado and Wyoming. It is a perennial, introduced in alfalfa seed, limited now in those two states, but potentially troublesome.

[4] Marsh, 1919.

Zygophyllaceae (Caltrop Family)

276. *Tribulus terrestris* L. Puncture vine, Caltrop, Ground burnut, Tackweed. Fig. 65, d–g.

Annual; reproducing by seeds. Pastures, roadsides, along railroad tracks, waste places and sometimes in cultivated fields. Occasional in the eastern states and Middle West. Locally common from Colorado south and westward to California, especially in warmer locations. Introduced from the Mediterranean region. June–August.

Description.—Stems prostrate, branching from the base to form dense mats of slender trailing branches from 3–20 dm. long, hairy. Leaves mostly opposite, pinnately compound with 4–8 pairs of oval leaflets. Flowers axillary on short peduncles; sepals mostly 5, hairy; petals 5, separate, about 1 cm. long, pale yellow; stamens 10, separate; pistil of 5 carpels united, 1–few-seeded, at maturity becoming hard and separating. Mature carpels about 6–8 mm. long, with 2–4 stiff spreading spines up to 7 mm. long. These bur-like carpels often lie in the soil for years before the seeds germinate. The spines are hard enough to stick into automobile tires.

Control.—Clean cultivation will control the puncture vine in fields. Plants in waste places and roadsides should be hoed off below the crown, piled and burned before seeds are matured.

Zygophyllum fabago L., Syrian bean caper, has been reported from fields and waste places in limited areas in Alamosa and Delta counties, Colorado.[1, 2]

Oxalidaceae (Wood Sorrel Family)

277. *Oxalis europaea* Jord. (*O. corniculata* L.). Yellow wood sorrel, Ladys sorrel. Fig. 65, a–c.

Perennial; reproducing by seeds and rootstocks. Cultivated fields, roadsides and waste places; on gravelly or sandy soils. Widespread and common in eastern North America to North Dakota and Colorado. Native; introduced into Europe from America. June–September.

Description.—Similar to 278 but the erect or decumbent stems from long, slender, perennial rootstocks. Leaves often purple. Capsules on spreading, but not recurved, pedicels; seeds similar to 278.

Control.—The same as for 278.

[1] Rogers, 1929.   [2] Thornton and Durrell, 1933.

278. *Oxalis florida* Salisb. (*O. filipes* Small). Yellow wood sorrel.
Perennial; reproducing by seeds and rootstocks. Gardens, cultivated fields, bare places in old fields and meadows; mostly on dry gravelly soils. Eastern United States west to Iowa and Arkansas. Native. May–August.

Description.—Stems erect or decumbent and rooting at the lower nodes, mostly 1–3 dm. high, covered with loose curly hairs. Leaves alternate, palmately compound, with 3 sessile obcordate leaflets on a slender petiole; stipules absent. Flowers perfect, regular, mostly 2–4 in an umbel on a long filiform peduncle; sepals 5, persistent; petals 5, yellow, about 1 cm. long, separate or slightly fused at the base; stamens 10, the filaments usually somewhat fused at the base; pistil 1, of 5 fused carpels, ovary superior, 5-celled, styles 5, distinct. Capsules on recurved pedicels, angular, hairy, 5-celled, 9–15 mm. long, gradually tapering toward the apex, style 2–4 mm. long, 2 or more seeds in each cell; seeds about 1 mm. long, flattened, obovate to elliptical, base obtuse, surface with 7–10 ridges on each side, light to chestnut-brown.

Control.—Clean cultivation (6). Hand weed or hoe the young weeds before seeds are formed (4, 5).

279. *Oxalis stricta* L. Upright yellow wood sorrel, Sour-grass.
Perennial; reproducing by seeds. Dry gravelly or stony fields and waste places. Widespread throughout North America, especially in the eastern United States. Native. May–September.

Description. Similar to 278, erect or sometimes rooting at the lower nodes, but without rootstocks. Leaves pale green, with stout petioles and firm oblong stipules. The stem, petioles and pedicels with appressed hairs. Capsules 15–25 mm. long, hairy, with a short abruptly pointed beak, seeds similar to 278.

Control.—The same as for 278.

GERANIACEAE (Geranium Family)

280. *Erodium cicutarium* (L.) L'Her. Storksbill, Heronsbill, Alfilaria, Pin-weed, Pin-grass. Fig. 66, d–h.

Annual or biennial; reproducing by seeds. Fields, pastures, lawns and waste places; on dry soil. Widespread across North America; abundant on the Pacific Coast. Introduced from Europe. April–June.

Description.—Stems low, spreading or prostrate, hairy. Leaves

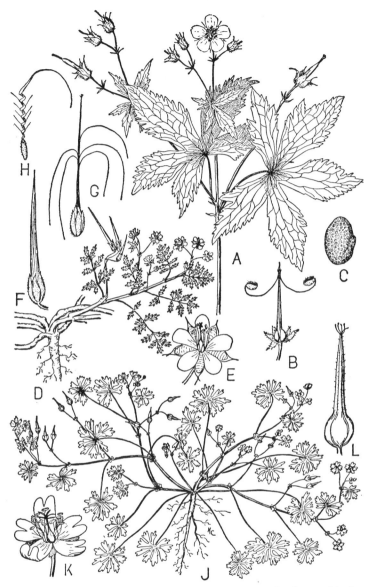

FIG. 66—**WILD CRANESBILL**, *Geranium maculatum:* **A**, shoot with flowers (× ½); **B**, fruit (× 1); **C**, seed (× 4).
STORKSBILL, *Erodium cicutarium:* **D**, part of plant showing habit (× ⅓); **E**, flower (× 2); **F**, unopened fruit (× 1); **G**, mature fruit splitting open (× 2); **H**, "seed" and style (× 2).
CRANESBILL, *Geranium molle:* **J**, plant (× ⅓); **K**, flower (× 2); **L**, unopened fruit (× 3).

alternate, pinnately compound, the sessile leaflets 1–2 pinnatifid, hairy, with acute stipules. Flowers perfect, regular, in umbel-like clusters; sepals 5, separate, bristle-tipped; petals 5, separate, pink to purple; anther-bearing stamens 5, sometimes 5 additional sterile filaments. Carpels 5, fused into a compound pistil; ovary superior, 5-lobed, 5-celled. In the mature capsule the 5 carpels, each bearing 1 seed, split apart and their long styles (tails) which are hairy on the inner surface, become spirally twisted; seed ovoid-lanceolate, minutely granular, orange-brown.

Control.—In lawns hoe or spud out the rosettes as soon as they can be recognized. Shallow cultivation of infested fields will induce the seeds to germinate before the crop is planted. Follow with a clean cultivated crop.

281. *Geranium carolinianum* L. Carolina cranesbill.

Annual or biennial; reproducing by seeds. Fields, pastures, lawns and waste places; in poor dry soils. Widespread throughout the United States and southern Canada. Native. May.

Description.—Stems erect, rather stout, with spreading branches, hairy. Leaves alternate, about 5-parted, the divisions cut and cleft into many oblong-linear segments. Flowers in crowded cymose clusters; sepals 5, ovate, persistent, awned, with hairy margins, about as long as the 5 pale pink or white notched petals; stamens 10; ovary similar to 282, hairy. Seeds ovoid, 1.6–2 mm. long, pitted, dark brown.

Control.—The same as for 282.

282. *Geranium columbinum* L. Long-stalked cranesbill.

Annual or biennial; reproducing by seeds. Dry fields and waste places. Northeastern United States; rare westward. Introduced from Europe. June–July.

Description.—Stems slender, decumbent, with forking branches, hairy. Leaves alternate, palmately parted, the segments parted into narrow lobes. Flowers perfect, regular, 1–3 on a slender peduncle; pedicels 3–10 cm. long; sepals 5, awned, about as long as the petals, persistent; petals 5, separate, 8–10 mm. long, rose-colored; stamens separate, 5 short ones opposite the petals and 5 long ones with glands at their bases alternate with the petals; carpels 5, fused into a 5-celled pistil with a deeply 5-lobed ovary with a long style. In the ripe pistil the carpels separate from below by curling upward into 5 1-seeded carpels with long tails which are glabrous on the inner surface;

ovary nearly glabrous, not wrinkled; seeds 2 mm. long, ovoid, pitted, gray-brown.

Control.—Destroy scattered plants in grasslands and lawns by hoeing or pulling before seeds are formed (5, 4). Plow badly infested fields and follow with a clean cultivated crop for a year (7, 6).

283. *Geranium maculatum* L. Wild cranesbill, Wild geranium, Spotted cranesbill, Alum-root. Fig. 66, a–c.

Perennial; reproducing by seeds. Gravelly or stony meadows, pastures, roadsides and woodlands. Maine to Alabama, westward to Manitoba and Kansas. Native. May–June.

Description.—Stems erect, 3–6 dm. high, sparingly branched, hairy, from a short, thick, perennial rootstock with fibrous roots. Leaves alternate, mostly radical, palmately veined and about 5-parted, the broad segments lobed and cut. Flowers perfect, regular, 1–3 on a long peduncle; sepals 5, slender, pointed; petals 5, separate, about 15 mm. long, light purple, bearded on the claw; stamens 10, separate. Carpels and ripe pistil similar to 282 but larger; seeds 2.5–3 mm. long, ovoid, reticulated, dark brown.

Control.—Destroy scattered plants in grasslands and lawns by hoeing or pulling (5, 4). Meadows should be mowed early to prevent seeds from maturing and being scattered with the hay. Plow badly infested fields and follow with a clean cultivated crop for a year (7, 6).

284. *Geranium molle* L. Cranesbill, Pigeon-foot, Dovesfoot, Culver-foot. Fig. 66, j–l.

Biennial; reproducing by seeds. Pastures, lawns and waste places. Nova Scotia to British Columbia and southward. Introduced from Europe. June–July.

Description.—Stems slender, weak, spreading with forking branches, hairy. Leaves alternate, palmately lobed and veined, the lobes crenate or cut. Flowers perfect, regular, 1–3 on a slender peduncle; pedicels 0.5–2 cm. long; sepals 5, ovate-oblong, persistent, awnless; petals 5, 4–5 mm. long, separate, rose-colored, notched; stamens 10, 5 long and 5 short. Carpels and ripe pistil similar to 282; ovary glabrous, transversely wrinkled; seeds about 1.5 mm. long, oblong, smooth or striate, orange-brown.

Control.—Hoe or spud the rosettes out of lawns as soon as they appear in the spring. Badly infested areas in lawns should be top-dressed and reseeded.

285. **Geranium pusillum** L. Small-flowered cranesbill.

Biennial; reproducing by seeds. Cultivated fields, pastures, lawns, and waste places; on sandy or gravelly soil. Widespread across the northern United States and southern Canada. Introduced from Europe. June–July.

Description.—Stems spreading or prostrate, branches forking, with scattered hairs. Leaves alternate, palmately veined and parted, the broad segments crenate or lobed. Flowers perfect, regular, 1–3 on a slender peduncle; sepals 5, persistent, awnless; petals 5, separate, purplish, about as long as the sepals (1 cm. long or less); stamens 5. Carpels similar to 282, the lobes of the ovary hairy; seeds like 284 but smooth.

Control.—The same as for 284.

Geranium pratense L., a perennial with glandular-pubescent pedicels and large deep purple petals, introduced from Europe, sometimes appears in meadows and grasslands from Quebec to Massachusetts and New York. The method of control is the same as for 283.

EUPHORBIACEAE (Spurge Family)

286. **Acalypha rhomboidea** Raf. (*A. virginica* L.). Three-seeded mercury, Wax balls, Copper-leaf, Mercury-weed.

Annual; reproducing by seeds. Meadows, pastures, abandoned fields and waste places. Widespread throughout eastern North America, west to Nebraska and Oklahoma. Native. August–September.

Description.—Stems branched near the base, erect or spreading, hairy. Leaves opposite below, alternate above, simple, with slender petioles, ovate to oblong-ovate, bluntly serrate or crenate. The entire plant often turns purple or copper-colored in late summer or autumn. Flowers imperfect, the staminate very small, clustered in small spikes; pistillate flowers with a large palmately 5–9-cleft leafy bract, 1–3 in the upper leaf-axils; calyx 3–5-parted; corolla absent; stamens 8–16, with short filaments which are fused at the base; carpels 3, fused into a 3-lobed pistil; styles 3. Ovary maturing into a 3-lobed capsule which splits into 3 globular 1-seeded 2-valved segments; seeds about 1.5 mm. long, obovoid, granular, dull, reddish-brown or gray with reddish-brown spots.

Control.—Clean cultivation (6). Hand pulling or hoeing before seeds are produced (4, 5).

Spurge Family

287. *Acalypha virginica* L.

Annual; reproducing by seed. Woods, fields, roadsides, and stream banks. Maine to Florida, west to Iowa and Texas. Native. June–October.

Description.—Similar to *Acalypha rhomboidea*. Stems erect, usually branched, with few to many long, spreading hairs, 1–6 dm. high. Leaf blades narrowly to broadly lanceolate, 2–8 cm. long, shallowly crenate; petioles one-third to one-half the length of the blade. Pistillate bracts with 9–15 lanceolate, sharply acute lobes. Staminate spikes as long as or longer than the bracts. Seeds 1.4–1.8 mm. long.

Control.—The same as for 286.

288. *Cnidoscolus stimulosus* (Michx.) Engelm. and Gray (*Jatropha stimulosa* Gray, *Bivonea stimulosa* Small). Spurge-nettle, Tread-softly.

Perennial; reproducing by seeds. Dry sandy soil; frequently a bad weed in cultivated fields and meadows in southern states. Virginia to Florida, west to Texas. Native. June–September.

Description.—Stems up to 12 dm. high, simple or branched above, from a long perennial root. Whole plant armed with stiff, stinging, bristle-like hairs, 3–6 mm. long. Leaves alternate, simple, long-petioled, roundish-cordate, deeply 3–5-lobed. Flowers white, fragrant, 1.8 cm. or more long, without corolla but calyx corolla-like, in few-flowered terminal cyme; stamens 10, unequal in length, monodelphous at base. Capsule bristly, 3-lobed, 3-seeded. Seeds cylindric, 6–8 mm. long.

Control.—Clean cultivation followed by hand hoeing or digging of scattered plants.

289. *Croton capitatus* Michx. Hogwort.

Annual; reproducing by seeds. Dry sandy soil and waste places. New York to Georgia, west to Iowa and Texas. Native. June–October.

Description.—Stems stout, erect, up to 2 m. high, branching, densely soft-woolly with stellate hairs. Leaves with long petioles, lanceolate-oblong to oblong or oval, entire, rounded or subcordate at base, 4–10 cm. long. Flowers in terminal clusters, 1–3 cm. long; staminate flowers with 5 sepals, 5 petals, and 7–14 stamens; pistillate flowers without petals, calyx 6–12 parted. Capsule globose, 7–9 mm. in diameter. Seeds 5 mm. long, lens-shaped, suborbicular.

Control.—The same as for 288.

Croton glandulosus L. is listed as a weed in the north central states.

290. *Eremocarpus setigerus* Benth. Turkey mullein, Woolly white drouth-weed, Dove-weed.

Annual; reproducing by seeds. Stubble fields, summer fallow land, waste places; in dry areas. Pacific Coast states. Native. July–August.

Description.—Stems much branched, low and spreading, or forming mats from 2–10 dm. in diameter. Leaves simple, broadly oval to almost circular, thick, with 3 prominent veins from the base, entire, petioled, the lower alternate, the upper opposite. Leaves and stems light gray in color and harsh to the touch due to a dense coating of branched bristly hairs. Flowers small, imperfect; staminate flowers in flat-topped clusters terminating the branches, with 5 or 6 sepals and 6 or 7 stamens; pistillate flowers solitary or in groups of 2–3 in the lower leaf-axils, composed of a solitary naked hairy pistil. Capsule 1-seeded, hairy; seeds about 3 mm. long, oval, with a ridge on one side, glossy, buff, mottled with brown or nearly black.

Control.—Harrowing or disking of stubble fields as soon as the crop is removed and again later will destroy the seedlings (8). Clean cultivation (6).

Euphorbia spp. The flowers in the genus Euphorbia are very much reduced, imperfect, without calyx and corolla and are grouped into small flower-like clusters. The several staminate flowers, each of which consists of a single stamen, and the one central pistillate flower, consisting of one 3-lobed pistil, are surrounded by a cup-like structure (involucre) resembling a calyx. These flower-like clusters are referred to as involucres in descriptions of weeds 291–301. The pistil, which matures into a capsule, has three 2-cleft styles and a 3-celled ovary which normally contains 3 seeds. The milky sap of most of the Euphorbias, on coming into contact with the skin, may produce severe irritation and blistering. Most of them are also poisonous when eaten, especially by cattle.

291. *Euphorbia corollata* L. Flowering spurge, White-flowered milkweed, Poison milkweed.

Perennial; reproducing by seeds and rootstocks. Dry sandy fields, old pastures and waste places. Widespread in the eastern and middle western states and in Ontario. Native; also introduced in the northeastern states. July–August.

Description.—Stems erect, simple or sparingly branched above, mostly glabrous, about 1 m. high, sap milky. Leaves simple, entire,

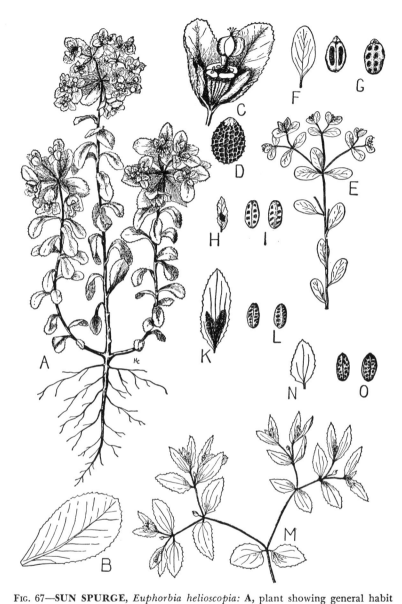

FIG. 67—**SUN SPURGE,** *Euphorbia helioscopia:* **A,** plant showing general habit (× ¼); **B,** leaf (× 1); **C,** flower-cluster (× 1); **D,** seed (× 6).
PETTY SPURGE, *Euphorbia peplus:* **E,** branch (× ⅓); **F,** leaf (× 1); **G,** seeds (× 6).
MILK PURSLANE, *Euphorbia supina:* **H,** leaf (× 1); **I,** seed (× 6).
SPOTTED SPURGE, *Euphorbia maculata:* **K,** leaf (× 1); **L,** seeds (× 6).
HAIRY SPURGE, *Euphorbia vermiculata:* **M,** branch (× ½); **N,** leaf (× 1); **O,** seeds (× 6).

sessile, ovate to linear-lanceolate, alternate below, opposite above, a whorl subtending the umbel. Flowers in a terminal umbel and in the upper leaf-axils; involucres on long, slender peduncles, with showy white petal-like appendages. Capsule smooth, on a slender pedicel; seeds about 2 mm. long, ovate, ash-gray or light brown mottled, with shallow coarse pits, a dark line on one side.

Control.—Dig out scattered plants (17). In pastures mow close to the ground before the blossoms appear (11). Infested fields should be plowed and planted with a clean cultivated crop for a year. Scattered plants in the cultivated crop should be pulled or hoed as soon as they appear above ground.

292. **Euphorbia cyparissias** L. (*Tithymalus cyparissias* Hill). Cypress spurge, Salvers spurge, Quack salvers-grass, Graveyard-weed. Fig. 68, a–d.

Perennial; reproducing mostly by creeping roots; in some localities also by seeds. Dry gravelly or sandy fields, pastures, roadsides, cemeteries, and waste places. Widespread and locally common in the northeastern and north central states; infrequent in the Pacific Northwest. Introduced as an ornamental from Europe. May–June.

Description.—Stems erect, simple or branched above, 1–4 dm. high, clustered from a crown or scattered from buds on the extensively creeping roots, glabrous, with milky sap. Leaves alternate, simple, linear, 1–2 mm. wide; the upper whorl of leaves subtending a terminal umbel of rays. Flowers similar to 291; involucral lobes with crescent-shaped glands. Capsule minutely granular; seeds oblong to nearly globose, about 1.5–2 mm. long, smooth with a dark line on one side, minutely pitted, light gray; with a yellowish-white thick caruncle attached.

Control.—The same as for 294.

Experimental plots of cypress spurge established by the writer, with the offspring from a single seed, did not produce seeds, while similar plots from the offspring of several seeds yielded an abundance. This apparent self-sterility would account for the failure of seed formation in some localities and the abundant seeding in other areas. In many localities, especially in western New York, where the cypress spurge has been introduced and spread by root propagation, all the plants in a neighborhood probably came from a single original individual. This explains why seeds are not produced under such conditions. In other places, such as in Orange County, New

York, and near Pittsfield, Massachusetts, where seeds are produced freely, the extensive areas probably represent the offspring from several or many seeds.

293. **Euphorbia dentata** Michx. (*Poinsettia dentata* Small). Toothed spurge.

Annual; reproducing by seeds. Cultivated fields, grain fields, meadows and waste places; on rich moist soil. Pennsylvania, west to Minnesota and Wyoming and south to Texas and Mexico. Native; also introduced northward. August–September.

Description.—Stems erect or ascending, branched, 3–8 dm. high. Leaves opposite, the lowest alternate, simple, variously shaped, coarsely dentate, hairy. Flowers in clusters of nearly sessile involucres terminating the stem and branches; involucres with 5 oblong toothed lobes and with 1 or a few short stalked glands. Capsules tubercled; seeds 2.5 mm. long, obovoid, slightly angled, a dark line on one side, tubercled, deep gray, with a flattened yellowish caruncle attached.

Control.—Harrow grain fields to destroy the weed seedlings (8). Clean cultivation (6) followed by hand hoeing (5).

294. **Euphorbia esula** L. (*E. virgata* Waldst. and Kit.). Leafy spurge, Faitours-grass. Fig. 68, e–h.

Perennial; reproducing by seeds and creeping roots. Meadows, pastures and cultivated fields. Widespread across the northern United States and southern Canada;[1] locally common in New York,[2] the northwestern states,[3,4] and Saskatchewan and Manitoba.[5] Introduced from Europe. June–July.

Description.—Stems erect, simple or branched above, 5–10 dm. high, frequently in clusters from a vertical root, glabrous, with milky sap; in late summer with several pinkish scaly buds just below the soil surface. Leaves alternate, simple, entire or slightly wavy along the margin, somewhat glaucous, linear-lanceolate, usually widest at or above the middle, 4–9 mm. wide. Flowers imperfect, very small, in greenish-yellow involucres grouped in small clusters surrounded by rounded greenish-yellow lobes with short horn-like brownish glands. Capsule smooth or granular; seeds obovoid-ovoid, 2 mm. long, nearly smooth, light gray or with brownish spots, with a dark line on one side, a yellowish caruncle often attached.

[1] Britton, 1921.
[2] Muenscher, 1930b.
[3] Hanson and Rudd, 1933.
[4] Barnett and Hanson, 1934.
[5] Batho, 1932.

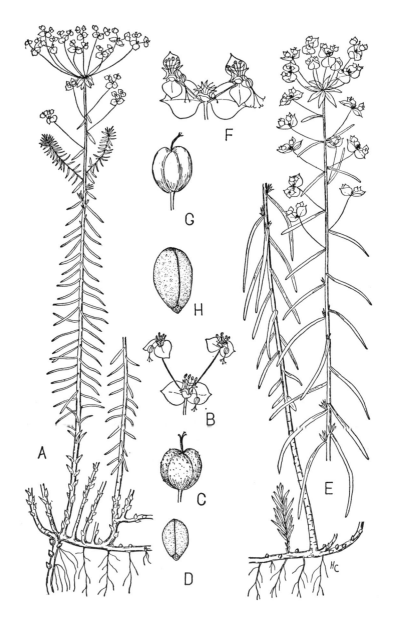

Fig. 68—**CYPRESS SPURGE**, *Euphorbia cyparissias:* **A**, plant showing habit (× ⅓); **B**, portion of a flower-cluster (× 1); **C**, capsule (× 3); **D**, seed (× 6).
LEAFY SPURGE, *Euphorbia esula:* **E**, plant showing habit (× ⅓); **F**, portion of a flower-cluster (× 1); **G**, capsule (× 3); **H**, seed (× 6).

Control.—Prevent seed production by mowing the leafy spurge plants as soon as the first blossoms appear (11). Large areas should be plowed and kept bare by summer fallow for a season and followed by a heavy smother crop before reseeding. Close pasturing with sheep will gradually run the leafy spurge out of a pasture.

295. *Euphorbia helioscopia* L. Sun spurge, Wart spurge, Wart-weed, Wart-grass, Cats-milk. Fig. 67, a–d.

Annual; reproducing by seeds. Cultivated fields and gardens; on rich, sandy or gravelly soils. Locally common in eastern Canada and the northeastern United States; infrequent on the Pacific Coast. Introduced from Eurasia. July–October.

Description.—Stems erect or spreading, branched at the base, glabrous, often purplish, with milky sap, 1–4 dm. high. Leaves alternate (except the terminal whorl), simple, finely serrate, obovate, rounded or notched at the apex, glabrous or nearly so. Involucres in a terminal umbel of about 5 stout 3-branched rays; umbel subtended by a whorl of leaves; lobes of involucre with stalked orbicular glands. Capsule smooth; seeds about 2 mm. long, ovoid-elliptical, with a ridge on one side, roughened by a raised fine meshed network, dark orange-brown.

Control.—Clean cultivation (6) followed by hoeing continued until late in the summer.

296. *Euphorbia lucida* Waldst. & Kit. Broad-leaved spurge, Shining spurge.

Perennial; reproducing by seeds and roots. Borders of fields, roadsides and waste places about abandoned buildings. Infrequent in the northeastern United States; local elsewhere; noxious (1954) in South Dakota. Introduced from Europe. June–July.

Description.—Stems erect, simple, stout, usually in clusters from the old crowns, glabrous, 5–10 dm. high, with milky sap. Leaves alternate, simple, entire, oblong to oblong-lanceolate, dark green, 18–25 mm. wide. Flowers similar to 294; involucral lobes with short hornlike glands. Capsule finely wrinkled; seeds 2.5 mm. long, ovoid with a dark line on one side, smooth, light gray, a yellowish-brown caruncle usually attached.

Control.—The same as for 294.

The broad-leaved leafy spurge seldom produces seed in the United States. Since it does not possess an extensive creeping root system like the leafy spurge, 294, it spreads very slowly.

297. *Euphorbia maculata* L. (*E. nutans* Lag., *E. preslii* Guss.). Spotted spurge, Eye-bright, Nodding spurge, Stubble spurge, Slobber-weed. Fig. 67, k–1.

Annual; reproducing by seeds. Fields, bare places in meadows and pastures, especially on dry gravelly soil. Widespread from Massachusetts to North Dakota and southward to Florida and Mexico; also introduced in California. Native. July–September.

Description.—Stems ascending, branched or sometimes nearly simple except at apex, about 2–8 dm. high, glabrous, often reddish, with milky sap. Leaves similar to 301 but larger, 1–3 cm. long, serrate, often reddish or with a red spot; stipules small, triangular. Involucres in small leafy cymes on long slender pedicels terminating the lateral branches; lobes of involucre with 4 appendaged glands. Capsule sharply angled, glabrous; seeds obovoid-oblong, about 1 mm. long, 3–4-angled, minutely pitted and transversely ridged, dark brown or black.

Control.—The same as for 301.

298. *Euphorbia marginata* Pursh. Snow-on-the-mountain.

Annual; reproducing by seeds. Fields, gardens, pastures and waste places. Minnesota to Montana, southward to Texas; also widely introduced and locally escaped in the eastern states. Native. June–August.

Description.—Stems erect, stout, much branched above, somewhat hairy, 3–10 dm. high. Leaves simple, the lower alternate, the uppermost opposite or whorled and with prominent white petal-like margins, sessile, ovate or oblong, entire. Flowers in umbels with 3 forked branches; involucres 5-lobed, with broad white appendaged glands. Capsule somewhat hairy; seeds 3–4 mm. long, obovate-globular, a dark line on one side, minutely pitted, tubercled, light gray to brown.

Control.—Clean cultivation (6). Pull or hoe scattered plants before seeds are formed (4, 5).

299. *Euphorbia peplus* L. Petty spurge. Fig. 67, e–g.

Annual; reproducing by seeds. Cultivated fields, gardens and yards; mostly on rich soils. Widespread in the northeastern and north central states and eastern Canada; also on the Pacific Coast. Introduced from Europe. August–October.

Description.—Stems erect, with spreading branches near the apex, with milky sap, 1–3 dm. high. Leaves alternate (except the terminal whorl), simple, entire, round-ovate, thin. Involucres in a terminal umbel of 3 branched rays; umbel subtended by a whorl of ovate leaves; lobes of involucre with long horn-like glands. Capsule with 2 crests on the back of each lobe; seeds obovoid-oblong, about 1.3 mm. long, slightly 4-angled, minutely pitted and marked with oval depressions, light gray or reddish-brown, a white caruncle usually attached.

Control.—The same as for 295.

300. *Euphorbia supina* Raf. (*E. maculata* L.). Milk purslane. Fig. 67, h–i.

Annual; reproducing by seeds. Dry, gravelly or sandy fields and sterile waste places. Widespread throughout the eastern and middle western states; infrequent on the Pacific Coast. Native; introduced on the Pacific Coast. July–October.

Description.—Stems prostrate, much branched, forming a dense mat, puberulent, often reddish, with milky sap. Leaves similar to 301, usually with a reddish-brown spot on the upper surface; stipules lanceolate. Involucres solitary in the branch axils or appearing in dense leafy lateral clusters; lobes of involucre with 4 appendaged glands. Capsule sharply angled, puberulent; seeds about 0.8 mm. long, 4-angled, transversely ridged, minutely pitted, gray-brown to reddish.

Control.—The same as for 301.

301. *Euphorbia vermiculata* Raf. (*E. hirsuta* Wieg.). Hairy spurge. Fig. 67, m–o.

Annual; reproducing by seeds. Dry, sandy or gravelly fields and waste places. Northeastern United States and eastern Canada, Indiana, Wisconsin, New Mexico, and Arizona. Native. July–September.

Description.—Stems procumbent or prostrate, much branched, forming a dense mat, hirsute, often reddish, with milky sap. Leaves opposite, simple, oblique at base, short-petioled, serrate, 8–15 mm. long, with small triangular stipules. Involucres solitary in the branch axils or appearing in terminal clusters; lobes of involucre with 4 appendaged glands. Capsule bluntly angled, smooth; seeds about 1 mm. long, 3–4-angled, pitted, dark gray to dark brown.

Control.—Clean cultivation begun early and continued as long as the crop will permit, followed by hand pulling of scattered plants before they produce seeds (6, 4). Harrowing grain fields while the grain is small will destroy many of the spurge seedlings (8).

ANACARDIACEAE (Cashew Family)

302. **Rhus radicans** L. (*R. toxicodendron* L., *Toxicodendron radicans* Ktze.). Poison ivy, Poison oak, Poison creeper, Three-leaved ivy, Picry, Mercury. Fig. 69.

Woody perennial; reproducing by seeds and creeping rootstocks. Dry rocky fields, pastures, fence-rows, banks and waste places, also in rich alluvial woodlands. Widespread throughout the United States and southern Canada. Native. June–July.

Description.—Stems woody, erect shrub or a vine climbing by aerial rootlets on fences, walls or trees; slender creeping rootstocks, from the basal stem nodes, sometimes run horizontally underground for several meters, sending up short leafy shoots from their nodes. Leaves alternate, compound, with 3 leaflets; leaflets glossy or dull green, glabrous or somewhat hairy, the margins entire, toothed or somewhat lobed. Flowers polygamous, in slender open axillary panicles; calyx with 5 lobes; corolla of 5 yellow-green petals; stamens 5, separate; pistil solitary, with a 1-celled ovary. Fruit a small drupe, white or cream-colored, almost dry, nearly globular, mostly 4–6 mm. in diameter, with a grayish, striped, 1-seeded stone about 3–4 mm. in diameter.

The treatment of poison ivy, as here recognized in a broad sense, allows for sufficient variation in habit and leaf form to include most, if not all, the trifoliate white-fruited forms occurring in the various sections of the United States. A number of these forms, based largely on whether their habit is erect or climbing and on differences in the texture, pubescence and degree of lobing of the leaflets, have been described as distinct species. Since the characters on which these species are based show such great variations under different field conditions, it has seemed best for the present purpose not to treat them as separate species until they are better understood. The more common of these forms, with their approximate ranges, are: *Rhus toxicodendron* L., with deeply lobed leaflets, Maryland to Texas; *R. microcarpa* Steud., a form with fruits 3–4 mm. in diameter, Atlantic states; *R. radicans* L. var. *rydbergii* Small, a thicker-leaved form, Great Plains and Rocky Mountains; *R. diversiloba* T. and G., with leaflets mostly crenate, on the Pacific Coast; *R. radicans* L., the common climbing form of the eastern states.

Control.—In fields, mow close to the ground in midsummer and

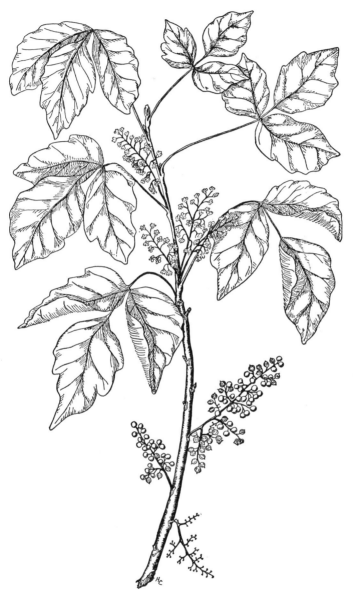

FIG. 69—**POISON IVY,** *Rhus radicans:* A branch showing leaves and flower-clusters on the new growth and clusters of berry-like fruits persisting on the growth of the preceding year ($\times \frac{1}{6}$).

follow by plowing and harrowing. The following spring plant a cultivated crop. Small areas on stony soil can be controlled by close grazing by sheep or goats. Among trees, shrubs and other plants, grub out the rootstocks of poison ivy.

The leaves, flowers, fruits and bark of stem and roots of poison ivy are very poisonous to the touch, frequently producing severe inflammation and blisters in susceptible persons.

For prevention of ivy poisoning after contact with the plant,[1, 2] wash the hands, face, or affected parts freely in a solution of 5 per cent iron chloride in a half-and-half mixture of alcohol and water. If this solution is applied either before or immediately after going into a region where poison ivy is known to grow, no harmful effects need be expected. This inexpensive and non-poisonous remedy can be obtained at almost any drug-store. Wash and rinse the skin thoroughly several times. Hot water and ordinary kitchen or laundry soap containing an excess of alkali is best. Do not use soaps containing oils, for the poison is soluble in oils and will, therefore, be spread over other parts of the skin. Make a heavy lather and rinse off repeatedly, at least three or four times.

As remedial measures when poisoning has begun, the following are of value. The application of baking soda, 1 or 2 teaspoonfuls to a cup of water, generally relieves the pain caused by the inflammation. Soaking in hot water also gives relief. For soothing pain and for preventing the general spread of the inflammation, fluid extract of Grindelia, diluted with 6 to 10 parts of water, may be applied with a clean bandage which should be kept moist and frequently changed. Ointments or other remedies containing fatty or oily substances should not be applied until after the poison has exhausted itself. The use of sugar of lead is discouraged since the lead may itself cause poisoning, especially on open pustules. Other remedies are employed with success, but when possible it is usually desirable to consult a physician, especially in bad cases of poisoning.

303. *Rhus vernix* L. (*Toxicodendron vernix* Ktze.). Poison sumac, Poison dogwood, Poison elder, Poison ash, Swamp sumac, Thunderwood. Fig. 70.

Woody perennial; reproducing by seeds. Swamps and boggy soils. Locally common from Maine to Ontario and Minnesota and southward. Native. June–July.

[1] Muenscher, 1934b.   [2] McNair, 1923.

Fig. 70—**POISON SUMAC,** *Rhus vernix:* A branch showing leaves and drooping clusters of berry-like fruits which may persist all winter ($\times \frac{1}{6}$).

Description.—Stems woody, erect, often in clumps, a large shrub or small tree with stout spreading branches and gray bark. Leaves alternate, pinnately compound with about 7–13 leaflets, petioles often reddish; leaflets entire, glabrous, frequently turned upward; the leaves of seedlings have only 3 leaflets. Flowers polygamous, in loose slender axillary panicles, similar to 302. Fruit like 302, white or pale yellow, somewhat flattened, especially the stone.

Control.—The bushes may be cut off near the surface of the ground, preferably in summer, and the stubs grubbed out to prevent new shoots from forming.

The entire plant is very poisonous to the touch, frequently producing severe inflammation and blisters similar to poison ivy. The treatment for poisoning is the same as for poison ivy.

MALVACEAE (Mallow Family)

304. *Abutilon theophrasti* Medic. Velvet-leaf, Indian mallow, Butter print, Velvet-weed, Butter-weed, Indian hemp, Cotton-weed. Fig. 71, i–j.

Annual; reproducing by seeds. Cultivated fields, gardens and waste places; on rich alluvial or somewhat sandy soils. Widespread in the eastern United States; also on the Pacific Coast. Most common in the warmer regions. Introduced from Asia. August–September.

Description.—Stems simple, erect, about 1 m. high, velvety. Leaves alternate, simple, cordate, with tapering apex and dentate margin, palmately veined, velvety. Flowers on short peduncles; sepals 5, fused at the base; petals 5, yellow, separate; stamens numerous, monadelphous, fused below with the petals. Carpels about 9–15, separate, hairy, beaked, arranged in a disk, when mature dehiscing to discharge the 3–9 seeds; seeds about 3 mm. long, ovate, notched, small hair-like prickles scattered over the minutely reticulated surface, grayish-brown. The seeds of velvet-leaf are very resistant; they have been known to retain their viability for more than fifty years when buried in the soil or when kept in dry storage. They will also germinate after having passed through a silo.

Control.—Clean cultivation continued as late as the crop will allow. Pull or hoe out scattered plants in the rows or hills of cultivated crops to prevent seed formation.

305. *Hibiscus trionum* L. Flower-of-an-hour, Bladder ketmia, Venice mallow, Modesty, Shoo-fly. Fig. 71, d–f.

Mallow Family. 311

Annual; reproducing by seeds. Cultivated fields, gardens, and waste places; on gravelly, or often somewhat limy, soils. Widespread and locally common in eastern North America; especially southward. Introduced from Europe; formerly grown extensively as an ornamental. July–September.

Description.—Stems erect or spreading, low branching, hairy, 2–5 dm. high. Leaves alternate, simple, pinnately veined, petioled; upper leaves 3-parted with lanceolate, pinnately lobed divisions, lower leaves pinnately lobed or parted, stipulate, hairy. Flowers solitary or 2 or 3 in the axils of upper leaves, subtended by several linear, hairy, involucral bracts, perfect, regular; calyx bladdery-inflated, of 5 fused sepals, soon becoming scarious, pale green with dark veining; corolla regular, of 5 sulfur-yellow petals, purplish in the center and with dark veinings, ephemeral; stamens numerous, monadelphous, united more than half their length, forming a column about pistil and united with bases of the petals; pistil with several cells in the ovary, styles united below. Fruit a many-seeded capsule inclosed in the bladdery calyx; seeds about 2 mm. long, obovoid to kidney-shaped or triangular in shape, thinner near the hilum, roughened by blunt yellow-tipped protuberances, dull, grayish-brown.

Control.—The same as for 304.

306. *Malva moschata* L. Musk mallow, Musk, Musk plant.

Perennial; reproducing by seeds. Dry grassy fields, meadows and waste places; especially on limy soils. Locally common in the northeastern United States and adjacent Canada; also in the Pacific Northwest; infrequent elsewhere. Introduced from Europe. June–July.

Description.—Stems erect, branching near the base, pubescent, 3–10 dm. high. Leaves alternate, simple, palmately lobed and veined, slender petioles about twice the length of leaf, stem-leaves mostly deeply parted and incised, the divisions cleft into linear lobes, somewhat pubescent on the veins, 3–5 cm. in length, linear stipules, basal leaves mostly rounded and lobed. Flowers clustered toward the ends of the branches, perfect, regular, subtended by 3 bracts, very showy, on short peduncles; calyx of 5 hairy sepals united at the base; corolla of 5 separate obcordate petals, rose-colored or white with pink veins; stamens numerous, monadelphous, united more than half their length, forming a column about the pistil 1 cm. or more high; pistil of 10–20 carpels, hairy, separating from the central axis when mature. Fruit consists of the indehiscent 1-seeded carpels; seeds

Fig. 71—**ROUND-LEAVED MALLOW**, *Malva neglecta:* **A,** plant showing general habit (× ¼); **B,** cheese-shaped fruit (× 2); **C,** seed (× 6).
FLOWER-OF-AN-HOUR, *Hibiscus trionum:* **D,** branch showing habit (× ⅓); **E,** capsule (× 1); **F,** seed (× 6).
PRICKLY SIDA, *Sida spinosa:* **G,** branch showing habit (× ⅓); **H,** fruit (× 2).
VELVET-LEAF, *Abutilon theophrasti:* **I,** branch showing general habit, flowers and seed capsules (× ⅓); **J,** seed (× 6).

about 2 mm. long, orbicular to kidney-shaped, sides concave, edges rounded, brown, usually surrounded by the hairy carpel.

Control.—Pull or spud scattered plants in meadows (4). Badly infested meadows should be mowed early before the mallow has produced seeds (11).

307. *Malva neglecta* Wallr. (*M. rotundifolia* L.) Round-leaved mallow, Cheeses, Low mallow. Fig. 71, a–c.

Annual or biennial; reproducing by seeds. Cultivated ground, new lawns, farmyards, and waste places. Widespread throughout North America. Introduced from Eurasia. June–October.

Description.—Stems procumbent, widely spreading from deep biennial root, pubescent and very leafy. Leaves alternate, simple, palmately veined, orbicular with cordate base, crenate and slightly lobed, 1–4 cm. in diameter, stipulate, with slender petioles several times the length of the leaf. Flowers single or clustered in the axils of the leaves, perfect, regular, subtended by 3 bracts, on long peduncles; calyx of 5 fused sepals, hairy; corolla of 5 separate notched petals, whitish and twice the length of the calyx; stamens numerous, monadelphous, united more than half their length, forming a column about the pistil; pistil of 10–20 hairy carpels, separating from the central axis when mature. Fruit consists of the indehiscent 1-seeded carpels arranged in a ring; carpel almost circular, flattened on two sides, light brown, slightly roughened, with radiating ridges, 1–2 mm. in diameter, containing a reddish-brown seed about 1.5 mm. long.

Control.—Hand weeding or hoeing when the seedlings are small (4, 5). Clean cultivation (6). Large fruiting plants in waste places or fields should be pulled and burned.

308. *Sida spinosa* L. Prickly sida, False mallow, Indian mallow, Spiny sida, Thistle mallow. Fig. 71, g–h.

Annual; reproducing by seeds. Cultivated fields, pastures, gardens and waste places. Widespread in the eastern United States, west to Nebraska; most abundant and troublesome in the southern states. Introduced from tropical America; possibly native in the southern states. June–September.

Description.—Stems erect, much branched, softly pubescent, 2–5 dm. high. Leaves alternate, simple, pinnately veined, petioled, stipulate, serrate, ovate-lanceolate or oblong, softly pubescent, 1–5 cm. long. Flowers solitary or clustered in the axils of the leaves, perfect, regular, without involucre, peduncled; calyx of 5 sepals united at the base; corolla of 5 separate entire pale yellow petals, about 1 cm. broad; stamens numerous, monadelphous, united more than half their length forming a column; pistil composed of 5 fused carpels. Fruit a 5-celled dehiscent capsule; carpels 1-seeded, splitting at the top into 2 beaks; seeds about 2 mm. long, broadly ovate, with a projection on one side, 3-angled, minutely granular, dull, dark reddish-brown.

Control.—The same as for 307.

Sida hederacea Dougl. White mallow or Alkali mallow, native to western North America, is locally common as a weed of cultivated crops in somewhat saline or sandy soils in the South and West. It is a low perennial with stems and leaves more or less yellow, due to an abundance of scurfy scales and forked yellow hairs. Control measures should consist of frequent clean cultivation. Badly infested fields should be plowed, thoroughly tilled, and planted with a heavy smother crop for a year. After the weeds have been weakened by the shading of the smother crop, the field should again be worked thoroughly before a cultivated crop is planted.

GUTTIFERAE (St. Johns-wort Family)

309. *Hypericum perforatum* L. St. Johns-wort, Tipton-weed, Klamath-weed, Goat-weed, Eola-weed, Amber, Rosin rose. Fig. 72.

Perennial; reproducing by seeds and short runners. Old meadows, pastures, along roadsides and waste places; usually on dry, gravelly or sandy soils. Newfoundland to British Columbia and south; locally very abundant and troublesome. Introduced from Europe. July–August.

Description.—Stems erect, slender, much branched above with short leafy shoots, somewhat 2-ridged, producing runners from the base, glabrous, with dark rings about the nodes, very leafy, 3–8 dm. high. Leaves opposite, simple, pinnately few-veined, sessile, entire, elliptic or linear-oblong, glabrous with pellucid dots, rarely 1 cm. broad. Flowers numerous in open leafy flat-topped cymes, perfect, regular, calyx of 5 lanceolate acute sepals, 5–7 mm. long; corolla of 5 deep yellow petals bearing black dots only on the margin, twice the length of the sepals, convolute in the bud; stamens numerous, in 3 clusters, showy; pistil compound, with 3 styles; ovary 1-celled, with parietal placentation. Fruit a 3-celled capsule with numerous seeds,

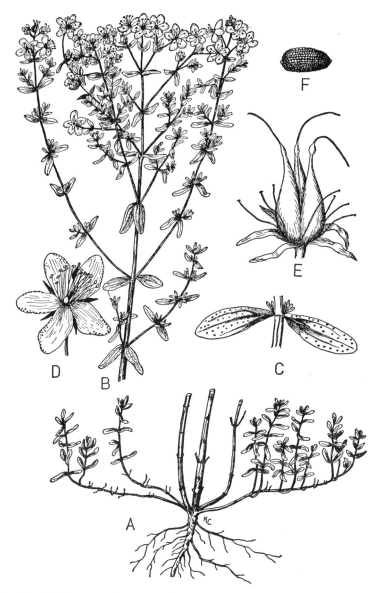

Fig. 72—**ST. JOHNS-WORT,** *Hypericum perforatum:* **A,** basal part of plant showing roots and runners (× ⅓); **B,** flowering shoot (× ⅓); **C,** opposite leaves showing translucent dots (× ⅓); **D,** flower (× 1); **E,** fruit (× 3); **F,** seed (× 10).

septicidal dehiscence, and persistent styles; seeds oblong, about 1 mm. long, slightly pointed at the ends, mottled with rows of minute depressions, glossy, dark brown.

Control.—Clean cultivation will soon destroy this weed. Pastures and waste lands should be mowed several times to prevent seed formation.

Under certain conditions cattle and sheep are poisoned by eating St. Johns-wort.[1]

310. *Hypericum spathulatum* (Spach) Steud. (*H. prolificum* L.) Shrubby St. Johns-wort, Broom brush.

Shrubby perennial; reproducing by seeds. Neglected fields, pastures and waste places; especially on sandy or stony soils. Locally common in the eastern United States and Ontario. Native. July–August.

Description.—Stems woody, much branched from the base, 3–15 dm. high; the branchlets 2-edged or ridged. Leaves opposite, simple, with short petioles or the upper sessile; blade entire, narrowly oblong, glabrous, with translucent dots; the axils of the larger leaves usually with tufts of smaller leaves. Flowers similar to 309, in terminal or axillary clusters, bright yellow; with numerous stamens, usually not in clusters. Capsule 3-celled; seeds about 1.5 mm. long, oblong, often slightly curved, ridged on one side, slightly glossy, marked with vertical rows of minute depressions, dark brown.

Control.—Grub out the bushes with a mattock.

VIOLACEAE (Violet Family)

311. *Viola arvensis* Murr. Wild pansy, Field pansy, Hearts-ease.

Annual; reproducing by seeds. Fields and bare places in meadows and new clover seedings; on gravelly or sandy soils. Infrequent in the northeastern states and eastern Canada. Introduced from Europe. June–September.

Description.—Stems spreading, irregularly branched or simple, leafy, angled and nearly glabrous, 1–3 dm. high. Leaves alternate, simple, pinnately veined, petioled; lower leaves orbicular to ovate, crenate; upper leaves oblong to oblanceolate and crenately serrate, somewhat pubescent, 2–5 cm. long; stipules large, leaf-like, pectinate at the base, the terminal lobe enlarged. Flowers axillary, solitary, nodding on long slender pedicels; calyx of 5 lanceolate acute sepals, auricled at base and persistent; corolla of 5 somewhat irregular

[1] Marsh and Clawson, 1930.

petals, seldom longer than the sepals, pale yellow or yellow with purple, the lower petal spurred; stamens 5, distinct, the 2 lower spurred; pistil compound of 3 carpels; style much enlarged upward into a globose hollow summit. Fruit a globose, 1-celled, 3-valved capsule with 3 parietal placentae; after opening, each valve as it dries folds together lengthwise, projecting the numerous seeds; seeds about 1 mm. long, obovoid, glossy, yellow to dark brown.

Control.—Clean cultivation, especially while the seedlings are small.

PASSIFLORACEAE (Passion-flower Family)

312. *Passiflora incarnata* L. Passion-flower, May-pop.

Perennial; reproducing by seeds and new shoots from roots. Cultivated fields and waste places. Maryland to Illinois, south to Florida and Texas. Native. May–August.

Description.—Stems climbing or trailing by means of axillary tendrils, 1–10 m. long, somewhat hairy. Leaves alternate, palmately veined and usually palmately 3-lobed; the lobes serrate, with acuminate apex. Flowers perfect, solitary, axillary, on long jointed peduncles, about 5 cm. in diameter; calyx of 5 sepals fused at base; corolla of 5 large white petals attached on the throat of the calyx and crowned with triple rows of long fringes which are pale purple with a lighter band near the center; stamens 5, fused by their filaments into a sheath around the long stalk of the ovary; pistil with 3 styles. Ovary 1-celled, with 3–4 placentae forming a many-seeded yellow berry about 2–4 cm. in diameter; seeds about 6 mm. long, ovate, flattened; surface dull, with hollow depressions; gray-brown to black.

Control.—Clean tillage of cultivated crops alternated with an annual smother crop of cowpeas or clover will usually control this weed. Deep plowing and deep cultivating with a knife type of cultivator will destroy the roots.

CACTACEAE (Cactus Family)

Various Opuntias, Prickly pears, of the Cactaceae or Cactus Family, are listed as weeds in the north central states. They grow on arid or sandy soils and are occasionally found east to New York and Massachusetts, west to British Columbia and Washington, and south to Florida, Texas, and California.

LYTHRACEAE (Loosestrife Family)

313. Cuphea petiolata (L.) Koehne (*Parsonsia petiolata* Rusby). Clammy loosestrife, Blue waxweed, Red stem, Tarweed, Sticky stem.

Annual; reproducing by seeds. Dry meadows, pastures and waste places. Widespread in the eastern United States, west to Iowa; most common in the South. Native; appearing as if introduced northward. August–September.

Description.—Stems erect, branching, very viscid-hairy, reddish, often sticky, 1–5 dm. high. Leaves opposite, simple, pinnately veined, petioled, entire, ovate-lanceolate, rough, viscid-hairy on midrib and veins, 3 cm. or less in length, often tinged with red. Flowers solitary or racemose, perfect, irregular, short-peduncled; calyx tubular, gibbous or spurred at the base on the upper side, 6-toothed, 12-ribbed, viscid-hairy, reddish; corolla of 6 unequal petals, ovate, deciduous, purplish; stamens 11 or 12 attached in two sets on throat of the corolla; stigma 2-lobed, style slender; ovary 1–2-celled and bearing a gland at the base next the calyx spur. Capsule oblong, few-seeded, rupturing early along one side, exposing the immature seeds attached to the placentae; seeds about 2.5 mm. long, broadly oval, flat, with a median ridge on each side, dull, granular, reddish-brown.

Control.—Infested meadows should be mowed again after the crop is removed and before the weeds produce seeds. Badly infested fields should be plowed and planted with an intertilled crop before reseeding.

314. Lythrum salicaria L. Purple loosestrife, Bouquet-violet.

Perennial; reproducing by seeds. Marshes, wet meadows, stream margins, and shores of lakes where it often chokes out native vegetation. Locally abundant Newfoundland, Quebec, New England and Minnesota, south to Virginia and Missouri; also in western Washington. Introduced from Europe. June–September.

Description.—Stem 5–12 dm. high, stout, erect; stem and foliage glabrous to downy pubescent. Leaves opposite or in whorls of threes, sessile, lanceolate, 3–10 cm. long, the larger leaves cordate at base. Flowers many in terminal spike-like panicles, trimorphic, showy, red-purple, 7–10 mm. long; spikes 1–4 dm. long, interspersed with lanceolate to ovate leafy bracts; bracts and calyx greenish, somewhat pubescent, the calyx-lobes much shorter than the subulate appendages; stamens twice as many as the petals.

Control.—The same as for 313.

ONAGRACEAE (Evening Primrose Family)

315. *Epilobium angustifolium* L. (*Chamaenerion angustifolium* Scop.). Fireweed, Willow-herb, Flowering willow, Burnt-weed, Indian wickup.

Perennial; reproducing by seeds and creeping rootstocks. Open fields, especially in recently cleared land, pastures and burned-over logged-off lands; mostly on gravelly soils. Widespread in the northern United States and Canada; abundant on the Pacific Coast. Native. Also native in Eurasia. July–August.

Description.—Stems erect, simple, somewhat stout and woody, reddish, usually glabrous, 5–20 dm. high, bearing scattered leaves. Leaves alternate, simple, pinnately veined, nearly sessile, distantly toothed or entire, lanceolate, glabrous, 5–20 cm. long; the lower leaves often opposite. Flowers in terminal racemes, perfect, symmetrical, regular, showy; calyx tubular with 4-cleft limb at top of ovary, purplish, fused with the ovary; corolla of 4 ovate, entire, magenta to purple petals, 1–2 cm. long; stamens 8, showy, declined, borne on summit of calyx-tube; ovary inferior, 4-celled; style long, declined, pink, bearing a 4-lobed stigma. Fruit a capsule, linear, many-seeded, splitting from the apex, reddish-brown, 5 cm. long or more; seeds about 1 mm. long, ovate-lanceolate, nearly smooth, comose with fine white hairs at apex, brown.

Control.—In newly cleared land under cultivation this weed will soon disappear provided the shoots are hoed off or pulled as soon as they sprout. Close grazing or mowing as soon as the flowers appear will gradually destroy the fireweed in pastures.

316. *Epilobium hirsutum* L. Willow-herb.

Annual, occasionally perennial; reproducing by seeds and short rootstocks. Low fields, along ditches and on the borders of swales in pastures and waste land; usually in gravelly somewhat limy soils. Locally common, especially in the river valleys of the northeastern states; also in Quebec and Ontario. Introduced from Europe. July–August.

Description.—Stems erect, much branched, somewhat stout and woody, densely covered with long fine hairs, 1–2 m. high. Leaves mostly opposite, simple, pinnately veined, sessile and clasping, serrulate, oblong, hairy, 1–6 cm. long. Flowers in corymbs or panicle-like clusters, a few in the upper leaf-axils, perfect, regular, showy; calyx tubular, prolonged a little beyond the ovary, 4-cleft, purplish, fused with the ovary; corolla of 4 ovate notched petals, magenta-colored, 1–2 cm. long; stamens 8, erect, borne on calyx-tube; ovary inferior, 4-celled; style long, erect, bearing a 4-lobed stigma. Capsule linear, many-seeded, dehiscing from apex downward, brown, pubescent, 3–7 cm. in length; seeds 1 mm. long, ovate, slightly flattened and grooved, reddish-brown, comose with fine white hairs at apex.

Control.—Improved drainage (21). Mow as soon as the first flowers appear (11).

317. *Gaura biennis* L. Biennial gaura.

Biennial; reproducing by seeds. Meadows, pastures, thickets, and waste places. Widespread throughout eastern North America, west to Minnesota and Missouri. Native. July–September.

Description.—Stems erect, much branched, stout, woody below, soft-hairy, 1–3.2 m. high; winter rosette on deep taproot. Leaves alternate, simple, pinnately veined, sessile, denticulate, oblong-lanceolate, glabrous above but pubescent on the lower surface, 4–10 cm. long. Flowers in wand-like spikes, sessile, nearly regular, perfect; calyx-tube obconical, 1 cm. or more in diameter, fused with but much prolonged beyond the ovary, deciduous, 4 reflexed lobes; corolla of 4 clawed unequal pinkish petals; stamens 8, declined; ovary inferior, 1-celled; style long, drooping, bearing a 4-lobed stigma. Fruit indehiscent, nut-like, fusiform, 4-angled, ribbed, downy, sessile, about 1 cm. long, 1-celled with 1 or several seeds; seeds naked, irregularly angular, gray or light brown, about 1 mm. long.

Control.—Hoe or spud the weeds while still in the rosette stage (12). Mow meadows before seeds are produced (11).

Gaura parviflora Dougl. Velvet-weed, a native annual in the western and southwestern states, sometimes encroaches upon dry fields and cultivated land. *G. villosa* Torr., *G. coccinea* Pursh, *G. odorata* Lag., and *G. sinuata* Nutt. were classed as noxious weeds in California in 1954. A rotation including an intertilled crop will destroy them.

318. *Oenothera biennis* L. Evening primrose, Tree primrose, Field primrose, Fever plant, Coffee plant.

Biennial; reproducing by seeds. Dry, gravelly, or sandy neglected fields and waste places. Widespread and locally common throughout eastern North America, west to North Dakota, also on the Pacific Coast. A variable species including several varieties. Native. July–September.

Description.—Stems erect, usually simple, stout, pubescent to hirsute, 6 to 18 dm. high; winter rosette of petioled lanceolate leaves on a deep fleshy tap-root. Leaves alternate, simple, pinnately veined, sessile, toothed, lanceolate, acuminate, velvety-pubescent. Flowers in a terminal spike, sessile, leafy-bracted, regular, perfect; calyx-tube terete, 2.5–3.5 cm. long, prolonged beyond the ovary, deciduous, with 4 reflexed lobes; corolla of 4 obovate yellow petals, 1.5–2.5 cm. long; stamens 8, inserted on top of calyx-tube; style with deeply 4-cleft stigma; ovary much elongated and 4-celled. Fruit a capsule, sub-cylindrical, velvety-pubescent, 4-celled, 4-angled, apex slightly dilated, many-seeded, 22–33 mm. long; seeds 1–1.6 mm. long, very irregular in shape, pyramidal with varying number of sides, dull, dark reddish-brown, granular, ridged and often with winged edges.

Control.—The same as for 317.

UMBELLIFERAE (Parsley Family)

The following characteristics common to the weeds in this family, numbers 320–329, are not repeated in the descriptions of the species. Flowers in simple or compound umbels, subtended by bracts called an involucre. Flowers small, perfect, regular; calyx tubular, 5-toothed, fused with the ovary; corolla of 5 separate petals; stamens 5, separate; pistil of 2 carpels each containing 1 seed; ovary inferior; styles 2, with a thickened base called a stylopodium. In the mature fruit the 2 carpels separate into 2 1-seeded mericarps. The walls of the mericarps, "seeds," are ribbed longitudinally and usually have oil-tubes in the intervals between the ribs.

319. *Aegopodium podagraria* L. Goutweed. Fig. 73.

Perennial; reproducing by seeds and new shoots from the creeping rhizomes. Cultivated in gardens and escaped into waste places and along roadsides. Newfoundland to Michigan, south to Nova Scotia, New England, Pennsylvania, and North Carolina; local elsewhere. Moist shaded places. Introduced from Eurasia. June–August.

Description.—Stems erect, glabrous, simple or sparingly branched, 3–11 dm. high, from creeping rhizomes. Lower leaves with long petioles, chiefly biternate with 9 leaflets but often irregular; leaflets elliptic to ovate, 2.5–9 cm. long, serrate or toothed; upper leaves reduced, short-petioled, largely once-ternate. Flowers in dense, long-peduncled umbels, 6–12 cm. wide, rising above the leaves; involucel none; pedicels numerous, short; sepals obsolete; petals white. Fruit ovate, flattened laterally, glabrous, 3–4 mm. long.

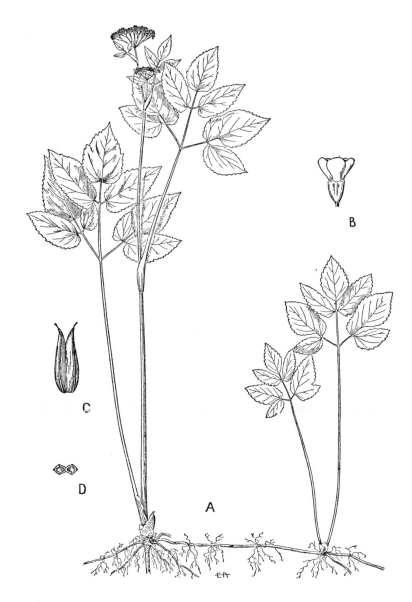

Fig. 73—**GOUTWEED**, *Aegopodium podagraria*: **A,** habit showing runners (× 1); **B,** flower (× 5); **C,** fruit (× 4); **D,** cross-section of seeds (× 4).

Control.—The same as for 322.

320. *Angelica atropurpurea* L. Purple-stem angelica, American angelica.

Biennial; reproducing by seeds. Meadows, pastures, roadsides, and waste places; on rich alluvial soils. Northeastern states, north central states, Quebec and Ontario. Native. June–July.

Description.—Stem erect, jointed and hollow, very stout, 2–3 m. high, glabrous, usually dark purple, from a thick tap-root which is often branched. Leaves alternate, 2–3-times ternately compound, with long petioles or the upper nearly sessile, the pinnate segments with 5–7 lanceolate to ovate leaflets with serrate margins and 2–4 cm. wide. Flowers in large compound umbels, without involucres; corolla small, white. Fruit flattened at right angles to the partition, with prominent primary ribs, stylopodium depressed; mericarps, "seeds," oblong, 5–8 mm. long, flattened on one side, the other with 5 winglike ribs; oil-tubes not distinct; pericarp free from the seed, yellow.

Control.—Cut the rosettes below the soil surface with a hoe or spud (12). The second season mow tops before seeds are formed (11).

321. *Carum carvi* L. Caraway.

Perennial; reproducing by seeds. In meadows and along roadsides; mostly on gravelly soils. Widespread and locally common, especially in the northeastern United States. Introduced from Europe. June–July.

Description.—Stems erect, glabrous, usually hollow, jointed, 6–9 dm. high. Leaves alternate or basal, petioled, pinnately decompound, leaflets filiform, glabrous. Flowers in compound umbels, involucral bracts entire; calyx with 5 small teeth; corolla white, or rarely pink; stylopodium short-conical. Mericarps, "seeds," narrow oblong, curved, 3–4 mm. long, flattened on one side, the rounded side appearing striped due to the 5 tan-colored corky ribs and the brown intervals each containing a solitary oil-tube.

Control.—Mow or pull the plants before seeds are formed.

322. *Cicuta maculata* L. Water-hemlock, Spotted-hemlock, Musquash root, Beaver poison, Spotted cowbane, Muskrat-weed, Childrens-bane. Fig. 74.

Perennial; reproducing by seeds and fleshy roots. Marshy ground, wet meadows and pastures, along ditches and streams. Widespread and locally abundant in northern and eastern United States and

FIG. 74—WATER-HEMLOCK, *Cicuta maculata*: **A**, upper part of plant showing habit (× ⅓); **B**, cluster of fleshy roots (× ⅓); **C**, vertical section through base of stem (× ⅓); **D**, flower (× 4); **E**, pistil (× 4); **F**, fruit (× 5); **G**, cross-section of a fruit showing oil-tubes (× 5); **H**, seed (mericarp), (× 5).

Parsley Family

in adjacent Canada, extending southwestward. Native. June–August.

Description.—Stems erect, branching, stout, hollow, jointed, streaked with purple, ridged, glabrous, 1–2.2 m. high; roots fusiform, 3–10 cm. long, fascicled. Leaves alternate or basal, petioled, ternately decompound with leaflets lanceolate to oblong-lanceolate, 3–12 cm. long, acuminate, deeply serrate, glabrous. Flowers in open spreading umbels having no involucre but involucels of slender bractlets, pedicels of the umbellets very unequal; calyx-teeth prominent; corolla white; stylopodium depressed. Mericarps, "seeds," broadly oval, about 2.5–3.5 mm. long, flat on one side, rounded and with 5 light ribs on the other side; oil-tubes solitary between the ridges; brown with yellow ribs.

Control.—Pull or dig the plants early in the spring while the ground is soft. Mow before seeds are formed.

The roots of water-hemlock and several other species of Cicuta which occur primarily in marshy areas are poisonous when eaten.[1]

323. **Conium maculatum** L. Poison-hemlock, Deadly-hemlock, Snake-weed, Poison parsley, Wode whistle, Poison stinkweed. Fig. 75.

Biennial; reproducing by seeds. Borders of fields, meadows, roadsides, and waste places; on rich, gravelly or loamy soils. Widespread in the northeastern and north central states and adjacent Canada; also in the Pacific Northwest. Introduced from Europe. July–August.

Description.—Stems erect, branching, stout, glabrous, purple-spotted, ridged, 6–15 dm. high; tap-root long, white, often branched. Leaves alternate or basal, petioled, ternately decompound, leaflets lanceolate, dentate and finely cut. Flowers in large open compound umbels, involucral bracts entire; corolla white. Mericarps, "seeds," oval, about 2–3 mm. long, granular, with conspicuous wavy ribs, without oil-tubes but with a layer of secreting cells next to the seed, grayish-brown.

Control.—The same as for 320.

According to tradition the poison-hemlock is the plant which furnished the "cup of death" given to Socrates in ancient Greece. It was formerly grown in the United States as a drug plant.

324. **Daucus carota** L. Wild carrot, Queen Annes lace, Birds-nest, Devils plague. Fig. 76, a–c.

Biennial; reproducing by seeds. Old meadows, pastures and waste

[1] Marsh and Clawson, 1914.

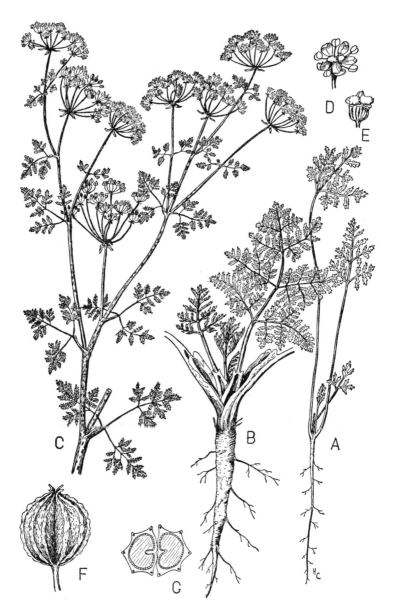

Fig. 75—**POISON-HEMLOCK**, *Conium maculatum:* **A**, seedling (× ⅓); **B**, root with leaf rosette at beginning of the second year (× ⅓); **C**, branch with umbels of flowers and fruits (× ⅓); **D**, flower (× 5); **E**, pistil (× 10); **F**, fruit (schizocarp), (× 6); **G**, cross-section of fruit showing two mericarps, each containing one seed (× 6).

places. Widespread throughout North America. Introduced from Eurasia. July–September.

Description.—Stems erect, branching, slender, hollow, ridged, bristly-hairy, 3–9 dm. high, bearing only scattered stem-leaves; taproot bearing a rosette of leaves the first season. Leaves alternate or basal, pinnately decompound, the segments narrow, often lobed, somewhat hairy; stem-leaves sessile with a sheathing base; basal leaves long-petioled. Flowers in compound flat-topped umbels which become concave as the fruits mature, involucral bracts cleft or pinnatifid; corolla white to pinkish; stylopodium depressed. Mericarps, "seeds," oblong, 2–3 mm. long, one side flattened, the other with 5 bristly primary ribs and 4 conspicuous secondary ribs which are winged and bear a row of barbed prickles; oil-tubes solitary under the secondary ribs and also 2 on the flat side; light grayish-brown.

Control.—The wild carrot causes most trouble in meadows after the hay has been removed. The second crop sometimes consists of more wild carrots than grass or clover. Such meadows should be mowed a second time as soon as the wild carrots begin to blossom. Badly infested meadows should be plowed under and planted to a clean cultivated crop for at least two years before reseeding. The frequent cultivation will induce the seeds already in the soil to germinate and destroy the new seedlings, thus preventing a fresh crop of seeds from contaminating the soil.

325. *Daucus pusillus* Michx. Rattlesnake weed.

Annual; reproducing by seeds. Dry woods, barrens, and hills. South Carolina, Kansas and British Columbia, south to Florida and northern Mexico; abundant on the Pacific Coast. Native. March–August.

Description.—Stems 1–8 dm. high, erect, retrorsely pubescent, from a slender taproot. Leaves and fruit as in *Daucus carota* but leaves more finely dissected. Umbels with long peduncles; terminal one usually less than 7 cm. wide; corollas white, or the central one larger and purple; bracts divided into lanceolate or linear segments 2–8 mm. long. Fruit ovoid-oblong, 3–5 mm. long, the wings parted into 1–8 flat prickles.

Control.—The same as for 324.

326. *Hydrocotyle sibthorpioides* Lam. (*H. rotundifolia* Roxb.). Lawn pennywort.

Perennial; reproducing by seeds and stems rooting at the nodes. Lawns, turfs, about greenhouses, cemeteries and in waste places.

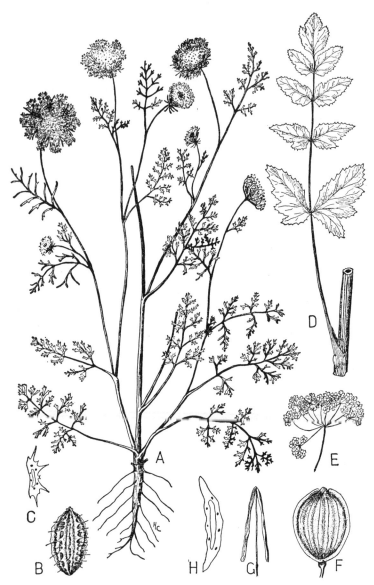

FIG. 76—**WILD CARROT**, *Daucus carota:* **A,** plant showing general habit (× ⅓); **B,** side view of "seed" (× 6); **C,** cross-section of "seed" (one-half of a fruit) (× 6).

PARSNIP, *Pastinaca sativa:* **D,** leaf showing attachment to stem (× ⅓); **E,** a flower-cluster (× ⅙); **F,** "seed" (× 4); **G,** fruit splitting into two parts, "seeds" (× 4); **H,** "seed" (one-half of a fruit) in cross-section (× 6).

Local from the middle Atlantic states westward to Kentucky[2] and southern Indiana.[3] Introduced from Asia; formerly somewhat used as an ornamental and ground-cover plant. June–July.

Description.—Stems slender, prostrate and creeping, rooting at the nodes, glabrous. Leaves alternate, nearly round or shield-shaped, 1–2 cm. in diameter, glabrous and glossy, with crenate margin and slender petiole. Flowers small, nearly sessile, in crowded head-like clusters; corolla of 5 small white petals. Fruit flattened at right angles to the partition, nearly orbicular; mericarps, "seeds," about 2 mm. long, with 5 primary ribs, 2 of which are often enlarged, oil-tubes none.

Control.—Hoe or dig out from lawns and fill in the bare places with sod or reseed.

327. *Pastinaca sativa* L. Wild parsnip, Birds-nest, Harts-eye, Madnip. Fig. 76, d–h.

Biennial; reproducing by seeds. Old fields, meadows, and waste places; mostly on rather rich heavy soils. Escaped and now widespread throughout northern United States and southern Canada. This is the wild form of the garden parsnip which was introduced from Europe and is grown extensively for its edible fleshy tap-roots. June–July.

Description.—Stems erect, branching, stout, hollow, grooved, glabrous, 6–12 dm. high; tap-root bearing a rosette of large leaves the first season. Leaves alternate or basal, pinnately compound, leaflets ovate to oblong, dentate, pubescent beneath; stem-leaves smaller, sessile and clasping; basal leaves with long grooved petioles. Flowers in compound umbels, involucral bracts few and entire or lacking; corolla yellow; stylopodium depressed. Mericarps, "seeds," oval, 4–7 mm. long, much flattened parallel to the partitions, ribs low, the 2 lateral extending into wings; oil-tubes small and solitary in the intervals, also 2–4 on the flat side; straw-colored to light brown.

Control.—The same as for 320.

The wet leaves of the parsnip, on coming into contact with certain individuals, produce a severe irritation and blistering of the skin and associated symptoms in a general way resembling those caused by poison ivy.

328. *Sium suave* Walt. (*S. cicutaefolium* Schrank). Water parsnip. Fig. 77.

[2] Garman, 1914.   [3] Hansen, 1921, 1923.

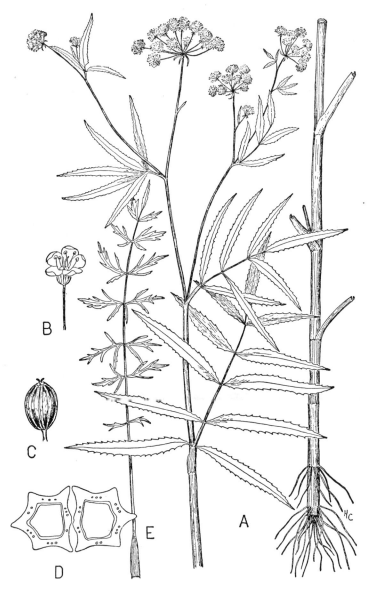

Fig. 77—**WATER PARSNIP,** *Sium suave:* **A,** plant showing general habit ($\times \frac{1}{3}$); **B,** flower ($\times 4$); **C,** fruit ($\times 4$); **D,** cross-section of fruit ($\times 8$); **E,** submerged leaf.

Heath Family

Perennial; reproducing by seeds. Marshy ground, sloughs and swales in pastures and meadows; also in shallow water along muddy shores. Widespread across North America. Native. July–September.

Description.—Stems stout, erect, branched above, glabrous and ridged. Leaves alternate, pinnately compound, with 3–8 pairs of lanceolate serrate leaflets, the lower submerged leaves often finely dissected. Flowers in compound umbels, with numerous narrow bracts; corolla white. Fruit ovate to oblong; stylopodium depressed; mericarps, "seeds," half-oval, 2.5–3 mm. long, with 5 prominent winglike ribs; oil-tubes 1–3 in each interval, dark brown with light brown or greenish ribs.

Control.—The same as for 322.

This plant has been reported as poisonous when eaten by stock.

329. *Torilis japonica* (Houtt.) DC. Japanese hedge-parsley.

Annual; reproducing by seeds. Open woods, fields, and waste places. New York to Iowa, south to Florida and Texas; also in the Pacific Coast states. Introduced from Eurasia. April–August.

Description.—Stems erect, slender, much branched, 3–8 dm. high; plants hispidulous throughout. Leaves in general outline triangular to ovate, ternately or pinnately once- or twice-compound, the ultimate segments linear or lanceolate, sharply serrate. Flowers in open loose umbels, peduncles 3–16 cm. long, white; primary rays 5–10, mostly 1–3 cm. long; pedicels short, strigose, mostly shorter than the bractlets. Fruits 1.5–4 mm. long, evenly covered with prickles.

Control.—Mow or pull the plants before seeds are formed.

ERICACEAE (Heath Family)

330. *Kalmia angustifolia* L. Sheep laurel, Lambkill, Narrow-leaved laurel, Dwarf laurel, Sheep poison, Calfkill, Wicky. Fig. 78, a.

Perennial shrub; reproducing by seeds and new shoots from the roots. Pastures, dry hillsides, sandy soils and in bogs; on acid soils. Eastern North America from Newfoundland to Georgia. Native. June–July.

Description.—Stems erect, low shrub, glabrous, rarely 1 m. high. Leaves mostly whorled or opposite, simple, petioled, entire, leathery and persistent, pale green to light brown beneath, glossy green above, glabrate, narrowly oblong, obtuse, flat, 2–5 cm. long. Flowers perfect, regular, in lateral corymbs, appearing later than the shoots of the season, slightly glandular; calyx of 5 fused sepals, persistent,

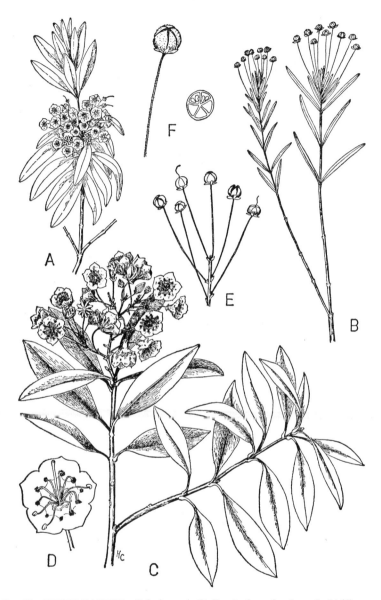

Fig. 78—**SHEEP LAUREL**, *Kalmia angustifolia:* **A**, flowering branch (× ⅓).
SWAMP LAUREL, *Kalmia polifolia:* **B**, flowering branch (× ⅓); **E**, cluster of capsules (× ⅓); **F**, capsule (× 1).
MOUNTAIN LAUREL, *Kalmia latifolia:* **C**, flowering branch (× ⅓); **D**, flower (× 1).

glandular, smaller than the capsule; corolla wheel-shaped, rarely 1 cm. broad, crimson, 5-lobed with 10 pouches receiving as many anthers; stamens 10, with long thread-like curving filaments; pistil compound; style 1; ovary 5-celled. Fruit a globose, 5-celled, many-seeded, depressed, nearly smooth capsule; seeds linear, about 1 mm. long, glossy, marked with fine vertical lines, straw-colored.

Control.—Grub out roots and all (17). Mow the tops with a brush scythe and burn.

The leaves of sheep laurel and the two following species are poisonous when eaten by sheep, cattle and horses.

331. *Kalmia latifolia* L. Mountain laurel, Calico bush, Spoonwood, Poison laurel, Ivy. Fig. 78, c–d.

Woody perennial; reproducing by seeds. Rocky or stony hillside woodlands and pastures; on acid soils. New Brunswick to Florida and Louisiana; most common from southern New York through the Appalachian Mountains to western Florida. Native. May–June.

Description.—Stems erect, shrub, much branched, 1–3 m. high, twigs nearly terete, glabrous. Leaves mostly alternate, simple, petioled, entire, leathery and persistent, bright glossy green on both sides, ovate-lanceolate or oblong, acute at each end, nearly flat, 5–12 cm. long. Flowers perfect, regular, in terminal corymbs; pedicels clammy-pubescent; calyx 5-parted, persistent, smaller than the capsules; corolla wheel-shaped, pink or white, 1.5–2.5 cm. in diameter, 5-lobed with 10 pouches receiving as many anthers; stamens 10, with long thread-like curved filaments; pistil compound; style 1; ovary 5-celled. Fruit a globular depressed, 5-celled, many-seeded, glandular capsule; seeds about 0.7 mm. long, oblong, often curved, surface reticulated, glossy, orange-brown.

Control.—Grub out small bushes or scattered clumps (17). In large areas cut the tops and burn them.

332. *Kalmia polifolia* Wang. Swamp laurel, Pale laurel, Alpine laurel. Fig. 78, b, e, f.

Woody perennial; reproducing by seeds. Pastures, woodlands and bogs; on acid soils. Common across Canada, southward into northern New York, Michigan, Minnesota and in the mountains of the Pacific Coast states. Native. June–July.

Description.—Stems erect or procumbent, forming a small straggly shrub, 1–6 dm. high, twigs 2-edged, glabrous. Leaves opposite, simple, nearly sessile, entire, leathery and persistent, glossy green above and glaucous-white beneath, oblong, 1–4 cm. long, margins revolute. Flowers perfect, regular, in few-flowered terminal corymbs; calyx 5-parted, persistent, smaller than the capsule; corolla wheel-shaped, rose-purple, 1–2 cm. broad, 5-lobed, with 10 pouches receiving as many anthers; 10 stamens with long thread-like curved filaments; pistil compound; style 1. Fruit an ovoid, 5-celled, many-seeded, glabrous capsule; seeds similar to 330 but about twice as long.

Control.—The same as for 330.

PRIMULACEAE (Primrose Family)

333. *Anagallis arvensis* L. Scarlet pimpernel, Poor-mans weatherglass, Red chickweed, Poison chickweed, Shepherds-clock, Eye-bright. Fig. 79, e–g.

Annual; reproducing by seeds. Gardens, cultivated fields, lawns and waste places. Widespread throughout North America; most troublesome in the middle Atlantic states and on the Pacific Coast. Introduced from Eurasia. May–September.

Description.—Stems prostrate, branched and spreading, glabrous, 4-angled, 1–5 dm. long. Leaves opposite or in whorls of 3, simple, sessile, entire, ovate, glabrous, glandular-dotted beneath, 2 cm. long or less. Flowers solitary in the leaf-axils, perfect, regular, about 1 cm. in diameter, borne on slender peduncles; calyx of 5 awl-shaped persistent sepals; corolla wheel-shaped, with almost no tube, 5-lobed, the lobes broadly obovate and minutely glandular-ciliate on the margin, reddish or rarely white, quickly closing at the approach of bad weather; stamens 5, opposite the petals; filaments bearded; pistil compound; style 1; ovary 1-celled, with a free-central placenta. Fruit a smooth, globose, many-seeded, 1-celled capsule, the top falling off as a lid when the capsule splits transversely; seeds 3-angled, elliptical, about 1.3 mm. long, finely pitted, brown.

Control.—Clean cultivation begun early and continued late (6). Hand pull or hoe scattered plants before the first blossoms appear (4, 5).

334. *Lysimachia nummularia* L. Moneywort, Creeping loosestrife, Yellow myrtle, Creeping jenny, Creeping charlie. Fig. 79, a–d.

Perennial; reproducing by seeds but more often by creeping stems. Low fields, gardens, lawns, and along ditches; especially on moist or shaded, rich or alluvial soils. Widespread and locally common Newfoundland to Ontario, south to Georgia, Missouri, and Kansas. In-

Fig. 79—**MONEYWORT,** *Lysimachia nummularia:* **A,** plant showing habit (× ⅓); **B,** flower (× 1); **C,** capsule (× 1); **D,** seed (× 6).
SCARLET PIMPERNEL, *Anagallis arvensis:* **E,** habit (× ⅓); **F,** capsule (× 2); **G,** seed (× 12).

troduced from Europe as an ornamental; still used as a ground-cover in some places. June–July.

Description.—Stems prostrate and creeping, branched, glabrous, rooting freely at the nodes. Leaves opposite, simple, short-petioled, nearly entire, obovate to orbicular, glabrous, glandular-dotted, 1–3 cm. in diameter. Flowers solitary on slender pedicels in the leaf-axils, perfect, regular, 2–3 cm. in diameter; calyx of 5 or 6 ovate overlapping sepals; corolla yellow, wheel-shaped, deeply 5-lobed, the lobes broadly obovate; stamens 5, opposite the petals, slightly fused by the filaments; pistil compound; style 1; ovary 1-celled, with a free-central placenta and bearing many seeds. Fruit a globose, 1-celled, several-seeded capsule, splitting vertically when mature; seeds 3-angled, elliptical, about 1 mm. long, dark brown to black, surface roughened by light-colored scaly ridges.

Control.—Improve the drainage (21). Dig out by hand, being careful to get every piece of stem (17). In lawns raise the runners with a rake and mow close (11).

APOCYNACEAE (Dogbane Family)

335. *Apocynum androsaemifolium* L. Spreading dogbane, Wandering milkweed, Honey-bloom, Milk ipecac. Fig. 80, j–k.

Perennial; reproducing by seeds and rootstocks. Old fields, meadows, roadsides and thickets; on dry, gravelly or sandy soils. Widespread throughout the northern states and southern Canada. Native; introduced in some localities. June–July.

Description.—Stems erect, 3–6 dm. high, with spreading branches, glabrous or nearly so, woody at the base, bark tough, fibrous, sap milky, creeping horizontal rootstocks. Leaves opposite, simple, pinnately veined, short-petioled, entire, broadly ovate, mucronate-tipped, dull dark green and smooth above, pale and somewhat pubescent beneath. Flowers in open paniculate clusters, both terminal and axillary, perfect, regular, short-pedicelled and nodding; calyx of 5 acute lobes, much shorter than the corolla-tube; corolla campanulate, with 5 revolute lobes, pale pink marked with reddish stripes, 6–7 mm. long, appendaged within; stamens 5, inserted on corolla, filaments shorter than the arrow-shaped convergent anthers, which slightly adhere to the stigma; carpels 2, separate except the 2 stigmas are fused. Fruit 2 long slender follicles, terete, curved, smooth, about 10 cm. long, containing numerous seeds; seeds about 1.5 mm. long,

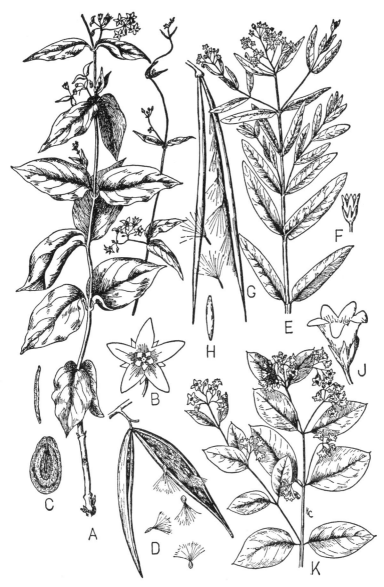

FIG. 80—**WHITE SWALLOW-WORT**, *Cynanchum vincetoxicum:* **A,** plant showing general habit (× ⅓); **B,** flower (× 2); **C,** surface and side view of seed (× 3); **D,** pod (follicle) shedding seed with hairs (× ½).
INDIAN HEMP, *Apocynum cannabinum:* **E,** flowering shoot (× ⅓); **F,** flower (× 2); **G,** pod (follicle), (× ⅓); **H,** seed (× 2).
SPREADING DOGBANE, *Apocynum androsaemifolium:* **J,** flower (× 2); **K,** flowering shoot (× ⅓).

linear, comose with a tuft of silky-white hairs at apex, reddish-brown.

Control.—Fields overrun with this weed should be plowed deep (7) and followed by a clean cultivated crop (18). The shoots sprouting after the last cultivation should be pulled or hoed off as soon as they appear. In waste places mow with a scythe or cut with a hoe several times during the summer (11).

336. *Apocynum cannabinum* L. Indian hemp, American hemp, Indian physic, Choctaw root, Bowmans root, Rheumatism-weed, Hemp dogbane. Fig. 80, e–h.

Perennial; reproducing by seeds and rootstocks. Gravelly fields, meadows, thickets, waste places and along streams. Widespread throughout the United States and southern Canada; represented by several varieties in different sections. Native. June–August.

Description.—Stems erect, branched, main axis exceeded by branches; woody at the base, bark tough, fibrous, sap milky, 5–25 dm. high. Leaves opposite, simple, pinnately veined, short-petioled or nearly sessile on the branches, entire, oblong-ovate to oblong-lanceolate, mucronate-tipped, pale green, glabrous or sparingly pubescent beneath. Flowers in terminal rather dense corymbose clusters, perfect, regular, short-pedicelled, mostly erect, the central cluster flowering first; calyx of 5 acute lobes equaling the corolla-tube in length; corolla campanulate, with 5 erect lobes, greenish-white, 2.5–4 mm. long, appendaged within; stamens 5, inserted on corolla, filaments shorter than the arrow-shaped convergent anthers which slightly adhere to the stigma; carpels 2, separate except the 2 stigmas are fused. Fruit 2 long slender follicles, terete, straight, dark reddish-brown, smooth, 8–12 cm. long, less than 4 mm. in diameter, containing numerous seeds; seeds about 5 mm. long, linear, comose with a tuft of silky-white hairs at the apex, brown.

Control.—The same as for 335.

337. *Vinca minor* L. Periwinkle, Myrtle.

Perennial; reproducing by runners; apparently only rarely by seeds. Moist rich soils bordering gardens, lawns, roadsides, cemeteries, and shaded waste places, in localities where it has been planted extensively as a ground-cover. Widespread, mostly in the northeastern states. Introduced from Europe. April–June.

Description.—Stems creeping, evergreen, branching, glabrous, rooting at the nodes. Leaves opposite, simple, pinnately veined, entire, ovate to oblong, narrowed at the base into petioles, firm, glossy, evergreen, 1.5–3 cm. long. Flowers solitary, axillary, perfect, regular, peduncled; calyx with 5 acuminate, lanceolate, glabrous lobes; corolla blue, the tube funnelform about 1 cm. long, the limb salverform with 5 truncate lobes, 1.5–2 cm. in diameter; stamens 5, inserted on the corolla, filaments short; carpels 2, nearly separate except the styles and stigma are fused. Fruit of 2 short cylindrical follicles about 2–4 cm. long, with several rough seeds.

Control.—In lawns raise the runners with a rake and mow them close (11). Dig out by hand (17). Clean cultivation (6).

ASCLEPIADACEAE (Milkweed Family)

338. *Ampelamus albidus* (Nutt.) Britt. (*Gonolobus laevis* Michx.). Honeyvine, Sandvine. Fig. 82.

Perennial; reproducing by seeds. River banks and rich alluvial or sandy thickets; often a troublesome weed in fields and low moist woods. Pennsylvania to Iowa, south to Florida and Texas. Native. July–September.

Description.—Stems twining, 1–4 m. high, smooth. Leaves 3.5–12 cm. wide, opposite, deeply cordate, acute, long-petioled. Flowers small, whitish, in axillary, pedunculate, raceme-like clusters, corolla about 6 mm. long.

Control.—Cultivation at frequent intervals.

339. *Asclepias speciosa* Torr. Showy milkweed.

Perennial; reproducing by seeds and rootstocks. Meadows, pastures and waste places; especially on alluvial soils. Minnesota to Arkansas and westward to the Pacific Coast. Native. June–August.

Description.—Similar to 340 in general habit but often less hairy. Corolla-lobes and crown longer, with a long appendage from the truncate summit of each hood. Seeds about 1 cm. long, flat, thin, distinctly margined with a ridge on one side, dull, brown, comose with a tuft of hairs at base.

Control.—The same as for 335.

340. *Asclepias syriaca* L. Milkweed, Silkweed, Cotton-weed. Fig. 81, a–e.

Perennial; reproducing by seeds and rootstocks. Fields, pastures and waste places; on rich sandy or gravelly loam. Common in eastern North America, west to Iowa and Kansas. Native. June–August.

Description.—Stems erect, commonly unbranched, mostly pubescent, with milky sap, 6–15 dm. high, usually in patches due to the

Fig. 81—**COMMON MILKWEED,** *Asclepias syriaca:* **A,** part of a plant showing flower-clusters (× ⅓); **B,** flower (× 3); **C,** flower, top view (× 3); **D,** surface and side views of seed (× 3); **E,** opening seed-pod (follicle), (× ⅓).
BUTTERFLY-WEED, *Asclepias tuberosa:* **F,** part of plant showing flower-clusters (× ⅓); **G,** opening seed-pod (follicle), (× ⅓); **H,** seed (× 3).

Milkweed Family

long horizontal creeping rootstocks. Leaves opposite or whorled, simple, closely pinnate-veined, with veins uniting before reaching the margin, entire, broadly oval to oblong-elliptic, green and glabrous above but pale and downy-pubescent beneath, 1-2 dm. long, with very short stout pubescent petioles. Flowers in simple many-flowered umbels, perfect, regular, with slender pubescent peduncles; calyx with 5 small persistent reflexed lobes; corolla deeply 5-parted, reflexed, purplish, 6-9 mm. long; crown of 5 yellowish hooded bodies with an incurved horn rising from the cavity of each hood; stamens 5, inserted on the base of the corolla, the filaments united, forming a tube about the pistil; anthers adhering to the stigma, each anther composed of 2 sac-like cells, the pollen forming waxy masses (pollinia); carpels 2, separate, fused by the stigmas. Fruit a hairy follicle, covered with soft spine-like projections, on a recurved pedicel, acuminate, 8-13 cm. long; seeds much flattened, narrowly obovate, wing-margined, dull brown, 6-8 mm. long, comose with a tuft of silky hairs at the base.

Control.—The same as for 335.

341. *Asclepias tuberosa* L. Butterfly-weed, Pleurisy root, Orange milkweed, Orange root, White root. Fig. 81, f-h.

Perennial; reproducing by seeds. Dry, sandy or loamy fields, pastures and waste places. Local in the eastern states, the Great Lakes region and the Mississippi Valley. Native. July-August.

Description.—Stems erect, branched near the top, stout, sap not milky, very rough, 3-9 dm. high; stems numerous, densely clustered. Leaves alternate, simple, inconspicuous venation, entire, nearly sessile, linear to narrowly oblong, rough-hairy on both surfaces, 4-10 cm. long. Flowers similar to 340 except the corolla is dull orange, the crown of hoods a bright orange. Fruit a follicle, narrowly acuminate, erect on deflexed pedicels, hoary, dark brown, 5-10 cm. long; seeds much flattened, 6-8 mm. long, brown, comose with a tuft of silky hairs at the base.

Control.—The same as for 335.

342. *Asclepias verticillata* L. Whorled milkweed.

Perennial; reproducing by seeds. Fields, prairies, roadsides, and upland woods, often in dry, sterile soil. Massachusetts, Ontario, and Saskatchewan, south to Florida and Texas. Native. April-September.

Description.—Stems from fibrous roots, slender, erect, 2-10 dm. high, mostly simple, leafy to the summit. Leaves narrowly linear,

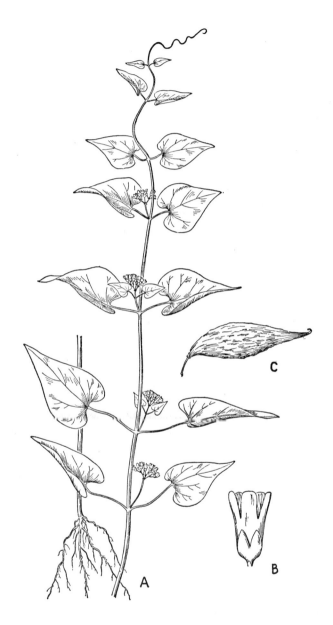

Fig. 82—**HONEYVINE**, *Ampelamus albidus*: **A**, habit (× ⅓); **B**, flower (× 2); **C**, follicle (× ⅓).

Milkweed Family

revolute, in whorls of 3–6. Umbels small, axillary and terminal; peduncles 1–3 cm. long; petals white or greenish-white, 4–5 mm. long; hoods greenish white, broadly ovate, about half as long as the incurved subulate horns. Pods smooth, slender, on erect pedicels, 4–5 cm. long.

Control.—The same as for 335.

Asclepias galioides HBK., Poison milkweed or Whorled milkweed, a native perennial of the ranges and dry hillsides from Kansas and Utah southward into Texas, California, and Mexico, appears in pastures, fields, along ditches and roadsides and in waste places. This weed spreads by seeds and creeping roots. It has whorls of linear leaves on smooth unbranched stems. This weed causes poisoning when eaten by livestock.[1, 2] The method of control is the same as for 335.

Asclepias labriformis Jones was listed as a noxious weed in Utah in 1954.

343. *Cynanchum nigrum* (L.) Pers. (*Vincetoxicum nigrum* Moench). Black swallow-wort, Climbing milkweed.

Perennial; reproducing by seeds. Old fields, pastures, fence-rows and waste places, chiefly in limestone regions. Local in the northeastern states and Ontario. Introduced from Europe. June–September.

Description.—Stems twining, mostly unbranched, nearly glabrous, 5–15 dm. high. Leaves opposite, simple, pinnately veined, petioled, entire, ovate-acuminate, thin, nearly glabrous, 8 cm. long or less. Flowers perfect, regular, in an axillary cymose cluster on a short peduncle; calyx small, 5-lobed; corolla wheel-shaped, with 5 spreading lobes, dark purple, pubescent within, crowned with a fleshy 5-lobed disk; stamens and carpels similar to 339. Fruit follicles, smooth, elliptic, very dark brown, about 5 cm. long, in pairs; seeds much flattened, obovate, wing-margined, brown, 5–6 mm. long, comose with a tuft of hairs at the base of the seed.

Control.—The same as for 335.

344. *Cynanchum vincetoxicum* (L.) Pers. (*Vincetoxicum officinale* Moench). White swallow-wort. Fig. 80, a–d.

Perennial; reproducing by seeds. Fields, meadows and fence-rows; chiefly on dry soils in limestone regions. Local in the northeastern states, west to Kansas. Introduced from Europe. June–September.

[1] May, 1920.   [2] Marsh, Clawson, Couch and Eggleston, 1920.

Description.—Stems twining or sub-erect, branching, somewhat pubescent, 5–15 dm. high. Leaves opposite, simple, pinnately veined, entire, ovate-lanceolate, slightly pubescent, 8 cm. long or less. Flowers similar to 343 except the corolla is usually greenish-white and glabrous within. The follicles and seeds similar to 343.

Control.—The same as for 335.

CONVOLVULACEAE (Morning-glory Family)

345. *Convolvulus arvensis* L. Bindweed, Wild morning-glory, Small-flowered morning-glory, European bindweed, Corn-bind, Bearbind, Creeping jenny, Green-vine. Fig. 83.

Perennial; reproducing by seeds and creeping roots. Cultivated land, grain fields and waste places; on rich, somewhat sandy or gravelly soils. Widespread throughout the United States and southern Canada; most abundant and troublesome in the western states. Introduced from Europe. June–July.

Description.—Stems slender, prostrate or twining, often 1–3 m. long, glabrous. Leaves alternate, simple, long-petioled, entire, ovate-oblong to hastate, with acute basal lobes, glabrous, 5 cm. long or less. Flowers perfect, regular, axillary, the long peduncles mostly 1-flowered, bracts minute and remote; calyx small, campanulate; corolla funnel-shaped, white or pinkish, 1.5–2 cm. long; stamens 5, attached to the corolla; pistil compound, stigmas 2, filiform; style filiform; ovary 2-celled. Fruit an ovate capsule, 2-celled, 2–4-seeded; seeds 3-angled, ovoid, 3–5 mm. long, flat on one or two sides, the other side rounded, coarsely roughened, dull dark gray-brown.

Control.—The bindweed is one of the most troublesome and difficult to eradicate on account of its very extensive root system which may penetrate the soil to a depth of 3–5 meters. The earlier methods of controlling this weed were based entirely on intensive cultivation and summer fallowing of the infested areas.[1] More recently a number of state experiment stations have been developing other methods of control based largely on the use of chemicals: California,[2, 8] Utah,[3] Kansas,[4] Washington,[5] Iowa,[6] New York,[7] Nebraska.[9]

[1] Cox, 1909.
[2] Gray, 1917.
[3] Stewart and Pittman, 1924.
[4] Latshaw and Zahnley, 1927, 1928.
[5] Schafer, Lee and Neller, 1929.
[6] Bakke, 1930.
[7] Muenscher, 1932.
[8] Crafts and Kennedy, 1930.
[9] Kiesselbach, Petersen and Burr, 1934.

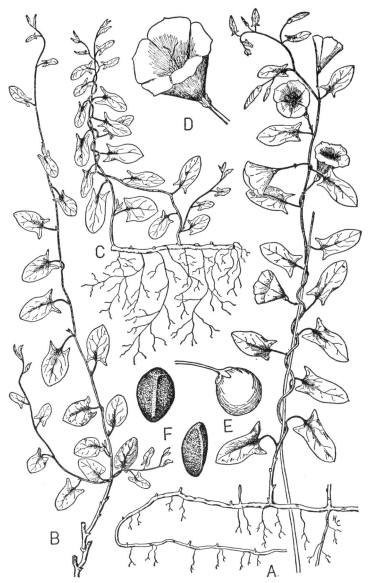

Fig. 83—BINDWEED, *Convolvulus arvensis:* **A,** part of a plant showing general habit (× ⅓); **B,** young shoot sprouting from underground stem (× ⅓); **C,** a small piece of root with new shoot and buds (× ⅓); **D,** flower (× ⅔); **E,** capsule (× 2); **F,** seed (× 3).

The control methods here suggested are general and must be selected or modified to suit the local conditions.

Clean cultivation or summer fallow, if performed thoroughly, will control bindweed in two years. The land must be kept black, that is no green shoots should be allowed to appear above ground. Cultivation should be at frequent intervals of about six days. Under certain conditions this time may be extended a few days and under other conditions it may have to be shortened to three- or four-day intervals. A cultivator with knife-like blades or a special knife weeder capable of cutting the bindweed from 3 to 6 inches below the soil surface gives best results. After the bindweed has been weakened or practically eradicated, sow an annual cover-crop such as millet, sorghum, or sudan-grass which produces a dense shade. Or, if the land is adapted for it, sow alfalfa, which on account of its extensive root-system, its shading effect, and the frequent cutting of the crop, will tend to prevent the bindweed from spreading. Small areas of bindweed can be destroyed by covering them with paper mulch (24).

346. *Convolvulus sepium* L. Wild morning-glory, Hedge bindweed, Devils-vine, Great bindweed, Bracted bindweed. Fig. 84.

Perennial; reproducing by seeds and creeping rootstocks. Cultivated fields, meadows and fence-rows; on rich moist lowlands on alluvial soils. Widespread Newfoundland to British Columbia, south to Florida, New Mexico, and Oregon. Native. July–August.

Description.—Stems high-twining or trailing extensively, glabrous or hairy. Leaves alternate, simple, long-petioled, entire, triangular-ovate, with angular-cordate base, acute, glabrous, 1 dm. long or less. Flowers axillary, peduncles elongated, 4-angled, with 2 large bracts at the base of the calyx; calyx of 5 overlapping sepals; corolla campanulate to funnelform, white to rose-colored, 3–6 cm. long; stamens 5, attached to the corolla; stigmas 2, oblong, style filiform; ovary 2-celled. Fruit a globose, 2-celled, 2–4-seeded capsule about 8 mm. in diameter, usually covered by the bracts and calyx; seed plump ovoid, 4–5 mm. long, one side nearly flat, the other side rounded, coarsely roughened, black or dark brown.

Control.—The same as for 345.

347. *Cuscuta campestris* Yuncker. Western field dodder.

Annual; reproducing by seeds. Parasitic on various herbs, includ-

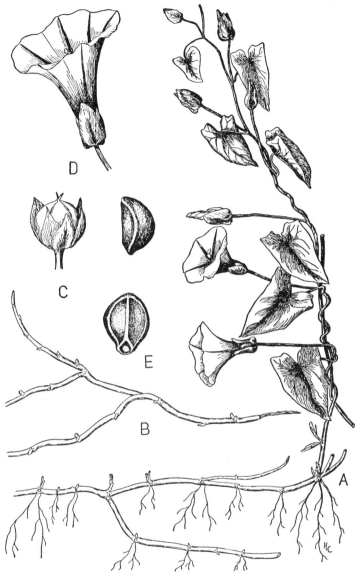

Fig. 84—**WILD MORNING-GLORY**, *Convolvulus sepium:* **A**, portion of a plant showing climbing habit (× ⅓); **B**, portion of underground stem (rhizome) with buds (× ⅓); **C**, capsule (× 1); **D**, flower (× ½); **E**, seed (× 9).

ing cultivated species. Washington to California, east across the continent. Native. June–October.

Description.—Stems light yellow, coarser than *Cuscuta pentagona*, with larger glomerules. Flowers often glandular, 2–3 mm. long, on short pedicels, in compact globular clusters; corolla-lobes spreading, triangular-ovate, tube campanulate, about the same length as the lobes; calyx-lobes broader than long; stamens borne at sinuses; anthers shorter than the filaments; styles slender, about equaling the ovary; scales reaching summit of corolla-tube, their longer fringe half as long as blade. Seeds 1–1.5 mm. long.

Control.—The same as for 355.

348. *Cuscuta coryli* Engelm. Hazel dodder. Fig. 85, e.

Annual; reproducing by seeds. Parasitic on various shrubs and herbs. Widely distributed in the eastern states and the Mississippi Valley. Native. July–September.

Description.—Similar to 355. Stigma capitate; calyx-lobes acute; corolla with 4 inflexed lobes. Capsule indehiscent; seeds about 1.5 mm. long, dark brown, scurfy, globular or compressed.

Control.—The same as for 355.

349. *Cuscuta epilinum* Weihe. Flax dodder, Hairweed, Devilshair. Fig. 85, c.

Annual; reproducing by seeds. Parasitic, mostly on flax, rarely on other hosts. In the flax sections of the northwestern states and Manitoba; also eastward. Noxious (1954) in all states. Introduced from Europe. June–August.

Description.—Similar to 355. Calyx-lobes obtuse and overlapping; style with linear-elongated stigma not longer than the ovary. Capsule opening by a lid; seeds 1–1.5 mm. long, ovoid to spherical, often in pairs; hilum minute, whitish; surface pitted, dull, yellow to brown.

Control.—The same as for 355.

350. *Cuscuta epithymum* Murray. Clover dodder, Thyme dodder. Fig. 85, d.

Annual; reproducing by seeds. Parasitic on clovers, alfalfa and other leguminous hosts. Widespread throughout North America. Introduced from Europe. June–September.

Description.—Similar to 355. Calyx-lobes acute, not overlapping; style with linear-elongated stigma longer than the ovary. Capsule capped by the withered corolla, opening by a lid; seeds about 1 mm. long, globular, often angular, dull, pitted, yellow to brown or gray.

Control.—The same as for 355.

351. *Cuscuta glomerata* Choisy.

Annual; reproducing by seeds. Parasitic on many coarse herbs. Ohio to Minnesota and South Dakota, south to Mississippi and Texas. Native. July–September.

Description.—Branches slender, floriferous, soon wound lightly about host. Flowers mostly sessile, 4–5 mm. long, in dense, twisted, cord-like masses. Bracts oblong or oval, as long as the calyx, with recurved tips; corolla lobes lanceolate to oblong, acutish, spreading, about 2 mm. long, shorter than the tube. Capsule flask-shaped, bearing withered corolla at summit. Seeds about 1.7 mm. long.

Control.—The same as for 355.

352. *Cuscuta gronovii* Willd. Common dodder, Goldthread vine, Onion dodder. Fig. 85, b.

Annual; reproducing by seeds. Parasitic on various plants; usually in low wet fields or in marshy thickets. Widespread and locally abundant throughout eastern North America. Native. July–September.

Description.—Similar to 355. Stigma capitate; corolla-tube campanulate, with 5 obtuse, erect or spreading lobes. Capsule globose-conical, indehiscent; seeds 1.5–2 mm. in diameter, spherical, angular or indented, light to dark brown, dull, minutely granular.

Control.—The same as for 355.

353. *Cuscuta indecora* Choisy. (*C. racemosa* var. *chiliana* Rydb.). Large-seeded alfalfa dodder, Chilean dodder.

Annual; reproducing by seeds. Damp sandy pinelands, openings and bottomland, where it is parasitic on a wide variety of herbs and shrubs. Florida to California, north to Virginia, Minnesota and Idaho. Native. July–September.

Description.—Stems relatively coarse, loosely matted. Flowers whitish, fleshy, 5-merous, 2–5 mm. long, on short pedicels, forming loose, irregular cymose clusters; calyx-lobes triangular-ovate, acute or subacute, much shorter than to equaling the broad corolla-tube; corolla campanulate, 3–5 mm. long, with ascending to erect triangular lobes with acute, inflexed tips; stamens about half as long as lobes of corolla; scales oblong, equal in length to corolla-tube, fringed. Capsule thickened at summit, globose, wrapped by the withering but persistent corolla. Seeds 1–1.7 mm. long.

Control.—The same as for 355.

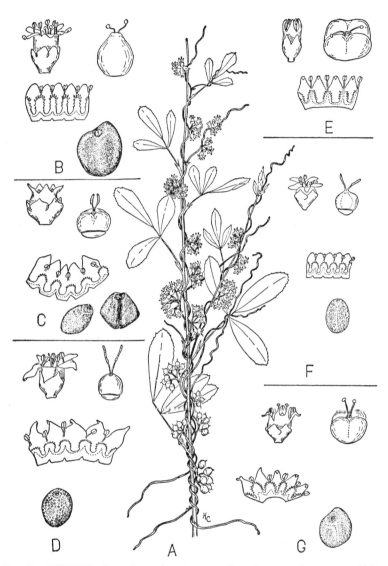

FIG. 85—**DODDER**, *Cuscuta* species: **A**, part of a plant growing on an alfalfa stem, showing general habit (× ⅓). In all the following, flower, corolla and capsule (× 3), seed (× 9). **B**, Common dodder, *Cuscuta gronovii;* **C**, Flax dodder, *Cuscuta epilinum;* **D**, Clover dodder, *Cuscuta epithymum;* **E**, Hazel dodder, *Cuscuta coryli;* **F**, Small-seeded alfalfa dodder, *Cuscuta planiflora;* **G**, Field dodder, *Cuscuta pentagona.*

354. **Cuscuta pentagona** Engelm. (*C. arvensis* Beyrich). Field dodder, Large-seeded dodder, Love-vine. Fig. 85, g.

Annual; reproducing by seeds. Parasitic on various herbaceous plants. Widely distributed throughout the United States. Native. April–October.

Description.—Similar to 355. Stigma capitate; calyx-lobes obtuse; corolla with 4–5 acute inflexed lobes. Capsule depressed-globose, indehiscent; seeds about the size of red clover seed, 1–1.5 mm. long, one side rounded, the other flattened and often with an obtuse ridge, minutely pitted, yellow to reddish-brown; hilum whitish, within a smooth circular area on the flattened side.

Control.—The same as for 355.

355. **Cuscuta planiflora** Tenore. Small-seeded alfalfa dodder. Fig. 85, f.

Annual; reproducing by seeds. Parasitic on alfalfa and other hosts. Widely distributed in the western states, from Washington and Wyoming to Mexico.[10] Noxious (1954) in all states. Introduced from Europe. June–August.

Description.—Stems thread-like, climbing or twining, yellow or reddish (without chlorophyll), leafless, with small suckers by which they parasitize the host plants upon which they grow. Leaves reduced to a few very minute scales. Flowers perfect, regular, in dense globular clusters, small, about 2 mm. long, white; calyx of 4–5 fused sepals, lobes broadly ovate, overlapping; corolla of 4–5 petals, fused below, with ovate lobes shorter than the narrow cylindrical tube; stamens 5 (rarely 4), each with a fringed scale-like appendage at its base; pistil with a 2-celled ovary and 2 slender styles longer than the ovary; stigma linear-elongated. Fruit a depressed-globose dehiscent capsule with the withered corolla persisting on its summit, 2-celled, 2–4-seeded; seeds about 1 mm. long, oval to oblong, angular, finely granular and more or less scurfy, light brown; hilum short, oblong.

Control.—Do not sow seeds infested with dodder; use only clean seed (1). Prevent seed formation by dodders in fence-rows, along ditches and on waste places by early cutting (3). Mow patches of dodder in clover and alfalfa fields; pile and burn on the spot to prevent scattering of seeds or pieces of vines (13). Harvest the first cutting of alfalfa or clover for hay or silage early enough so that no seeds

[10] *Yuncker*, 1921, 1932.

are matured. Plow badly infested fields and follow with a cultivated crop for a year.

356. *Cuscuta polygonorum* Engelm.

Annual; reproducing by seeds. Parasitic on Polygonum but also on other herbs of low ground, especially Lycopus and Penthorum. Quebec to Minnesota, south to New England, Maryland, Tennessee and Texas. Native. July–September.

Description.—Stems slender, orange. Flowers in small dense clusters, mostly 4-merous, 2–2.5 mm. long, sessile or subsessile; calyx-lobes ovate, obtuse, as long as or longer than the corolla-tube; corolla-lobes triangular, acute, ascending or erect, commonly as long as or longer than the tube; scales oblong, toothed or 2-cleft; included stamens borne from the sinuses; 2 styles shorter than the depressed-globose ovary. Capsule subglobose, with the withered corolla persistent at base, 2.5–3 mm. broad. Seeds yellowish-brown, roundish, about 1.3 mm. long.

Control.—The same as for 355.

357. *Cuscuta suaveolens* Ser.

Annual; reproducing by seed. On alfalfa, *Medicago sativa*. Local, Maine to South Dakota, south to Texas and California. Noxious (1954) in all states. Introduced from South America. August–October.

Description.—Stem slender, straw-colored. Flowers in loose cymes; pedicels exceeding the flowers; calyx-lobes triangular, short; corolla bell-shaped with short spreading or reflexed lobes which are less than half the length of corolla tube; scales short and fringed; ovary shorter than style. Capsule slightly flattened, spherical.

Control.—The same as for 355.

Cuscuta spp. A comprehensive taxonomic treatment of the American dodders has been presented by Yuncker.[11] The relation of the dodders to farm seeds has been discussed by Hillman.[12]

358. *Ipomoea coccinea* L. Morning-glory.

Annual; reproducing by seeds. Fields, gardens and waste places; especially on alluvial soils. Southern New York to Michigan, south to Florida and Arizona. Introduced from tropical America. July–October.

Description.—Stems twining or trailing extensively, glabrous, often reddish and ridged. Leaves alternate, simple, long-petioled, entire, angular-cordate, acuminate, glabrous, 4–6 cm. long. Flowers

[11] Yuncker, 1921, 1932.   [12] Hillman, 1907.

Morning-glory Family

perfect, regular, 2 to several on a long axillary peduncle; calyx with 5 awn-pointed lobes; corolla salverform, limb about 16 mm. broad, with a very slender tube 2–3 cm. long, scarlet; stamens 5, exserted; stigma capitate; style exserted, ovary 2-celled. Fruit a globular capsule, 4–6-seeded, 2–4-valved; seeds 3–4 mm. long, 3-angled, ovoid, one face convex, the other two sloping to the edges from a central ridge, granular, brownish-black; hilum pubescent, in a notch.

Control.—Hand pulling and hoeing (4, 5). Clean cultivation (6).

359. *Ipomoea hederacea* (L.) Jacq. Ivy-leaved morning-glory.

Annual; reproducing by seeds. Gardens, fields and waste places. New England to North Dakota, south to Florida and Arizona; chiefly in the southeastern states. Introduced from tropical America. July–September.

Description.—Stems twining or trailing, with recurved hairs. Leaves alternate, simple, 4 cm. long or less, palmately veined and 3-lobed, the lobes entire, acute or acuminate, hairy. Flowers 1–3 on an axillary peduncle, perfect, regular, showy; calyx densely hairy below, with 5 acuminate lobes; corolla funnelform or nearly campanulate, 2.5–4.5 cm. long, white and purple or pale blue; stamens 5, included; stigma capitate, style included, ovary 2-celled. Fruit a globular capsule, 4–6-seeded, 2–4-valved; seeds similar to 358 but larger, 5–6 mm. long, granular, pubescent, dark brown to black.

Control.—The same as for 358.

360. *Ipomoea purpurea* (L.) Roth. Morning-glory.

Annual; reproducing by seeds. Waste places and gardens. Southeastern states, extending elsewhere. Introduced from tropical America. July–September.

Description.—Stems twining or trailing, hairy, sometimes 2–3 m. long. Leaves alternate, simple, long-petioled, entire, broadly cordate, acuminate, pubescent, 8 cm. long or less. Flowers perfect, regular, 3–5 borne on a long axillary peduncle; calyx bristly-hairy below, with 5 acute lobes; corolla funnelform, 4–7 cm. long, purple, blue or white; stamens 5, attached on the corolla, included; stigma capitate, style included, ovary 2-celled. Fruit a globular capsule, 4–6-seeded, 2–4-valved; seeds similar to 358 but longer, 4–5 mm., granular, pubescent, brownish-black; hilum glabrous.

Control.—The same as for 358.

Ipomoea pandurata (L.) G. F. W. Mey. is listed (1954) as a weed in the north central and the northeastern states.

POLEMONIACEAE (Phlox Family)

361. *Navarretia intertexta* Hook.

Annual; reproducing by seeds. Pastures, roadsides and waste places; mostly on dry soils. Native to British Columbia and the Pacific Coast states; sparingly introduced in the eastern states. June–August.

Description.—Stems erect, much branched and spreading above, 5–15 cm. high, pubescent, but neither glandular nor with disagreeable odor. Leaves alternate, 1- or 2-pinnately cleft into linear or spinelike segments, glabrous or nearly so. Flowers perfect, regular, in crowded short spikes with prominent bracts; calyx tubular, villous, the 5 lobes ribbed, cleft into spiny segments; corolla tubular, 5-lobed, 5–6 mm. long, pale blue; stamens 5, separate, exserted. Capsule 1–3-celled, containing 3–4 seeds in each cell; seeds 2–3 mm. long, ovate, wrinkled, orange-brown.

Control.—Clean cultivation.

362. *Navarretia squarrosa* H. & A. Skunkweed.

Annual; reproducing by seeds. Pastures, roadsides and waste places. Native to British Columbia and the Pacific Coast states; sparingly introduced eastward. June–July.

Description.—Stems erect or spreading, much branched, 1–3 dm. high, glandular-sticky and with a disagreeable "skunk-like" odor. Leaves mostly alternate, pinnately parted and the segments cleft or parted into sharp-pointed lobes; the upper leaves and bracts of the spike spine like. Flowers perfect, regular, in short clusters on the ends of the branches; calyx tubular, with lobes mostly entire and long, spiny-tipped; corolla tubular, 5-lobed, 8–10 mm. long, pale blue; stamens 5, included. Capsule elliptical, 3-celled, with 8–12 seeds in each cell, dehiscent; seeds about 1 mm. long, ovoid, angular, granular, dark purplish-brown.

Control.—Clean cultivation.

HYDROPHYLLACEAE (Waterleaf Family)

363. *Ellisia nyctelea* L. Ellisia. Fig. 86, a–c.

Annual; reproducing by seeds. Grain fields and meadows; alluvial woods, stream banks, and waste places. Manitoba to Alberta and Idaho, south to Louisiana, Texas, and New Mexico; adventive in East from New England to North Carolina; local elsewhere. Native. April–July.

Description.—Stems weak, 1-4 dm. long, often branched from the base, glabrous or minutely rough-hairy; entire plant with a rank, disagreeable odor. Leaves petioled, deeply pinnately divided into

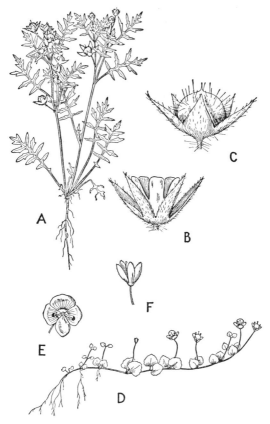

FIG. 86—**ELLISIA,** *Ellisia nyctelea:* **A,** plant showing habit (× 1); **B,** flower (× 2); **C,** fruit (× 2).
CREEPING VERONICA, *Veronica filiformis:* **D,** plant showing habit (× 1); **E,** flower (× 3); **F,** calyx (× 3).

7-13 spreading, oblong, entire or coarsely toothed lobes. Flowers mostly solitary on slender axillary peduncles, similar to those of *Phacelia purshii* but corolla-lobes not fringed and calyx enlarging in fruit. Capsule pendulous, 2-celled, 4-seeded.

Control.—Use a short rotation including clean cultivated crops.

Harrow infested grain fields early in the spring to destroy the seedlings (8).

364. *Phacelia purshii* Buckley. Miami mist, Scorpion weed.

Annual or biennial; reproducing by seeds. Rich woods, thickets, waste places, clearings, gardens, and fields. Pennsylvania to Minnesota, south to Alabama and Oklahoma; casually adventive in the Northeast. Native. April–June.

Description.—Stem erect or ascending, simple to much branched, 1.5–5 dm. high. Leaves pinnately divided. Flowers small, blue or white, in one-sided terminal clusters; calyx 5-lobed; corolla campanulate, about 1 cm. broad, with 5 fringed lobes; stamens 5, on the corolla-tube; pistil with a 2-celled ovary forming a many-seeded capsule.

Control.—Plow the infested land and plant a cultivated crop for a year. Plants missed by the cultivator should be pulled or hoed before they produce seed.

BORAGINACEAE (Borage Family)

365. *Amsinckia intermedia* Fisch. & Mey. Tarweed, Buckthornweed, Finger-weed, Fiddle-necks, Fireweed, Yellow burnweed, Yellow forget-me-not. Fig. 89, a–d.

Annual; reproducing by seeds. Grain fields, gardens, orchards, meadows and waste places. Native to the Pacific Coast; introduced locally in the central and eastern states. May–July.

Description.—Stems erect or with spreading branches, 3–8 dm. long, covered with stiff white bristly hairs. Leaves alternate, lanceolate to linear, rather thick and covered with short bristly hairs. Flowers perfect, regular, in a dense, one-sided, coiled, leafy-bracted raceme which becomes much elongated in fruit; calyx with 5 linear, acute, very densely bristly lobes; corolla salverform, 5-lobed, yellow, 5–6 mm. long; stamens 5, separate, inserted on the corolla-tube; ovary deeply 4-lobed, in fruit separating into 2 or 4 seed-like nutlets which are attached above the base. Nutlets about 2.5 mm. long, ovoid, angular, apex curved, with a scar near the base, somewhat winged on one angle, unarmed but roughened by wrinkles, gray to dark brown or nearly black. A variable species represented by several forms which are sometimes separated as species.

Control.—Infested grain fields should be harrowed in the spring to destroy the weed seedlings (8). Badly infested meadows should be plowed and followed with a clean cultivated crop for a year (6).

366. *Amsinckia lycopsoides* Lehm. Tarweed.

Annual; reproducing by seeds. Cultivated fields and waste places. Pacific Coast. Native. May–July.

Description.—Stems decumbent, branched. Leaves alternate, ovate to lanceolate, somewhat toothed, sparsely bristly. Flowers similar to 365 but the calyx-lobes narrowly ovate and only sparsely bristly. Seeds similar to 365, 2–2.5 mm. long, irregularly wrinkled and finely granular, brownish-black.

Control.—Clean cultivation (6). Plants in waste places should be hoed or mowed off and burned if they already bear seeds (11).

367. *Cynoglossum officinale* L. Hounds-tongue, Dog bur, Sheep lice, Woolmat. Fig. 87, a–d.

Biennial; reproducing by seeds. Pastures and waste fields; mostly on gravelly somewhat limy soils. Widespread and locally common in eastern North America; infrequent west of the Great Plains. Introduced from Europe. June–July.

Description.—Stems erect, simple, 3–9 dm. high, branching near the top, stout, ridged, hairy and leafy to the top. Root thick, black, and crowned the first year with tufted, coarse, hairy, oblong leaves. Leaves 10–30 cm. long, pointed, shaped like a hound's tongue, entire and long-petioled; stem-leaves simple, alternate, lanceolate, 4–15 cm. long, softly hairy, sessile or clasping. Flowers in terminal simple or branched racemes, perfect, regular; calyx 5-lobed, hairy, enlarging in fruit; corolla funnelform, reddish-purple, 5-lobed, less than 15 mm. broad; stamens 5, borne on the corolla; pistil with a deeply 4-lobed ovary and a simple style, in fruit separating into 4 obovoid nutlets. Nutlets about 6 mm. long, attached laterally and spreading, flat on the upper surface, with a scar on one side near the base, covered with short barbed prickles, brown or grayish-brown.

Control.—The young rosettes should be cut below the crown with a spud or hoe in autumn or early spring (12). Mow the flowering stems close to the ground before seeds are formed (11). Badly infested fields should be plowed and followed with a clean cultivated crop for a year (7).

368. *Echium vulgare* L. Blue thistle, Blue devil, Blue-weed, Vipers bugloss, Snake flower. Fig. 87, e–j.

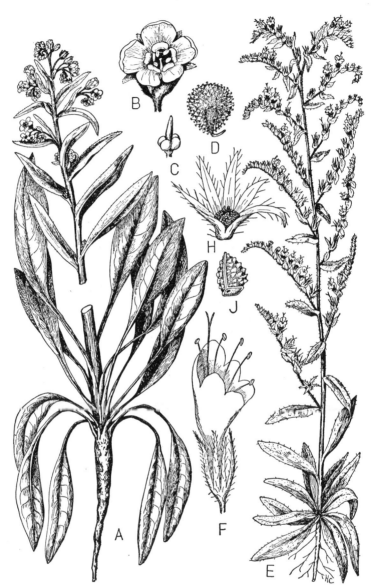

Fig. 87—**HOUNDS-TONGUE**, *Cynoglossum officinale:* **A,** plant showing habit (× ¼); **B,** flower (× 2); **C,** pistil showing nutlets (× 2); **D,** mature nutlet (× 2).
BLUE THISTLE, *Echium vulgare:* **E,** plant showing habit (× ¼); **F,** flower (× 2); **H,** nutlets and calyx (× 2); **J,** nutlet (× 4).

Biennial; reproducing by seeds. Dry, gravelly or stony pastures, meadows, and fields; especially in limestone regions. Widespread in eastern North America; locally common, especially northward; rare west of the Mississippi River. Introduced from Europe. June–July.

Description.—Stems erect, simple or branched below, stout, 3–9 dm. high, bristly when young, the bristles becoming prickles at maturity; these bristles springing from red tubercles which speckle the stem. First year's growth forming a rosette of coarse bristly-hairy leaves crowning a long, thick, dark tap-root. Leaves linear-oblong, 7–15 cm. long, entire and petioled; stem-leaves alternate, simple, linear-lanceolate, entire, sessile and speckled with bristle-bearing tubercles. Flowers perfect, irregular; flower-spike compound, formed of one-sided curving spikelets borne in the axils of the upper leaves; calyx 5-toothed, bristly; corolla funnelform, pink in the bud, bright blue when open, 1–2 cm. long, showy, with 5 unequal rounded and spreading lobes; stamens 5, borne on the corolla, unequal in length and mostly protruding beyond the corolla; anthers red; pistil with a deeply 4-lobed ovary and a thread-like style. Fruit of 4 small, 3-angled, obovate nutlets, attached at the base, rough and wrinkled but not prickly, about 2–2.5 mm. long, with conspicuous flattened base, gray-brown.

Control.—The same as for 367.

369. *Hackelia floribunda* (Lehm.) I. M. Johnston (*Lappula floribunda* Greene). Stickseed.

Perennial; reproducing by seeds. Pastures, grasslands and waste places; damp thickets and shores. Ontario to British Columbia, south to New Mexico and California. Native. June–July.

Description.—Similar to *Lappula echinata,* stickseed, but taller (0.3–1.5 m. high); pedicels reflexed in fruit; nutlets much larger (4–6 mm. long) and margined with a row of flat, lanceolate prickles.

Control.—The same as for 367.

370. *Hackelia virginiana* (L.) I. M. Johnston (*Lappula virginiana* Greene). Wild comfrey.

Biennial; reproducing by seeds. Old pastures, neglected fields, rich woods. Widespread throughout eastern North America. Native. June–September.

Description.—Same general appearance as *Lappula echinata.* The stem is 3–12 dm. high; pedicels are recurved in fruit; the nutlets are attached near the base; the pistil is shorter than the nutlets; and the nutlets are subequally prickly over the whole back and along the margins.

Control.—The same as for 367.

371. *Heliotropium curassavicum* L. Wild heliotrope, Chinese pusley, Seaside heliotrope, White-weed, Devil-weed, Alkali heliotrope. Fig. 89, e–g.

Perennial or sometimes biennial; reproducing by seeds. Fields, pastures, and waste places; especially on alkaline soils in the West. Widespread from Delaware southward and westward to the Pacific Coast. Native; apparently introduced in some localities. May–September.

Description.—Stems prostrate or ascending, much branched, 3–7 dm. long, fleshy, glabrous, and glaucous. Leaves alternate, simple, oblong to linear, the lower petioled, the upper sessile, with entire or wavy margins; herbage ashy-green, turning purplish-brown on drying. Flowers perfect, in long, bractless, coiled, terminal, usually paired, one-sided spikes; calyx 5-parted; corolla regular, tubular, white with a yellow or purple center, 5-lobed; stamens 5, inserted on the corolla-tube; ovary globular, in fruit separating into 2–4 nutlets. Nutlets 2 mm. long, ovate-oblong, one side convex and grooved, the other side with a dark circular scar near the middle, surface uneven, straw-colored.

Control.—Clean cultivation (6). Improved drainage (21).

372. *Lappula echinata* Gilib. Stickseed, Burweed, Blue bur, Sheep bur. Fig. 88, c g.

Annual; reproducing by seeds. Old dry fields, pastures and waste places. Widespread but usually not very abundant. Introduced from Eurasia. June–July.

Description.—Stems erect, simple, slender, 3–6 dm. high, branching at the top, leafy to the top, gray with appressed hairs. Leaves alternate, simple, sessile, oblong to linear, 1–3 cm. long and 2–5 mm. wide, entire, grayish with soft hairs. Flowers perfect, regular, in terminal leafy-bracted racemes, the pedicels short, stout and erect in fruit; calyx 5-pointed, hairy, enlarging as the fruit matures; corolla regular, salverform, blue, short, less than 5 mm. broad, with 5 stamens inserted on its tube; pistil with a deeply 4-lobed ovary and a simple style which is longer than the nutlets. Fruit of 4 small erect nutlets, attached laterally all along the side; nutlets 2–3 mm. long, with a narrow scar along the inner side, the outer side rounded and

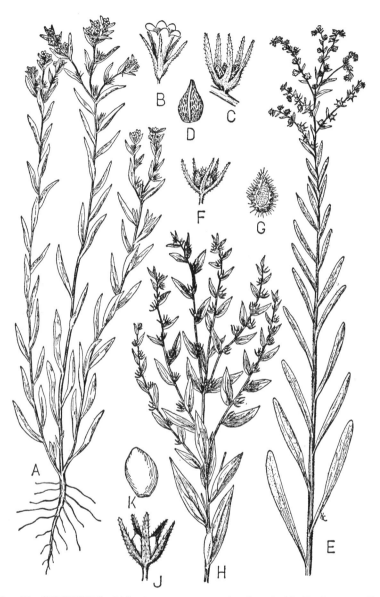

FIG. 88—**GROMWELL,** *Lithospermum arvense:* **A,** plant (× ⅓); **B,** flower (× 2); **C,** nutlets surrounded by calyx (× 2); **D,** seed (nutlet) (× 4).
STICKSEED, *Lappula echinata:* **E,** plant showing habit (× ⅓); **F,** nutlets surrounded by calyx (× 2); **G,** seed (nutlet), (× 4).
GROMWELL, *Lithospermum officinale:* **H,** upper part of plant showing habit (× ⅓); **J,** shiny white nutlets surrounded by calyx (× 2); **K,** nutlet (× 4).

Fig. 89—**TARWEED,** *Amsinckia intermedia:* **A,** plant showing habit (× ⅓); **B,** flower (× 4); **C,** fruiting calyx (× 4); **D,** nutlets (× 4).
WILD HELIOTROPE, *Heliotropium curassavicum:* **E,** plant showing habit (× ⅓); **F,** flower (× 4); **G,** nutlets (× 4).

having around the margin and back a double row of barbed prickles which break off easily, gray-brown.

Control.—The same as for 367.

373. **Lithospermum arvense** L. Gromwell, Red-root, Wheat thief, Stone seed, Puccoon, Pigeon-weed. Fig. 88, a–d.

Winter annual or biennial; reproducing by seeds. Grain fields, meadows and waste places. Widespread and locally common throughout the United States and southern Canada. Introduced from Europe. May–June.

Description.—Stems erect, slender, simple or branching at the base, 2–7 dm. high, minutely roughened and hairy, leafy to the top. Leaves alternate, simple, entire, sessile, lanceolate to linear, 1–3 cm. long, hairy on both sides. Flowers perfect, regular, nearly sessile in the axils of leafy bracts of the terminal racemes; calyx 5-pointed, hairy; corolla funnelform, with 5 rounded spreading lobes, cream-colored, about the length of the calyx (5–7 mm.), with 5 short stamens inserted on its tube; pistil with a deeply 4-lobed ovary and simple style. Fruit of 4 small erect nutlets, conical, ovoid, about 3 mm. long, one side angled, the other convex, base with 2 small white tubercles and scar, roughened with tubercles and ridges, dull, grayish-tan.

Control.—This weed does not persist long under cultivation. It is most troublesome where winter wheat or rye is grown year after year. Use a system of rotation in which the land is cultivated for at least one year after each grain crop (20).

374. **Lithospermum officinale** L. Gromwell, Pearl plant, Gray mile, Little wale. Fig. 88, h–k.

Perennial; reproducing by seeds. Stony or gravelly pastures, neglected fields and waste places; in limestone regions. Locally common from Quebec to New Jersey, and westward to Minnesota. Introduced from Europe. June–August.

Description.—Stems erect, stout, much branched, 3–10 dm. high, rough-hairy and leafy to the top. Leaves alternate, simple, broadly lanceolate, pointed at both ends, entire, thinnish, rough above, downy beneath, sessile and with a few distinct veins. Flowers perfect, regular, almost sessile in the axils of the upper leaves; calyx of 5 long hairy segments elongating as the fruit matures; corolla funnelform, greenish-white to pale yellow, 5-lobed with 5 hairy crests in the throat, less than 5 mm. long; stamens 5, inserted on the tube; pistil with a deeply 4-lobed ovary and simple style. Fruit of 4 very smooth shiny ivory-white nutlets, attached at the base; nutlets about 3 mm. long, ovoid, base truncate with 2 small tubercles and a scar, glossy, white or slightly yellowish with a brownish base.

Control.—The same as for 367.

375. *Lycopsis arvensis* L. Small bugloss.

Annual; reproducing by seeds. Local in fields and waste places; especially on dry sandy soil. Quebec to Ontario and southward. Introduced from Europe. July–August.

Description.—Stems erect or with spreading branches, slender, bristly-hairy, 2–6 dm. long. Leaves alternate, lanceolate, entire or wavy-margined, with short, margined petioles or the upper sessile, bristly-hairy. Flowers in leafy, raceme-like curved clusters; calyx with 5 narrow lobes, as long as the corolla-tube; corolla funnel-shaped but the tube curved, the throat closed by 5 obtuse bristly scales opposite the lobes, pale blue; stamens 5, inserted on the corolla; ovary 4-parted, in fruit separating into 4 nutlets. Nutlets 3–4 mm. long, ovoid, with prominent concave, basal scar, surface granular and ridged, dull, light to dark brown.

Control.—The same as for 366.

376. *Symphytum officinale* L. Comfrey, Healing herb, Backwort, Bruisewort, Asses-ears.

Perennial; reproducing by seeds. Moist meadows and waste places; on moist, rich, somewhat limy soils. Widespread but local in northeastern North America, south to Georgia and Louisiana. Introduced from Europe as a medicinal herb. June–August.

Description.—Stems erect, branched, stout and hairy, 3–9 dm. high. Root thick, mucilaginous, covered with thin black bark. Leaves alternate, simple, pinnately veined, the upper leaves decurrent upon the stem in long wedge-shaped wings, the lower leaves narrowed at the base to long winged petioles, entire, long-ovate to ovate-lanceolate, thick, rough hairy on both surfaces, 6–20 cm. long. Flowers in nodding raceme-like clusters, perfect, regular, pedicelled; calyx of 5 lanceolate segments, acute, rough, hairy; corolla tubular with 5 spreading lobes, the throat crested below the lobes, yellowish-white, rarely pink or purplish, 10–15 mm. long; stamens 5, included, inserted on corolla-tube; pistil with a deeply 4-lobed ovary and thread-like style. Fruit of 4 erect nutlets, attached at the base; nutlets ovoid, 3–4 mm. long, the base hollowed and finely toothed on its margin, nearly smooth, glossy, brownish-black.

Control.—The same as for 367.

VERBENACEAE (Vervain Family)

377. *Verbena bracteata* Lag. and Rodr. (*V. bracteosa* Michx.). Bracted vervain, Prostrate vervain. Fig. 90, g–h.

Annual; reproducing by seeds. Grasslands, lawns and waste places; mostly on dry soils. Native from Ohio westward and southward; locally introduced in the northeastern states and on the Pacific Coast. June–August.

Description.—Stems widely spreading, with some prostrate and some ascending branches, 1–4 dm. long, rough-hairy, somewhat 4-angled. Leaves opposite, simple, wedge-shaped in outline, but cut-pinnatifid and much toothed, 2–8 cm. long, rough-hairy, short-petioled. Flowers perfect, sessile, on a thick leafy-bracted spike, few in bloom at any time; bracts stiff, hairy, longer than the calyx, 1–2 cm. long, nearly concealing the flowers; calyx short, tubular, 5-toothed; corolla purplish-blue, irregular, tubular, salverform, 5-cleft; stamens 2 long and 2 short, included; style slender, terminal; ovary 4-celled, not 4-lobed but in fruit splitting into 4 1-seeded indehiscent nutlets, "seeds." Nutlet about 2 mm. long, linear-oblong, margin slightly winged, the convex surface with longitudinal ridges forming a network on the apical third, with a white scar near the base, dark brown.

Control.—Plow badly infested fields and plant a clean cultivated crop for a year. Pull or hoe plants in lawns.

378. *Verbena hastata* L. Blue vervain, Wild hyssop, Ironweed, Purvain. Fig. 90, a–d.

Perennial; reproducing by seeds and short rootstocks. Moist meadows and pastures; on gravelly or heavy loam soils. Common in the eastern states and Mississippi Valley, rare westward. Native. June–September.

Description.—Stems erect, angled and coarsely grooved, simple and branching near the top, 6–20 dm. high, sparingly rough-pubescent. Leaves opposite, simple, lanceolate or oblong, taper-pointed, doubly serrate, rough, dark green above and grayish-green beneath, prominently pinnately veined, 4–12 cm. long. Flowers sessile in erect slender

Fig. 90—**BLUE VERVAIN,** *Verbena hastata:* **A**, part of plant (× ⅓); **B**, flower (× 3); **C**, mature calyx (× 6); **D**, seed (× 9).
WHITE VERVAIN, *Verbena urticifolia:* **E**, part of plant (× ⅓); **F**, seed (× 9).
BRACTED VERVAIN, *Verbena bracteata:* **G**, part of plant (× ⅓); **H**, seed (× 9).

linear spikes, borne in conspicuous terminal panicles; bracts shorter than the calyx; flowers and fruit similar to 377; corolla blue or rarely pink, about 3–4 mm. in diameter. Nutlets linear-oblong, 1.5–2 mm. long, the convex surface with longitudinal ridges, reddish-brown.

Control.—Plow badly infested fields and plant a clean cultivated crop for a year. Scattered patches in pastures should be mowed or grubbed out.

379. *Verbena stricta* Vent. Hoary vervain, Mullein-leaved vervain, Woolly vervain.

Perennial; reproducing by seeds. Native prairies, pastures, old fields, and waste places; mostly on dry or gravelly soils. Native from Ontario and Ohio westward to the Rocky Mountains; sparingly introduced in the northeastern states. June–September.

Description.—Stems erect, simple or with a few branches above, 3–9 dm. high, velvety with soft whitish hairs. Leaves opposite, simple, obovate or oblong, 2–8 cm. long, doubly serrate, sessile, downy with whitish hairs, pinnately veined. Flowers in dense spikes; bracts hairy, shorter than the calyx; flower and fruit similar to 377 but the corolla purple and about 1 cm. in diameter. Nutlet about 3 mm. long, similar to 378, dark or grayish-brown.

Control.—The same as for 378.

380. *Verbena urticifolia* L. White vervain, Nettle-leaved vervain. Fig. 90, e–f.

Perennial; reproducing by seeds and short rootstocks. Old fields, pastures and waste places; on rather rich heavy soils. Widespread and common throughout eastern North America to North Dakota, Texas and Mexico; rare on the Pacific Coast. Native. July–September.

Description.—Stems erect, simple, 5–15 dm. high, with a few ascending branches, angled, slender with scattered hairs or nearly glabrous. Leaves opposite, simple, oval or oblong-ovate, 6–15 cm. long, unevenly, coarsely toothed, long-pointed, thin, often splotched with a white mildew. Flowers in open long slender spikes which are grouped into a loose open panicle; bracts shorter than the calyx; flower and fruit similar to 377 except that the corolla is white, about 2 mm. in diameter. Nutlet 1.5 mm. long, similar to 377 but more nearly oval, dark reddish-brown.

Control.—The same as for 378.

Verbena officinalis L. European vervain, an annual introduced from Europe, has become established locally, chiefly from the middle

Atlantic states southward; also in California. It has pinnatifid leaves and small purplish flowers on slender open spikes. Control the same as for 377.

Lippia cuneifolia Steud. Fog-fruit is a native perennial from the north central states and westward. It has procumbent jointed branches from a woody base and opposite wedge-shaped leaves. Flowers purplish, in head-like clusters on axillary peduncles; calyx short, 2-lipped. Control the same as for 377.

LABIATAE (Mint Family)

Some of the characteristics common to the weeds belonging to the mint family (381–402) are recorded here and are not repeated in the descriptions of the species. Stems 4-sided; leaves opposite, simple, more or less aromatic or strongly scented. Flowers perfect, irregular; corolla more or less 2-lipped; stamens 2 long and 2 short, or only a single pair; pistil with a deeply 4-lobed ovary which in fruit forms 4 seed-like nutlets, attached by their bases around the single style, in the bottom of the persistent calyx; each nutlet contains 1 exalbuminous seed with a straight embryo.

381. *Dracocephalum parviflorum* Nutt. Dragonhead.

Annual or biennial; reproducing by seeds. Meadows, fields, gardens, and waste places; on gravelly or stony soils, often in recent clearings or near woods. Apparently native in southern Canada, the Lake states, and the northwestern states; sparingly introduced in the northeastern states. June–August.

Description.—Stems erect, often branched below and forming clusters, bluntly 4-angled, 2–10 dm. high. Leaves pinnately veined, ovate-lanceolate, coarsely serrate, petioled. Flowers in whorls on the upper nodes; whorls with leafy bristly-toothed bracts, crowded into a terminal head or spike; calyx tubular, 5-toothed, 13–15-nerved, open in fruit; corolla bluish, strongly 2-lipped, the upper lip notched, the lower lip 3-lobed and the middle lobe largest and notched; stamens 4, the upper pair longer than the lower pair. Nutlets 2–2.5 mm. long, oval, with a light kidney-shaped basal scar on the angular side, granular, brownish-black.

Control.—Clean cultivation, followed by hand hoeing and weeding (6). In waste places mow the plants before seeds are formed (11).

Mint Family

382. Galeopsis tetrahit L. Hemp nettle, Dog nettle, Bee nettle, Wild hemp, Flowering nettle, Ironweed. Fig. 94, a–d.

Annual; reproducing by seeds. Cultivated fields, truck lands, pastures, and waste places; mostly in areas formerly timbered. Locally common in Canada and the northeastern United States, west to Iowa, introduced from Europe. July–September.

Description.—Stems erect or spreading, much branched, stout, 4-angled, bristly-hairy, 3–9 dm. high. Leaves pinnately veined, with long bristly petioles, coarsely serrate, ovate-acuminate, bristly-hairy, 2–12 cm. long. Flowers in dense, very short, leafy-bracted, terminal spikes and in dense clusters in the upper leaf-axils; calyx tubular, with 5 long spiny-pointed teeth, persistent; corolla 2-lipped, tubular, the upper lip stiff, arched and entire, the lower lip 3-cleft and spreading, white to pink or purple, about twice the length of the calyx; stamens 4, the upper pair shorter than the lower pair, inserted on the corolla. Nutlets about 2 mm. long, obovate, one side convex, the other angular and with scar at base, brown mottled.

Control.—Clean cultivation, especially while the seedlings are small (6).

383. Glechoma hederacea L. (*Nepeta hederacea* Trev.) Ground ivy, Gill-over-the-ground, Creeping charlie, Cats-foot, Field balm. Fig. 91, h–j.

Perennial; reproducing by seeds and creeping stems. Lawns, gardens, orchards, and waste places; especially on damp rich or somewhat shaded soils. Widespread throughout the northern United States and southern Canada; infrequent in the southern states. Introduced from Eurasia. April–June.

Description.—Stems creeping or trailing, with numerous erect flowering branches, 4-angled, glabrous or nearly so, rooting at the nodes. Leaves palmately veined, petioled, orbicular-reniform, crenate, glabrous and green on both sides, 1–3 cm. in diameter. Flowers in axillary clusters; calyx tubular, with 5 equal teeth, pubescent, persistent; corolla purplish, tubular, 2-lipped, the upper lip erect, rather concave, 2-cleft, the lower lip 3-lobed; stamens 4, the upper pair much longer than the lower pair; ovary deeply 4-lobed. Nutlets ovoid, 1.5–2 mm. long, granular, brown with a small whitish hilum at base.

Control.—Plow infested fields and follow with a clean cultivated crop for a year. In lawns raise the runners in early spring and mow close.

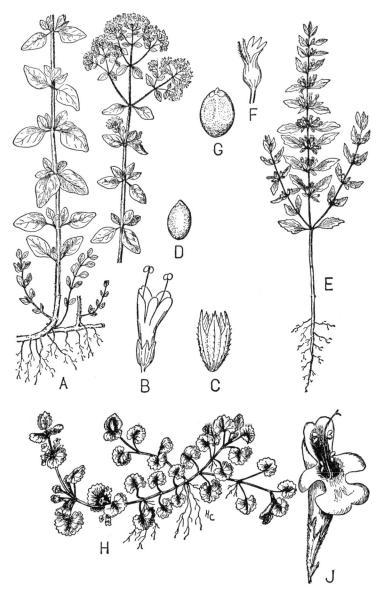

Fig. 91—**WILD MARJORAM,** *Origanum vulgare:* **A,** portion of plant showing habit ($\times$ 1/4); **B,** flower ($\times$ 3); **C,** fruiting calyx ($\times$ 6); **D,** seed ($\times$ 12).
AMERICAN PENNYROYAL, *Hedeoma pulegioides:* **E,** plant showing habit ($\times$ 1/3); **F,** calyx ($\times$ 3); **G,** seed ($\times$ 12).
GROUND IVY, *Glechoma hederacea:* **H,** plant showing habit ($\times$ 1/3); **J,** flower ($\times$ 3).

384. *Hedeoma pulegioides* Pers. American pennyroyal, Mock pennyroyal, Squaw mint, Stinking balm, Mosquito plant. Fig. 91, e–g.

Annual; reproducing by seeds. Meadows and pastures; on dry stony or sandy soils. Widespread in eastern North America, west to Minnesota and Kansas. Native. July–August.

Description.—Stems erect, much branched, slender, mostly 4-angled, pubescent, 1–3 dm. high. Leaves pinnately veined, petioled, entire or slightly serrate, oblong-ovate, nearly glabrous but dotted with small glands, 1–3 cm. long. Flowers in loose axillary whorls; calyx tubular, 2-lipped, 13-nerved, upper lip 3-toothed, lower lip with 2 bristle-like teeth; corolla tubular, nearly regular, slightly 2-lipped, bluish, upper lip notched, nearly flat, lower lip spreading and 3-cleft; stamens 2 sterile and 2 bearing anthers, the latter exserted. Nutlets obovate, about 1 mm. long, minutely granular, dark brown or black.

Control.—Badly infested meadows and pastures should be fertilized to stimulate the grass to smother this weed. Plowing followed by a cultivated crop before reseeding will destroy it.

385. *Lamium amplexicaule* L. Dead nettle, Henbit, Blind nettle, Bee nettle. Fig. 92, c–d.

Winter annual or biennial; reproducing by seeds and rooting stems. Gardens and cultivated fields; on rich soils. Widespread and locally common throughout eastern North America; also on the Pacific Coast. Introduced from Eurasia. April–June; September.

Description.—Stems decumbent, with numerous ascending branches, slender, 4-angled, nearly glabrous, 1–4 dm. high, frequently rooting at the lower nodes. Leaves palmately veined, the upper sessile and clasping, the lower petioled, doubly crenate-lobed, orbicular, pubescent, 1–2 cm. long. Flowers in axillary whorls; calyx tubular, with 5 nearly equal awl-shaped teeth, mostly 5-nerved; corolla pink to purplish, tubular, slender, 2-lipped, the upper lip entire, erect, arched, bearded, the lower lip 3-lobed (the middle lobe broadest and notched), spotted, 1–1.5 cm. long; stamens 4, the upper pair shorter than the lower pair. Nutlets nearly 2 mm. long, obovate-oblong, sharply 3-angled, apex truncate, grayish-brown, speckled with silvery-gray granules.

Control.—This weed is most troublesome during cool weather in early spring and again in autumn. Early spring plowing followed by surface tillage will destroy the spring crop of weeds. The fall crop of

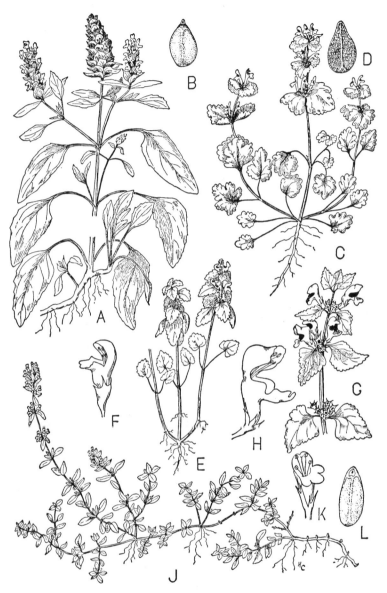

FIG. 92—**HEAL-ALL,** *Prunella vulgaris:* **A,** portion of plant (× 1/3); **B,** seed (× 6).
DEAD NETTLE, *Lamium amplexicaule:* **C,** plant (× 1/3); **D,** seed (× 6).
RED DEAD NETTLE, *Lamium purpureum:* **E,** plant (× 1/3); **F,** flower (× 2).
SPOTTED DEAD NETTLE, *Lamium maculatum:* **G,** portion of plant showing habit (× 1/3); **H,** flower (× 2).
CREEPING THYME, *Thymus serpyllum:* **J,** plant showing creeping habit (× 1/3); **K,** flower (× 2); **L,** seed (× 6).

seedlings should be destroyed by shallow surface cultivation as soon as the crop is removed.

386. *Lamium maculatum* L. Spotted dead nettle. Fig. 92, g–h.

Perennial; reproducing by seeds and rooting stems. Old gardens, roadsides, and waste places; on rich moist soil. Widespread throughout the northern United States; most common in the northeastern states. Introduced from Eurasia. April–May.

Description.—Stems decumbent, with numerous ascending branches, 4-angled, nearly glabrous, 1–5 dm. high, rooting at the lower nodes. Leaves petioled, crenate, ovate to cordate, pubescent, 1–3 cm. long, often with a pale spot on the upper surface. Flowers similar to 385 but the purplish to white corolla larger, 1.5–2.5 cm. long. Nutlets similar to 385 but larger, 2–3 mm. long and not so prominently speckled or angled.

Control.—Dig out small patches early in the spring. Large areas should be plowed and planted with a clean cultivated crop or a smother crop.

387. *Lamium purpureum* L. Red dead nettle. Fig. 92, e–f.

Winter annual or biennial; reproducing by seeds. Gardens and cultivated fields on rich soil. Widely distributed in eastern North America; infrequent in the western states. Introduced from Eurasia. April–June.

Description.—Stems decumbent, with numerous ascending branches, slender, 4-angled, glabrous, 1–3 dm. high, rooting at the lower nodes. Leaves palmately veined, all petioled, crenate-dentate, ovate to cordate, pubescent, 0.5–2 cm. long, the upper usually purplish and crowded. Flowers similar to 385, crowded in whorls in the axils of the upper leaves; corolla less than 1.5 cm. long. Nutlets similar to 385, with a yellowish-white caruncle at base.

Control.—The same as for 385.

388. *Leonurus cardiaca* L. Motherwort, Lions-tail, Lions-ear, Cowthwort. Fig. 93, e–g.

Perennial; reproducing by seeds. Neglected yards, gardens, meadows and roadsides; on rich soils. Locally common throughout eastern North America, south to North Carolina; infrequent westward to the Pacific Coast. Introduced from Europe. July–September.

Description.—Stems very erect and stiff, with only a few branches, prominently 4-angled, nearly glabrous, 6–15 dm. high. Leaves pal-

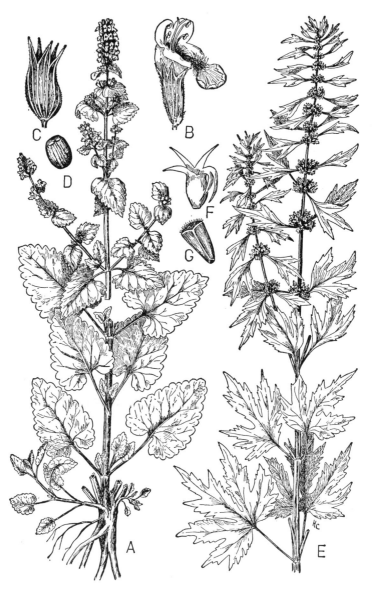

FIG. 93—**CATNIP,** *Nepeta cataria:* **A,** plant showing habit (× ⅓); **B,** flower (× 3); **C,** calyx (× 3); **D,** seed (nutlet), (× 3).
MOTHERWORT, *Leonurus cardiaca:* **E,** branch showing general habit (× ⅓); **F,** calyx (× 3); **G,** seed (nutlet), (× 6).

Mint Family

mately veined, long-petioled, the upper leaves 3-cleft with cuneate base and acuminate tip; lower leaves palmately cleft or incised, roughened by a few scattered hairs. Flowers in dense axillary whorls; calyx top-shaped, with 5 nearly equal teeth which are rigid and spine-tipped, strongly nerved and firm in texture, persistent; corolla white, pink or purplish, 2-lipped, the upper lip entire, somewhat arched and bearded, the lower lip spreading, 3-lobed, the middle lobe largest; stamens 4, nearly equal, inserted on the corolla. Nutlets about 2 mm. long, obovoid-oblong, 3-angled, hairy at apex, dark brown.

Control.—Clean cultivation (18). In waste places pull or dig the clumps before seeds are formed (17).

389. *Lycopus americanus* Muhl. Water horehound, Cutleaf bugleweed.

Perennial; reproducing by seeds, rootstocks and runners. Meadows, pastures, and waste places; on low moist ground. Widespread throughout the United States and southern Canada. Native. July–September.

Description.—Stems erect or spreading, usually branched above, 4-angled, slightly hairy, often purplish below, 2–4 dm. high, from short, tuberous, jointed rootstocks and slender runners. Leaves narrowly ovate, tapering into a short petiole or sessile, coarsely toothed, the lower pinnatifid, green or purplish. Flowers in dense axillary clusters; calyx bell-shaped, open, with 4–5 awl-shaped teeth, which are longer than the nutlets; corolla tubular, with erect lobes, pale purple to nearly white; perfect stamens 2, the upper pair rudimentary or wanting. Nutlets about 1 mm. long, obovate, 3-angled, toothed at apex, yellow to dark brown.

Control.—Improve the drainage (21). Mow close to the ground every time the new shoots begin to blossom (11). Plow when practicable and plant a clean cultivated crop for two years before reseeding.

390. *Lycopus uniflorus* Michx. Bugleweed.

Perennial; reproducing by seeds, rootstocks and runners. Meadows, pastures, and waste places. Widespread throughout the northern United States and southern Canada. Native. July–September.

Description.—Plant similar to 389 but the leaves toothed but not pinnatifid. Calyx-teeth broadly triangular and shorter than the nutlets. Nutlets about 1.5 mm. long, narrowly obovate, 3-angled, apex truncate and toothed, yellow-brown.

Control.—The same as for 389.

391. **Marrubium vulgare** L. Horehound, Houndsbane, Marrube, Marvel.

Perennial; reproducing by seeds. Old fields, pastures, waste places, and ranges; on dry upland soils. Widespread across the United States and southern Canada; most troublesome on the Pacific Coast. Introduced from Europe. June–August.

Description.—Stems erect, stout, usually in clumps, bluntly 4-angled, densely white-woolly, 3–10 dm. high. Leaves ovate to oblong or cordate, green above and white-woolly beneath, deeply irregularly crenate on the margin, petioled. Flowers in crowded axillary clusters; calyx tubular, with 10 spiny teeth which become recurved in fruit; corolla small, white, 2-lipped, the upper erect, notched, the lower spreading and 3-cleft; stamens 4. Nutlets obovate, 2 mm. long, grayish-brown, with black or dark brown granules scattered over the surface.

Control.—This weed does not persist under clean cultivation. In lands that cannot be plowed, mow or hoe the plants close to the ground as soon as they begin to grow in the spring and again each time new shoots start.

Horehound leaves and flower-stalks are used in the preparation of certain cough medicines and for flavoring candy. The entire plant has a characteristic bitter taste so that sheep and other animals do not eat it. The spiny-hooked teeth of the calyx containing the nutlets frequently catch in the wool of the sheep, thus helping in the scattering of the plant and also lowering the value of the wool.

392. **Mentha arvensis** L. Wild mint, Corn mint, Field mint.

Perennial; reproducing by rootstocks and seeds. Meadows, pastures, along ditches, and waste places; mostly on moist soils. Common throughout the northern United States, the Pacific Coast and Canada. Native; also native to Eurasia. July–September.

Description.—Stems erect, freely branching, 4-angled, somewhat retrorse-pubescent, 1–8 dm. high, from creeping rootstocks. Leaves pinnately veined, petioled, serrate, ovate to oblong, minutely glandular-pubescent, 1–5 cm. long, strongly aromatic-scented. Flowers in dense axillary whorls; calyx tubular, with 5 teeth, glandular-pubescent, strongly nerved; corolla tubular, nearly regular, upper lip nearly erect and not arched, white, pink or lavender; stamens 4, nearly equal, inserted on the corolla. Nutlets ovoid, 0.5 mm. long, smooth, light-brown, with an irregular dark line on the convex side.

Mint Family

A variable species represented by several varieties in the United States.

Control.—Improve drainage (21). In pastures mow close to the ground each time that new shoots appear (11). When practicable plow the field in autumn, drag the rootstocks out with a spring-tooth harrow and remove them, follow with a clean cultivated crop for two years before reseeding (7, 6).

393. *Mentha gentilis* L. Creeping whorled mint, Wild bergamot.

Perennial; reproducing by creeping rootstocks; rarely by seeds. Rich moist fields, pastures, and waste places. Locally common in the northeastern states; infrequent elsewhere. Introduced from Europe. July–August.

Description.—Stems erect, freely branching, 4-angled, nearly glabrous, 3–10 dm. high, often reddish in color. Leaves pinnately veined, petioled, serrate, ovate, glandular-dotted, pubescent on the veins, often blotched with white, aromatic-scented. Flowers in dense axillary whorls; calyx tubular with 5 teeth, glandular-pubescent; corolla tubular, nearly regular, upper lip nearly erect and not arched, pink; stamens 4, nearly equal, inserted on the corolla. Mature nutlets not observed; apparently only rarely produced.

Control.—The same as for 392.

394. *Mentha piperita* L. Peppermint, Lamb mint, Brandy mint.

Perennial; reproducing by creeping rootstocks and runners. Low wet fields, pastures, roadsides, and waste places. Locally common; especially in the eastern United States and adjacent Canada. Introduced from Europe. This plant is also cultivated for the oil of peppermint, principally in Michigan and on the Pacific Coast.[1] July–September.

Description.—Stems erect, branching, 4-angled, glabrous, 3–12 dm. high, reddish-purple or green. Leaves pinnately veined, petioled, serrate, ovate-lanceolate to ovate-oblong, glandular-dotted, nearly glabrous, dark green, aromatic-scented. Flowers in thick, oblong, terminal, spike-like heads, each flower on a pedicel; calyx tubular, glabrous, with 5 hirsute teeth, glandular-dotted, purplish-colored; corolla tubular, nearly regular, pink or lavender; stamens 4, equal, inserted on the corolla; ovary deeply 4-lobed. Mature nutlets rarely, if ever, produced in the United States.

Control.—The same as for 392.

[1] Sievers, 1929.

395. *Mentha spicata* L. Spearmint, Lamb mint, Garden mint, Our ladys mint.

Perennial; reproducing by creeping rootstocks. Moist pastures and waste places. Locally common; especially in the northeastern and north central states; also on the Pacific Coast. Introduced from Eurasia. This plant is also cultivated in certain localities for the oil of spearmint.[2] July–September.

Description.—Stems erect, branching, 4-angled, nearly glabrous, 3–8 dm. high, slightly reddish near the base. Leaves pinnately veined, sessile or nearly so, serrate, ovate-lanceolate to ovate-oblong, glandular-dotted, nearly glabrous, pale green, aromatic-scented. Flowers in slender terminal spikes; calyx tubular, glabrous, with 5 hirsute teeth, glandular-dotted, green; corolla tubular, nearly regular, white to pinkish-lavender; stamens 4, equal, inserted on the corolla; ovary deeply 4-lobed. Mature nutlets rarely produced in the United States.

Control.—The same as for 392.

396. *Nepeta cataria* L. Catnip, Catmint. Fig. 93, a–d.

Perennial; reproducing by seeds and short rootstocks. Neglected yards, pastures, and waste places; mostly on rich soils. Widespread throughout the northern United States and southern Canada; infrequent in the southern states. Introduced from Eurasia. July–September.

Description.—Stems erect, branched, 4-angled, downy-pubescent, 3–10 dm. high, purplish below. Leaves long slender-petioled, cordate-oblong, dentate-crenate, pubescent above and whitish-downy beneath, 1–5 cm. long. Flowers in terminal and upper-axillary clusters; calyx tubular with 5 equal teeth, pubescent, persistent; corolla tubular, 2-lipped, the upper lip erect, rather concave, 2-cleft, the lower lip 3-lobed, the middle lobe largest and with crenate margin, whitish, spotted with purple, and twice the length of the calyx; stamens 4, the upper pair much longer than the lower pair; ovary deeply 4-lobed. Nutlets broadly ovoid, 1–1.5 mm. long, with 2 white spots near the base, dark reddish-brown.

Control.—The same as for 391.

397. *Origanum vulgare* L. Wild marjoram, Winter sweet, Organy, Pot marjoram. Fig. 91, a–d.

Perennial; reproducing by seeds and short rootstocks. Pastures, meadows, old fields, lawns, and waste places; mostly on dry soils.

[2] Sievers, 1929.

Mint Family

Locally common from New England to Ontario, southward to North Carolina. Introduced from Europe. July–August.

Description.—Stems erect or ascending, branched above, 4-angled, pubescent, 2–6 dm. high, often purplish. Leaves pinnately veined, petioled, nearly entire, round-ovate, pubescent; the upper leaf axils usually bear small leafy shoots. Flowers in corymbose clusters of crowded cylindrical spikes, imbricated, with large purplish bracts; calyx ovoid-tubular, 5-toothed, hairy; corolla tubular, 2-lipped, the upper lip rather erect and slightly notched, the lower lip longer, of 3 nearly equal spreading lobes, purplish in color; stamens 4, exserted. Nutlets ovoid, about 0.7 mm. long, slightly flattened, dull reddish-brown with a gray scar at base.

Control.—Plow badly infested meadows and pastures and plant a clean cultivated crop for a year before reseeding. In pastures where plowing is impracticable the addition of fertilizer and early spring harrowing followed by reseeding with grass and white clover will stimulate the sod to crowd out this weed. In lawns small patches should be dug out.

398. *Prunella vulgaris* L. Heal-all, Self-heal, Carpenters-weed. Fig. 92, a–b.

Perennial; reproducing by seeds and short rootstocks. Pastures, meadows, lawns and waste places. Widespread throughout North America. Native to North America and Europe; apparently introduced in many sections of the United States. June–September.

Description.—Stems erect, ascending or prostrate, branching, 4-angled, pubescent, becoming nearly glabrous when older, 0.5–4 dm. high. In lawns or in closely grazed pastures the stems are prostrate and frequently root at the nodes. Leaves pinnately veined, long-petioled, entire or irregularly dentate, ovate-oblong, pubescent or glabrous. Flowers sessile in a close thick spike, 3 in the axils of each rounded membranaceous bract; calyx tubular, slightly 2-lipped, veiny, purplish; corolla pink to purple, tubular, 2-lipped, the upper lip erect, arched, entire, the lower lip spreading, 3-cleft and denticulate; stamens 4, the upper pair shorter than the lower pair, exserted. Nutlets obovate, about 1.5 mm. long, slightly flattened, brown with dark vertical lines, the base tapering to a pointed white caruncle, slightly roughened and glossy. Somewhat variable and represented by several varieties.

Control.—The same as for 397.

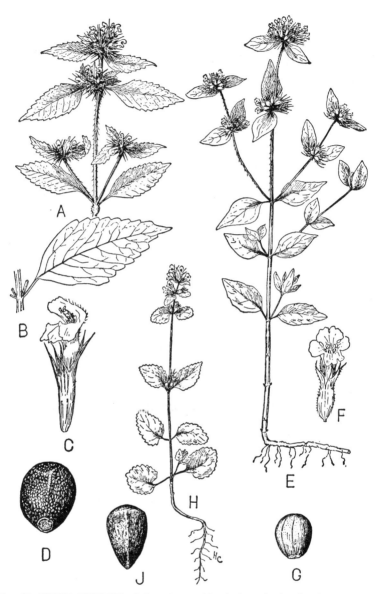

FIG. 94—**HEMP NETTLE**, *Galeopsis tetrahit:* **A,** branch showing habit (× ⅓);
B, lower leaf (× ⅓); **C,** flower (× 2); **D,** seed (× 6).
WILD BASIL, *Satureja vulgaris:* **E,** plant (× ¼); **F,** flower (× 2); **G,** seed (× 12).
HEDGE NETTLE, *Stachys arvensis:* **H,** plant (× ¼); **J,** seed (× 12).

Mint Family

399. *Satureja vulgaris* Fritsch. Wild basil, Field basil, Calamint, Horse thyme, Stone basil, Basil-weed, Dog mint. Fig. 94, e–g.

Perennial; reproducing by seeds and short creeping stems. Old pastures, meadows and borders of open woods. Common, chiefly from the northeastern states south to North Carolina and Tennessee, west to Minnesota and New Mexico; also in eastern Canada. Native; also native to Eurasia. July–August.

Description.—Stems erect, simple or with a few branches, 4-angled, hairy, 2–6 dm. high. Leaves pinnately veined, mostly short-petioled, nearly entire, ovate to ovate-lanceolate, pubescent, 1.5–5 cm. long. Flowers in dense many-flowered, axillary clusters, sessile, with a conspicuous bristly-bracted involucre; calyx tubular, hairy, 5-toothed; corolla white to reddish-purple, tubular, straight, slightly 2-lipped, the upper lip entire, the lower lip 3-lobed, slightly longer than the calyx; stamens 4, the upper pair shorter than the lower pair. Nutlets broadly ovate, about 1 mm. long, smooth, chestnut-brown, with a gray scar at base, minutely granular.

Control.—The same as for 397.

400. *Stachys arvensis* L. Hedge nettle, Low nettle. Fig. 94, h–j.

Annual; reproducing by seeds. Cultivated ground and waste places. Infrequent, chiefly on the Atlantic Coast. Introduced from Europe. May–October.

Description.—Stems prostrate or creeping, branched, 4-angled, hirsute, 1–5 dm. long. Leaves palmately and pinnately veined, petioled, crenate, ovate, hirsute, 1–3 cm. long, the uppermost sessile. Flowers in terminal spikes of 3-several flowered axillary whorls; calyx tubular with 5 equal lanceolate teeth, hirsute and strongly nerved, 4.5–6 mm. long; corolla 2-lipped, the upper lip erect and nearly entire, the lower lip longer and 3-lobed, purple, scarcely longer than the calyx; stamens 4, the upper pair shorter than the lower pair. Nutlets ovate, about 1.5 mm. in diameter, obscurely angled, smooth, rounded at the apex, granular, dark brown or black.

Control.—Surface cultivation early in the spring and in late autumn will kill the seedlings (8). Clean cultivation followed by hand pulling of scattered plants (6).

Stachys annua L., an annual introduced from Europe, occurs sparingly in cultivated soils in the northeastern United States. It has erect stems and yellow or cream-colored flowers. Control the same as for 400.

Stachys palustris L., Woundwort, a rank smelling perennial, occurs in northeastern United States and adjacent Canada; noxious (1954) in Maine.

401. *Thymus serpyllum* L. Creeping thyme, Wild thyme, Hillwort, Penny mountain. Fig. 92, j–l.

Perennial; reproducing by seeds and creeping stems. Pastures, lawns, and old fields; most troublesome on dry stony soils and in limestone regions. Locally common from New England and eastern New York to North Carolina; infrequent elsewhere. Introduced from Europe. June–August.

Description.—Stems prostrate or creeping, slightly 4-angled, pubescent, rooting at the nodes, usually only 1–2 dm. high. Leaves pinnately veined, very short-petioled, entire, ovate to elliptical, glabrous, glandular-dotted, 1 cm. long or less. Flowers in terminal heads consisting of crowded axillary whorls; calyx tubular, slightly 2-lipped, with ciliate awl-shaped teeth, strongly nerved and tinged with purple; corolla lavender, 2-lipped, the upper lip straight and notched at the apex, the lower lip 3-cleft; stamens 4, straight, exserted; ovary deeply 4-lobed. Nutlets nearly elliptical, with a minute white point at the base, minutely granular, light to dark brown.

Control.—The same as for 397.

402. *Trichostema dichotomum* L. Blue curls, Bastard pennyroyal.

Annual; reproducing by seeds. Old fields, pastures and waste places; mostly on dry sandy or gravelly soil. Native northeastern states, south to Florida and west to Michigan and Texas; sparingly introduced westward to the Pacific Coast. July–September.

Description.—Stems erect, slender and stiff, much branched above, minutely pubescent and somewhat sticky. Leaves short-petioled, narrowly-lanceolate, pointed at both ends, entire, finely pubescent. Flowers mostly on 1- or 2-flowered pedicels on the axillary branches, forming an open panicle-like cluster; calyx bell-shaped, deeply 5-cleft, the upper 3 teeth elongated and partly united, the 2 lower very short; corolla blue or pink, with a very slender tube and 5 declined lobes, the lower lobe oblong, the other 4 broad; stamens 4, with very long upcurved filaments, violet. Nutlets obovoid, about 2.5 mm. long, slightly flattened and with a large scar near the base, coarsely reticulated, pubescent at apex, brown.

Control.—Plow, fertilize and turn under a green-manure crop to add humus to the soil.

Nightshade Family

Salvia reflexa Hornem. Lance-leaved sage, is listed (1954) as a weed in the north central states.

Several additional native perennials belonging to the mint family may occur as weeds in moist meadows and grasslands bordering marshes or in dry stony fields and pastures near areas covered with native vegetation and waste places. The methods listed under 392 may give suggestions for the control of such weeds on moist land and the methods given for 397 may be used on dry situations. For the identification of these weeds a manual or flora covering the locality will be helpful.

SOLANACEAE (Nightshade Family)

403. *Datura stramonium* L. Jimson-weed, Jamestown-weed, Thorn-apple, Mad-apple, Stinkwort. Fig. 95, e–g.

Annual; reproducing by seeds. Cultivated fields and waste places; mostly on rich alluvial or gravelly soils. Widespread across the United States and into southern Canada, especially troublesome southward. Introduced from the Tropics. July–September.

Description.—Stems erect, stout, 3–15 dm. high, with spreading branches above, glabrous, green or purple. Leaves alternate, simple, ovate, unevenly toothed, glabrous, dark green above, 7–20 cm. long, on stout petioles, strong-scented. Flowers perfect, solitary on short peduncles in the axils of the branches; calyx 5-toothed and 5-ridged, inclosing the lower part of the corolla-tube; corolla funnelform, 6–15 cm. long, with a spreading 5-toothed plaited border, white or purple; stamens 5, separate; inserted on the corolla; ovary 4-celled or 2-celled above; style with a 2-lobed stigma. Fruit a hard, prickly, ovate, 4-celled, many-seeded capsule splitting from above into 4 valves, 3–5 cm. long, subtended by the basal part of the calyx; seeds about 3 mm. long, kidney-shaped, flat, pitted and wrinkled, black or dark brown.

Control.—Clean cultivation followed by hand weeding or hoeing. Mow the large plants before seeds are formed. Burn plants with mature seed-pods.

This weed is poisonous when eaten and in some individuals contact with it causes a skin eruption. The purple-flowered forms have been treated as a separate species, *Datura tatula* L., by some authors.

404. *Nicandra phvsalodes* (L.) Pers. Apple-of-Peru.

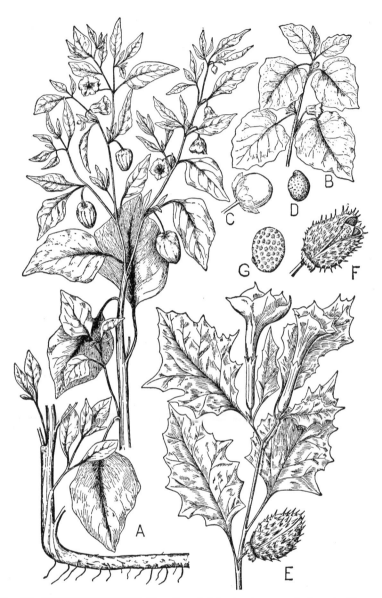

Fig. 95—**GROUND-CHERRY,** *Physalis subglabrata:* **A,** parts of plant (× ⅓).
GROUND-CHERRY, *Physalis heterophylla:* **B,** branch showing hairy leaves and flowers (× ⅓); **C,** berry with part of husk removed (× 1); **D,** seed (× 4).
JIMSON-WEED, *Datura stramonium:* **E,** branch showing flowers and immature capsule (× ⅓); **F,** ripe capsule (× ⅓); **G,** seed (× 4).

Annual; reproducing by seeds. Fields, gardens, and waste places. Southeastern United States; infrequent northward to Ontario. Introduced from South America as an ornamental. July–September.

Description.—Stems stout, glabrous, angled, 3–12 dm. high, erect, with spreading branches above. Leaves alternate, ovate, thin, with wavy or angular margin, petiole grooved. Flowers solitary on terminal or axillary peduncles, perfect, regular; calyx 5-parted, 5-angled, becoming enlarged and bladder-like in fruit; corolla rotate, 5-lobed or nearly entire, pale blue; stamens 5, erect, separate; ovary superior, 3–5 celled, developing into a globular many-seeded berry, becoming dry with age. Seeds about 2 mm. long, short kidney-shaped to obovate, sides flattened, dull, coarsely pitted, orange-brown.

Control.—Clean cultivation, followed by hand weeding or hoeing. Pull and burn plants that have gone to seed.

405. **Physalis alkekengi** L. Chinese lantern plant.

Perennial; reproducing by seeds and by the creeping rootstocks. Grown for its ornamental fruit and occasionally escaping locally into waste places near gardens. Introduced from Eurasia. Throughout the summer.

Description.—Stems mostly smooth, angled, erect and unbranched, 4–6 dm. high. Leaves ovate, firm, with petioles broadening toward the top. Corolla white, about 15 mm. wide, deeply lobed. Fruiting calyx firm, conic-ovoid, pointed, about 5 cm. long, deeply sunken at base, blood red, somewhat persistent.

Control.—Dig out plants before seeds are formed.

406. **Physalis heterophylla** Nees (including var. *ambigua* Rydb.). Ground-cherry, Husk-tomato. Fig. 95, b–d.

Perennial; reproducing by seeds and rootstocks. Meadows, pastures, and old fields; mostly on gravelly or stony soils. Widespread throughout eastern North America, westward to Saskatchewan and Texas. Native. June–August.

Description.—Stems erect, branched or widely spreading, puberulent or tomentose, usually viscid or glandular, ridged, about 3–8 dm. high. Leaves alternate, simple, broadly ovate, rounded or cordate base, viscid-pubescent, wavy or bluntly toothed. Flowers perfect, solitary in the leaf-axils; calyx 5-cleft, reticulate and enlarging after flowering, at length much inflated and inclosing the berry; corolla wheel-shaped to short funnelform, 1.5–2.2 cm. in diameter, 5-angled, 5-toothed; stamens 5, inserted on the corolla. Fruit a 2-celled, many-seeded, globose, yellow berry surrounded by the inflated calyx; seeds about 2 mm. in diameter, flattened, obovate to semi-circular, dull, granular, light orange to straw-colored.

Control.—A rotation which includes a clean cultivated crop every few years will usually destroy this weed, provided the scattered plants appearing after the last cultivation are kept hoed or pulled out (20). Scattered patches should be kept mowed close to the ground.

407. *Physalis subglabrata* Mackenz. and Bush. Ground-cherry. Fig. 95, a.

Perennial; reproducing by seeds and rootstocks. Pastures, meadows, and cultivated fields; on rich, moist, gravelly or alluvial soils. Eastern North America, extending south and west. Native. July–September.

Description.—Stems erect, with spreading branches, glabrous or nearly so, not glandular, ridged, 5–10 dm. high. Leaves alternate, simple, ovate-oblong, oblique base, entire or sparingly unevenly toothed. Flowers perfect, solitary in the leaf-axils; calyx 5-cleft, reticulated, enlarging after flowering, finally inflated and inclosing the berry; corolla rotate or short tubular, yellowish or violet-spotted in the center. Berry orange-red to purple, about 1–2 cm. in diameter, many-seeded; seeds about 1.5 mm. long, flattened, short kidney-shaped, dull, granular, light yellow.

Control.—The same as for 406.

Physalis ixocarpa Brot., Ground-cherry or Tomatilla, and *Physalis pubescens* L., Husk-tomato, two common native annual weeds of the southern states, frequently persist or escape in fields and gardens in the North where they are sometimes grown for their edible fruits. *P. ixocarpa* is nearly glabrous and has large purple berries which burst the calyx. *P. pubescens* is pubescent and produces small yellow berries inclosed in the enlarged husk-like calyx. *P. lobata* Torr. Purple-flowered ground-cherry, and *P. longifolia* Nutt., Perennial ground-cherry, are western weeds. These species can be controlled by clean cultivation.

408. *Solanum carolinense* L. Horse nettle, Bull nettle, Apple-of-Sodom, Wild tomato, Sand brier. Fig. 96, f–g.

Perennial; reproducing by seeds and by creeping rootstocks and roots. Old meadows, pastures, cultivated crops, and waste places. Native in the southern states; introduced northward; locally abundant and spreading in the central and eastern states; occasional in the west, and in southern Canada. June–August.

Description.—Stems erect, stout, loosely branched, 2–6 dm. high,

Nightshade Family 387

beset with stout yellowish spines and short stiff 4–8 rayed hairs; rootstocks deep and spreading. Leaves alternate, simple, oblong to ovate in outline, irregularly wavy-toothed or lobed, covered with stellate hairs; veins, midrib and petiole prickly. Flowers perfect, in open cymose racemes on prickly peduncles; calyx 5-lobed, hairy, persistent at base of fruit; corolla pale violet or white, wheel-shaped, 5-lobed, about 2–3 cm. in diameter; stamens 5, fused by their anthers, opening by pores at the tips; ovary 2-celled. Fruit an orange berry, smooth, globular, 1–2 cm. in diameter, full of juicy pulp and many seeds; seeds about 2 mm. long, obovate, flattened, glossy, granular, yellow to light brown.

Control.—The same as for 406.

The herbage and berries of this plant are reputed to be poisonous when eaten.

409. *Solanum dulcamara* L. European bittersweet, Blue nightshade, Woody nightshade, Poison berry, Climbing nightshade, Scarlet berry. Fig. 96, h–k.

Woody perennial; reproducing by seeds. Rich moist soil; fencerows, waste places in yards, thickets, banks of streams and ditches. Widespread; most common in the eastern and north central states; local on the Pacific Coast. Introduced from Eurasia. June–August.

Description.—Stems slender, reclining, climbing or twining, woody at the base, sometimes 2–3 m. long, slightly pubescent or smooth when older. Leaves alternate, thin, ovate-heart-shaped, the upper halberd-shaped or with 2 lobes or leaflets at the base, entire, smooth, dark green or with a purplish color, leaves and stems with a strong disagreeable odor. Flowers in small cymose clusters usually terminating short shoots; calyx with 5 short lobes, persistent at base of fruit, but not inclosing berry; corolla purple or bluish, wheel-shaped, with 5 pointed lobes, about 1 cm. in diameter; stamens and ovary similar to 408. Fruit an ovoid or elliptic pulpy bright red berry, with a thin transparent skin and many seeds; seeds about 2–2.5 mm. in diameter, nearly circular, flattened, coarsely pitted, dull, light yellow.

Control.—This weed does not persist under cultivation. When this is not practicable cut the stem close to the ground with a brush scythe, or brush hook, and grub out the roots.

The herbage and the berries of *Solanum dulcamara* contain toxic alkaloids which make them mildly poisonous. There is still some difference of opinion concerning the poisonous properties of the berries, some maintaining that they are poisonous while others claim that

Fig. 96—**BLACK NIGHTSHADE,** *Solanum nigrum:* **A,** branch showing habit (× ⅓); **B,** flower (× 2); **C,** berry (× 1).
BUFFALO BUR, *Solanum rostratum:* **D,** branch (× ⅓); **E,** fruit (× 1).
HORSE NETTLE, *Solanum carolinense:* **F,** branch (× ⅓); **G,** flower (× ½).
EUROPEAN BITTERSWEET, *Solanum dulcamara:* **H,** branch showing habit (× ⅓); **J,** flower, view from above (× ½); **K,** flower, side view (× ½).

they are harmless.[1, 2] It appears that not all stages are equally toxic.

410. *Solanum elaeagnifolium* Cav. White horse nettle, Bull nettle, Trompillo, Silver-leaf nightshade.

Perennial; reproducing by seeds and creeping roots. Meadows, pastures, and cultivated fields. Native to the middle western states; most troublesome from Missouri south-westward to Arizona; introduced into California,[3] noxious (1954) in several southern states. May–August.

Description.—Stems from deep-seated rootstocks and creeping roots, erect, branching, usually armed with slender prickles. Leaves alternate, varying from oblong to linear, petioled, entire or wavy-margined, usually armed with prickles, silvery-white with a dense covering of star-shaped clusters of hairs. Flowers in cyme-like clusters; calyx with 5 narrow acute hairy lobes; corolla rotate, 5-lobed, violet or blue; stamens 5, with tapering anthers; ovary white-hairy, developing into a globular, smooth, many-seeded, yellow or orange berry; seeds about 3 mm. long, obovate to nearly circular, flattened, finely granular, yellow to dark brown.

Control.—Clean cultivation followed by hand hoeing of scattered plants (6).

411. *Solanum nigrum* L. Black nightshade, Deadly nightshade, Poison berry, Garden nightshade. Fig. 96, a–c.

Annual; reproducing by seeds. Fields and waste places on loamy or gravelly soils; frequently in the shade. Widespread from southern Canada throughout the United States to Mexico, but apparently nowhere very abundant. Introduced from Europe; apparently some forms of this species are also native to North America. July–September.

Description.—Stems low, much branched and often spreading, nearly glabrous, angular, 2–10 dm. high. Leaves alternate, simple, ovate, wavy-toothed or nearly entire. Flowers perfect, in small umbel-like lateral drooping clusters; calyx of 5 small lobes, persistent; corolla white, similar to 408 except smaller, 3–8 mm. in diameter; stamens 5. Fruit a many-seeded globular black berry, 0.5 to 2 cm. in diameter; seeds about 1.5 mm. in diameter, flattened, minutely wrinkled, dull, pitted, yellow to dark brown.

Control.—Clean cultivation.

Poisonous when eaten, although certain cultivated forms are grown,

[1] Thomson and Sifton, 1922.   [2] Esser, 1910.   [3] Ball, 1932.

under such names as "garden huckleberry" or "wonder berry," for the berries which are used in preserves.

412. *Solanum rostratum* Dunal. Buffalo bur, Beaked nightshade, Sandbur, Colorado bur, Texas thistle. Fig. 96, d–e.

Annual; reproducing by seeds. Meadows, pastures, fields, and waste places; also introduced about farm buildings. Native to the Great Plains from South Dakota to Mexico; introduced locally but increasingly serious, in the eastern states and on the Pacific Coast. July–August.

Description.—Stems erect, stout, 3–6 dm. high, much branched, covered with many stout yellow spines. Leaves alternate, simple, bluntly lobed or 1–2 pinnatifid, very irregular in size, covered with stellate hairs, the veins, midribs and petioles very prickly. Flowers perfect, in open racemose clusters, on prickly peduncles; calyx densely prickly, becoming the bur which incloses the fruit; corolla yellow, about 2–3 cm. in diameter, wheel-shaped, with 5 lobes; stamens 5, declined, with unequal anthers. Fruit a berry, 1 cm. in diameter, inclosed by the spiny calyx, 5–20 mm. long; seeds about 2.5 mm. in diameter, flattened, coarsely pitted, dull, brownish-black.

Control.—Clean cultivation (6). Include cultivated crops in the rotation (20). Mow waste places (11).

413. *Solanum triflorum* Nutt. Cut-leaved nightshade, Three-flowered nightshade, Wild tomato.

Annual; reproducing by seeds. Originally limited to native prairies, but as these were brought under cultivation, this weed also encroached on cultivated fields, pastures, gardens, pea fields, and waste places.[4] Western Ontario to British Columbia, southward to Oklahoma and California, occasional east to the Atlantic Coast. Native. July–September.

Description.—Stems erect or much branched and spreading, 2–8 dm. high, glabrous or somewhat hairy. Leaves alternate, pinnatifid or pinnately lobed, the lobes rounded or pointed. Flowers 1–3 in axillary clusters; calyx 5-lobed; corolla rotate, 5-lobed, white, about 1 cm. in diameter; stamens 5. Fruit a smooth round berry, green or yellow; seeds about 2 mm. long, obovate, flattened, minutely pitted, dull, yellow to light brown.

Control.—The same as 412.

The berries of *Solanum triflorum* are poisonous.[5]

[4] Swingle, Morris and Jahnke, 1920.   [5] Thomson and Sifton, 1922.

414. *Solanum villosum* Mill. Hairy nightshade.

Annual; reproducing by seeds. Local in waste places. Massachusetts, to the south and west; noxious (1954) in Florida, North Carolina, South Carolina, and Utah. Adventive from Eurasia. May–October.

Description.—Similar to *Solanum nigrum* but cinerous and densely villous or pilose; leaves smaller; umbellate flowers 1–5, on slender, club-shaped, stiff pedicels; calyx densely pubescent; berries yellow or red; seeds 1.8–2.2 mm. long.

Control.—The same as for 412.

SCROPHULARIACEAE (Figwort Family)

415. *Chaenorrhinum minus* (L.) Lange (*Linaria minor* Desf.). Small snapdragon. Fig. 97, c.

Annual; reproducing by seeds. Sterile, sandy or gravelly fields, waste places, and on cinders, especially along railroad tracks. Widespread along the north Atlantic Coast and westward to Ontario and Michigan. Introduced from Europe. June–September.

Description.—Stems low branched, glandular-pubescent, 1–3 dm. high. Leaves alternate, simple, spatulate-linear, entire, sessile, 1–3 cm. long. Flowers in leafy-bracted racemes on long slender pedicels, similar to 416; corolla irregular, purplish, 5–9 mm. long. Capsule similar to 416; seeds about 0.7 mm. long, ovoid, with several acute vertical ridges, dull, dark brown to black.

Control.—Clean cultivation. Hoe or mow waste places before seeds are formed.

416. *Cymbalaria muralis* Gaertn., Mey. and Scherb. (*Linaria cymbalaria* Mill.). Kenilworth ivy, Coliseum ivy. Fig. 97, e.

Annual; reproducing by seeds. In gardens, lawns, waste places, and about greenhouses. Local in the southeastern United States; infrequent northward to New York. Introduced from Eurasia. June–September.

Description.—Stems prostrate or trailing, much branched, glabrous, sometimes rooting at the nodes. Leaves mostly alternate, simple, round to kidney-shaped, palmately veined and bluntly lobed, glabrous, petioles slender. Flowers perfect, solitary, axillary on long slender pedicels which become recurved in fruit; calyx 5-parted, the lobes lanceolate; corolla of 5 fused petals, 2-lipped, the upper lip covering the lower in the bud, spurred at the base, purplish-white; stamens 4, 2 long and 2 short, inserted on the corolla; pistil compound, 2-celled.

Fig. 97—**BUTTER-AND-EGGS,** *Linaria vulgaris:* **A,** plant showing general habit with creeping roots (× ⅓); **B,** flower (× 1).
SMALL SNAPDRAGON, *Chaenorrhinum minus:* **C,** branch showing flowers and leaves (× 1).
TOADFLAX, *Kickxia elatine:* **D,** upper part of creeping stem (× 1).
KENILWORTH IVY, *Cymbalaria muralis:* **E,** upper part of a creeping stem (× 1).

Capsule thin, many-seeded, opening below the apex by 1 or more pores; seeds about 1 mm. in diameter, globular, irregularly wrinkled, brownish-black.

Control.—Clean cultivation. Hoe off in waste places. Rake the runners from lawns.

417. *Digitalis purpurea* L. Foxglove.

Biennial; reproducing by seeds. Rich meadows, pastures, and waste places. Locally common on the Pacific Coast from British Columbia to northern California; rare elsewhere. Introduced from Europe as an ornamental. June–July.

Description.—Stems erect, stout, simple, pubescent, 1–2 m. high. Leaves alternate, ovate to lanceolate, the lower petioled, the upper sessile and often decurrent on the stem, hairy, entire or coarsely serrate. Flowers perfect, in long racemes; calyx 5-parted; corolla tubular, irregular, inflated, drooping, purple or white, spotted, about 4–5 cm. long; stamens 4, included in the corolla; pistil 2-celled, forming a 2-celled, ovate, many-seeded capsule. Seeds about 0.8 mm. long, linear, truncate at the ends, angled and pitted, dull, yellowish-brown.

Control.—Hoe or spud the rosettes in early spring. Mow close to the ground as soon as the flower-stalks appear.

418. *Kickxia elatine* (L.) Dumort. (*Linaria elatine* Mill.) Toadflax, Sharp-pointed fluellin. Fig. 97, d.

Annual; reproducing by seeds. Old fields, pastures, gardens; mostly on sandy or gravelly soils. Locally abundant, Massachusetts to Indiana, south to Georgia and Kansas. Introduced from Europe. June–September.

Description.—Stems prostrate or trailing, hairy. Leaves alternate, simple, entire, hastate, hairy, on very short petioles. Flowers perfect, solitary, axillary on slender peduncles, similar to 417; corolla 0.5–1 cm. long, irregular, purplish with a yellow spur. Seeds 1.2 mm. in diameter, nearly globular, with a network of raised wavy ridges, dark brown.

Control.—Clean cultivation followed by hand hoeing.

419. *Linaria canadensis* (L.) Dumont. Old-field toadflax. Fig. 101, a–c.

Annual or biennial; reproducing by seeds and trailing offshoots which form overwintering rosettes. Dry sandy or sterile soil. Nova Scotia to South Dakota, south to Florida and Texas; occasional on the

Pacific Coast from British Columbia to central California. Native. April–September.

Description.—Stems erect, 1–8 dm. high, very slender, glabrous, with several procumbent or widely spreading sterile basal shoots. Leaves narrowly linear, 1–3 cm. long, those of the erect stems scattered, alternate; those of the sterile shoots shorter and wider, crowded, often opposite or whorled. Flowers in nearly glabrous racemes which become elongated; corolla blue, violet, or rarely white, 5–10 mm. long, the lips much longer than the tube, spur 2–6 mm. long, lower lip bearing two short white ridges. Capsule 2–3.5 mm. long, 3–4 mm. in diameter, equaling or exceeding calyx. Seeds slenderly obpyramidal, wingless, nearly smooth, with sharp angles.

Control.—The same as for 416.

Linaria canadensis (L.) Dumont var. *texana* (Scheele) Pennell is widely distributed in the southern and western states and on the Pacific Coast. Corolla 10–14 mm. long, deeper violet, the spur 5–10 mm. long; seed more rugose and with more rounded angles.

Control.—The same as for 416.

420. *Linaria vulgaris* Mill. Butter-and-eggs, Yellow toadflax, Ramsted, Flaxweed, Wild snapdragon, Eggs-and-bacon, Jacobs-ladder. Fig. 97, a–b.

Perennial; reproducing by seeds and creeping roots. Cultivated land, grain fields, meadows and pastures; mostly on gravelly or sandy soils. Common throughout eastern North America; also local on the Pacific Coast. Introduced from Eurasia. June–October.

Description.—Stems erect, sparingly branched, glabrous or somewhat glandular above; often in clumps from creeping roots. Leaves mostly alternate, simple, linear, sessile and entire, pale green. Flowers in racemes; sepals 5; petals 5, fused into an irregular, spurred, 2-lipped, yellow corolla with orange throat, the upper lip covering the lower lip in the bud; stamens 4, 2 long and 2 short; pistil compound, 2-celled. Capsule 2-celled, many-seeded, opening by 2–3 pores or slits just below the apex; seeds 1.5–2 mm. in diameter, nearly circular, flattened, with notched wing, the body minutely warted, dark brown or black.

Control.—Plant a crop in check rows and cultivate until late in the season. Follow with an annual smother crop before reseeding.

421. *Penstemon digitalis* Nutt. (*P. laevigatus* Ait. var. *digitalis* Gray). Beard-tongue, Wild foxglove. Fig. 100.

Figwort Family

Perennial; reproducing by seeds and short rootstocks. Meadows and pastures; mostly on moist gravelly soils. Maine to South Dakota, south to Virginia and Texas. Native. June–July.

Description.—Stems erect, 5–15 dm. high, usually simple and reddish-colored, glabrous up to the inflorescence. Leaves opposite, simple, ovate-oblong to lanceolate, 3–15 cm. long, sparingly toothed, the upper ones sessile with clasping bases, the lower ones tapering to margined petioles, sessile. Flowers showy, in terminal racemose panicles, the peduncles with scattered hairs; calyx 5-parted; corolla about 2 cm. long, somewhat irregular, slightly 2-lipped, tubular, the tube dilated and with 5 lobes; stamens 5, inserted on the corolla-tube, 4 fertile with purplish anthers, the sterile filament, "tongue," bearded on its upper side. Fruit an ovoid capsule, 0.5–1 cm. long, with persistent style, 2-celled, containing many seeds; seeds about 1.5 mm. long, irregular, angular, winged, truncate at apex, dull, dark brown with light brown margins.

Control.—Hoe or mow scattered plants in meadows and pastures. Badly infested meadows should be turned under and cultivated for a year.

422. *Penstemon gracilis* Nutt.

Perennial; reproducing by seeds. Prairies and dry open woods. Wisconsin and western Ontario, west to Alberta and New Mexico; occasional eastward. Native. June–July.

Description.—Stems slender, 2–6 dm. high, finely puberulent to glabrous. Basal leaves short-petioled, oblanceolate; cauline leaves lanceolate to linear, usually denticulate, 5–10 cm. long, 0.2–1 cm. wide, glabrous or puberulent. Lateral branches of slender inflorescence erect; pedicels glandular-puberulent; sepals 4–9 mm. long, in fruit almost as long as the capsule; corolla slender, 1.5–2 cm. long, the lower lip pale violet.

Control.—The same as for 421.

423. *Scrophularia marilandica* L. Figwort, Pilewort.

Perennial; reproducing by seeds. Meadows, pastures, and waste places. Eastern United States and the Mississippi Valley. Native. July–September.

Description.—Stems erect, slender, 4-angled, much branched above, nearly glabrous except for the glandular hairs on the branched terminal panicles; roots knotted. Leaves opposite, simple, ovate, 1–3 dm. long, serrate, with long petioles. Flowers in long open leafless panicles; calyx deeply 5-cleft; corolla inflated, nearly globular, with 4 erect lobes and 1 spreading lobe; stamens 4 fertile and the fifth rudimentary, brownish-purple. Capsule subglobose, with a short conical summit, 2-celled, many-seeded, opening at the top; seeds about 1 mm. long, linear-ovoid, marked with obtuse vertical ridges, minutely reticuled, dull, dark brown.

Control.—Mow close to the ground or spud out the plants early in the spring.

Scrophularia lanceolata Pursh., with habit and appearance similar to 423, has a greenish-yellow sterile stamen and more coarsely serrate leaves. It is native to the northeastern United States but frequently grows in rather dry meadows and pastures throughout the United States and southern Canada. Control the same as 423.

424. *Verbascum blattaria* L. Moth mullein. Fig. 98, d–h.

Biennial; reproducing by seeds. Pastures, meadows, old fields and waste places; mostly on dry gravelly soils. Widespread throughout the United States; most common eastward. Introduced from Eurasia. June–September.

Description.—Stems erect, slender, 6–15 dm. high, stiff, simple or sometimes branched, round, glabrous or with glandular hairs near the top. Basal leaves tufted, forming a rosette about 2–6 dm. in diameter, oblong, doubly toothed or pinnately lobed, with prominent veins; upper leaves alternate, simple, pointed, sessile and partly clasping. Flowers perfect, in long terminal open racemes, on slender pedicels in the axils of bracts, in buds 5-pointed, yellow tinged with orange; calyx 5-parted; corolla about 2 cm. in diameter, rotate, 5-lobed, nearly regular, pale yellow or pinkish-white; stamens 5, of unequal length, inserted on corolla, the filaments conspicuously bearded with purple hairs. Capsule 2-celled, many-seeded, globose, 5–10 mm. in diameter; seeds brown, truncate, shaped like a piece of corn-cob with all 6 rows of kernels removed, about 0.8 mm. long, columnar, slightly angular, base truncate, often oblique, apex slightly rounded, marked with vertical rows of deep pits, those in contiguous rows alternating, dark brown.

Control.—Mow close to the ground (11). Hoe or spud out the rosettes below the crown in autumn or early spring (5, 12). Badly infested fields should be plowed, fertilized and planted with a cultivated crop before reseeding.

425. *Verbascum lychnitis* L. White mullein.

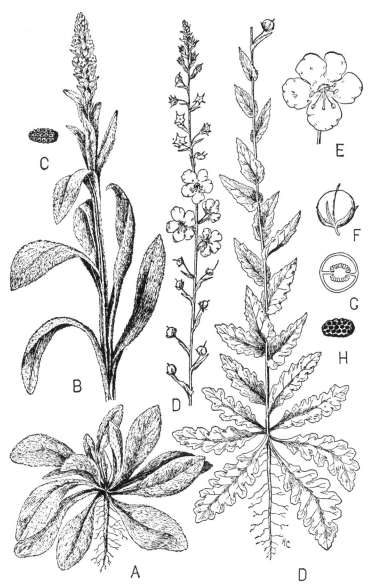

Fig. 98—**MULLEIN,** *Verbascum thapsus:* **A,** rosette (× ⅙); **B,** flowering shoot (× ⅙); **C,** seed (× 7).
MOTH MULLEIN, *Verbascum blattaria:* **D,** plant showing habit (× ⅓); **E,** flower (× ⅔); **F,** capsule (× 1); **G,** cross-section of capsule (× 1); **H,** seed (× 7).

Biennial; reproducing by seeds. Roadsides and fields. Local, Massachusetts to Ontario, Virginia, and West Virginia; also occasional elsewhere. Adventive from Eurasia. June–October.

Description.—Stems erect, up to 1.5 m. high, usually much branched above, white-tomentose. Leaves ovate or oblong-oblanceolate, entire or distantly toothed, acute, sessile, not decurrent, canescent beneath, green above. Flowers yellow or white, 10–15 mm. broad, in large pyramidal panicles or interrupted racemes, flowers pedicelled, several at each node; filaments white-woolly. Capsule ellipsoid, twice as long as calyx.

Control.—The same as for 424.

426. *Verbascum thapsus* L. Mullein, Velvet dock, Big taper, Candle-wick, Flannel-leaf, Torches, Jacobs-staff. Fig. 98, a–c.

Biennial; reproducing by seeds. Pastures, meadows, old fields, and waste places; mostly on dry, gravelly, or stony soils. Widespread throughout the United States and southern Canada. Introduced from Eurasia. June–September.

Description.—Stems erect, very stout, 5–25 dm. high, simple or with a few upright branches near the top, woolly throughout, angled and winged by the decurrent leaf-bases. Basal leaves tufted, forming a large thick rosette, 2–6 dm. in diameter, oblong, tapering to the thick base, densely woolly with branched and interlacing hairs; upper leaves alternate, simple, narrower and more pointed, their bases prolonged and encasing the stem downward to the next leaf, making the stem 4-winged. Flowers perfect, nearly sessile, in long dense terminal cylindrical spikes; calyx 5-lobed and very woolly; corolla 2–3 cm. in diameter, rotate, 5-lobed, nearly regular, sulfur-yellow; stamens 5, with white filaments, the 3 upper ones shorter and bearded. Capsule 2-celled, many-seeded, globular, about 6 mm. in diameter, downy; seeds similar to 424, but marked with wavy ridges alternating with deep grooves.

Control.—The same as for 424.

Verbascum phlomoides L., Mullein, occurs as a weed locally from Maine to Minnesota and south to North Carolina, Kentucky, and Iowa.

Veronica spp.

The flower of the Veronicas (427–435) is perfect, contains a 4–5-parted calyx, a rotate or short salverform, nearly regular, 4-lobed, white or pale blue to lavender corolla, 2 exserted stamens, and a 2-

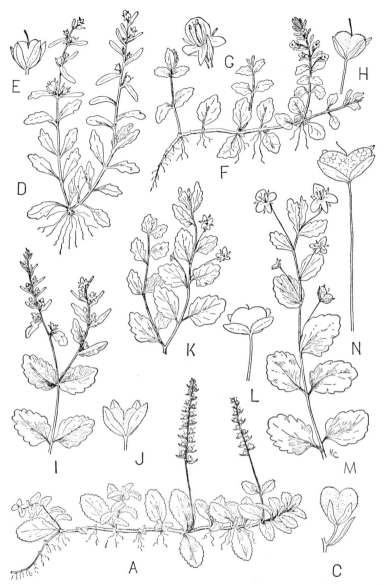

Fig. 99—**SPEEDWELL**, *Veronica officinalis:* **A**, part of plant (× ¼); **C**, capsule (× 2).
PURSLANE SPEEDWELL, *Veronica peregrina:* **D**, plant (×⅔); **E**, capsule (× 2).
THYME-LEAVED SPEEDWELL, *Veronica serpyllifolia:* **F**, plant showing creeping habit (× ½); **G**, flower (× 3); **H**, capsule (× 2).
CORN SPEEDWELL, *Veronica arvensis:* **I**, tip of plant (× ⅔); **J**, capsule (× 2).
FIELD SPEEDWELL, *Veronica polita:* **K**, tip of plant (⅔); **L**, capsule (× 2).
SPEEDWELL, *Veronica persica:* **M**, tip of plant (× ⅔); **N**, capsule (× 2).

celled several-seeded pistil with a single style and stigma. Its fruit is a 2-celled many-seeded capsule.

427. **Veronica arvensis** L. Corn speedwell, Rock speedwell, Wall speedwell. Fig. 99, i–j.

Annual or winter annual; reproducing by seeds. Dry gravelly fields, pastures, gardens and waste places. Throughout the United States and southern Canada. Introduced from Eurasia; perhaps also native in some sections of North America. April–August.

Description.—Stems simple or diffusely branched, erect, 0.5–4 dm. high, hairy. Leaves mostly opposite, simple, thin and rugose, the lower petioled, ovate, crenate, hairy, 1 cm. long or less; the uppermost sessile, lanceolate to entire. Flowers in terminal racemes or spikes; corolla bright blue. Capsules flattened, with 2 rounded lobes; seeds about 1 mm. long, oval, one side convex, the other concave, with a scar in the center, minutely granular, yellow.

Control.—Clean cultivation. Harrow grain fields and meadows early in the spring to destroy the speedwell seedlings. Shallow surface cultivation in autumn will destroy the fall crop of seedlings.

428. **Veronica chamaedrys** L. Germander speedwell, Birds-eye.

Perennial; reproducing by seeds and creeping stems. Meadows, pastures, and lawns. Infrequent in the northeastern states and adjacent Canada. Introduced from Europe. May–June.

Description.—Stems prostrate or creeping and rooting at the nodes, very slender and hairy, with erect or ascending branches. Leaves opposite, simple, ovate or cordate, hairy, 2 cm. long or less, crenate, petioled, the upper sessile. Flowers in flexuous loosely-flowered axillary racemes, the pedicels slightly longer than the calyx; corolla pale blue, with stripes. Capsule triangular-obcordate, flattened.

Control.—The same as for 431.

429. **Veronica filiformis** Sm. Creeping veronica. Fig. 86, d–f.

Perennial; reproducing by creeping stems. Introduced into gardens as an ornamental but spreads freely into adjacent lawns and gardens. Well established locally in the northeastern states. Introduced from Eurasia. Early summer.

Description.—Similar to *Veronica polita* but densely matted; leaves small, reniform; pedicels elongate; corolla 1 cm. broad or less; style 2–3 mm. long.

Control.—Pull out plants when ground is soft. Be sure all parts of every plant are removed and burned. Inspect frequently for new plants.

430. *Veronica hederaefolia* L. Ivy-leaved speedwell.

Annual; reproducing by seeds. Fields, grassy slopes, woodlands and waste ground. Local, New York and Ontario to South Carolina, Ohio, Washington. Introduced from Europe. March–June.

Description.—Stems prostrate to loosely ascending, much branched. Principal leaves rotund to reniform, 8–15 mm. long, obtuse, shallowly 3–5-lobed. Flowers solitary in leaf axils, pedicels 8–15 mm. long. Calyx-lobes ovate, long-ciliate, truncate to cordate at base; corolla lilac to blue, 2–5 mm. wide. Capsule turgid, glabrous. Seeds 2 in each locule, rugose, 2.5–3 mm. long, 1.9–2.4 mm. wide.

Control.—The same as for 427.

431. *Veronica officinalis* L. Speedwell, Gipsy-weed, Fluellin, Pauls betony, Ground-heal. Fig. 99, a–c.

Perennial; reproducing by seeds and creeping stems. Old fields, pastures, and open woodlands; mostly on stony or gravelly, somewhat acid soils. Common from eastern Canada southward to Georgia and Tennessee, westward to South Dakota. Native; also native to Eurasia. June–July.

Description.—Stems prostrate or creeping, rooting at the base, with erect mostly simple branches, stout, pubescent. Leaves opposite, simple, obovate-elliptical, obtuse, serrate, hairy and more or less rugose, 2–4 cm. long, short-petioled. Flowers in dense many-flowered axillary racemes, the pedicels shorter than the calyx; corolla pale blue, marked with darker lines, about 6 mm. in diameter. Capsule obovate-triangular, 3–4 mm. in diameter, flattened, broadly notched; seeds about 1 mm. long, oval to elliptical, flattened, minutely granular, lemon-yellow.

Control.—Clean cultivation (6). Harrow meadows and pastures in the spring to destroy the runners (8). In lawns raise the runners with a rake and mow close.

432. *Veronica peregrina* L. Purslane speedwell, Neckweed, Winter purslane. Fig. 99, d–e.

Annual or winter annual; reproducing by seeds. Gardens, cultivated fields, new lawns, and waste places; especially in gravelly or sandy soils. Widespread and common in the eastern and central United States; Alaska to Oregon. Native. April–October.

Description.—Stems erect, usually much branched, 1–3 dm. high, glabrous and rather stout, frequently rooting at the lower nodes. Lower leaves opposite, short-petioled, dentate, glabrous or nearly so, somewhat fleshy; upper leaves alternate, oblong-linear, entire, sessile or nearly so, much longer than the flowers in their axils. Lower flowers solitary and axillary; upper flowers in terminal spike-like racemes; corolla whitish. Capsule orbicular, flat, slightly notched; seeds about 0.8 mm. long, oval, flattened, with a scar on one side and a vertical ridge on the other, minutely granular, translucent, glossy, orange-yellow.

Control.—The same as for 427.

433. *Veronica persica* Poir. (*V. tournefortii* C. C. Gmel.). Speedwell. Fig. 99, m–n.

Annual or winter annual; reproducing by seeds. Gardens, cultivated fields, lawns, and waste places; on rich soils. Widespread throughout North America; locally common in the eastern states. Introduced from Eurasia. April–June.

Description.—Stems prostrate or ascending, slender, diffusely branched, pubescent, 1–3 dm. long, often rooting at the lower nodes. Leaves alternate, simple, round or ovate, 0.5–2.5 cm. long, dentate, slightly hairy, short-petioled to sessile, the lower opposite. Flowers solitary, on long pedicels in the axils of ordinary leaves; corolla 10–12 mm. in diameter, much longer than the calyx, pale blue with deep blue lines; style 3 mm. long. Capsule obcordate or reniform, broadly notched, lobes widely spreading, reticulated, pedicels in the fruit 15–25 mm. long; seeds about 1.2 mm. long, obovate-oval, one side deeply concave, the other convex and irregularly cross-ridged, pale yellow or nearly white.

Control.—Disking or harrowing early in the spring will kill the plants before they produce seeds. Follow with a clean cultivated crop. Rake the creeping stems from lawns and mow close.

434. *Veronica polita* Fries. Field speedwell. Fig. 99, k–l.

Annual or winter annual; reproducing by seeds. Gardens, lawns, and waste places; on rather gravelly or sandy soils. Infrequent from New York to Michigan, south to Florida and Texas. Introduced from Eurasia. April–May.

Description.—Stems prostrate or ascending, diffusely branched, slightly pubescent, 0.5–2 dm. long. Leaves alternate, simple, broadly ovate, crenate-serrate, rugose, slightly hairy, 0.5–1.5 cm. long, the

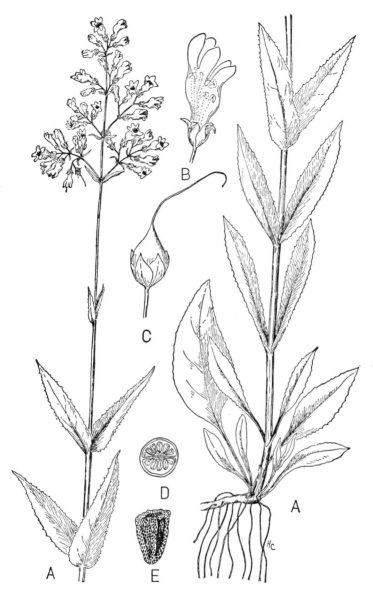

FIG. 100—**BEARD-TONGUE**, *Penstemon digitalis:* **A,** plant showing general habit (× ⅓); **B,** flower (× 1); **C,** capsule (× 2); **D,** cross-section of capsule (× 2); **E,** seed (× 9).

lower opposite. Flowers solitary on long pedicels in the axils of ordinary leaves; corolla 4–6 mm. in diameter, violet-blue; style 1 mm. long. Capsule long-pedicelled, broadly orbicular, with 7–10 seeds in each cell, lobes not spreading, not reticulated, pedicels in the fruit rarely 10 mm. long; seeds about 1.5 mm. long, elliptical, one side convex and ridged horizontally, the other side concave, orange-yellow.

Control.—The same as for 433.

435. *Veronica serpyllifolia* L. Thyme-leaved speedwell, Creeping speedwell. Fig. 99, f–h.

Perennial; reproducing by seeds and creeping stems. Lawns, pastures, and old fields; especially on moist soils. Common throughout North America. Native; apparently also introduced from Eurasia. April–October.

Description.—Stems prostrate or creeping, rooting at the nodes, branched, forming dense mats, branches simple and ascending, about 1 dm. high. Leaves opposite, simple, ovate to oblong, 1.5 cm. long or less, crenate or entire, rather thick and mostly glabrous, the lowest petioled and rounded, the upper alternate, sessile, narrow, becoming bracts. Flowers in terminal racemes; corolla bluish-white with darker stripes, 3–4 mm. in diameter, longer than the calyx. Capsule obcordate, broader than long, 3–4 mm. broad, flat; seeds about 0.9 mm. long, oval, concave on one side, slightly convex on the back, minutely granular, orange.

Control.—This weed does not persist under cultivation. It causes most concern in lawns and pastures. Badly infested areas should be plowed under and cultivated for a year before reseeding. Small patches can be hoed out of lawns.

Veronica agrestis L. Garden speedwell or Germander chickweed, a small introduced annual similar to 434, occurs in fields, gardens and waste places in eastern North America, especially near the coast. It has bright green finely serrate leaves, white or pale blue corollas, and 4–6 seeds in each cell of the capsule. Control the same as for 433.

BIGNONIACEAE (Bignonia Family)

436. *Campsis radicans* (L.) Seem. (*Tecoma radicans* (L.) Juss., *Bignonia radicans* Juss.). Trumpet-creeper, Cow-itch. Fig. 101, d–f.

Perennial; reproducing by seeds. Alluvial woods, roadsides, thickets, and fence rows. New Jersey to Florida, westward to Iowa and

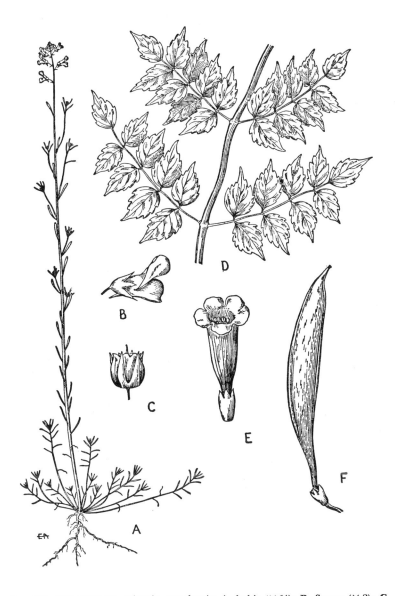

FIG. 101—**TOADFLAX,** *Linaria canadensis:* **A,** habit (× ⅓); **B,** flower (× 2); **C,** fruit (× 3).
TRUMPET-CREEPER, *Campsis radicans:* **D,** stem showing leaf arrangement (× ⅓); **E,** flower (× ⅓); **F,** fruit capsule (× ⅓).

Texas; planted as an ornamental vine outside its natural range. Native. July–September.

Description.—Stems woody, climbing by aerial rootlets. Leaves opposite, pinnately-compound, with 5–13 ovate or lanceolate leaflets with serrate margins. Flowers in corymbs; calyx bell-shaped, 5-toothed; corolla funnelform, 5-lobed, nearly regular, orange to scarlet, 5–8 cm. long; stamens 4, included; pistil forming a 2-celled, somewhat flattened, spindle-shaped capsule, 10–15 cm. long with many winged seeds about 15 mm. long.

Control.—Cut and burn. Dig out roots and destroy capsules before seeds mature.

OROBANCHACEAE (Broom-rape Family)

437. *Orobanche ramosa* L. Broom-rape, Hemp broom-rape, Tobacco broom-rape.

Annual; reproducing by seeds. Hemp and tobacco fields; parasitic on the roots of hemp, tobacco, and occasionally on tomato and other species. New York to Illinois, southward to North Carolina and Tennessee. Introduced from Europe. June–August.

Description.—Stems erect, much branched, slender, brown or straw-colored, 1–2 dm. high, parasitic on the roots of host plants. Leaves reduced to yellowish scales destitute of chlorophyll. Flowers in a spike, perfect, each flower with 3 bracts at the base of the calyx; calyx 4-lobed; corolla tubular, 2-lipped, yellow or pale blue, 1–1.5 cm. long; stamens 4, 2 long and 2 short, inserted on the corolla-tube; pistil with a superior 2-celled ovary, a long style and large stigma. Capsule 1-celled, with 4 placentae, many-seeded; seeds about 0.3 mm. long, ovate, finely ridged, dull, yellowish-brown.

Control.—Clean cultivation followed by hand weeding or hoeing the broom-rape plants before they produce seeds. Badly infested fields should be planted for two or three years with crops that are not parasitized by the broom-rape.

Orobanche ludoviciana Nutt., Louisiana broom-rape, native to sandy soils, chiefly in the Mississippi Valley, parasitizes the roots of tobacco and other plants. In general habit and appearance it is similar to 437, but it is stouter and each flower has 1–2 bracts at the base of the calyx. Its flowers are in a dense spike and its corolla-lobes are acute. Control as for 437.

Orobanche minor J. E. Smith. Clover broom-rape, parasitizing

Plantain Family

clover plants, has been introduced from Europe into the Atlantic states from New Jersey to North Carolina and rarely westward. It is similar to the Louisiana broom-rape but it has more open spikes and its corolla-lobes are rounded. Control as for 437.

PLANTAGINACEAE (Plantain Family)

438. *Plantago aristata* Michx. Bracted plantain, Rat-tail plantain, Clover choker, Western buckhorn, Western ripple-grass. Fig. 102, k–l.

Annual or winter annual; reproducing by seeds. Clover fields, meadows, pastures, and waste places; on rather dry soils. Most common in the middle western United States, where it is native; introduced locally in the eastern states and on the Pacific Coast. June–July.

Description.—Stems erect, simple, leafless, hairy, terminated by the flower-spike. Leaves basal, linear, loosely hairy or becoming glabrous. Flowers in long terminal spikes, with long bracts 2–8 times as long as the flowers, similar to 440. Capsule 2-celled, 2-seeded, opening by a lid; seeds 2–2.5 mm. long, boat-shaped, broadly oval, convex side with a transverse groove near middle, concave side with 2 white-margined scars and a white ellipse, dull, light brown.

Control.—A rotation including a green-manure crop plowed under and followed by clean cultivated crops for two seasons.

439. *Plantago indica* L. (*P. arenaria* Waldst. and Kit.) Whorled plantain, Sandwort plantain. Fig. 102, m–n.

Annual; reproducing by seeds. Waste places and gardens, especially about cities and towns. Local, mostly in the eastern and north central states. Introduced from Europe. July–September.

Description.—Stems erect or spreading, much branched, jointed, hairy, 1–4 dm. high. Leaves whorled or opposite on the upper nodes, simple, linear, entire, sessile, hairy. Flowers in numerous, slender-peduncled, short, axillary spikes; sepals 4, persistent, with dry margins; petals 4, fused into a short tubular corolla; stamens 4, with long filaments; pistil 2-celled, with a long style. Capsule opening transversely by a lid, 2-seeded; seeds about 2 mm. long, often slightly wider at one end, concave side with a scar in the center, convex side with a faint groove near the middle, nearly smooth, somewhat glossy, dark brown.

Control.—Clean cultivation and hand weeding or hoeing.

Large quantities of seeds of several species of Plantago have been

Fig. 102—**BUCKHORN**, *Plantago lanceolata:* **A**, plant (× ¼); **B**, seed (× 5).
RUGELS PLANTAIN, *Plantago rugelii:* **C**, plant showing general habit (× ¼); **D**, capsule (× 3); **E**, side and end view of seed (× 5); **F**, flower (× 3).
BROAD-LEAVED PLANTAIN, *Plantago major:* **G**, capsule (× 3); **H**, side and end view of seed (× 5).
BRACTED PLANTAIN, *Plantago aristata:* **K**, flower-spike and leaf (× ¼); **L**, seed (× 5).
WHORLED PLANTAIN, *Plantago indica:* **M**, branch (× ¼); **N**, seeds (× 5).

Plantain Family

imported from Europe. These mucilaginous seeds, usually known as "Psyllium," are used as laxatives and demulcents. Many of the "French Psyllium" samples consist of *Plantago indica* seed.[1] This explains the appearance of this plant about many cities on waste places, dumping grounds, and in the sludge taken from the settling basins of some sewer systems.

440. *Plantago lanceolata* L. Buckhorn, Buckhorn plantain, English plantain, Narrow-leaved plantain, Rib-grass, Ribwort, Buck plantain, Black-jacks. Fig. 102, a–b.

Perennial; reproducing by seeds and by new shoots from roots. Clover fields, meadows, lawns, and waste places. Widespread throughout the United States and Canada. Introduced from Eurasia. June–September.

Description.—Stems erect, simple, leafless, terminated by a short spike of flowers. Leaves in a basal rosette, lanceolate, hairy or nearly glabrous, with several prominent veins. Flowers in short cylindrical spikes or heads; sepals 4, persistent, with dry margins; petals 4, fused into a short tubular corolla; stamens 4, with long filaments; pistil 2-celled, with a long style. Capsule 2-seeded, opening transversely by a lid; seeds 1.5–2.5 mm. long, boat-shaped, concave side with a scar in the center, glossy, light to dark brown.

Control.—Badly infested meadows should be plowed under, fertilized and planted with a clean cultivated crop for two years before reseeding. In lawns scattered plants should be dug out with a spud (12). In badly infested cases it is easier to remake the lawn after a thorough cultivation for a summer than to eradicate the plantain from the sod.

441. *Plantago major* L. Broad-leaved plantain, Dooryard plantain. Fig. 102, g–h.

Perennial; reproducing by seeds and by new shoots from roots. Pastures, meadows, lawns, dooryards and waste places; most common on rich, somewhat moist, soils. Throughout North America. Introduced from Europe; probably also native northward. June–September.

Description.—Stems erect, simple, leafless, glabrous or hairy, terminated by a long slender spike. Leaves ovate or nearly orbicular, with several prominent veins, petiole long, with prominent veins and sometimes purplish. Flowers similar to 440. Capsule ovoid,

[1] Youngken, 1934.

2-celled, 6–20-seeded, opening by a lid near the middle; seeds 1–1.5 mm. long, variable in shape, angular on one side, with ridges radiating from the scar, glossy, minutely netted, light to dark brown.

Control.—The same as for 440.

442. *Plantago media* L. Hoary plantain, Lambs-tongue.

Perennial; reproducing by seeds. Lawns and waste places. Northeastern United States and eastern Canada; rare in the western states. Introduced from Eurasia. June–September.

Description.—Plant similar to 440 but the leaves ovate to oblong, canescent, coarsely toothed. Seeds 1.5–2.5 mm. long, flattened or slightly concave on one side, granular, dull, dark brown or black.

Control.—The same as for 440.

443. *Plantago purshii* R. and S. Purshs plantain, Woolly plantain.

Annual; reproducing by seeds. Dry fields, meadows, pastures, and waste places. Native from Ontario to Illinois, westward and southward; sparingly introduced eastward. May–August.

Description.—Stems erect, slender, 5–30 cm. high, leafless, woolly or silky-hairy. Leaves basal, ascending, linear, tapering at the apex and at the base narrowed into winged petioles, 3-nerved, 3–8 mm. wide, entire or nearly so. Spike dense, cylindrical, 3–15 cm. long, hairy; bracts rigid, as long as or slightly exceeding the flowers; flowers perfect; sepals 4, oblong with scarious margins; petals 4, fused below, broadly ovate; stamens 4. Capsule opening transversely near the middle, 2-seeded; seeds about 2 mm. long, boat-shaped, convex side with a transverse groove at middle, concave side with 2 white margined scars and a white ellipse, minutely netted, dull, light brown.

Control.—The same as for 438 and 440.

444. *Plantago rugelii* Dcne. Rugels plantain, Broad-leaved plantain, Pale plantain, Silk plant, White mans foot, Purple-stemmed plantain. Fig. 102, c–f.

Perennial; reproducing by seeds and by shoots from roots. Fields, lawns, and waste places; on rich soil. Native to eastern North America; introduced locally west of the Mississippi Valley. June–September.

Description.—Plant similar to 441 but usually less hairy, somewhat stouter, the leaves usually wavy-toothed and with a purplish petiole. Capsule nearly cylindrical, 2-celled, 4–10-seeded, opening by a lid much below the middle; seeds 1–2 mm. long, irregularly ovate, angular, with a scar near center of flat side, minutely granular, dull, dark brown or black.

Control.—The same as for 440.

445. *Plantago virginica* L. Hoary plantain.

Annual or biennial; reproducing by seeds. Dry, open, or sandy soil, especially pastures and fallow fields. Widespread in many parts of the United States. Native. April–June.

Description.—Leaves in basal rosettes, oblanceolate to obovate, usually 5–12 but occasionally up to 19 cm. long, obtuse, entire or undulate or shallowly dentate, villous. Species polygamodioecious. Scape up to 25 cm. long, long-villous; spikes dense, commonly 3–8 cm. (up to 30 cm.) long, 6–8 mm. in diameter. Sepals oblong-obovate, rounded at tip, 2–2.5 mm. long; petals acute, erect and converging at the summit after flowering, 2–3 mm. long. Seeds 2–4, pale brown to nearly black, 1.3–2 mm. long.

Control.—The same as for 438.

RUBIACEAE (Madder Family)

446. *Diodia teres* Walt. Button-weed.

Annual; reproducing by seeds. Cultivated fields, grain fields, and waste places; mostly on dry or sandy soil. Common in the southern states and frequent northward along the coast to Connecticut. Native. July–September.

Description.—Stems branched, spreading, hairy, 1–8 dm. long. Leaves opposite, entire, linear-lanceolate, sessile, stiff; stipules fused, forming several long bristles. Flowers perfect, 1–3 in a leaf-axil; calyx fused to the ovary, 4-toothed; corolla whitish, funnel-shaped, 5 mm. long, with 4 short lobes; stamens 4, attached to the corolla; pistil 2- or 3-celled, at maturity turbinate, splitting into 2 or 3 indehiscent carpels, "seeds." "Seeds" 3–4 mm. long, hairy, light or grayish-brown, with a few ribs, crowned with 3–4 short calyx-teeth.

Control.—Clean cultivation (6). When the crop is removed, sow a fall cover-crop and plow it under the next spring (22).

447. *Galium aparine* L. Cleavers, Goose-grass, Scratch-grass, Grip-grass, Catch-weed. Fig. 103, o–r.

Annual; reproducing by seeds. Bare places in meadows, especially near woodlands and in thickets; mostly on moist, rich, sandy, or alluvial soils. Widespread throughout North America. Native; probably also introduced from Eurasia. May–July.

Description.—Stems decumbent, slender, wiry, 4-angled, jointed, glabrous, with short bristly hooks along the ridges, 6–15 dm. long.

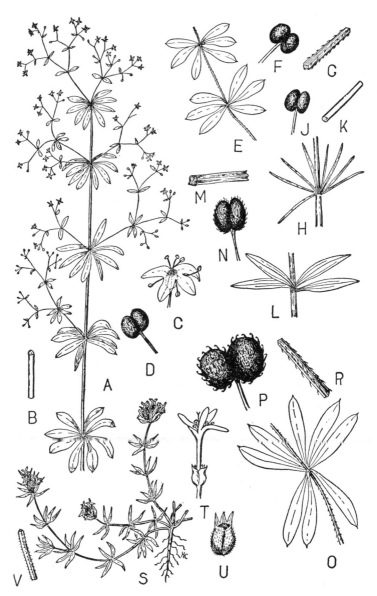

FIG. 103—**CLEAVERS**, *Galium mollugo:* **A**, a flowering stem (× ½); **B**, section of a main stem (× 2); **C**, flower (× 4); **D**, fruit (× 4). **ROUGH BEDSTRAW**, *G. asprellum:* **E**, leaf whorls (× ⅔); **F**, fruit (× 4); **G**, section of a main stem (× 2). **YELLOW BEDSTRAW**, *G. verum:* **H**, leaf whorls (× ⅔); **J**, fruit (× 4); **K**, section of a main stem (× 2). **NORTHERN BEDSTRAW**, *G. boreale:* **L**, leaf whorl (× ⅔); **M**, section of a main stem (× 2); **N**, fruit (× 4). **CLEAVERS**, *G. aparine:* **O**, leaf whorl (× ⅔); **P**, fruit (× 4); **R**, section of a main stem (× 2). **BLUE FIELD MADDER**, *Sherardia arvensis:* **S**, part of plant (× ⅓); **T**, flower (× 4); **U**, half fruit (× 4); **V**, section of a main stem (× 2).

Madder Family

Leaves in whorls of 6–8 at a node, simple, entire, oblanceolate to linear, rough, sparsely hairy, 1-veined. Flowers perfect, axillary, 1–3 on a peduncle; calyx obsolete; corolla rotate with 4 blunt lobes, white; stamens 4, on the corolla; styles 2; ovary 2-celled, separating when mature into 2 nearly globular 1-seeded indehiscent carpels or "seeds." "Seeds" warted and bristly, 2–3 mm. in diameter, gray-brown.

Control.—Clean cultivation (6). Rake the cleavers from hedgerows and thickets before seeds are formed and burn them.

448. *Galium asprellum* Michx. Rough bedstraw, Cleavers, Kidney-vine. Fig. 103, e–g.

Perennial; reproducing by seeds. Fields, waste places and borders of woods; mostly on moist rich alluvial soils. Widespread from eastern Canada to South Carolina and westward to Nebraska. Native. July–September.

Description.—Stems erect, decumbent or nearly prostrate, 5–12 dm. long, branched below, rough, with hooked bristles along the 4 angles, jointed. Leaves whorled, mostly 6 at a node, simple, entire, oblanceolate, cuspidate, 1-veined. Flowers in terminal panicles, similar to 447; corolla white; ovary glabrous. "Seeds" globular, about 1.5 mm. in diameter, granular, glabrous, light brown.

Control.—Mow meadows early before seeds are produced (11). Close pasturing will prevent the bedstraw from spreading (10). Plow badly infested meadows and plant a clean cultivated crop before reseeding.

449. *Galium boreale* L. Northern bedstraw, Chicken-weed. Fig. 103, l–n.

Perennial; reproducing by seeds and rootstocks. Old meadows and fields; mostly on rocky or gravelly soils. Widespread throughout the northern United States and Canada. Native; also native to Eurasia. June–July.

Description.—Stems erect, branched, 4-angled, glabrous, 5–10 dm. high, usually in clumps. Leaves mostly 4 in a whorl, lanceolate, blunt, 3-nerved. Flowers in dense terminal panicles, clear white; ovary hairy. "Seeds" oval, 1.3 mm. long, hairy, dark brown.

Control.—The same as for 448.

450. *Galium mollugo* L. Cleavers, Wild madder, White hedge bedstraw, Whip-tongue. Fig. 103, a–d.

Perennial; reproducing by seeds and rootstocks. Meadows, pastures, lawns, roadsides, and waste places; on gravelly or sandy loam soils. Locally common in the northeastern United States, also Newfoundland to Ontario; spreading considerably in recent years. Introduced from Europe. June–August.

Description.—Stems erect or decumbent and spreading, branched, wiry, 4-angled, 4–10 dm. high, usually in dense clumps from rootstocks, glabrous or nearly so. Leaves mostly 8 in a whorl on the main stem, 5 or 6 in a whorl on the branches, oblanceolate to linear-lanceolate, 1-nerved. Flowers in terminal panicles; corolla white; ovary glabrous. "Seeds" 1–1.5 mm. long, kidney-shaped to globular, glabrous, wrinkled, dark brown. This is a variable species apparently represented by several varieties in eastern North America.

Control.—The same as for 448.

451. *Galium verum* L. Yellow bedstraw, Ladys bedstraw, Cheeserennet, Curdwort, Bedflower, Yellow cleavers. Fig. 103, h–k.

Perennial; reproducing by seeds and rootstocks. Fields, pastures, and lawns; on rich, gravelly or sandy soils. Locally common in the northeastern United States, Ontario and Quebec, west to North Dakota and Kansas. Introduced from Europe. June–August.

Description.—Stems erect or spreading, rather stout, bluntly 4-angled, somewhat hairy. Leaves mostly 8 in a whorl on the main stem, 6 in a whorl on the branches, linear, cuspidate, 1-veined, rough. Flowers in dense terminal panicles; corolla yellow; ovary glabrous. "Seeds" kidney-shaped to nearly globular, about 1 mm. in diameter, glabrous, wrinkled, dark brown to black.

Control.—The same as for 448.

452. *Sherardia arvensis* L. Blue field madder, Spurwort, Herb sherard. Fig. 103, s–v.

Annual; reproducing by seeds. Lawns, meadows and waste places. Northeastern and north central states; eastern Canada, south to North Carolina, Tennessee, and Missouri; also on the Pacific Coast. Introduced from Europe. June–July.

Description.—Stems prostrate or creeping, with slender, erect, 4-angled, hairy branches, 1–3 dm. high. Leaves in whorls of 4–6, simple, entire, lanceolate, acuminate at apex, with scattered hairs. Flowers perfect, in terminal clusters, surrounded by an involucre of fused bracts; calyx of 4 lanceolate sepals; corolla regular, tubular, 4-lobed, pink to bluish; stamens 4; pistil with a 2-cleft style and 2-lobed ovary, at maturity separating into two 1-seeded, indehiscent carpels. "Seeds"

1–2 mm. long, ovate or obovate, dark brown, with short stiff light yellow hairs below, usually crowned with 3 stiff light yellow keeled calyx-lobes.

Control.—Cultivation will destroy this weed (6). In meadows and clover fields mow early to prevent seed formation. (11).

CAPRIFOLIACEAE (Honeysuckle Family)

453. *Lonicera japonica* Thunb. Japanese honeysuckle.

Twining or trailing shrubby perennial; reproducing by seeds and by creeping stems rooting at the nodes. Fields, thickets, orchards, gardens and waste places. Massachusetts to Indiana, south to Florida and Texas; most troublesome from Maryland southward. Introduced from eastern Asia as an ornamental. June–July.

Description.—Stems woody, climbing or trailing, hairy. Leaves opposite, simple, entire, short-petioled, ovate or oblong, hairy, mostly persistent throughout the winter, especially southward. Flowers in pairs on the summit of solitary axillary peduncles; sepals 5, fused with the ovary; petals 5, fused into a yellow or white tubular 2-lipped corolla; stamens 5, attached on the corolla; pistil solitary, 2–3-celled. Fruit a purplish-black berry with 2–3 seeds; seeds ovate to oblong, 2–3 mm. long, flattened, 3-ridged on the back, flat or concave on the inner face, dark brown or gray-brown.

Control.—When the honeysuckle occurs in fields, mow the vines, rake them into piles and burn. Follow by plowing and harrowing to remove as many as possible of the rootstocks from the soil. Plant with an intertilled crop, such as corn, and keep it clean. When this crop is removed, plow again and sow an annual smother crop the next spring. Close pasturing with sheep or goats will hold the honeysuckle in check.

VALERIANACEAE (Valerian Family)

454. *Valerianella olitoria* (L.) Poll. (*V. locusta* Betcke). Corn-salad, Lambs lettuce, Milk-grass, Fetticus. Fig. 104, d–f.

Winter annual or biennial; reproducing by seeds. Meadows, pastures, old fields and waste places. Infrequent in the eastern United States, also in the western states. Introduced from Europe as a salad plant. May–July.

Description.—Stems erect, with long internodes, simple, or with a few broadly forked branches, glabrous, somewhat succulent, 1–4 dm.

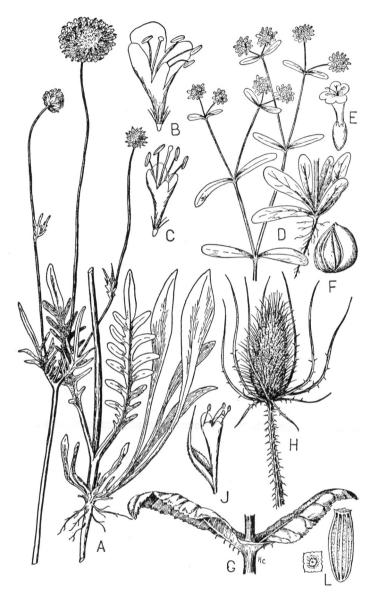

Fig. 104—**FIELD SCABIOUS**, *Knautia arvensis:* **A,** part of a plant showing general habit (× ⅓); **B,** marginal flower (× 2); **C,** flower from center of head (× 2).
CORN-SALAD, *Valerianella olitoria:* **D,** plant showing habit (× ⅓); **E,** flower (× 6); **F,** fruit (× 6).
TEASEL, *Dipsacus sylvestris:* **G,** part of a stem with a pair of opposite leaves (× ⅓); **H,** head of flowers (× ⅓); **J,** flower with bract (× 2); **L,** seed (achene) surface and end view (× 6).

Teasel Family

high. Leaves opposite, simple, obovate to oblanceolate, entire or the upper coarsely dentate or lobed, glabrous. Flowers perfect, in small slender-peduncled bracted cymes; calyx fused with the ovary; corolla short, tubular, 5-lobed, nearly regular, bluish; stamens 2 or 3; ovary 3-celled, only 1 cell producing a seed. Fruit 1-seeded but usually with 1 or 2 narrow empty cells, indehiscent, glabrous or slightly hairy, triangular in cross-section, about 3–4 mm. long.

Control.—Clean cultivation.

DIPSACACEAE (Teasel Family)

455. *Dipsacus sylvestris* Huds. Teasel, Card teasel, Venus-cup, Card thistle, Gipsy-combs. Fig. 104, g–l.

Biennial; reproducing by seeds. Pastures, old fields and waste places; on damp rich soils. Frequent or locally common in the northeastern United States; south to North Carolina, Tennessee, and Missouri; infrequent westward to the Pacific Coast. Introduced from Europe. July–September.

Description.—Stems erect, stiff, stout, branched above, angled, coarsely prickly, 1–3 m. high. Leaves opposite, the upper pairs fused by the bases into a cup, oblong-lanceolate, irregularly toothed, prickly along the edge; rosette leaves usually very spiny and prominently veined. Flowers perfect, in conical heads surrounded by an involucre of long ascending bracts exceeding the head; the middle flowers in the head opening first; receptacle chaffy, with long-tapering flexible straight-pointed bristles; calyx fused with the ovary, pappus none; corolla tubular, 4-cleft, nearly regular, pale lavender; stamens 4, separate, on the corolla; pistil with a slender style and 1-celled ovary developing into an achene. Achene linear, 3–4 mm. long, 4-angled, with longitudinal ridges, truncate at apex, pubescent, grayish-brown.

Control.—Clean cultivation (6). Badly infested fields should be plowed in autumn to destroy the teasel rosettes (7). Mow off the flowering stalks in waste places to prevent seed formation (11).

Dipsacus laciniatus L., with pinnatifid leaves, appears locally from Massachusetts to Michigan.

456. *Knautia arvensis* (L.) Duby (*Scabiosa arvensis* L.). Field scabious. Fig. 104, a–c.

Perennial; reproducing by seeds. Old fields, meadows, pastures and waste places; mostly on dry, gravelly or stony soils. Locally common

Newfoundland to North Dakota, south to New England and Pennsylvania. Introduced from Europe. June–July.

Description.—Stems erect, sparingly branched above, hairy, 4–12 dm. high. Leaves opposite, pinnately compound, pinnately lobed or pinnatifid. Flowers in short hemispherical heads; receptacle hairy but not chaffy; involucre of many narrow bracts; calyx with 8 awns; corolla funnelform, 4-lobed, pale lilac to pink; stamens 4, separate, on the corolla; pistil with a slender style, and a 1-celled ovary developing into an achene. Achene 4–5 mm. long, lanceolate, with truncate apex, a white knob at the base, covered with long bristly hairs, yellow or greenish.

Control.—Dig out scattered plants. Mow infested meadows early before seeds are formed, plow soon after hay crop is removed and keep the soil stirred up until autumn. Next spring plant an intertilled crop like corn or potatoes.

457. *Succisa australis* (Wulf.) Reichenb. Devils-bit, Southern scabious.

Perennial; reproducing by seeds. Moist, sandy, or gravelly meadows on somewhat limy soils. Widespread but very local in the northeastern United States. Introduced from Europe. July–September.

Description.—Stems erect, mostly 3-forked, hairy or smooth, 4–10 dm. high. Leaves opposite, simple, lanceolate, entire. Flowers in small subcylindric heads; involucre with many long bracts; receptacle chaffy; calyx fused with the ovary, minutely 5-toothed, corolla funnelform, 4-lobed, light blue; stamens 4, on the corolla; pistil with a long style and a 1-celled ovary developing into an achene. Achenes 3–4 mm. long, vase-shaped, with 8 prominent longitudinal ridges, apex truncate with a crenate border, olive-green or brown.

Control.—Dig out scattered plants. Mow infested meadows early before seeds are formed, plow soon after the hay crop is removed and keep the soil stirred up until autumn. Next spring plant an intertilled crop like corn or potatoes. Improve the drainage.

CUCURBITACEAE (Gourd Family)

458. *Echinocystis lobata* (Michx.) T. and G. Wild cucumber, Balsam-apple, Mock-apple.

Annual; reproducing by seeds. Fence-rows, thickets, and waste places; on damp rich soil, especially in river valleys. Widespread

New Brunswick to Saskatchewan, south to Florida and Texas. Native; also planted for arbors and escaping. August–September.

Description.—Stems climbing by 3-forked tendrils, angular, nearly glabrous. Leaves alternate, palmately lobed and veined, the 5 lobes sharply pointed. Flowers imperfect, the staminate numerous, in axillary panicles, the pistillate few or solitary in the same axils; corolla of sterile flower 6-parted; anthers 3, more or less fused, stigmas 2 or 3. Fruit fleshy, prickly, 2-celled, 4-seeded, opening at the apex, the inner part fibrous-netted; seeds about 2 cm. long, 0.8 cm. wide, spindle-shaped, flattened, rough, brown or black mottled with brown.

Control.—Clean cultivation when the seedlings appear, followed by hand hoeing or weeding of scattered plants appearing later.

459. *Sicyos angulatus* L. One-seeded bur cucumber, Nimble kate.

Annual; reproducing by seeds. Fence-rows, thickets, and waste places; on rich alluvial soils. Widespread in eastern North America, westward to the Mississippi Valley. Native. August–September.

Description.—Stems climbing by 3-forked tendrils, angular, clammy-hairy. Leaves alternate, palmately lobed and veined, the 5 lobes blunt or pointed. Flowers imperfect, the staminate numerous in axillary corymbs, the pistillate from the same axils, in a dense cluster on a long slender peduncle; petals 5, fused below into a shallow bell-shaped corolla; anthers fused; pistil with a 1-celled ovary, slender style and 3 stigmas. Fruit bristly barbed, 1-seeded, dry, indehiscent, broadly ovate, flattened; seeds about 12 mm. long, ovate, flattened, roughened with ridges and warts, hairy, black or dark brown.

Control.—The same as for 458.

Cucurbita foetidissima HBK., Wild gourd, is a weed in the north central states.

CAMPANULACEAE (Bluebell Family)

460. *Campanula rapunculoides* L. Bellflower, Creeping bellflower. Fig. 105, j–l.

Perennial; reproducing by seeds and roots. Meadows, old fields, and rich gravelly roadsides. Newfoundland to North Dakota, south to West Virginia and Missouri. Introduced from Eurasia as an ornamental. June–August.

Description.—Stems erect, slender, glabrous or nearly so, 5–10 dm. high, usually in clusters from thick fleshy roots; sap milky. Leaves alternate, simple, unevenly toothed, the lower long-petioled, cordate,

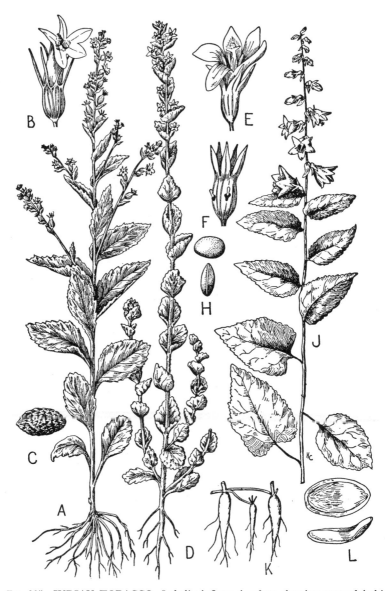

Fig. 105—**INDIAN TOBACCO,** *Lobelia inflata:* **A,** plant showing general habit ($\times \frac{1}{3}$); **B,** flower ($\times 2$); **C,** seed ($\times 18$).
VENUS LOOKING-GLASS, *Specularia perfoliata:* **D,** plant showing habit ($\times \frac{1}{3}$); **E,** flower ($\times 2$); **F,** capsule with calyx attached ($\times 2$); **H,** seed ($\times 18$).
BELLFLOWER, *Campanula rapunculoides:* **J,** part of plant showing flowers ($\times \frac{1}{3}$); **K,** roots ($\times \frac{1}{3}$); **L,** seed ($\times 18$).

the upper sessile or nearly so, ovate-lanceolate. Flowers perfect, in terminal racemes; calyx 5-cleft, fused with the ovary; corolla bell-shaped, 5-lobed, 2–3 cm. long, blue; stamens 5, separate, free from the corolla; pistil with 3-celled ovary and 3 styles. Fruit a capsule, 3-celled, many-seeded, opening by basal pores; seeds oval, about 1 mm. long, convex on one side, with a low obtuse ridge on the other side, finely striate, glossy, yellow-brown.

Control.—Small patches can be eradicated by digging the roots out. Extensive areas in fields should be plowed under and planted with a cultivated crop for a year, followed with an annual smother crop.

461. *Specularia perfoliata* (L.) A. DC. Venus looking-glass, Clasping bellwort. Fig. 105, d–h.

Annual; reproducing by seeds. Dry, gravelly or sandy fields, pastures, open woods and waste places. Widespread, Quebec to British Columbia, south to Florida and Texas. Native. June–August.

Description.—Stems erect, simple or branched at the base, hairy, 1–8 dm. high; sap milky. Leaves alternate, simple, nearly round or ovate, clasping, toothed, hairy. Flowers sessile, 1–3 in the leaf-axils; calyx 4–5-lobed, fused with the ovary; corolla rotate, 5-lobed, blue or purplish; stamens 5, separate; pistil with 3 stigmas and a 3-celled ovary. Fruit a short, ellipsoid, many-seeded capsule, opening below the middle by 3 valves; seeds 0.5 mm. long, ovoid, with convex sides, smooth, glossy, dark reddish-brown.

Control.—Clean cultivation (6). Increase the organic content of the soil by plowing under a green-manure crop.

LOBELIACEAE (Lobelia Family)

462. *Lobelia inflata* L. Indian tobacco, Lobelia, Eye-bright, Gagroot, Asthma-weed, Emetic-weed, Bladder-pod, Puke-weed. Fig. 105, a–c.

Annual or winter annual; reproducing by seeds. Meadows, pastures, grain fields, cultivated fields, and open woodlands. Eastern Canada southward to Georgia; less common westward to Saskatchewan, Nebraska and Kansas. Native. July–October.

Description.—Stems erect, densely branched, hairy, 3–8 dm. high; sap milky. Leaves alternate, simple, ovate to oblong, toothed, the upper reduced to leafy bracts. Flowers in loose racemes; calyx fused to the ovary, 5-cleft, becoming inflated in fruit; corolla tubular, irregular, 2-lipped, 5-lobed, pale blue; stamens 5, fused by the anthers; pistil with 1 style and a 2-celled ovary. Capsule 2-celled, many-seeded, opening at the apex; seeds about 0.6 mm. long, fusiform, marked with a network of ridges, slightly glossy, yellow-brown.

Control.—Clean cultivation followed by hand hoeing and weeding (6, 5). Hand mowing or pulling in pastures (4, 5). Mow meadows a second time so as to destroy the weeds before they mature seed (11).

Lobelia inflata is poisonous because of alkaloids which it contains. It is also used in certain medicines.

COMPOSITAE (Composite Family)

The flowers of the Compositae are small and grouped into heads on a receptacle surrounded by bracts forming an involucre. The bracts among the flowers on the surface of the receptacle are called chaff. A receptacle without bracts among the flowers is naked. The calyx, modified and reduced, usually in the form of bristles, awns or scales, is called the pappus. The corolla, composed of 5 fused petals, may be tubular as in the disk-flowers or strap-shaped (ligulate) in the ray-flowers. Stamens 5, attached on the corolla and mostly united by their anthers into a tube. Pistil 1-celled, 1-seeded, ripening into an achene which is often crowned by the pappus. Seeds exalbuminous, with a straight embryo.

463. *Achillea millefolium* L. Yarrow, Milfoil, Thousand-leaf, Bloodwort. Fig. 131, a–c.

Perennial; reproducing by seeds and rootstocks. Meadows, pastures, lawns, and waste places. Widespread throughout North America. Native; probably also introduced in some localities. June–September.

Description.—Stems erect, simple or sometimes branched above, hairy or nearly glabrous, 3–10 dm. high. Leaves alternate, sessile, bipinnately dissected into fine divisions, pubescent, 2–14 cm. long. Heads in dense flat-topped compound corymbose clusters, numerous, many-flowered, 3–5 mm. in diameter; involucral bracts imbricated, hairy, scarious-margined, 3–5 mm. long; receptacle flattish, chaffy; ray-flowers 5–10, white to reddish, pistillate, 1.5–2.5 mm. long; disk-flowers white, perfect. Achenes oblong, 2–3 mm. long, flattened, margined, finely striate longitudinally, white to gray; pappus wanting.

Control.—Clean cultivation will destroy yarrow. In meadows mow early before seeds are matured to avoid scattering them in the hay. In lawns where yarrow is kept cut close to the ground it will make a

464. *Ambrosia artemisiifolia* L. (*A. elatior* L.). Ragweed, Wild tansy, Hog-weed, Bitterweed, Mayweed, Hay-fever weed, Blackweed. Fig. 106, a–b.

Annual; reproducing by seeds. Cultivated fields, grain fields, old meadows, pastures and waste places. Widespread throughout North America; most common in the eastern and north central states. Native; probably introduced east of the Allegheny Mountains. August–September.

Description.—Stems erect, much branched, rough-pubescent, 3–15 dm. high. Leaves mostly alternate, bipinnately parted, variable, nearly glabrous, thin, 5–10 cm. long. Heads unisexual; staminate heads in single or panicled crowded racemes near the top of the plant, 5–20-flowered, about 2 mm. in diameter with the involucre cup-shaped, of 7–12 united bracts; pistillate heads axillary, sessile, 1–3 together, involucre obovoid, closed, becoming woody, resembling an achene and inclosing a single flower, and later a single achene; receptacle usually chaffy; disk-flowers greenish-white; ray-flowers wanting. Mature involucre 2–3 mm. long, top-shaped, with an apical beak surrounded by 5–10 tubercles, longitudinally ridged, straw-colored to brown; pappus wanting.

Control.—Clean cultivation of intertilled crops begun as soon as the seedlings appear and continued as late as possible. Mow waste places and pastures before the ragweed flowers open. Grain fields and meadows may be mowed again if ragweeds appear after the crop is removed.

The inhaled pollen of the ragweeds, 464–466, is responsible for many cases of autumnal hay fever.[1]

465. *Ambrosia psilostachya* DC. Perennial ragweed, Western ragweed. Fig. 106, f–g.

Perennial; reproducing by seeds and creeping roots. Fields, meadows, and waste places. Widespread in the Mississippi Valley and the Great Plains states and in the prairie provinces of Canada; locally established both east and west. Native. June–September.

Description.—Stems erect, branched, hispid, 6–18 dm. high; often in dense patches from creeping roots. Leaves mostly alternate, 1-

[1] Gahn, 1933.

FIG. 106—**RAGWEED**, *Ambrosia artemisiifolia*: **A,** plant showing general habit (× ¼); **B,** seed (achene), (× 3).
GREAT RAGWEED, *Ambrosia trifida*: **C,** upper portion of a plant showing flowers (× ¼); **D,** fruiting branch (× ¼); **E,** seed (achene), (× 1½).
PERENNIAL RAGWEED, *Ambrosia psilostachya*: **F,** a single leaf (× ⅓); **G,** portion of creeping root system with buds (× ⅙).

pinnatifid, lobes acute, rough. Similar to 464, but mature involucres pubescent and often less tubercled.

Control.—The same as for 464.

466. *Ambrosia trifida* L. Great ragweed, Kinghead, Giant ragweed, Crown-weed, Wild hemp, Horse-weed, Bitterweed, Tall ambrosia. Fig. 106, c–e.

Annual; reproducing by seeds. Fields and waste places; mostly on rich or alluvial soils. Quebec to British Columbia, south to Florida and Arizona; most common in the Middle West. Native. July–September.

Description.—Stems erect, branched above, stout, rough-hairy, very rank and coarse, 1–6 m. high. Leaves opposite, petioles margined; lower leaves deeply 3-lobed, rough-hairy; upper leaves variable, 3-lobed or simple, ovate-lanceolate, dentate or entire. Heads unisexual, similar to 464 except the staminate racemes are longer, up to 3 dm. long; pistillate heads larger, 2–4 mm. in diameter; receptacle naked. Mature involucre with achene 6–12 mm. long, plus a beak 2–3 mm. long, with 5 short stout tubercles, each terminating a stout longitudinal ridge, mottled, brown or gray; pappus wanting.

Control.—The same as for 464.

Ambrosia bidentata Michx., Lance-leaved ragweed, is listed [2] (1954) as a weed in the north central states. Other Composites included in the list, and not otherwise mentioned in "Weeds" are *Aster pilosus* Willd., White heath aster; *Coreopsis tinctoria* Nutt.; *Erigeron divaricatus* Michx., Dwarf fleabane; *Eupatorium serotinum* Michx., Thoroughwort; *Gutierrezia dracunculoides* (DC.) Blake, Broomweed; *Helianthus maximiliani* Schrad., Sunflower; *Rudbeckia laciniata* L., Tall cone-flower; *Senecio aureus* L., Golden ragwort; *Silphium perfoliatum* L., Cup-plant; *Solidago nemoralis* Ait., Gray goldenrod; *Vernonia altissima* Nutt., Tall ironweed; and *Vernonia baldwinii* Torr., Western ironweed.

467. *Anaphalis margaritacea* (L.) C. B. Clarke. Pearly everlasting, Silver leaf, Moonshine, Cotton-weed, Silver button, Poverty-weed. Fig. 121, a–c.

Perennial; reproducing by seeds and short rootstocks. Old pastures and meadows; especially on dry, gravelly, or stony soils. Common throughout the northern United States and Canada, south to Colorado and California. Native; also native to Asia. August–September.

---

[2] Weeds of the North Central States.

Description.—Stems erect, simple or corymbose at the summit, very leafy, white-woolly, 2–9 dm. high. Leaves alternate, simple, clasping, entire, linear-lanceolate, tomentose on both surfaces. Heads numerous, in terminal corymb, many-flowered and mostly dioecious; involucral bracts very numerous, imbricated, papery-white, finely striate, spreading at maturity; receptacle convex, naked; disk-flowers whitish; corolla of staminate flowers filiform, of pistillate flowers slender, tubular, 5-pointed; ray-flowers wanting. Achenes cylindrical, less than 1 mm. long, yellowish-brown; pappus capillary.

Control.—Plow infested areas and plant a cultivated crop before reseeding. In pastures cut close before seeds are formed (11).

468. *Antennaria plantaginifolia* (L.) Hook. Ladies tobacco, Early everlasting, Plantain-leaved everlasting, White plantain, Pussy-toes, Mouse-ears. Fig. 121, h.

Perennial; reproducing by seeds and rootstocks. Dry sterile pastures, old fields, and meadows on stony hilly soils. Widespread in the eastern United States, westward to Minnesota. Native. April–May.

Description.—Stems decumbent, simple, slender, white-woolly, 1–3 dm. high, leaves mostly in a basal rosette. Stem-leaves alternate, simple, sessile, entire, lanceolate, white-woolly; basal leaves simple, broadly obovate to spatulate, 3-nerved, tapering into long petioles, entire, tomentose. Heads in small corymbose clusters, dioecious, many-flowered, about 6 mm. in diameter; involucral bracts imbricated, scarious, white or sometimes purplish, 5–8 mm. high; receptacle convex, naked; disk-flowers whitish; corolla of staminate flowers filiform, of pistillate flowers slender, tubular, 5-toothed; ray-flowers wanting. Achenes cylindrical or somewhat flattened; pappus capillary.

Control.—The same as for 467.

Several other species of Antennaria sometimes act as weeds in old meadows and pastures on infertile soils.

469. *Anthemis arvensis* L. var. *agrestis*. (Wallr.) DC. Corn chamomile, Field chamomile. Fig. 107, c–d.

Biennial; reproducing by seeds and rooting stems. Cultivated fields, new seedings, old fields and waste places. Widespread throughout the northern United States, southern Canada, and south to Georgia. Introduced from Europe. May–August.

Description.—Stems erect or partly decumbent, with spreading branches, pubescent, 2–5 dm. high; lower branches frequently root-

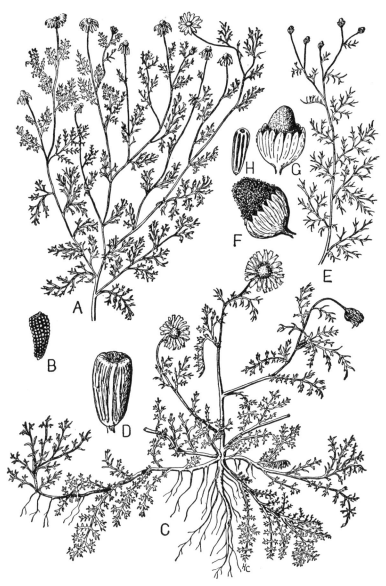

Fig. 107—**DOG FENNEL**, *Anthemis cotula*: **A**, branch showing general habit (× ½); **B**, seed (achene), (× 10).
CORN CHAMOMILE, *Anthemis arvensis* var. *agrestis*: **C**, plant showing general habit (× ⅓); **D**, seed (achene), (× 10).
PINEAPPLE-WEED, *Matricaria matricarioides*: **E**, branch showing general habit (× ⅓); **F**, head of flowers (× 2); **G**, disk of head after seeds have fallen (× 2); **H**, seed (achene), (× 10).

ing at the nodes. Leaves alternate, 1–2-pinnately divided, sessile, grayish-green, pubescent, scented. Heads solitary, terminal, numerous, many-flowered, 12–30 mm. in diameter; involucre hemispherical, of many small imbricated scarious bracts; receptacle conical, chaffy throughout; ray-flowers 15–20, white, pistillate, spreading, about 1 cm. long; disk-flowers numerous, yellow, perfect. Achenes about 2 mm. long, ridged, truncate at apex, glabrous, usually straw-colored; pappus forming a minute border.

Control.—Clean cultivation of intertilled crops followed by hand weeding or hoeing (6). Mow waste places and roadsides before seeds are formed (11). Pull or hoe scattered plants in new seedings of clover and alfalfa.

470. *Anthemis cotula* L. Dog fennel, Mayweed, Stink-weed, Dogs chamomile, Dill-weed, Stinking daisy, Hogs fennel, Fetid chamomile. Fig. 107, a–b.

Annual or winter annual; reproducing by seeds. Cultivated fields, gardens, farmyards and waste places; mostly on rich gravelly soils. Widespread throughout North America. Introduced from Europe. June–October.

Description.—Stems erect, branching, pubescent, 2–6 dm. high. Leaves alternate, sessile, 3-pinnately dissected, yellowish-green, pubescent, strong scented. Heads solitary, terminal, numerous, many-flowered, 12–30 mm. in diameter; involucre hemispherical, of many small imbricated scarious bracts; receptacle conical, chaffy only near the middle; ray-flowers 15–20, white, spreading, neutral, about 1 cm. long; disk-flowers numerous, yellow, perfect. Achenes less than 2 mm. long, with 10 tubercle-bearing ridges, rough, light brown; pappus wanting.

Control.—Clean cultivation of intertilled crops followed by hand weeding or hoeing (6). Mow waste places and roadsides before seeds are formed (11). Harrow grain fields while the dog fennel seedlings are small (8).

471. *Arctium lappa* L. Great burdock, Beggars-buttons, Cocklebutton. Fig. 108, e–f.

Biennial; reproducing by seeds. Roadsides, fence-rows, and waste places; especially about neglected farmyards; mostly in rich soils. Widespread and locally common in eastern Canada, northeastern and north central states; local on the Pacific Coast. Introduced from Europe. July–October.

Fig. 108—**SMALLER BURDOCK,** *Arctium minus:* **A,** branch and basal leaf (× ¼); **B,** head (× 1); **C,** a single flower (× 2); **D,** achene (× 3).
GREAT BURDOCK, *Arctium lappa:* **E,** head (× 1); **F,** achene (× 3).

Description.—Stems erect, branching, rough-hairy, stout, ridged, 1–3 m. high; from a large tap-root bearing a rosette of large leaves. Leaves alternate, simple, pinnately veined, petioles solid; margins slightly ruffled, entire to dentate, round-ovate, mostly cordate, rough, tomentose and paler beneath; lower leaves often 3 dm. in length. Heads in corymb-like axillary clusters, each head 3–5 cm. in diameter, many-flowered; involucre nearly globular, bur of imbricated coriaceous bracts with hooked tips, the outer bracts about as long as the inner ones; receptacle chaffy with bristles; disk-flowers pinkish-purple, perfect, 10–13 mm. long; ray-flowers wanting. Achenes oblong, 4–6 mm. long, 3–5-angled, straight or curved, slightly ridged, glabrous, mottled brown, pappus of bristly scales, deciduous.

Control.—Plowing, disking, or in inaccessible and waste places hoeing, as soon as the seedlings appear will destroy them. Rosettes in the second year stage should be cut with a spud or spade deep below the soil surface; otherwise the tap-roots will send up new crowns. Plants with flower-stalks should be mowed before seeds are formed.

472. *Arctium minus* (Hill) Bernh. Smaller burdock, Clotbur, Cuckoo-button, Cockle-button. Fig. 108, a–d.

Biennial; reproducing by seeds. Waste places and about neglected farmyards; in rich soils. Throughout North America. Introduced from Europe. July–October.

Description.—Plants similar to 471 but mostly smaller and with lower petioles usually hollow. Heads in raceme-like axillary clusters; each head 1.5–3 cm. in diameter; involucre with outer bracts successively shorter than the inner; flowers and achenes similar to 471.

Control.—The same as for 471.

473. *Artemisia absinthium* L. Wormwood, Absinthe, Madderwort, Warmot.

Perennial; reproducing by seeds. Pastures, roadsides and waste places; on dry soils. Eastern Canada and northeastern United States; rare elsewhere. Introduced from Europe; frequently planted as a garden herb.

Description.—Stems erect, in clumps, branched, somewhat woody at base, silky-hoary, 5–10 dm. high. Leaves alternate, 2–3-pinnately parted, the divisions mostly entire, the lower petioled, the upper sessile, gray-woolly. Heads less than 5 mm. across, nodding, sessile, in short axillary spikes or clusters in a large loose panicle; involucral bracts few, imbricated, margins dry and scarious; receptacle flat, hairy;

disk-flowers yellowish or purplish, perfect or the outer ones pistillate, about 1 mm. long. Achenes obovoid, about 1 mm. long, somewhat flattened, finely striate, light gray-brown; pappus wanting.

Control.—Dig out scattered plants or plow badly infested land (17). In pastures and waste land mow close to the ground several times in a summer (11).

474. *Artemisia annua* L. Annual wormwood. Fig. 109, a–c.

Annual; reproducing by seeds. Waste places, gardens and cultivated fields; on dry, sandy or gravelly soils. Locally common in the north central states and Ontario; infrequent east and south. Introduced from Eurasia. July–September.

Description.—Stems erect, much branched near the top, glabrous, ridged, pale green, 6–12 dm. high. Leaves alternate, 2–3-pinnately divided, dentate, nearly glabrous, 5–14 cm. long, the upper sessile, the lower petioled, sweet-scented. Heads less than 5 mm. in diameter, in open spreading panicles, composed of disk-flowers only; involucral bracts few, imbricated, margins dry and scarious; receptacle flat, naked; disk-flowers perfect or sometimes the outer ones pistillate, greenish, about 1 mm. long. Achenes oval, similar to 475, light yellow; pappus wanting.

Control.—Clean cultivation (6). Hand pulling (4). Close mowing in waste places (11).

475. *Artemisia biennis* Willd. Biennial wormwood, False tansy, Bitterweed. Fig. 109, d–f.

Biennial or annual; reproducing by seeds. Old fields, roadsides, waste places, neglected lots; mostly on dry gravelly soil. Native in north central and northwestern states; introduced eastward and on the Pacific Coast. August–September.

Description.—Stems erect, simple or with short nearly erect branches, glabrous, 6–15 dm. high. Leaves alternate, 1–2-pinnately parted, coarsely dentate, the lower leaves petioled and the upper sessile, dark green, glabrous. Heads less than 5 mm. in diameter, composed of disk-flowers only, in short axillary spike-like clusters, crowded in a slender leafy panicle, erect, sessile; involucral bracts few, imbricated, margins dry and scarious; receptacle flat, naked; flowers perfect or the outer ones pistillate, yellowish or purplish, about 1 mm. long. Achenes obovoid, 1–1.5 mm. long, with a small summit, somewhat flattened, 3–4-angled, brown, wrinkled, glabrous; pappus wanting.

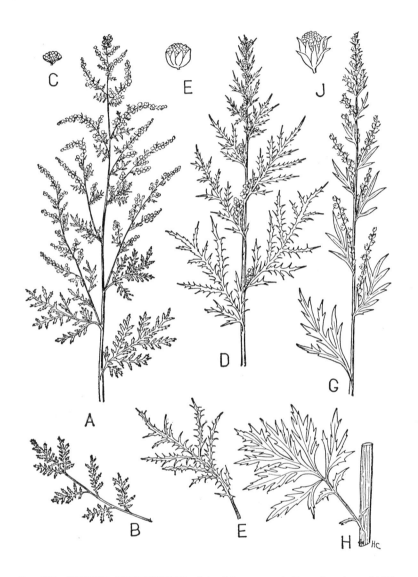

Fig. 109—**ANNUAL WORMWOOD**, *Artemisia annua:* **A,** flowering shoot (× ⅓); **B,** lower stem-leaf (× ⅓); **C,** head of flowers (× 2).
BIENNIAL WORMWOOD, *Artemisia biennis:* **D,** flowering shoot (× ⅓); **E,** lower leaf (× ⅓); **F,** head (× 2).
MUGWORT, *Artemisia vulgaris:* **G,** flowering shoot (× ⅓); **H,** lower leaf (× ⅓); **J,** head (× 2).

Control.—The same as for 474.

476. *Artemisia tridentata* Nutt. Sage-brush.

Shrubby perennial; reproducing by seeds. High plains and slopes of the foothills and lower mountains. Widespread in arid regions from the western edge of the Great Plains westward to British Columbia and California. Native. July–September.

Description.—Stems woody, much branched, 1–3 m. high, usually covered with silvery-gray hair. Leaves alternate, simple, wedge-shaped, with 3–5 blunt teeth at the apex, 2–4 cm. long, silvery-gray, aromatic-scented. Heads sessile, in crowded spikes in the leaf-axils and terminating the branches; terminal clusters frequently forming large panicles; heads composed of perfect, fertile, tubular flowers; receptacle flat, hairy. Achenes obovoid, about 1.5 mm. long, hairy, dull, light yellow; pappus wanting.

Control.—Grub out the bushes and burn them. Plowing the infested areas will destroy the roots.

This is the commonest of the several species of "sage-brush" so general over extensive arid sections of western North America.

477. *Artemisia vulgaris* L. Mugwort, Wormwood. Fig. 109, g–j.

Perennial; reproducing by seeds and short rootstocks. Waste places, fields and pastures; especially on limy soils. Widespread across the northern and western United States. Introduced eastward; native in the western states where it is variable and represented by several forms which are sometimes designated as varieties or species. July–September.

Description.—Stems erect, branching, ridged, often reddish, 3–10 dm. high. Leaves alternate, deeply pinnatifid and cleft, the lobes usually entire, acute, green-glabrous above, white-woolly beneath, the upper sessile, the lower petioled. Heads numerous, in terminal or axillary spike-like clusters; involucre bell-shaped, with several lanceolate scarious-margined bracts; receptacle naked, flat; disk-flowers only, small, greenish-yellow. Achenes oblong, about 15 mm. long, with narrow base, ridged, with a crown of minute bristles at the apex, brown; pappus wanting.

Control.—The same as for 473.

478. *Aster simplex* Willd. (*A. paniculatus* Lam.). White field aster, Frost flower.

Perennial; reproducing by seeds and rootstocks. Neglected fields, pastures, and cultivated fields on recently plowed land; mostly on heavier clay soils. Native from Quebec to Saskatchewan, southward to Virginia and Kansas. This variable species is most abundant in the northeastern states. August–September.

Description.—Stems erect, much branched, nearly glabrous, 5–20 dm. high. Leaves alternate, simple, sessile, serrate or entire, oblong to oblanceolate, tapering to a narrow base, glabrous. Heads paniculate, on erect or ascending branches, each head about 2 cm. broad, many-flowered, on leafy peduncles; involucral bracts in 3–5 series, linear-attenuate, green-tipped, unequal, 5 mm. long or less; receptacle flat, naked; ray-flowers 14–50 in a single series, white or bluish, pistillate, 6–8 mm. long; disk-flowers fewer than the ray-flowers, perfect. Achenes linear, about 2 mm. long, flattened, hairy, yellow-brown; pappus capillary, white.

Control.—Mow pastures and waste places before the asters blossom. Infested fields that have been plowed should be followed by an intertilled crop for two years before sowing grain or grass.

Several other native species of aster frequently overrun neglected fields and pastures, especially in the eastern United States. Their general habits are similar to 478 but they are not as persistent when the land is plowed and put under cultivation for a year.

479. *Bellis perennis* L. English daisy, Lawn daisy.

Perennial; reproducing by seeds. Lawns, pastures and waste places. Chiefly in the Pacific Northwest; local in the northeastern states and southward; naturalized to some extent in Newfoundland. Introduced from Europe as an ornamental and also with grass seed. April–June; September–November.

Description.—Stems simple, leafless, 1–2 dm. high, terminated by a solitary head of flowers. Leaves obovate or spatulate, entire or wavy-margined, forming a basal tuft or rosette. Heads solitary, 3–5 cm. in diameter, many-flowered; involucral bracts herbaceous, in about 2 rows; receptacle conical, naked; ray-flowers numerous, pistillate, white, pink or rose; disk-flowers yellow. Achene linear, about 2 mm. long, finely striate, yellow-brown; pappus bristly, falling with the corolla.

Control.—In lawns English daisy can be removed by digging the rosettes when the ground is loose (17).

480. *Bidens bipinnata* L. Spanish needles, Cuckold. Fig. 110, f–g.

Annual; reproducing by seeds. Cultivated fields, gardens, and waste places. Massachusetts to Florida, westward to Kansas and Mexico;

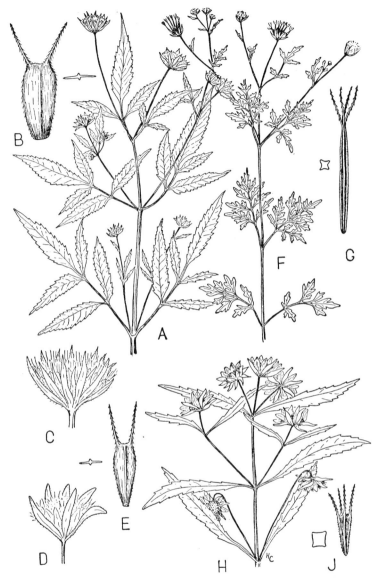

FIG. 110—**TALL BEGGAR-TICKS,** *Bidens vulgata:* **A,** branch showing flowers (× ⅓); **B,** seed (achene), and cross-section (× 2); **C,** head of flowers (× 1).
BEGGAR-TICKS, *Bidens frondosa:* **D,** head of flowers (× 1); **E,** achene (× 2).
SPANISH NEEDLES, *Bidens bipinnata:* **F,** branch showing flower-heads (× ⅓); **G,** seed (achene), (× 2).
STICK-TIGHTS, *Bidens cernua:* **H,** branch with flower-heads (× ⅓); **J,** achene (× 2).

most troublesome in the South. Native, or introduced northward. July–September.

Description.—Stems erect, branched, glabrous, slender, somewhat ridged and angled. Leaves opposite, 1–3-pinnately parted, petioled, leaflets ovate-lanceolate, dentate. Heads axillary, numerous, on long slender peduncles, few-flowered; involucre of 2 series of bracts, the outer leaf-like; receptacle flat, chaffy; ray-flowers very short, pale yellow, neutral, sometimes wanting; disk-flowers yellow, perfect. Achenes 1–2 cm. long, linear, spindle-shaped, 4-angled, nearly smooth or hairy at the apex; pappus of 3–4 downwardly barbed awns.

Control.—Clean cultivation continued until late in the season (6). Pull scattered plants. Mow the weeds in waste places whenever the flowers appear (11).

481. *Bidens cernua* L. Stick-tights, Smaller bur-marigold, Pitchforks. Fig. 110, h–j.

Annual; reproducing by seeds. Wet meadows, pastures, and along ditches. Widespread throughout the United States and southern Canada. Native. July–October.

Description.—Stems erect, sparingly branched, nearly glabrous, slender. Leaves opposite, simple, sessile, chiefly connate, coarsely serrate, linear-lanceolate. Heads axillary or in cyme-like clusters, many-flowered, and on short erect peduncles which nod later; involucre with 2 series of bracts, the outer ones leaf-like and longer than the head; receptacle flat, chaffy; ray-flowers large, about twice the length of bracts, yellow, neutral, sometimes wanting; disk flowers yellow, perfect. Achenes wedge-shaped, 4–8 mm. long, 4-angled, usually roughened by scattered barbed tubercles, brown; pappus of 3–4 downwardly barbed awns.

Control.—Improve the drainage. Mow or hoe the plants along ditches and in pastures before seeds are produced.

482. *Bidens frondosa* L. Beggar-ticks, Stick-tights, Devils bootjack, Bur-marigold, Pitchfork-weed. Fig. 110, d–e.

Annual; reproducing by seeds. Pastures, roadsides, gardens, cultivated fields, and waste places; especially on moist soil. Widespread throughout the United States and southern Canada; most common in the eastern and north central states. Native. August–September.

Description.—Stems erect, with spreading branches, glabrous. Leaves opposite, compound, with 3 or 5 divisions; leaflets lanceolate, serrate, glabrous, pinnately veined. Heads similar to 481, 1.5 cm.

long or less; involucre of 2 series of bracts, outer involucral bracts 4–8; receptacle flat, chaffy; ray-flowers orange-yellow, neutral, sometimes wanting; disk-flowers orange, 5-toothed, perfect. Achenes 6–10 mm. long, wedge-shaped, flat, with a strong nerve on each side, rough or nearly smooth, dull, dark brown to nearly black; pappus of 2 downwardly barbed awns.

Control.—Clean cultivation early in the season followed by hand pulling or hoeing later. Mow or pull plants in waste places along roadsides and ditches before seeds are formed. Improving the drainage on moist land usually helps to control this weed.

483. *Bidens polylepis* Blake (*B. involucrata* (Nutt.) Britt.). Beggarticks. Fig. 111.

Annual or biennial; reproducing by seeds. Wet places, shores, old meadows, fields, and swampy pastures. Indiana to Colorado, south to Tennessee and Texas; infrequently naturalized eastward. Native. August–October.

Description.—Stems 3–10 dm. high, branched. Leaves pinnately-compound, the leaflets linear-lanceolate, incised to pinnatifid. Heads 3–5 cm. in diameter; rays 10–12, large, golden yellow; outer involucral bracts 12–25. Achenes flat, obovate, ciliate with 2 teeth.

Control.—The same as for 482.

484. *Bidens vulgata* Greene. Tall beggar-ticks, Stick-tights. Fig. 110, a–c.

Annual; reproducing by seeds. Cultivated fields, neglected farmyards, roadsides, and waste places; on rich heavy soils. Widespread throughout the northern United States and southern Canada; south to North Carolina and California; most common in the north central states. Native. August–September.

Description.—Stems erect, branching, glabrous. Leaves opposite, pinnately 3–5-divided, slender-petioled, nearly glabrous; leaflets lanceolate, serrate. Heads and involucre similar to 481; outer involucral bracts 10–16, unequal; receptacle flat, chaffy; ray-flowers pale yellow, neutral, sometimes wanting; disk-flowers straw-colored. Achenes similar to 481 but somewhat broader and often larger.

Control.—The same as for 482.

Bidens connata Muhl., Swamp beggar-ticks, and *Bidens comosa* Wiegand, Leafy-bracted tickseed, are sometimes troublesome along ditches and in wet pastures and meadows in the northeastern and middle western states. Like *Bidens cernua,* 481, they are native an-

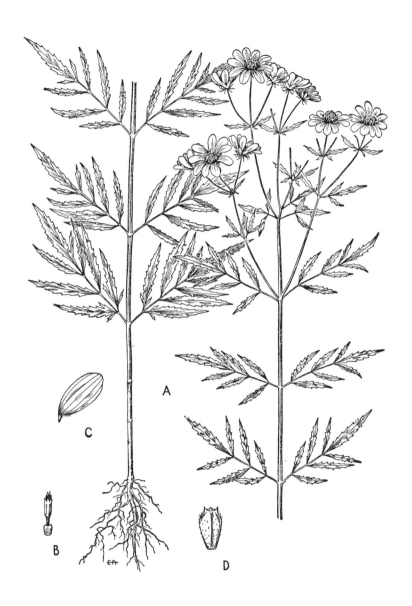

FIG. 111—**BEGGAR-TICKS,** *Bidens polylepis:* **A,** upper and lower parts of plant, showing habit (× ⅓); **B,** disk-flower (× 2); **C,** ray-flower (× 1); **D,** achene (× 2).

nuals with simple leaves but with ray-flowers wanting or very small. *Bidens connata* has orange 5-toothed corollas, exserted stamens, and achenes 4–6 mm. long. *B. comosa* has straw-colored 4-toothed corollas, included stamens, and achenes 8–11 mm. long. Control the same as for 481.

485. *Carduus acanthoides* L. Plumeless thistle.

Annual or biennial; reproducing by seeds. Pastures, meadows, and waste places. Local in the northeastern and north central states. Introduced from Europe. June–September.

Description.—Stems erect, branched, about 1 m. high, winged and spiny. Leaves alternate, pinnatifid, decurrent and spiny. Heads 1 or several at the ends of spiny winged branches, 2–4 cm. in diameter, with many disk-flowers; involucre hemispherical, 1–2.5 cm. in diameter, bracts numerous, imbricated, linear, the outer somewhat herbaceous and spreading; receptacle densely bristly; disk-flowers perfect, rose-purple; corolla about 18 mm. long. Achenes oblong, 3–3.5 mm. long, flattened, straight or curved, with an apical tubercle, finely striate; pappus capillary, not plumose, about 1 cm. long, deciduous.

Control.—Clean cultivation (6). Cut the rosettes below the crown with a spud or hoe (12). Mow badly infested pastures as soon as the first blossoms appear (11).

486. *Carduus nutans* L. Musk thistle, Plumeless thistle, Nodding thistle.

Biennial; reproducing by seeds. Pastures, meadows and waste places; mostly on dry or gravelly soils. Northeastern and north central states; south to Missouri; also in eastern Canada. Introduced from Europe. June–October.

Description.—Stems erect, branched above, 1–1.5 m. high, pubescent, spiny-winged. Leaves alternate, pinnatifid, decurrent, very spiny, much coarser than 485. Heads solitary on the ends of long nearly naked peduncles, 3–5 cm. in diameter, nodding, containing many disk-flowers; involucre 3–4 cm. in diameter, bracts numerous, imbricated, lanceolate, mostly spiny-tipped; receptacle densely bristly; disk-flowers perfect, purple or rarely white, about 2 cm. long. Achenes oblong, about 4 mm. long; apex truncate, tipped with a whitish tubercle in the center surrounded by a crown; striated, glossy, yellow-brown or olive-brown; pappus capillary, not plumose, about 2 cm. long, deciduous.

Control.—The same as for 485.

Carduus crispus L. Welted thistle, a biennial introduced from Europe, is occasionally found in meadows and pastures in the eastern and north central states. It is very similar to 485 but has several small heads, which are sessile or nearly so, crowded in clusters at the ends of the spiny-winged branches; involucre ovoid, 1–1.3 cm. in diameter, and with linear bracts, the outer rigid and less spreading.

487. *Centaurea calcitrapa* L. Caltrops, Purple star-thistle, Maize thorn.

Annual or biennial; reproducing by seeds. Grain fields, pastures, and waste places. Widespread from the Atlantic states to the Pacific Coast, but very local. Introduced from Europe. June–September.

Description.—Stems erect or spreading, 3–6 dm. high, much branched, the branches becoming hard and stiff, somewhat hairy. Leaves alternate, slightly hairy, the lower pinnately divided into narrow irregularly toothed lobes, with a slightly winged petiole, the upper narrow, sessile, somewhat clasping, with minute spines on the narrow lobes. Heads terminal or axillary, sessile or on short peduncles, about 1–2 cm. in diameter, many-flowered; involucral bracts tipped with stout, sharp, light yellow, spreading spines, 1–2 cm. long and with 1–3 pairs of small spines at their base; receptacle bristly, flat; flowers all with tubular purple corollas, the outer row sterile, the others perfect and fertile. Achenes obovoid, 2–3 mm. long; apex truncate, with a small central tubercle; base with a lateral scar; smooth, glossy, pale yellow mottled with brown; pappus wanting.

Control.—Use a system of rotation which includes an intertilled crop to permit clean cultivation. Mow scattered plants in pastures and waste places before seeds are formed.

488. *Centaurea cyanus* L. Bachelors-button, Corn-flower, Bluebottle, French pink.

Annual or winter annual; reproducing by seeds. Grain fields, new meadows, and waste places. Widespread, escaped from cultivation in many localities; most abundant in the Pacific Northwest. Introduced from Europe as an ornamental. June–September.

Description.—Stems erect, with many slender branches, 4–10 dm. high, softly hairy, gray-green. Leaves alternate, softly hairy, the lower narrow, often toothed or pinnatifid, the upper linear and entire. Heads solitary on long slender peduncles, blue, violet, pink or white; involucre ovoid, with many imbricated bracts in about 4 unequal series; bracts greenish-straw-color with margin darker or fringed with teeth; receptacle bristly, flat; flowers all tubular; the central ones small, slender, perfect, and fertile; the outer row with much longer, funnel-shaped, spreading, deeply notched corollas, neutral. Achenes oblong, 3–4 mm. long, similar to 487, pappus bristly, orange-brown, 2–3 mm. long.

Control.—In waste places and meadows pull or mow before seeds are formed. In badly infested fields include an intertilled crop in the rotation, and do not sow grain after grain (20).

489. *Centaurea diffusa* Lam.

Annual or biennial; reproducing by seeds. Fields, roadsides, waste places; rapidly spreading. Local in Massachusetts, Michigan, Iowa, Washington; probably elsewhere. Introduced from Europe. July–September.

Description.—Stems diffusely branched, 1–6 dm. high, thinly grayish-pubescent, angled, rigid, many-headed, paniculate-corymbose. Basal leaves bipinnatifid, cauline small, pinnatifid and deciduous below, mostly entire above, grayish-pubescent. Flowers in numerous narrow, sessile heads on short branchlets; involucre ellipsoid-cylindric, 8–10 mm. high, outer bracts pectinate or spinose-ciliate and tipped with a slender spine 1.5–4 mm. long; flowers few, roseate to creamy, marginal ones not enlarged; pappus deciduous.

Control.—The same as for 491.

490. *Centaurea iberica* Trev. Iberian star-thistle.

Biennial; reproducing by seeds. Fields and waste places. Local in California.[3] Noxious in Nevada and Oregon (1954). Introduced from Asia. June–July.

Description.—Stems erect, with spreading branches from the base, 5–10 dm. high, somewhat hairy. Leaves alternate, often with glandular dots below, hairy; the basal leaves pinnatifid or lobed; the cauline leaves sessile with few lobes, or the upper narrow and nearly entire. Heads solitary on the ends of branches, many-flowered, 1–1.5 cm. in diameter; involucre with imbricated bracts terminating in rigid straw-colored spines, each 1–2 cm. long and bearing 1–3 basal prickles; receptacle bristly-hairy; flowers all tubular, purple, pink or rarely white. Achenes 3–4 mm. long, apex truncate, base with a lateral scar, straw-colored mottled with gray or dark brown; pappus a crown of about 3 rows of bristles, shorter than the achene.

Control.—The same as for 493.

[3] Ball, Robbins and Bellue, 1933.

491. *Centaurea jacea* L. Star-thistle, Brown knapweed, Rayed knapweed, Brown centaury, Horse-knobs, Hard-heads. Fig. 112, h–m.

Perennial; reproducing by seeds. Dry gravelly fields, meadows, pastures, and roadsides. Locally common in the northeastern states; and more recently appearing in the north central states and the Pacific Northwest. Introduced from Europe. July–August.

Description.—Stems erect, branching near the top, rough-pubescent, 3–8 dm. high. Leaves alternate, simple, slightly dentate or lobed, lanceolate or oblanceolate, rough-pubescent, the upper sessile, the lower petioled. Heads solitary, terminal, about 3 cm. in diameter, many-flowered; involucral bracts numerous, imbricated, apex pectinate, brown, enlarged; receptacle bristly, flat; flowers all tubular, purple or pink, perfect, the marginal ones enlarged, neutral. Achenes 2–3 mm. long, obovoid, flattened, scar of attachment in a notch to one side of the round base, apex truncate with a small tubercle in the middle, cream-color to brown, striated, with scattered hairs; pappus wanting.

Control.—Plow badly infested fields and plant a clean cultivated crop before reseeding. In pastures and waste places mow the weeds close to the ground or spud them out before seeds are formed.

Several varieties of this species occur in various parts of the United States. These differ primarily in the shape of the involucral bracts, and are sometimes designated as distinct species.

492. *Centaurea maculosa* Lam. Spotted knapweed. Fig. 112, a–g.

Biennial; reproducing by seeds. Dry, sterile, gravelly or sandy pastures and old fields and roadsides. Northeastern states, north central states, and the Pacific Northwest. Introduced from Europe, spreading in recent years. August–September.

Description.—Stems erect or ascending, with slender branches, rough-pubescent, 3–10 dm. high. Leaves alternate, pinnatifid with narrow divisions, rough-pubescent, the upper often linear. Heads terminal and axillary, numerous and clustered, 1–2 cm. in diameter, many-flowered; involucral bracts numerous, imbricated, the apex pectinate, brown, not enlarged; receptacle bristly, flat; flowers all tubular, pink to purple, perfect, the marginal ones enlarged, neutral. Achenes similar to 491; pappus bristly, scanty, 1–2 mm. long, persistent.

Control.—The same as for 491.

493. *Centaurea melitensis* L. Napa thistle, Tocalote, Rayless centaury. Fig. 112, s–w.

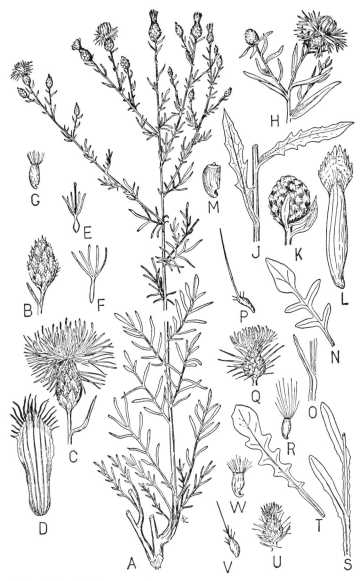

Fig. 112—**SPOTTED KNAPWEED,** *Centaurea maculosa:* **A,** plant showing habit (× ⅓); **B,** flower-bud (× 1); **C,** head (× 1); **D,** involucral bract (× 3); **E,** fertile flower (× 1); **F,** sterile flower (× 1); **G,** seed (achene), (× 3).
STAR-THISTLE, *Centaurea jacea:* **H,** branch with head (× ⅓); **J,** leaves (× ⅓); **K,** head of flower-buds (× 1); **L,** involucral bract (× 3); **M,** achene (× 3).
BARNABYS THISTLE, *Centaurea solstitialis:* **N,** basal leaf (× ⅓); **O,** stem leaf (× ⅓); **P,** involucral bract (× 1); **Q,** head (× ½); **R,** achene (× 3).
NAPA THISTLE, *Centaurea melitensis:* **S,** stem leaves (× ⅓); **T,** basal leaf (× ⅓); **U,** head (× ⅓); **V,** bract (× 1); **W,** achene (× 3).

Annual; reproducing by seeds. Grain fields, new meadows, pastures, and waste places. Locally common on the Pacific Coast; infrequent in the northeastern and north central states. Introduced from Europe. June–September.

Description.—Stems erect, much branched, rough-pubescent, winged, 3–10 dm. high. Leaves alternate, simple, slightly decurrent, dentate, oblanceolate, rough-pubescent. Heads terminal, usually solitary, 1–1.5 cm. in diameter, many-flowered, nearly sessile; involucral bracts imbricated, numerous, the tips bearing rigid, purplish, pinnately branched spines, 6–9 mm. long; receptacle bristly, flat; flowers all tubular, yellow. Achenes similar to 491 but light gray; pappus bristly, scanty, white, 1–2 mm. long, often persistent.

Control.—A rotation including an intertilled crop to permit clean cultivation. Pull or spud scattered plants out of grain fields and new seedings. Hoe or mow waste places before seeds are formed.

494. *Centaurea repens* L. (*C. picris* Pall.). Russian knapweed, Turkestan thistle.

Perennial; reproducing by seeds and creeping roots.[4] Grain fields, alfalfa, cultivated fields, meadows, and waste places. Locally common from North Dakota to Missouri,[5] westward to the Pacific Coast,[6] spreading east to Michigan and south to Texas. Introduced from Asia with alfalfa seed. June–August.

Description.—Stems erect, 4–10 dm. high, branched near the base, striate, tomentose to nearly glabrous, from creeping roots. Leaves alternate, simple, the basal lobed or pinnatifid, gradually narrowed to a petiole, the upper entire, linear. Heads solitary on the ends of leafy branchlets; involucre oblong, about 1–2 cm. long, bracts roundish to oblong, with green base and whitish, entire or nearly entire appendages; flowers numerous, all tubular, rose to purple or blue. Achenes 2–3 mm. long, similar to 491 but with basal scar not oblique; pappus capillary, in one series, dehiscent.

Control.—Plow and plant an intertilled crop that can be kept free from weeds. Summer fallow badly infested fields and follow with an annual smother crop or alfalfa (19, 22).

495. *Centaurea solstitialis* L. Barnabys thistle, Yellow star-thistle. Fig. 112, n–r.

Annual; reproducing by seeds. New clover and alfalfa seedings, cultivated fields, and waste places. Widespread and local from On-

[4] Rogers, 1928. [5] Blake, 1922. [6] Ball and Robbins, 1931.

tario to Iowa, south to Florida and California. Introduced from Europe. July–September.

Description.—Stems with rigid spreading branches from the base, gray-pubescent, winged, 3–10 dm. high. Leaves alternate, strongly decurrent, gray-pubescent; the upper small, entire or dentate, linear to lanceolate; the lower deeply lobed, serrate. Heads terminal, solitary, 2–3 cm. in diameter, many-flowered; involucral bracts imbricated, numerous, the tips bearing rigid yellowish spines, 1–2 cm. long; receptacle bristly, flat; flowers all tubular, yellow. Achenes similar to 491; pappus bristly, 3–5 mm. long, persistent or deciduous.

Control.—The same as for 493.

496. *Centaurea vochinensis* Bernh.

Perennial; reproducing by seeds. Meadows, fields, roadsides, and waste places. Maine to Ontario, south to Virginia and Missouri. Introduced from Europe. June–September.

Description.—Stems harsh, branching, up to 10 dm. high. Leaves of rosette oblanceolate, petiolate, often lobed; cauline leaves smaller, mostly sessile, dentate or entire. Involucre of several unequal series; corollas rose-purple, the marginal enlarged and falsely radiate; outermost involucral bracts ovate, short, with blackish, acute tip bearing 5–7 pairs of long cilia; middle ones elongate-lanceolate, terminated by a dilated, blackish, pectinate appendage; innermost elongate, with brightly colored or dark erose or lacerate appendage.

Control.—The same as for 491.

497. *Chondrilla juncea* L. Skeleton-weed, Gum succory.

Biennial; reproducing by seeds. Dry fields, pastures, and waste places. Chiefly from southern New York to Virginia; rare elsewhere. Introduced from Europe. July–September.

Description.—Stems very slender, stiff, much branched, 3–10 dm. high, nearly leafless, bristly-hairy below, from a thick tap-root, with milky juice. Leaves of the first-year rosette very irregular, like a dandelion leaf, leaves of the second year's growth alternate, reduced to narrow or awl-shaped bracts. Heads in groups of 2–3, on short peduncles terminating the branches, or sessile and lateral on nearly naked branches; involucre cylindrical, of several linear equal bracts and a row of small bractlets at the base; flowers few, all ligulate, corollas yellow. Achenes cylindrical, with 5 vertical ribs, with a slender beak, roughened toward apex by small scaly teeth-like projec-

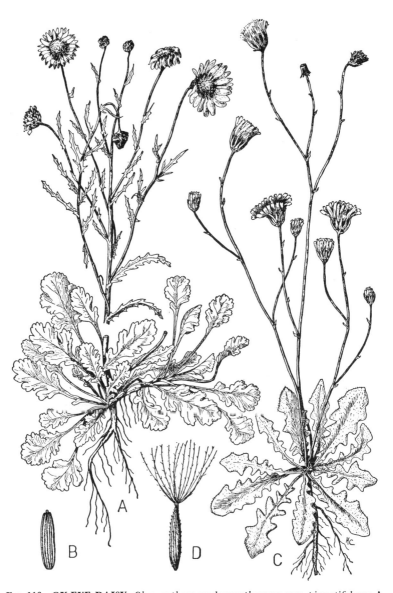

Fig. 113—**OX-EYE DAISY**, *Chrysanthemum leucanthemum* var. *pinnatifidum:* **A**, plant showing general habit ($\times \frac{1}{3}$); **B**, achene ($\times 7$).
GOSMORE, *Hypochoeris radicata:* **C**, plant showing general habit ($\times \frac{1}{3}$); **D**, achene ($\times 3$).

tions, light yellow; pappus capillary, white, about 5 mm. long, often deciduous.

Control.—Spud out the rosettes or mow the flower-stalks in pastures before seeds are formed. Plant an intertilled crop when practicable.

498. *Chrysanthemum leucanthemum* L. var. *pinnatifidum* Lecocq and Lamotte. Ox-eye daisy, White daisy, White-weed, Field daisy, Marguerite, Poorland flower. Fig. 113, a–b.

Perennial; reproducing by seeds and short rootstocks. Meadows, pastures, old fields and waste ground. Widely distributed throughout the United States. Introduced from Europe. June–July.

Description.—Stems erect, simple or forked near the top, glabrous; plants in patches due to short rootstocks. Stem-leaves alternate, simple, sessile, dentate, narrowly oblong or oblanceolate, subpinnatifid at the base, glabrous; basal leaves pinnately lobed, in rosettes. Heads terminal, solitary, 3–5 cm. in diameter, many-flowered; involucral bracts imbricated, numerous, narrow, brown-margined; receptacle flattish, naked; ray-flowers 20–30, white, pistillate, 10–15 mm. long; disk-flowers numerous, yellow, perfect. Achenes narrowly obovate, 1–1.5 mm. long; bearing a tubercle at the apex, black with 8 or 10 light gray ribs; pappus wanting.

Control.—Mow infested meadows early, as soon as the first flowers appear. A short rotation including a clean cultivated crop at least once every three years will control this weed.

499. *Cichorium intybus* L. Chicory, Succory, Blue sailors, Blue daisy, Coffee-weed, Bunk. Fig. 128, a–b.

Perennial; reproducing by seeds and roots. Old meadows and waste places; most troublesome in limestone regions. Locally common from Nova Scotia to British Columbia; southward to North Carolina and California. Introduced from Europe. July–August.

Description.—Stems erect, much branched, round, hollow, pubescent, 5–20 dm. high, juice milky. Leaves alternate, sessile, clasping, rough, hairy; the upper entire to dentate, oblong-lanceolate, auriculate, 3–7 cm. long; the basal and lower leaves pinnatifid, dentate, 1–2 dm. long. Heads axillary, 1–3 clustered or solitary and sessile on short branches, 2–3 cm. in diameter, several-flowered; involucral bracts in 2 series, the inner 8–10, the outer of 5 shorter bracts, spreading, margins minutely spiny; receptacle chaffy; flowers all ligulate, bright blue, perfect. Achenes obovate, 2–3 mm. long, 4- or

5-angled, straight or slightly curved, apex truncate, striated, light brown, mottled; pappus a small crown of many pale bristle-like scales.

Control.—Scattered plants in grasslands and waste places should be cut well below the crown (12). If area is extensive, plow it deep in autumn and plant a clean cultivated crop for two years before reseeding to grass. If a short rotation is used chicory is soon suppressed. Close grazing by sheep will control the chicory in pastures.

Chicory is grown in certain European countries and in Michigan for the roots which are used as a coffee substitute.[7, 8]

500. *Cirsium altissimum* (L.) Spreng. Tall thistle.

Biennial; reproducing by seeds. Pastures, wet meadows, roadsides, and waste places. Eastern United States, westward to North Dakota and Texas. Native. August–September.

Description.—Stems erect, branching above, 1–3 m. high, downy, from a fleshy tap-root. Leaves alternate, rough-hairy above and white-woolly beneath; the lower often pinnatifid with broad lobes and short petioles; the upper sinuate-toothed or with weak prickles, sessile and somewhat clasping but not decurrent. Heads solitary, terminal on the branches, about 5–7 cm. in diameter, with many disk-flowers; involucre 2–3 cm. high, the outer bracts spine-tipped and with a short dark glandular line on the back; the inner bracts soft, with spineless but serrulate tips; disk-flowers purple. Achenes oblong, about 5 mm. long, apex truncate, with a conspicuous rounded tubercle in the center, striate, slightly glossy, yellow-brown with a light brown band at apex; pappus plumose, about 2 cm. long, deciduous.

Control.—The same as for 503.

501. *Cirsium arvense* Scop. (*Carduus arvensis* Robson). Canada thistle, Creeping thistle, Small-flowered thistle, Perennial thistle, Green thistle. Fig. 114.

Perennial; reproducing by seeds and creeping roots. Grain fields, pastures, meadows, cultivated fields and waste places; mostly on rich or rather heavy soils. Common throughout southern Canada and northern United States, southward to Virginia and northern California. Introduced from Eurasia. July–October.

Description.—Stems erect, rigid, branched above, nearly glabrous or slightly hairy, ridged, 4–12 dm. high, from short rootstocks or

[7] Kains, 1900. [8] Cormany, 1927.

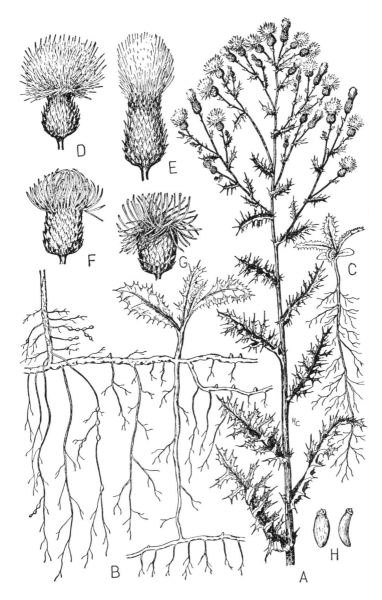

Fig. 114—**CANADA THISTLE**, *Cirsium arvense:* **A,** flowering shoot (× ⅓); **B,** lower part of stem and creeping root system (× ⅓); **C,** seedling (× ⅓); **D,** head of flowers from a fertile (pistillate) plant (× 1); **E,** fertile head one week later (× 1); **F,** head of flowers from a sterile (staminate) plant (× 1); **G,** sterile head one week later (× 1); **H,** seed (achene), (× 4).

from creeping roots. Leaves alternate, oblong or lanceolate, very irregularly lobed and spiny-toothed on the margin, the upper sessile but only slightly decurrent, hairy beneath or when mature often glabrous or nearly so, dark green. Heads dioecious, in corymbose clusters, terminal and axillary, with many rose-purple, lavender or white disk-flowers, 2.5 cm. in diameter or less; involucral bracts numerous, imbricated, spineless; receptacle bristly-chaffy. The evidence of the dioecious nature of the Canada thistle as well as the differences in the heads and flowers of staminate and pistillate plants has been recorded by Detmers.[9] Staminate heads oblong; corollas long; anthers fully developed and shedding pollen at the time of opening of the corolla; styles short, branches not separated; stigmatic surface not developed; pappus shorter than the flower and not in evidence after the heads have blossomed; achenes begin to shrivel soon after blossoming. Pistillate heads ovoid or flask-shaped; corolla short; stamens short, shrivelled and blackened at time of opening of corolla, pollen none; styles long, exserted, style branches spreading, stigmatic surface exposed on the inner surface; pappus long; achenes normal. Achenes oblong, 2.5–3.5 mm. long, flattened, curved or straight, apex truncate with a tubercle in the center, smooth, light to dark brown; pappus plumose, about 2 cm. long, deciduous.

The Canada thistle is one of the most feared weeds in the United States, having been declared noxious in the seed laws of thirty-seven states (1935), forty-three states in 1954. The noxiousness of the Canada thistle is due to its creeping root system, every piece of which can give rise to a new plant,[10] and to the numerous seeds which are easily scattered by the wind and which have a great longevity. Along the southern border of its range, the Canada thistle does not produce seed as freely as farther north. In certain localities it is commonly believed that it does not seed.[11] This belief, as has been pointed out by Detmers,[12] is due to several reasons. Since this species is dioecious, that is the staminate and pistillate heads are borne on different plants, two individuals are necessary for seed production. If a patch of Canada thistle is started from one seed and spreads or its roots are dragged throughout a field, it still represents the offspring from one seed. If this happens to be a staminate plant no seed can ever be produced by its offspring; if it is a pistillate plant, its

[9] Detmers, 1927. [10] Rogers, 1928. [11] Hansen, 1918.
[12] Detmers, 1927.

offspring cannot produce seed unless other staminate plants are near enough to supply pollen for pollination.[13]

Another reason why Canada thistle appears not to produce seed is due to the attacks of certain insects such as *Dasyneura gibsoni*, Canada thistle midge, and *Trypeta florescentiæ*, whose larvae feed on and destroy the undeveloped achenes.

In localities where attacks by these and other insects are severe most, if not all, the Canada thistle plants may be prevented from maturing seeds.

Weather conditions unfavorable for pollination may reduce the set of seed somewhat, but since the flowering period extends over a considerable time this would seldom prevent plants from producing any seeds at all.

The leaves of Canada thistle vary greatly in the amount of hairiness, degree of lobing and spininess. Wimmer and Grabowski [14] long ago described four varieties based on these variations:

Var. *integrifolium*—leaves smooth, entire, setose-spinulose, glabrous beneath.

Var. *vestitum*—leaves denticulate, white-tomentose beneath.

Var. *mite*—stem-leaves sinuate-pinnatifid, branch-leaves nearly entire or dentate, minutely spiny.

Var. *horridum*—all leaves undulate, with narrow, strongly spiny, pinnate lobes.

Observations in America indicate that, although all four of these so-called varieties may be present, they are not constant but represent only such leaf variations as can be found in many Canada thistle patches.

Var. *integrifolium*, Uncut-leaved Canada thistle, and var. *vestitum* have been reported locally from Quebec through New England, New York and Ohio [15] to Iowa.[16]

Control.—Small patches of Canada thistle should be destroyed promptly by digging out the roots (17), or by covering with heavy mulch paper (24). In pastures persistent cutting with a scythe or hoe will gradually destroy the thistle. The first cutting should be made just before the plants blossom. At this stage the reserve foods stored in the roots have been much reduced and no seeds have yet been formed. Additional cuttings should be made as soon as new shoots

[13] Hayden, 1934.
[15] Detmers, 1927.
[14] Wimmer and Grabowski, 1829.
[16] Pammel and King, 1925.

appear. On large areas clean cultivation is the most practical method of controlling Canada thistle. A short rotation of a grain crop (one year), clover (one year) and cultivated crop (two years) will permit of sufficient cultivation to control the Canada thistle under most conditions. As preventive measures, no seed or manure containing viable Canada thistle seeds should be put on land. Avoid scattering the roots from small patches of thistles to clean parts of a field with cultivators. Allow no thistles to produce seed.

502. *Cirsium pumilum* (Nutt.) Spreng. Pasture thistle, Fragrant thistle. Fig. 115, a.

Biennial; reproducing by seeds. Old fields and pastures; mostly on dry, sandy, or gravelly, somewhat acid soils. Chiefly in the northeastern states, southward to South Carolina. Native. July–September.

Description.—Stems erect, sparingly branched above, stout, ridged, hairy, very leafy, 3–10 dm. high, from a fleshy tap-root. Leaves alternate, lanceolate-oblong, pinnatifid and prickly on the lobes, partly clasping but not decurrent, nearly glabrous; leaves of the first year forming a large flat rosette. Heads solitary, terminal, 4–9 cm. in diameter, with many disk-flowers; involucral bracts imbricated, numerous, appressed, the outer spine-tipped, the inner soft; receptacle flat, bristly; disk-flowers pinkish-purple or rarely white, perfect, fragrant. Achenes 3–4 mm. long, similar to 501, yellow with a lighter band at apex; pappus plumose, white, about 3 cm. long, deciduous.

Control.—The same as for 503.

503. *Cirsium vulgare* (Savi) Tenore (*C. lanceolatum* Hill). Bull thistle, Spear thistle, Plume thistle. Fig. 115, b–c.

Biennial; reproducing by seeds. Pastures, old fields, and waste places; on rich rather moist soils. Widespread throughout the United States and southern Canada. Introduced from Eurasia. July–October.

Description.—Stems erect, branched, stout, ridged, woolly, prickly-lobed, winged, very leafy, 6–12 dm. high, from a fleshy tap-root. Leaves alternate, lanceolate, strongly decurrent, pinnatifid, the lobes spiny, rough above, hairy beneath; leaves of the first year forming a large flat rosette. Heads solitary, mostly on the ends of branches, 3–6 cm. in diameter, with many disk-flowers; involucral bracts imbricated, numerous, all tipped with spreading spines; receptacle flat, bristly; disk-flowers pinkish-purple, perfect. Achenes 3–4 mm. long,

Fig. 115—**PASTURE THISTLE,** *Cirsium pumilum:* **A,** flowering shoot showing general habit (× ⅙).
BULL THISTLE, *Cirsium vulgare:* **B,** flowering shoot (× ⅓); **C,** seed (achene), (× 4).

similar to 501, straw-colored with brown or black vertical lines; pappus plumose, about 2 cm. long, deciduous.

Control.—This weed does not persist under cultivation. In pastures or meadows cut the rosettes below the crown early in the spring with a spud or hoe. Badly infested fields and waste places should be mowed before seeds have an opportunity to ripen.

504. *Cnicus benedictus* L. Blessed thistle. Fig. 116.

Annual; reproducing by seeds. Roadsides, sandy fields, and waste places. Sparingly established, New Brunswick to Illinois, south to Georgia and Alabama; rare elsewhere. Adventive from Europe. April–September.

Description.—Stems hirsute or villous, branching, 1.5–8 dm. high. Leaves mostly 5–15 cm. long and 5 cm. wide, coarsely reticulate, the lower with petioles, the upper somewhat clasping, hardly, if at all, decurrent. Flowers all tubular, yellow, marginal ones sterile and shorter than the others, in many-flowered heads. Involucre 3–4 cm. high, of broad, spinescent bracts, generally surpassed by the closely subtending foliage leaves. Achenes about 8 mm. long, terete, short, many-ribbed, crowned with 10 short and horny teeth, 10 rigid bristles, and 10 shorter alternating bristles in an inner row.

Control.—The same as for 491.

505. *Crepis capillaris* (L.) Wallr. Hawks-beard.

Annual or biennial; reproducing by seeds. Meadows, pastures, lawns, and waste places. Local from Connecticut to New Jersey and Pennsylvania; also on the Pacific Coast; increasing in range. Introduced from Europe. July–September.

Description.—Stem erect or ascending, slender, branched above, slightly glandular-pubescent, leafy, 3–9 dm. high; juice milky. Leaves alternate, simple, mostly clasping with sagittate-auriculate base, lanceolate-spatulate to pinnatifid, dentate or the uppermost entire, nearly glabrous; basal leaves numerous, pinnatifid, with backward pointed lobes. Heads numerous, in an open corymbose cluster, 10–15 mm. in diameter, few to many-flowered, on long slender peduncles; involucral bracts in a single row, equal, erect, pubescent, 5–7 mm. long; receptacle flat, naked; flowers all ligulate, yellow, perfect. Achenes fusiform, 2–3 mm. long, straight or curved, 10-ribbed, minutely granular, brown; pappus capillary, white, soft, simple, persistent.

Control.—Plow badly infested meadows and plant a clean culti-

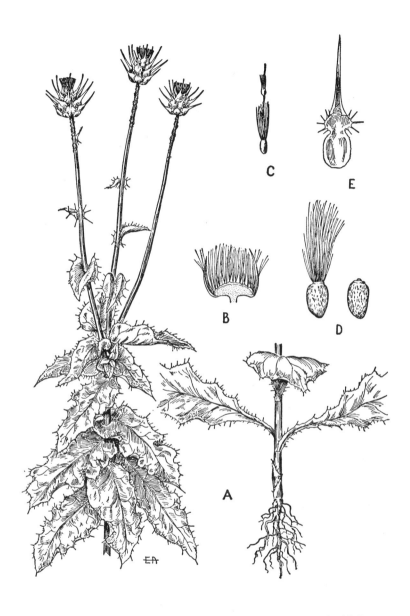

Fig. 116—**BLESSED THISTLE**, *Cnicus benedictus*: **A**, habit sketch (× ⅓); **B**, cross-section of receptacle (× ⅔); **C**, flower (× 1); **D**, achene with and without pappus (× 1½); **E**, bract of involucre (× 1).

vated crop. Spud or hoe scattered plants out of lawns or pastures as soon as they can be recognized. Mow waste places before seeds are formed.

506. *Crepis tectorum* L. Hawks-beard.

Annual or biennial; reproducing by seeds. Meadows, pastures, and waste places. Local, Ontario to North Carolina, westward to Nebraska. Introduced from Europe. July–September.

Description.—Similar to 505 in general habit and appearance but somewhat smaller. Achenes fusiform, about 3 mm. long, straight or curved, with 10 upwardly-scabrous ribs, chestnut-brown; pappus capillary, copious, persistent.

Control.—The same as for 505.

Crepis biennis L., a biennial introduced from Europe, occurs locally in fields and waste places in the northeastern states. It is similar to 505 but its achenes have 13 ribs instead of 10.

507. *Dyssodia papposa* (Vent.) Hitchc. (*Boebera papposa* Rydb.). Fetid marigold, False mayweed, Stinkweed. Fig. 119, c–f.

Annual; reproducing by seeds. Fields, new meadows, waste places. Native, Ontario to Illinois, southward to Mexico; also spreading west to Montana and Arizona; introduced eastward. June–August.

Description.—Stems erect or spreading, much branched from the base, smooth or hairy, glandular-dotted, 1–5 dm. high. Leaves opposite, pinnately parted into narrow toothed lobes, glandular-dotted, with a fetid odor. Heads numerous, on short peduncles, many-flowered, about 1 cm. in diameter, involucre of one row of purplish bracts united into a cup with a few separate bracts at the base; receptacle flat, with short bristles; ray-flowers few, pistillate, very short; disk-flowers many, with yellow corollas, perfect and fertile. Achenes 3–4 mm. long, slender, 4-angled, with a truncate apex and calloused base, hairy, dark brown to black; pappus a row of scales dissected into rough bristles.

Control.—Clean cultivation while the seedlings are small (6). Pull, hoe or mow waste places before seeds are formed (4, 5).

508. *Erechtites hieracifolia* (L.) Raf. Fireweed, Pilewort. Fig. 127, h–k.

Annual; reproducing by seeds. Old fields, pastures, recent clearings, and burned-over woodlands; in rather moist soil. Widespread in eastern North America, and west to Minnesota, in the regions that were formerly woodlands. Native. August–September.

Description.—Stems erect, simple or branched above, coarse, glabrous or hairy, grooved, 1–2 m. high. Leaves alternate, simple, sharply dentate, oblanceolate to pinnatifid, thin, glabrous; the lower tapering to margined petioles; the upper with clasping auricled bases. Heads numerous, 15–20 mm. long, with many disk-flowers, in an open terminal panicle; involucral bracts in a single row, linear, acute; receptacle flat, naked; flowers greenish-white, perfect, the marginal flowers pistillate. Achenes linear, flattened, tapering to each end, 10-ribbed, minutely hairy, the apex with a small white tubercle in the center, dark brown; pappus capillary, white, glistening, copious.

Control.—Clean cultivation (6). Pull, hoe or mow plants before seeds are formed (4, 5).

Erechtites prenanthoides (A. Rich.) DC. Australian burnweed, was considered noxious in 1954 in Oregon.

509. *Erigeron annuus* (L.) Pers. White-top, Daisy fleabane, Whiteweed, Sweet scabious. Fig. 117, b–c.

Annual or biennial; reproducing by seeds. Meadows, pastures, and waste places. Common from Nova Scotia to Manitoba, southward to Georgia and Missouri. Native. June–September.

Description.—Stems erect, branched above, 2–15 dm. high, glabrous or with scattered stiff hairs. Leaves alternate, simple, coarsely and sharply dentate, thin, nearly glabrous; lower leaves ovate, tapering into margined petioles; upper leaves lanceolate, acute and entire at both ends, mostly sessile; the uppermost leaves often entire, linear. Heads in a corymbose cluster, numerous, 1.5–2 cm. in diameter, many-flowered, on naked peduncles; involucral bracts in 1–2 series, narrow, equal, slightly hairy; receptacle flat, naked; ray-flowers about 50–75, white or tinged with purple, much longer than disk-flowers, 1 mm. wide, pistillate; disk-flowers numerous, yellow, perfect. Achenes about 1 mm. long, obovate, flattened, with minute appressed hairs, straw-colored; pappus double, the outer a crown of short scales, the inner of deciduous bristly hairs, usually wanting in the ray-flowers.

Control.—Plow infested meadows in the spring and follow with a cultivated crop or spring grain. Close pasturing with sheep will destroy the white-top rosettes in pastures and meadows. New meadows infested with white-top should be clipped as soon as the flower-stalks of the weeds appear. If the meadow has not been clipped, the hay crop should be cut early before the white-top has formed seeds.[17]

[17] Pipal, 1918.

FIG. 117—**ROUGH DAISY FLEABANE,** *Erigeron strigosus:* **A,** plant showing general habit (× ⅓); **E,** leaf (× ⅓); **F,** seed (achene), (× 6).
WHITE-TOP, *Erigeron annuus:* **B,** leaf (× ⅓); **C,** seed (achene), (× 6).
PHILADELPHIA FLEABANE, *Erigeron philadelphicus:* **D,** plant showing general habit (× ⅓).

Composite Family 459

510. *Erigeron canadensis* L. (*Leptilon canadense* Britton, *Conyza canadensis*). Fleabane, Horse-weed, Bitterweed, Hog-weed, Marestail, Blood stanch. Fig. 118, a–e.

Annual; reproducing by seeds. Cultivated fields, pastures and meadows; mostly on rather dry soils. Common throughout North America. Native; also introduced in some localities. July–October.

Description.—Stems erect, wand-like, simple or rarely branching from the base, bristly-hairy, 2–20 dm. high. Leaves alternate, simple, hairy; the lower spatulate, dentate, sometimes cut-lobed, tapering to petioles; the upper lanceolate to linear, mostly entire, sessile. Heads cylindrical, in panicled clusters, very numerous, about 5 mm. in diameter, many-flowered, on naked peduncles; involucral bracts in 1–2 series, narrow, equal, hairy; receptacle flat, naked; ray-flowers about as long as the disk-flowers, over 100 in a head, greenish-white, less than 1 mm. wide, pistillate; disk-flowers numerous, perfect, yellow. Achenes similar to 509.

Control.—Plow in the autumn and plant a clean cultivated crop next spring. Pull or hoe scattered plants from grain fields and meadows. Mow the hay crop early in badly infested meadows.

511. *Erigeron philadelphicus* L. Philadelphia fleabane, Daisy fleabane, Skevish. Fig. 117, d.

Perennial; reproducing by seeds and short rootstocks. Moist meadows and pastures. Widespread throughout North America. Native. June–August.

Description.—Stems erect, simple or branched above, slightly hairy, 3–9 dm. high. Leaves alternate, simple, entire or sparingly dentate, glabrous to somewhat hairy; the lower spatulate, narrowing to short margined petioles; the upper oblong, mostly entire, with cordate and clasping base. Heads in a terminal corymbose cluster, numerous, 1.5–2 cm. in diameter, many-flowered, on naked peduncles; involucral bracts in 1–2 series, narrow, equal, pubescent; receptacle flat, naked; ray-flowers much larger than the disk-flowers, 100–150 in a head, pink, less than 1 mm. wide, pistillate; disk-flowers numerous, perfect, yellow. Achenes similar to 509.

Control.—The same as for 509.

512. *Erigeron pulchellus* Michx. Robins plantain, Blue spring daisy, Rose petty. Fig. 118, f.

Perennial; reproducing by seeds and short rootstocks. Meadows, pastures and open woodlands. Widespread from Maine to Ontario

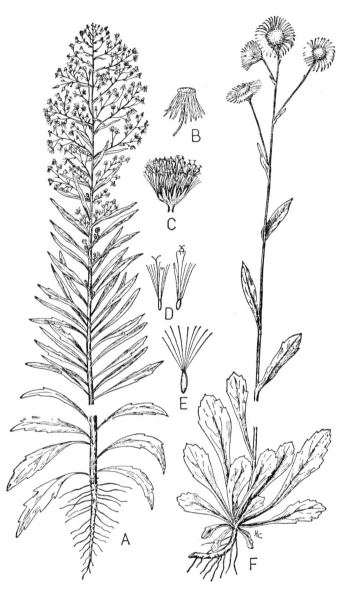

Fig. 118—**FLEABANE,** *Erigeron canadensis:* **A,** plant showing general habit ($\times \frac{1}{3}$); **B,** receptacle after shedding seeds ($\times 3$); **C,** head of flowers ($\times 3$); **D,** ray-flower (left) and disk-flower (right), ($\times 4$); **E,** seed (achene) with pappus ($\times 3$).
ROBINS PLANTAIN, *Erigeron pulchellus:* **F,** plant showing general habit ($\times \frac{1}{3}$).

and Minnesota, southward to Florida, Louisiana, and Kansas. Native. May–June.

Description.—Stems erect, simple, softly hairy, 2–5 dm. high, producing short rootstocks. Leaves alternate, simple, mostly entire, hairy; basal leaves obovate and spatulate, with dentate tips, base tapering to short margined petioles; cauline leaves lanceolate-oblong, sessile or partly clasping. Heads few in a terminal cluster, 2.5–3.5 cm. in diameter, many-flowered, on slender naked peduncles; involucral bracts in 1–2 series, narrow, equal, hairy; receptacle flat, naked; ray-flowers much larger than the disk-flowers, about 50 in a head, bluish-purple, 1 mm. wide, pistillate; disk-flowers numerous, perfect, greenish-yellow. Achenes 1–2 mm. long, similar to 509 but more glossy and light brown.

Control.—The same as for 509.

513. *Erigeron strigosus* Muhl. (*E. ramosus* BSP.) Rough daisy fleabane, White-top. Fig. 117, a; e–f.

Annual or biennial; reproducing by seeds. Meadows, abandoned fields, and waste places. Widespread throughout the United States and southern Canada; most common in the northeastern states. Native. June–September.

Description.—Stems erect, branched above, appressed-hairy, 3–10 dm. high. Leaves alternate, simple, mostly entire, pubescent; the lower and basal leaves oblong to spatulate, tapering to a slender petiole, somewhat dentate; upper leaves lanceolate, sessile. Heads, flowers and achenes similar to 509.

Control.—The same as for 509.

514. *Eupatorium capillifolium* (Lam.) Small. Dog-fennel. Fig. 119, a–b.

Annual; reproducing by seeds. Old fields, pastures, roadsides, borders of woods, and clearings. New Jersey to Tennessee, south to Florida and Texas. Native. September–November.

Description.—Stems several, slender, erect, up to 3 m. high, much branched, scabrous-hirsute, resembling Artemisia. Leaves delicate, crowded, glabrous, largely alternate, 2–10 cm. long, larger ones pinnately divided into thread-like segments; inflorescence leaves mostly simple. Flowers in a tall wand-like panicle with many small, 3–6 flowered heads, 2–3.5 mm. long; tubular corolla greenish-white, 1.5–2.5 mm. long. Achenes glabrous, 1 mm. long.

Control.—The same as for 516.

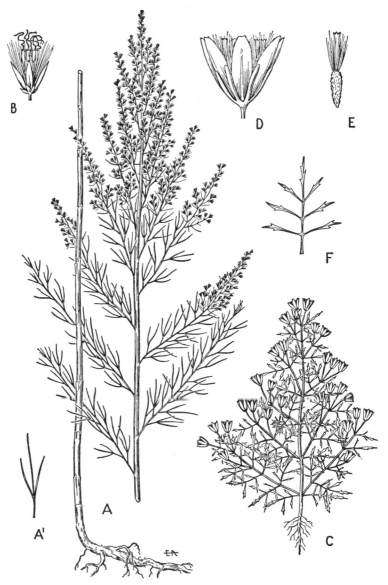

Fig. 119—**DOG-FENNEL**, *Eupatorium capillifolium:* **A,** habit (× 1); **A′,** leaf (× 3); **B,** flower head (× 9).
FETID MARIGOLD, *Dyssodia papposa:* **C,** habit (× 1); **D,** flower head (× 6); **E,** flower (× 6); **F,** leaf (× 3).

515. **Eupatorium maculatum** L. (*E. purpureum* in part, of Gray's Manual). Joe-pye weed, Purple boneset, Tall boneset, Trumpet-weed.

Perennial; reproducing by seeds. Old fields, wet meadows, pastures, and borders of swampy woods. Common from Newfoundland to Michigan, southward to North Carolina and New Mexico; infrequent in British Columbia and northwestern Washington. Native. August–September.

Description.—Stems erect, simple, stout, ridged, glabrous, usually speckled with dark purple, 5–30 dm. high. Leaves in whorls of 3–6, simple, petioled, coarsely serrate, oblong-ovate to lanceolate, puberulent beneath, 8–16 cm. long. Heads in a large compact flat-topped compound corymb, cylindrical, 9–15-flowered; involucral bracts numerous, unequal, purplish; receptacle flat, naked; disk-flowers perfect, corolla purple, 5 mm. long. Achenes similar to 516 but larger; pappus bristles 4–5 mm. long.

Control.—The same as for 516.

516. **Eupatorium perfoliatum** L. Boneset, Thoroughwort, Agueweed, Fever-weed, Sweating plant.

Perennial; reproducing by seeds. Wet fields, pastures and waste places; along swamps and ditches. Widespread in Quebec to Manitoba, south to Florida and Texas. Native. August–September.

Description.—Stems erect, simple or rarely branching above, stout, ridged, hairy, 5–15 dm. high. Leaves opposite, simple, very veiny, serrate, acuminate, connate-perfoliate, united at the base around the stem, wrinkled above, pubescent beneath, 1–2 dm. long. Heads in large compact corymbs, 10–40-flowered, 5–7 mm. in diameter; involucral bracts in several imbricated rows, linear-lanceolate, unequal, greenish, hairy; receptacle naked; disk-flowers dull white or rarely blue, perfect. Achenes 2–2.5 mm. long, linear, 5-angled, apex truncate with a tubercle, black or dark brown, with scattered yellow dots; pappus capillary, about 2 mm. long.

Control.—Boneset does not persist under cultivation. When plowing is not practicable grub the plants out, roots and all, as soon as they appear. In extensive areas mow the weeds close to the ground several times during the summer before seeds are formed. Improved drainage will often help to control boneset.

517. **Eupatorium rugosum** Houtt. (*E. urticaefolium* Reichard). White snake-root, Richweed, White sanicle, Indian sanicle, Deerwort, Squaw-weed. Fig. 120, a–d.

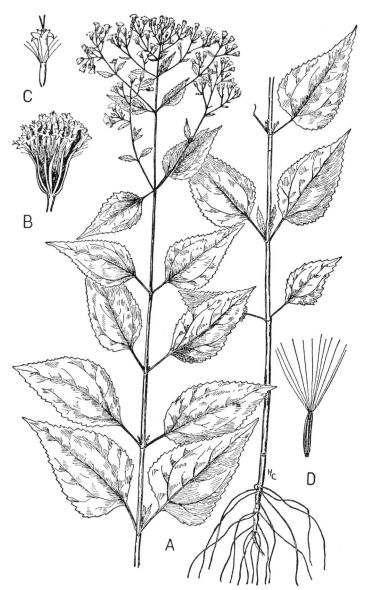

Fig. 120—**WHITE SNAKE-ROOT,** *Eupatorium rugosum:* **A,** plant showing general habit (× ⅓); **B,** head of flowers (× 3); **C,** flower (× 4); **D,** seed (achene) with pappus attached (× 6).

Perennial; reproducing by seeds. Woodlands, banks, damp or shady pastures and fields; mostly on gravelly or limy soils. Widespread from New Brunswick to Saskatchewan, south to Florida and Texas. Native. August–September.

Description.—Stems erect, simple or branching above, slender, glabrous, 3–12 dm. high. Leaves opposite, simple, long-petioled, 3-nerved, coarsely dentate, broadly ovate, pointed, thin, glabrous, 7–12 cm. long. Heads in dense compound corymbs, with 8–30 perfect disk-flowers; corolla white, 5–7 mm. in diameter; involucral bracts nearly equal, usually in one row, green; receptacle flat, naked. Achenes similar to 516.

Control.—The same as for 516.

White snake-root is the plant that causes the disease known as trembles in cattle or milk sickness in humans, in the eastern states. The poisonous principle, tremetol, may be passed in the milk of animals that have grazed on this plant to persons and suckling animals and cause a disease.[18] This disease was formerly rather common, especially from the Ohio Basin southward to Georgia.

518. *Franseria discolor* Nutt. (*Gaertneria discolor* Kuntze). Silverleaf poverty-weed, White-leaved franseria, Bur-ragweed, White-weed.

Perennial; reproducing by seeds and creeping roots. Meadows, pastures, cultivated fields and waste places; in dry regions, also in irrigated fields. Western Great Plains and Rocky Mountain states. Native. July–September.

Description.—Stems erect or ascending, 2–6 dm. high, branching from the base, white-woolly. Leaves alternate, mostly bipinnately lobed, white-woolly beneath, green or grayish above, lobes narrow, irregularly serrate, petiole winged. Heads monoecious; the staminate heads small, drooping, solitary or in several terminal racemes; bracts of staminate involucre fused; pistillate heads 1 or 2 in the axils of leaves. Mature involucre bur-like, inclosing 1 or more achenes, obovoid, about 5 mm. long, armed with 2 or more rows of stout hooked spines, light brown.

Control.—The same as for 501.

Franseria tomentosa Gray, Woolly-leaved poverty-weed, a native perennial with the general habit and distribution of 518, occurs in similar situations but it is more troublesome in moister soils. It is a

[18] Couch, 1933.

taller plant with leaves white-woolly on both sides and usually only 1-pinnately lobed and the terminal lobe is much larger than the others.

Franseria tenuifolia Torr. and Gray, Poverty-weed, is a noxious weed (1954) in Arizona.

519. *Galinsoga ciliata* (Raf.) Blake (*G. parviflora* Cav. var. *hispida* of authors). Galinsoga, French-weed. Fig. 132, d–g.

Annual; reproducing by seeds and rooting stems. Gardens, cultivated fields and waste places; mostly on rich moist soils. Widespread and becoming common from Maine and Ontario to Georgia and westward to Oregon and Mexico. Introduced from tropical America. July–September.

Description.—Stems erect or spreading, much branched, slender, green, pubescent, 3–6 dm. high. Leaves opposite, simple, petioled, crenate-serrate, ovate, glabrous. Heads numerous, terminal and axillary, several-flowered, less than 1 cm. in diameter; involucral bracts 4–5, ovate, thin, green; receptacle conical, chaffy; ray-flowers 4–5, corolla white, 3-toothed, pistillate, scarcely longer than the disk-flowers; disk-flowers perfect, with yellow corolla. Achenes about 1.5 mm. long, wedge-shaped, 4-sided, finely white-hairy, dark brown to black; pappus a fringe of chaffy scales.

Control.—Plow in the spring after the seeds have germinated and plant a crop that can be cultivated at frequent intervals while the weed seedlings are small. Pull large plants before seeds are formed. Mow waste places.

520. *Gnaphalium macounii* Greene (*G. decurrens* Ives). Clammy everlasting, Clammy cudweed. Fig. 121, g.

Annual or biennial; reproducing by seeds. Upland pastures, old fields and waste places. Locally common, Quebec to British Columbia, southward to West Virginia and Arizona. Native. August–September.

Description.—Stems erect, simple or sparingly branched, 3–10 dm. high, woolly, glandular-viscid, fragrant. Leaves alternate, simple, clasping and decurrent, entire, linear-lanceolate, somewhat scabrous above, densely white-woolly beneath. Heads in dense corymbose clusters, numerous, many-flowered, about 5 mm. high; involucral bracts numerous, imbricated, appressed, scarious, yellowish; receptacle flat, naked; disk-flowers cream-white to straw-color, the outer pistillate and very slender, the inner perfect. Achenes about 0.7 mm.

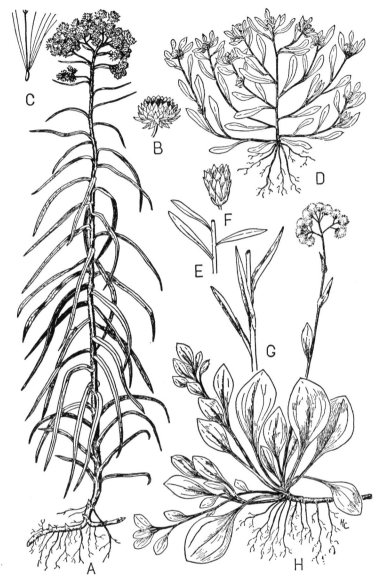

FIG. 121—**PEARLY EVERLASTING,** *Anaphalis margaritacea:* **A,** plant showing general habit (× ⅓); **B,** head (× 2); **C,** achene (× 3).
LOW CUDWEED, *Gnaphalium uliginosum:* **D,** plant showing habit (× ⅓).
FRAGRANT EVERLASTING, *Gnaphalium obtusifolium:* **E,** leaves (× ⅓); **F,** head (× 2).
CLAMMY EVERLASTING, *Gnaphalium macounii:* **G,** leaves (× ⅓).
LADIES TOBACCO, *Antennaria plantaginifolia:* **H,** plant showing habit (× ⅓).

long, oval, flattened or somewhat angled, pale brown; pappus capillary, 3–4 mm. long, deciduous.

Control.—This weed does not persist if the land can be plowed and cultivated for a year. In permanent pastures mow close to the ground before seeds are formed.

521. *Gnaphalium obtusifolium* L. (*G. polycephalum* Michx. of Gray's Manual, ed. 7). Fragrant everlasting, Old field balsam, Cudweed, Chafe-weed. Fig. 121, e–f.

Annual or winter annual; reproducing by seeds. Old fields, pastures and waste places; on moist or dry soils. Widespread in eastern North America, westward to Manitoba, Kansas and Texas. Native. August–September.

Description.—Stems erect, simple or sparingly branched, 3–10 dm. high, white-woolly, slightly glandular, fragrant. Leaves alternate, simple, sessile, undulate or entire, lanceolate, tapering to the base, glabrous above, woolly beneath. Heads numerous, clustered at tips of panicled corymbose branches, about 5 mm. high, many-flowered; involucral bracts numerous, imbricated, appressed, scarious, white; receptacle flat, naked; disk-flowers white, the outer pistillate and very slender, the inner perfect. Achenes similar to 520.

Control.—The same as for 520.

522. *Gnaphalium uliginosum* L. Low cudweed, Marsh cudweed, Wartwort, Mouse-ear. Fig. 121, d.

Annual; reproducing by seeds. Low meadows, pastures, and cultivated fields; common on land that is subject to inundation. Widespread and abundant throughout the northern United States and Canada. Native; also native to Europe. July–September.

Description.—Stems low, much branched and spreading on the ground, very white-woolly, 5–30 cm. high. Leaves alternate, simple, sessile, entire, spatulate-lanceolate to linear, densely white-woolly except the lowest, which are often glabrous. Heads numerous, in terminal capitate clusters subtended by leaves, several-flowered, about 2 mm. high; involucral bracts numerous, slightly imbricated, appressed, scarious, brownish; receptacle flat, naked; disk-flowers white, the outer pistillate and slender, the inner perfect. Achenes similar to 520 but usually smaller, 0.4–0.6 mm. long.

Control.—Improved drainage so that the water does not stand in low areas will induce grass to grow where the cudweed usually thrives. Plowing, followed by clean cultivation, will control cudweed.

523. *Grindelia squarrosa* (Pursh) Dunal. Gumweed, Tarweed, Gum-plant, Rosin-weed. Fig. 122, a–b.

Perennial or biennial; reproducing by seeds. Dry pastures, roadsides, banks, and waste places. Native from Illinois to Manitoba, southward to Texas and Mexico; locally introduced in the northeastern states and westward. August–September.

Description.—Stems erect, branched, glabrous, often reddish, 2–6 dm. high. Leaves alternate, simple, sessile, often clasping, spinulose-serrate, spatulate to oblong, rigid, slightly scarious. Heads solitary and terminal on the branches, many-flowered, about 2.5 cm. in diameter; involucral bracts numerous, imbricated, in several rows, recurved, subulate, glutinous; receptacle conical, naked; ray-flowers pistillate and fertile, with yellow corolla; disk-flowers perfect, corolla yellow. Achenes 2–4 mm. long, variable, short and thick, somewhat 4-angled, often curved, smooth or ridged longitudinally, gray or light brown; pappus of 2–8 deciduous awns.

Control.—Plow and follow with a clean cultivated crop. In grasslands and waste places cut or hoe the gumweed plants before seeds are formed.

524. *Helenium autumnale* L. Sneeze-weed, Staggerwort, Yellow star, Swamp sunflower.

Perennial; reproducing by seeds. Wet pastures, meadows, along ditches, streams and swamps. Widespread in eastern North America, west to Minnesota and Missouri. Native. August–October.

Description.—Stems erect, branched above, angular, winged, glabrous, greenish, 4–20 dm. high. Leaves alternate, simple, sessile and decurrent, dentate, lanceolate to ovate-oblong, glabrous. Heads terminal, solitary, numerous, many-flowered, 2–4 cm. in diameter; involucral bracts small, linear, hairy, reflexed; receptacle globose, naked; ray-flowers spatulate, 3-toothed, drooping, pistillate and fertile, golden-yellow; disk-flowers perfect, greenish-yellow. Achenes about 2 mm. long, top-shaped, angular, hairy-ribbed, yellow or grayish-brown; pappus of 5–8 thin, awned scales, persistent.

Control.—Improve the drainage. Cultivation will destroy sneezeweed. In permanent pastures and meadows or where plowing is not practicable pull or mow the plants as soon as the flowers appear.

FIG. 122—**GUMWEED**, *Grindelia squarrosa*: **A,** part of plant showing habit (× ⅓); **B,** achene (× 4).
ELECAMPANE, *Inula helenium*: **C,** flowering shoot and basal leaf (× ¼); **D,** achene with pappus (× 6).

The sneeze-weeds, 524–526, are poisonous when eaten by stock. Dairy products from cows that have eaten sneeze-weeds become tainted by them.

525. *Helenium nudiflorum* Nutt. Purple-headed sneeze-weed.

Perennial; reproducing by seeds. Wet meadows, pastures and borders of marshes and ditches. Native from Virginia to Florida, westward to Texas and Missouri; introduced northward to New England and through the north central states. June–September.

Description.—Stems erect, branched above, angular, winged, glabrous, brownish, 3–10 dm. high. Leaves alternate, simple, sessile and decurrent, entire, lanceolate to linear, puberulent; basal leaves spatulate and dentate. Heads similar to 524, usually smaller, 2–3 cm. in diameter; receptacle globose, naked; ray-flowers spatulate, 3-toothed, pistillate and sterile, yellow or sometimes brownish, sometimes wanting; disk-flowers perfect, brown. Achenes similar to 524, about 1 mm. long, dark brown.

Control.—The same as for 524.

526. *Helenium tenuifolium* Nutt. Bitterweed, Fennel, Yellow dog fennel, Sneeze-weed.

Annual; reproducing by seeds. Meadows, pastures, yards, waste places, and roadsides. Native from Virginia to Kansas and southward to Florida and Texas; locally introduced northward to Michigan and Massachusetts. August–October.

Description.—Stems erect, much branched above, glabrous, very leafy, 1–8 dm. high. Leaves alternate, simple, sessile, entire, linear-filiform, glabrous. Heads similar to 524, smaller, 1.5–2 cm. in diameter; ray-flowers few, wedge-shaped, 3-toothed, drooping, pistillate and fertile, yellow; disk-flowers perfect, yellow. Achenes similar to 524, 1–1.5 mm. long, reddish-brown, with appressed hairs.

Control.—Bitterweed does not persist if the land can be cultivated. In permanent pastures, where this weed causes most trouble, avoid too early grazing. If the grass is allowed to get a good start before stock is turned in, the bitterweed seedlings will be smothered out. Badly infested pastures should be mowed off as soon as the flowers appear and again when new shoots are formed. Several cuttings a season for two or three years will control most of the bitterweed in a pasture.

In certain localities in the South bitterweed is the cause of serious losses to dairymen, as bitter unmarketable milk is produced by cows that graze the plant heavily.

527. *Helianthus annuus* L. Sunflower, Wild sunflower. Fig. 123, d–g.

Annual; reproducing by seeds. Cultivated fields, grain fields, fencerows, roadsides and waste places; mostly on rich lowland soils. Native from Minnesota to Saskatchewan, southward to Missouri and Texas; frequently introduced eastward and westward. August–September.

Description.—Stem erect, usually branching above, stout, coarse, rough and pubescent, 0.5–2 m. high. Leaves mostly alternate, simple, petioled, with 3 main veins, ovate or the lower cordate, serrate. Heads solitary and terminal, or also axillary, 6 cm. or more in diameter; involucral bracts imbricated, lanceolate, rough-pubescent; receptacle flat, chaffy, 4 cm. or more in diameter; ray-flowers mostly neutral, bright yellow; disk-flowers numerous, perfect, with tubular brownish corolla. Achenes about 5 mm. long, ovate to wedge-shaped, slightly 4-angled and flattened, white, gray or dark brown with lighter stripes or gray mottled; pappus of 2 thin scales, chaffy, deciduous.

Control.—Cultivate thoroughly while the seedlings are small and follow with hoeing and hand weeding. In grasslands and waste places mow or pull the weeds before seeds are matured. A heavy stand of grass or alfalfa will usually control the sunflower.

528. *Helianthus tuberosus* L. Jerusalem artichoke, Earth-apple, Girasole. Fig. 129, a–c.

Perennial; reproducing by seeds and tubers. Cultivated fields, gardens and waste places; on rich moist soils. Apparently native from western New York to Minnesota and southward to Georgia and Arkansas, although in some of these regions it may have been introduced by the Indians for its edible tubers; also cultivated and becoming established eastward, northward and on the Pacific Coast. August–September.

Description.—Stems erect, branching above, stout, coarse, rough-hairy, 1–3 m. high; rootstocks short, bearing tubers at the apex. Leaves simple, petioled, coarsely serrate, ovate or subcordate to oblong-lanceolate, acuminate, scabrous above, gray-pubescent beneath, 1–2 dm. long, the lower opposite, the upper alternate. Heads solitary or in corymbs, many-flowered; involucral bracts imbricated, lanceolate, attenuate, hirsute-ciliate; receptacle flat, chaffy, 1.5–2

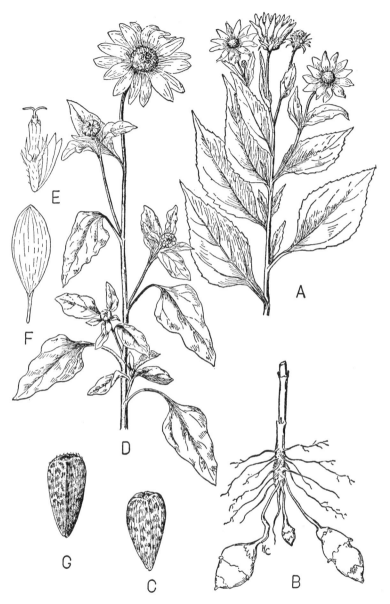

Fig. 123—**JERUSALEM ARTICHOKE,** *Helianthus tuberosus:* **A,** flowering shoot (× ⅕); **B,** tubers (× ¼); **C,** achene (× 4).
SUNFLOWER, *Helianthus annuus:* **D,** flowering shoot (× ¼); **E,** a single disk-flower (× 2); **F,** a single ray-flower (× 1); **G,** achene (× 4).

cm. in diameter; ray-flowers 12–20, neutral, deep yellow; disk-flowers numerous, perfect, with tubular yellow corolla. Achenes about 6–8 mm. long, oblong to wedge-shaped, flattened or obtusely angled, glabrous, striate, gray or brown, often mottled with black; pappus similar to 527.

Control.—Clean cultivation. Frequent and close cutting begun about midsummer. Hogs like the tubers and will dig them if turned into a small area at a time.

Helianthus ciliaris DC., Blue-weed, a native perennial of the grasslands of the Southwest, causes considerable trouble in cultivated land throughout the south. This is a small plant, usually less than 10 dm. high, with gray-green or blue-green stem and leaves. It spreads by seeds and also by an extensive creeping root system. Plowing about 2 dm. deep at intervals of two months controls this weed. For larger areas a combination of deep plowing followed by smother crops or clean cultivation is recommended.[19]

Helianthus petiolaris Nutt., Prairie sunflower, a native annual from Minnesota to Manitoba and Texas and westward, also locally introduced in the eastern states, has the same general habits and appearance as 527. It is a more slender plant with leaves mostly entire; disk of the receptacle about 2 cm. in diameter; achenes hairy.

529. *Hieracium aurantiacum* L. Orange hawkweed, Orange paintbrush. Fig. 124, d.

Perennial; reproducing by seeds and runners. Old fields, pastures, meadows and lawns; on dry, sterile or gravelly, mostly acid soils. Very abundant from New Brunswick to Ontario, southward to New Jersey and Virginia, increasingly troublesome in the north central states and rare in the Pacific Northwest. Introduced from Europe. June–August.

Description.—Stems scapose, glandular-hirsute, 2–6 dm. high, bearing 1 or 2 leaves or bracts near the base; juice milky; stolons numerous, slender. Leaves basal, simple, sessile, entire, oblanceolate, hirsute. Heads in a crowded corymb, terminating the scape, about 2 cm. in diameter, bearing 12–many perfect orange-red ligulate flowers; involucral bracts imbricated, in 2 or 3 series, lanceolate, hairy; receptacle convex, naked. Achenes 1.5–2 mm. long, cylindrical, vertically ridged, glabrous, minutely granular, dark brown or black; pappus a single row of capillary bristles, rough, fragile, 2–4 mm. long.

[19] Karper, 1922.

Control.—Mow infested meadows early, before seeds are formed, and as soon as the hay is removed plow shallow and keep well cultivated until autumn. Plant a cultivated or annual smother crop the next spring. Stony pastures and fields should be harrowed, fertilized and reseeded.

530. *Hieracium florentinum* All. King devil. Fig. 124, f.

Perennial; reproducing by seeds and new shoots from creeping roots. Meadows, old fields, pastures and waste places; especially on dry, sandy or gravelly soil. Very abundant from Newfoundland and Ontario to Virginia and Iowa. Introduced from Europe. June–July.

Description.—Stems scapose, sparingly bristly-hairy, bearing 1 or 2 bracts, 3–8 dm. high, from short stout rootstocks or creeping roots.

Leaves basal, simple, sessile, entire, narrowly oblanceolate or spatulate, glabrous or nearly so, glaucous. Heads and achenes similar to 529 but the ligulate flowers yellow.

Control.—The same as for 529.

531. *Hieracium floribundum* Wimm. and Grab. Hawkweed, Yellow devil.

Perennial; reproducing by seeds and stolons. Pastures, old fields, and waste places; mostly on dry, gravelly or sandy soils. Eastern Canada and New England, west to Ohio. Introduced from Europe. June–July.

Description.—Stems very similar to 530 in general appearance but producing numerous slender leafy stolons at the base. Leaves basal, glaucous, glabrous above and with scattered bristly hairs beneath. Heads similar to 529, but with yellow ligulate flowers. Achenes similar to 529.

Control.—The same as for 529.

532. *Hieracium pilosella* L. Mouse-ear hawkweed, Felon herb, Mouse bloodwort. Fig. 124, e.

Perennial; reproducing by seeds and rootstocks. Lawns, pastures, and waste places; on rather dry, sandy or gravelly soils. Newfoundland to Minnesota, south to North Carolina and Ohio; most troublesome northward. Introduced from Europe. June–July.

Description.—Stem scapose, glandular-pubescent, 0.5–2 dm. high; juice milky; bearing at its base several slender stolons. Leaves basal, simple, nearly sessile, spatulate, bristly-hairy above, white-tomentose beneath. Heads solitary or 1–3 on the scape, 2–3 cm. in diameter,

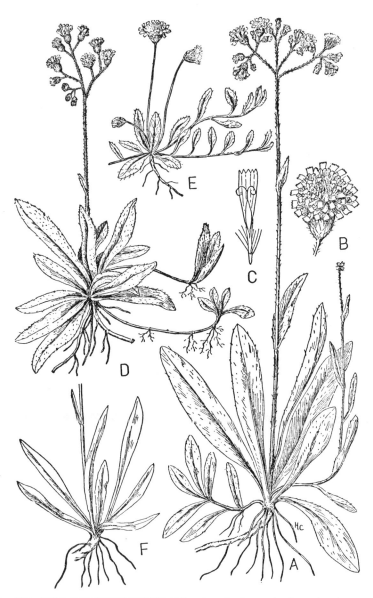

Fig. 124—**YELLOW PAINTBRUSH,** *Hieracium pratense:* **A,** plant showing general habit (× ⅓); **B,** head (× 1); **C,** flower (× 3).
ORANGE HAWKWEED, *Hieracium aurantiacum:* **D,** plant showing habit (× ⅓).
MOUSE-EAR HAWKWEED, *Hieracium pilosella:* **E,** plant showing habit (× ¼).
KING DEVIL, *Hieracium florentinum:* **F,** base of plant showing habit (× ¼).

bearing many perfect golden-yellow ligulate flowers, otherwise similar to 529. Achenes similar to 529.

Control.—The same as for 529.

533. *Hieracium praealtum* Gochnat. King devil.

Perennial; reproducing by seeds. Grasslands and pastures. Locally abundant, Newfoundland to New York. Introduced from Europe. June–July.

Description.—Stems soft, glaucous, scapose or nearly so, arising from a basal rosette, 0.4–1 m. high. Rhizome short and thick with many slender, leafy, spreading or ascending branches arising from among basal leaves. Leaves narrowly oblanceolate, somewhat hispid on both surfaces and stellate-pubescent on lower surface. Heads of yellow ray flowers 7 to many on setose scape.

Control.—The same as for 529.

534. *Hieracium pratense* Tausch. Yellow paintbrush, Devils paintbrush, Yellow devil, Field hawkweed. Fig. 124, a–c.

Perennial; reproducing by seeds and runners. Old fields, meadows, pastures, and lawns; on dry, sterile, gravelly, somewhat acid soils. Very abundant from Quebec and Ontario to North Carolina and Tennessee; infrequent westward to Michigan. Introduced from Europe. June–July.

Description.—Stems scapose, hirsute, 4–8 dm. high, bearing 1–3 well developed leaves; juice milky; stolons few, slender. Leaves basal, simple, nearly sessile, entire or sparingly dentate, narrowly oblanceolate, bristly-hairy, green, not glaucous, 1–2.5 dm. long. Heads in a loose open corymb, terminating the scape, about 2 cm. in diameter, bearing 12–many perfect yellow ligulate flowers. Achenes and pappus similar to 529.

Control.—The same as for 529.

535. *Hypochoeris radicata* L. Gosmore, Cats-ear, Flatweed, Coast dandelion, False dandelion. Fig. 113, c–d.

Perennial; reproducing by seeds. Lawns, meadows, pastures, and waste places; mostly on sandy or gravelly soils. Very abundant on the Pacific Coast, locally common in eastern Canada and the northeastern and north central states. Introduced from Europe. June–September.

Description.—Stems scapose, stout, branching several times, glabrous, 2–4 dm. high; juice milky. Leaves basal, lanceolate, with irregular or rounded lobes, hirsute. Heads solitary, terminating the

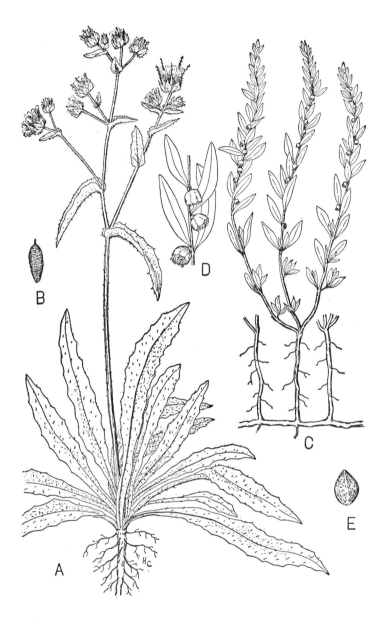

FIG. 125—**BRISTLY OX-TONGUE,** *Picris echioides:* **A,** plant showing habit (× ⅓); **B,** seed (achene), (× 3).
POVERTY-WEED, *Iva axillaris:* **C,** plant showing habit (× ⅓); **D,** portion of flowering branch (× 2); **E,** seed (achene), (× 3).

branches of the scape, 2–4 cm. in diameter, with numerous perfect yellow ligulate flowers; involucral bracts imbricated, in several series, smooth, appressed; receptacle flat, chaffy. Achenes 4–5 mm. long, spindle-shaped, with long ridges, beak barbed, about 5 mm. long, glabrous, reddish-brown; pappus of plumose bristles, 8–10 mm. long, persistent.

Control.—Plow badly infested fields and cultivate for one or two years before reseeding (6). Scattered plants in grasslands and lawns should be spudded out below the crown in early spring as soon as the leaves appear (12).

536. *Inula helenium* L. Elecampane, Horseheal, Horse-elder, Yellow starwort, Scabwort, Elf-dock, Elf-wort. Fig. 122, c–d.

Perennial; reproducing by seeds. Old fields, pastures, roadsides, and waste places; mostly in wet, somewhat gravelly, soils. Northeastern United States, westward to Minnesota and Missouri, rare in the Pacific Northwest. Introduced from Eurasia. July–September.

Description.—Stems erect, very stout, simple or sometimes branched, pubescent, 1–1.5 m. high. Leaves alternate, simple, sessile and partly clasping, ovate, dentate, glabrous above, white-woolly beneath; basal leaves with stout hairy petioles. Heads solitary, terminal, 5–10 cm. in diameter, on stout hairy peduncles; involucre hemispherical, of whitish imbricated bracts, the outer leaf-like; receptacle naked; ray-flowers numerous, pistillate, narrow and yellow; disk-flowers numerous, perfect, yellow. Achenes linear, 4–6 mm. long, flattened, 4–5-sided, vertically striate, light brown or olive-brown; pappus capillary, bristles simple, unequal.

Control.—This weed does not persist if the land can be cultivated for a year (6). In pastures and meadows cut the crowns below the ground with a spud or spade (12).

537. *Iva axillaris* Pursh. Poverty-weed, Death-weed, Devils-weed, Small-flowered marsh elder. Fig. 125, c–e.

Perennial; reproducing by seeds, creeping roots and erect rootstocks. Grain fields, meadows, cultivated crops, and waste places; common in alkaline or saline soils. Manitoba to British Columbia, southward to Mexico. Native. July–September.

Description.—Stems erect or ascending, mostly branched, smooth or hairy, 2–6 dm. high. Leaves opposite, or the upper alternate, sessile, entire, narrowly oblong, usually 3-nerved, rather thick and crowded, stiff and rough to the touch. Heads small, drooping, axillary in the upper leaves, with several disk-flowers, pistillate and staminate flowers in the same head; involucral bracts 5, united into a 5-lobed cup; receptacle chaffy; flowers greenish-yellow, the 4 or 5 marginal pistillate and fertile, the others perfect but sterile. Achenes obovoid to lens-shaped, about 2 mm. long, scurfy, often with scattered resinous dots, gray-brown; pappus wanting.

Control.—A short rotation including a clean cultivated crop every two or three years. When the soil is adapted for alfalfa a thick stand will tend to crowd out the poverty-weed. Mow waste places several times a season to prevent seed formation.

538. *Iva xanthifolia* Nutt. Marsh elder, Highwater shrub, False ragweed, Red-river-weed, False sunflower. Fig. 126, f–h.

Annual; reproducing by seeds. Grain fields, meadows, cultivated fields, farmyards, and roadsides; usually on rich soils. Native from Michigan and Manitoba westward to Alberta and southwestward to Mexico; introduced in the northeastern states and on the Pacific coast. July–September.

Description.—Stems erect, branched, stout, mostly glabrous below, rough-hairy above, 1–3 m. high. Leaves mostly opposite, simple, long-petioled, doubly dentate, broadly ovate to cordate, pubescent. Heads in panicled spikes, terminal and axillary, drooping, sessile, numerous, with several disk-flowers; pistillate and staminate flowers in the same head; involucral bracts few, rounded; receptacle flat, chaffy; flowers greenish-white, the 5 marginal pistillate and fertile, the others perfect but sterile. Achenes obovoid, 1.5–2.5 mm. long, slightly flattened, ribbed longitudinally, glabrous or scurfy, gray-brown to purplish-black; pappus wanting.

Control.—Clean cultivation begun early and followed by hand pulling or hoeing of scattered plants. Mow waste places to prevent seed production.

539. *Lactuca muralis* (L.) Gaertn.

Annual or biennial; reproducing by seeds. Moist land, roadsides, and waste places. Local, Quebec and New York, west to Michigan. Introduced from Europe. July–September.

Description.—Stems slender, glabrous, often glaucous, 3–9 dm. high. Leaves thin, lower surface more or less glaucous, lyrate-pinnatifid with large terminal segment; upper leaves few and much reduced. Flowers 5, bright yellow, in several to many narrow heads in an open panicle; involucre slenderly cylindric, 9–11 mm. high in fruit.

Fig. 126—**COCKLEBUR,** *Xanthium orientale:* **A,** seedling (× ⅓); **B,** leaf and fruiting branch showing habit (× ⅓); **C,** mature bur (× ⅔); **D,** vertical section of bur (× ⅔); **E,** two seeds removed from bur (× 1).
MARSH ELDER, *Iva xanthifolia:* **F,** basal leaf and root (× ⅓); **G,** flowering branch (× ⅓); **H,** seed (achene), (× 6).

Achenes black or dark red, oblanceolate, 5–7 ribbed on each face, about 0.5 mm. wide and 4 mm. long, including the very short beak.

Control.—The same as for 541.

540. **Lactuca pulchella** (Pursh) DC. Blue lettuce, Showy lettuce, Wild lettuce, Milkweed.

Perennial; reproducing by seeds and creeping roots. Meadows, pastures, grain fields, and waste places. Michigan and Ontario, westward to Washington and southward; rarely adventive in the east. Native. July–September.

Description.—Stems erect, simple or branched above, pale or glaucous, about 1 m. high; juice milky. Leaves alternate, sessile, oblong to linear-lanceolate, entire or the lower lobed, glabrous. Heads few and large, 2–3 cm. broad, in racemose clusters, on bracted peduncles, each with several to many blue ligulate flowers; receptacle naked; involucre cylindrical, the bracts imbricated, in 3–4 ranks. Achenes 4–6 mm. long, elliptical, with long tapering apex and short beak, 3–4 vertical ridges on each side, glabrous, reddish to dark brown; pappus capillary, white, copious, deciduous.

Control.—Plow infested fields and plant a cultivated crop for one or two years. Sow alfalfa where the land is adapted to this crop. Mow waste places, fence-rows and ditch banks several times during the summer.

541. **Lactuca scariola** L. Wild lettuce, Prickly lettuce, Compass plant, Milk thistle, Horse thistle, Wild opium. Fig. 128, c–d.

Annual or winter annual; reproducing by seeds. Grain fields, cultivated fields, and waste places; usually on light or dry soils. Widespread and locally abundant throughout the northern United States and southern Canada. Introduced from Europe. July–August.

Description.—Stems erect, simple or with short branches, mostly glabrous, ridged, stiff, 6–15 dm. high; juice milky. Leaves alternate, simple, clasping, prickly-denticulate, midrib white and prickly; the upper oblong; the lowest pinnatifid, sagittate, glabrous, thin, tending to turn the blade into a vertical position. Heads numerous, on short pedicels, in a large open terminal panicle, 4–10 mm. in diameter; with 6–30 perfect pale yellow ligulate flowers; receptacle naked; involucral bracts imbricated, in 2 or more sets of unequal length. Achenes about 3 mm. long, obovate, flattened, with a long slender beak, with 6 vertical ridges on each side, gray-brown; pappus capillary, white, deciduous.

FIG. 127—**YELLOW GOATS-BEARD,** *Tragopogon pratensis:* **A,** plant showing habit ($\times \frac{1}{4}$); **B,** a single flower ($\times 1$); **C,** seed (achene), ($\times 2$).
NIPPLE-WORT, *Lapsana communis:* **D,** part of plant ($\times \frac{1}{6}$); **E,** head showing persistent bracts ($\times 3$); **F,** a single flower ($\times 3$); **G,** seed (achene), ($\times 6$).
FIREWEED, *Erechtites hieracifolia:* **H,** part of a plant showing habit ($\times \frac{1}{3}$); **J,** achene with pappus ($\times 2$); **K,** seed (achene), ($\times 6$).

Control.—Badly infested areas should be planted with an intertilled crop for a year or two. Plants in waste places should be mowed close to the ground as soon as the first flowers appear and repeated later for new shoots. Scattered plants can be destroyed by cutting the rosettes below the crown with a hoe or spud.

Lactuca canadensis L., Wild lettuce, with yellow flowers, and *Lactuca biennis* (Moench) Fern., with pale blue flowers, two biennials native chiefly in the eastern states and adjacent Canada, sometimes encroach upon meadows and neglected fields. They have smooth leaves without prickles. Control the same as for 541.

542. *Lapsana communis* L. Nipple-wort, Succory dock, Ballogan. Fig. 127, d–g.

Annual; reproducing by seeds. Gardens, orchards, fields and waste places. Local in the northeastern United States, south to Virginia and west to Michigan and Missouri; also common in the Pacific Northwest. Introduced from Europe. July–August.

Description.—Stems erect, slender, much branched, nearly glabrous, 3–12 dm. high, juice milky. Leaves alternate, simple, sessile, dentate, ovate to lanceolate, glabrous; the lower lyrate, dentate to undulate, petioled. Heads numerous, 4–7 mm. in diameter, in a loose panicle, with 8–12 perfect yellow ligulate flowers; receptacle naked; involucre cylindric, of 8 smooth linear bracts. Achenes 3–4 mm. long, spindle-shaped, with about 20 vertical ridges, glossy, silver-gray to light brown; pappus wanting.

Control.—The same as for 541.

543. *Leontodon autumnalis* L. (*Apargia autumnale* Hoffm.). Fall dandelion, Lions-tooth, Hawkbit. Fig. 130, d–g.

Perennial; reproducing by seeds and short rootstocks. Meadows, pastures, roadsides, and waste places; mostly on light gravelly soils. Locally common in the northeastern United States and adjacent Canada. Introduced from Europe. July–September.

Description.—Stems 1–8 dm. high, simple or sparingly branched, leafless but with scaly bracts above, glabrous, with milky juice. Rootstock short, thick, horizontal. Leaves basal, simple, sessile, laciniate-dentate or pinnatifid, slightly pubescent. Heads solitary or a few terminating the stem, erect, 2–3 cm. in diameter, bearing only ligulate flowers; involucral bracts 15–20, linear, scarcely imbricated; receptacle naked; flowers many, perfect, with bright yellow corollas. Achenes cylindrical to fusiform, 3–5 mm. long, 5-ribbed, with rows

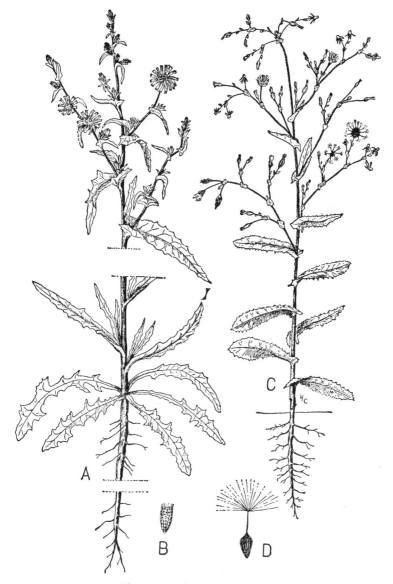

Fig. 128—**CHICORY,** *Cichorium intybus:* **A,** plant showing general habit (× ⅙); **B,** seed (achene), (× 3).
WILD LETTUCE, *Lactuca scariola:* **C,** plant showing general habit (× ⅙); **D,** seed (achene), (× 2).

of tubercles, brown; pappus capillary, plumose, tawny, persistent.

Control.—Mow infested meadows early before seeds are produced and when the crop is removed, plow deep and follow with a clean cultivated crop the next year.

544. *Lygodesmia juncea* (Pursh) D. Don. Skeleton-weed, Rush pink, Wild asparagus, Devils shoestring.

Perennial; reproducing by seeds and roots. Meadows, fields, and waste places; mostly in light dry soils. From the north central states to Saskatchewan, southward to Missouri and Arizona. Native. June–August.

Description.—Stems erect, usually tufted, stiff, much branched, finely grooved, somewhat glaucous, appearing leafless and rush-like, 2–6 dm. high; juice milky. The basal leaves lanceolate to linear, rigid; the upper alternate, awl-shaped and minute. Heads numerous, terminal, with 5 rose-purple ligulate flowers; involucre cylindrical, of 5–8 linear bracts in a single row and several small bractlets at the base. Achenes 4–8 mm. long, linear, with 8–10 vertical ridges on each side, dull; pappus capillary, light brown, soft, about 8–10 mm. long.

Control.—A short rotation including an intertilled crop at least once every three years. Summer fallow badly infested fields.

545. *Madia glomerata* Hook. Tarweed.

Annual; reproducing by seeds. Fields, roadsides, pastures, and waste places; mostly on dry soils. Native on the Pacific Coast; local east to Minnesota, infrequent and locally introduced in the eastern states. June–September.

Description.—Stems erect, simple or with erect branches, very leafy above, hairy, 3–10 dm. high. Leaves alternate or the lower opposite, linear or linear-lanceolate, ascending, scabrous and hirsute, entire or somewhat toothed. Heads in terminal racemose clusters; involucre laterally compressed; receptacle chaffy at the margin; ray-flowers few and pistillate, yellow, or none; disk-flowers 1–5, perfect, their corollas pubescent. Achenes about 5 mm. long, narrowly obovate, straight or curved, with 4 obtuse angles or flattened, bronze-brown mottled with black; pappus wanting.

Control.—The same as for 546.

546. *Madia sativa* Molina var. *congesta* T. and G. Tarweed, Common madia.

Annual; reproducing by seeds. Fields, meadows, pastures, ranges, and waste places; mostly in dry areas. Pacific Coast states and Mexico; infrequent eastward. Introduced from South America. June–September.

Description.—Stems erect, stout, simple or with erect branches, sticky, pubescent and very glandular, 3–10 dm. high. Leaves alternate or the lower opposite, lanceolate or the upper linear, sessile, entire or somewhat toothed. Heads in narrow, dense, panicled clusters; receptacle chaffy at the margin; ray-flowers 5–12, pistillate, with pale yellow corollas; disk-flowers 1–5, perfect. Achenes similar to 545 but usually somewhat smaller, black or brown mottled with black.

Control.—Mow infested fields and waste places before seeds are formed. Plants with ripe seed should be mowed, piled and burned. Pull or spud out scattered plants in lawns.

547. *Matricaria matricarioides* (Less.) Porter (*M. suaveolens* Buchenau). Pineapple-weed, Rayless chamomile. Fig. 107, e–h.

Annual; reproducing by seeds. Grain fields, farmyards, waste places, and roadsides. Native on the Pacific Coast and eastward to Montana and Wyoming. Introduced in the northeastern states and spreading rapidly, especially about cities and towns. May–September.

Description.—Stems erect or spreading, much branched, glabrous, 1–4 dm. high; odor of bruised plant suggesting pineapple. Leaves alternate, 2–3-pinnately parted into short linear divisions. Heads solitary, terminal, numerous, on short peduncles, with many greenish-yellow perfect disk-flowers, about 1 cm. in diameter or smaller; involucral bracts imbricated, shorter than the disk, oval, margins white-scarious; receptacle conical, naked. Achenes obovate to oblong, 1–1.5 mm. long, 3–5 ribbed, apex oblique, with a small tubercle in the center, glabrous, yellow or gray, often with 2 red stripes; pappus of a minute crown or wanting.

Control.—Badly infested fields should be plowed in the spring and planted with an intertilled crop (6). Mow roadsides and waste places before seeds are formed (11).

548. *Parthenium hysterophorus* L. Parthenium. Fig. 129.

Annual; reproducing by seeds. Waste places, roadsides, cultivated ground. Florida to Texas, north locally to Massachusetts, Michigan, Missouri, and Kansas. Adventive from tropical America where it is widespread. July–October.

Description.—Stems usually branched, up to 1 m. high, glabrous to sometimes puberulent above and hirsute below. Leaves mem-

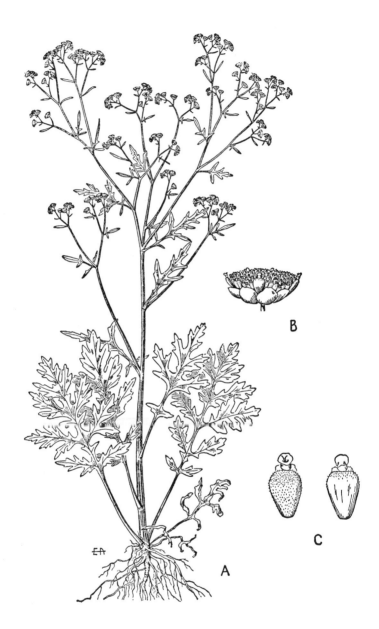

FIG. 129—**PARTHENIUM,** *Parthenium hysterophorus:* **A,** plant showing habit ($\times \frac{1}{3}$); **B,** flower head ($\times 3$); **C,** ray flower in ventral and dorsal view ($\times 5$).

branous, bipinnatifid. Flowers in small, numerous, slender-stalked heads, in loose corymb-like cymes, the disk 3–5 mm. wide; rays tiny, whitish; pappus of 2 large orbicular scales. Achenes obovate, black, up to 2 mm. long.

Control.—Mow close to the ground as soon as the first flowers appear and repeat later for new shoots. Scattered plants can be destroyed by cutting the rosettes below the crown with a hoe or spud.

549. *Picris echioides* L. Bristly ox-tongue, Bugloss. Fig. 125, a–b.

Annual or biennial; reproducing by seeds. Fields, pastures, and waste places. Local in the northeastern and north central states; also on the Pacific Coast. Introduced from Europe. July–September.

Description.—Stems erect or spreading, much branched, bristly with stiff hairs, 3–10 dm. high; juice milky. The basal leaves spatulate, irregularly toothed; the upper leaves alternate, mostly entire, sessile and clasping; all bristly-hairy. Heads on short peduncles, in corymbose clusters, each about 2 cm. broad, with many perfect yellow ligulate flowers; involucre with outer bracts loose and spreading, broadly ovate, leaf-like and prickly hairy, the inner bracts membranous, narrow and pointed. Achenes about 3 mm. long, narrowly obovate-oblong, apex prolonged into a beak; surface roughened by transverse broken wavy ridges, light orange; pappus plumose, deciduous.

Control.—Clean cultivation (6). Mow infested meadows before the ox-tongue seeds are mature (11).

550. *Picris hieracioides* L. Ox-tongue, Bugloss. Fig. 130, a–c.

Annual or biennial; reproducing by seeds. Meadows, grain fields, cultivated fields, and waste places. Locally common in the northeastern and north central states. Introduced from Europe. July–September.

Description.—Stems erect, branched, rough-bristly, 3–7 dm. high; juice milky. The basal leaves spatulate, narrowed to margined petioles; stem-leaves alternate, simple, lanceolate, clasping, irregularly dentate; all bristly-hairy. Heads on short peduncles, in spreading corymbose clusters, 1–2 cm. in diameter, with many perfect yellow ligulate flowers; involucral bracts unequal, the outer narrow and spreading, bristly. Achenes about 3 mm. long, fusiform, with a short beak, straight or curved, with 5–10 vertical ribs and transverse ridges between them, reddish-brown; pappus plumose, deciduous.

Control.—The same as for 549.

Fig. 130—**OX-TONGUE**, *Picris hieracioides*: **A**, plant showing habit (× ⅓); **B**, achene with pappus (× 3); **C**, seed (achene), (× 6).
FALL DANDELION, *Leontodon autumnalis*: **D**, plant showing general habit (× ⅓); **E**, flower (× 2); **F**, achene with pappus (× 3); **G**, seed (achene), (× 6).

551. **Ratibida pinnata** (Vent.) Barnh. (*Lepachys pinnata* T. & G.), Cone-flower, Cone-headed daisy.

Perennial; reproducing by seeds. Meadows, pastures, neglected fields, and waste places. Native from the Great Lake states southward to Texas and Arizona; also locally introduced eastward. June–July.

Description.—Stems erect, slender, branched above, grooved, hoary with minute appressed hairs, 5–15 dm. high. Leaves alternate, pinnately compound with 5–7 long, narrow, acute leaflets. Heads solitary on long slender peduncles, many-flowered; involucral bracts few, small and spreading; receptacle columnar, with truncate chaff; ray-flowers few, neutral, with long, drooping, light yellow or brownish corollas; disk-flowers numerous, perfect, much shorter than the ray-flowers. Achenes 2.5–4 mm. long, obovate-oblong, angular, apex truncate with a small projection in the hollowed center, granular, striate, grayish-black; pappus wanting.

Control.—Plow infested areas and plant a clean cultivated crop for a year. Scattered plants and waste areas should be cut each time when flowers begin to appear.

552. **Rudbeckia serotina** Nutt. (*R. hirta* L.) Black-eyed Susan, Darky-head, Ox-eye daisy, Nigger-head, Yellow daisy, Cone-flower. Fig. 131, d–e.

Biennial; reproducing by seeds. Meadows, pastures, and old fields. Native from western New York and Ontario south to Georgia and Alabama and west to Colorado and Texas; also introduced and locally abundant eastward to the Atlantic Coast. June–August.

Description.—Stems erect, simple or sparingly branched near the base, rough-hairy, 3–8 dm. high. Leaves alternate, simple, 1-nerved, sessile, nearly entire, oblong, rough-hairy; lower spatulate, triple-nerved, petioled. Heads solitary, terminal, many-flowered, 5–10 cm. in diameter; involucral bracts in 1–2 rows, lanceolate, rough-hairy, spreading; receptacle conical, chaffy; ray-flowers 14–16, orange-yellow, neutral, 3–4 cm. long; disk-flowers perfect, numerous, dark purplish-brown. Achenes fusiform, about 2 mm. long, 4-angled, with fine vertical ridges, glabrous, black; pappus wanting.

Control.—This weed does not persist when the land can be plowed and cultivated for a year (6). Infested meadows should be mowed early, before seeds are formed (11). Scattered plants should be pulled or hoed out (4).

553. **Rudbeckia triloba** L.

FIG. 131—**YARROW,** *Achillea millefolium:* **A,** plant showing general habit ($\times \frac{1}{3}$); **B,** one cluster (head) showing several small flowers ($\times$ 3); **C,** seed (achene), ($\times$ 12). **BLACK-EYED SUSAN,** *Rudbeckia serotina:* **D,** plant showing general habit ($\times \frac{1}{3}$); **E,** seed (achene), ($\times$ 12).

Annual or biennial; rarely a short-lived perennial; reproducing by seeds. Open woods, rocky slopes, thickets, and fields, especially in calcareous soil. Native; adventive or naturalized in the Northeast. New York to Minnesota and Kansas, south to Florida and Texas. June–October.

Description.—Stems 0.3–1.6 m. high, hispid, hirsute or glabrous, usually with leafy branches. Juvenile and lower cauline leaves sometimes simple and cordate, long-petioled, 3-lobed to pinnately 5–7-parted; middle and upper cauline leaves ovate to lanceolate, entire to coarsely toothed, short-petioled or sessile, acuminate. Heads with short peduncles; bracts of the involucre narrow; rays 6–12, deep yellow but often with the base orange to brown; 15–35 mm. long; disk usually purple–black, depressed but becoming ovoid, 5–13 mm. broad. Achene quadrangular, with pappus or minute crown. Highly variable.

Control.—The same as for 552.

554. *Senecio jacobaea* L. Stinking willie.

Biennial or winter annual, occasionally perennial; reproducing by seeds. Dry soil in fields, pastures, roadsides, and waste places. Newfoundland throughout much of the Maritime Provinces, south locally to Massachusetts, New Jersey and Ontario; Pacific Coast. Avoided by grazing animals. Naturalized from Europe. July–October.

Description.—Stems coarse, hard, solitary, erect, from a short taproot, simple up to the inflorescence, cobwebby but quickly glabrate, 0.2–1.2 m. high. Leaves evenly distributed, cobwebby beneath, mostly 2–3-pinnatifid, 0.4–2.3 dm. long and 2–11 cm. wide, the lower petioled, often deciduous, the uppermost sessile. Heads several to many in a showy, terminal, broad corymb; the disk 5–10 mm. wide; involucre about 4 mm. high, with 13 bracts, rays 4–10 mm. long. Receptacle naked, pappus capillary. Achenes minutely pubescent.

Control.—The same as for 552.

555. *Senecio vulgaris* L. Groundsel, Grimsel, Simson, Bird-seed, Ragwort. Fig. 132, a–c.

Annual or winter annual; reproducing by seeds. Gardens, truck lands, nurseries, and waste places; mostly on moist rich land. Widespread throughout the northern United States and Canada, south to Texas and California. Introduced from Europe. April–October.

Description.—Stems erect or decumbent, much branched, glabrous, 1–5 dm. high, frequently rooting at the lower nodes. Leaves alter-

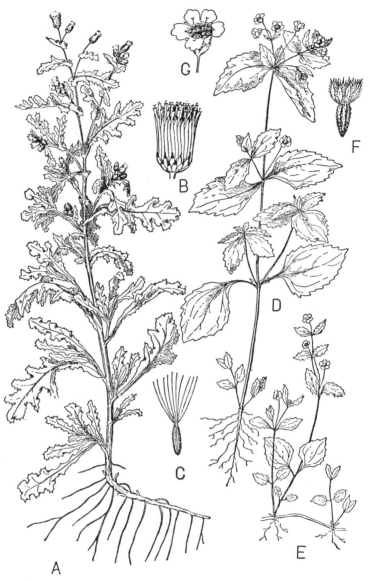

Fig. 132—**GROUNDSEL,** *Senecio vulgaris:* **A,** plant showing general habit (× ⅓); **B,** head of flowers (× 2); **C,** achene (× 3).
GALINSOGA, *Galinsoga ciliata:* **D,** plant showing general habit (× ⅓); **E,** branch showing roots at nodes (× ⅓); **F,** achene (× 6); **G,** head of flowers (× 2).

nate, pinnatifid, dentate, clasping, slightly auricled, glabrous; the lower tapering to a petiole. Heads mostly in corymbose clusters, about 1 cm. in diameter, with many perfect yellow disk-flowers but no ray-flowers; involucral bracts linear, green with black tips, in a single row with a few bractlets at the base; receptacle flat, naked. Achenes 3–4 mm. long, spindle-shaped, apex truncate, with about 10 vertical ridges, minutely hirsute, reddish-brown to gray-brown; pappus of soft capillary bristles, deciduous.

Control.—This weed is most troublesome during cool wet weather in autumn and spring. If infested land is harrowed several times in autumn, many of the seeds will germinate and the seedlings will be destroyed. Plowing early next spring will destroy another growth of seedlings before the crop is planted.

556. *Solidago canadensis* L. Canada goldenrod, Tall goldenrod.

Perennial; reproducing by seeds and rootstocks. Meadows, old fields, fence-rows and waste places; mostly on rich soils. Newfoundland to Manitoba, south to Virginia and Colorado. Native. August–October.

Description.—Stems erect, slender, mostly simple, glabrous or pubescent above, 5–15 dm. high; from short rootstocks. Leaves narrowly lanceolate, essentially uniform from base to summit of stem, mostly serrate, 3-nerved, glabrous above and pubescent beneath, at least on the nerves. Heads in crowded recurved racemes and forming a broad panicle; receptacle naked; involucre 2–3 mm. high; otherwise similar to 558. Achenes 1 mm. long, similar to 558.

Control.—The same as for 558.

557. *Solidago gigantea* Ait. (*S. serotina* Ait.). Goldenrod.

Perennial; reproducing by seeds and rootstocks. Moist meadows, pastures and waste places. Widespread throughout eastern North America; infrequent westward to British Columbia, Oregon, and Texas. Native. July–September.

Description.—Stems erect, simple, glabrous and often glaucous, 5–25 dm. high. Leaves alternate, simple, sessile, lanceolate, sharply serrate except at the base, 3-nerved, nearly glabrous. Heads numerous, in a broad pubescent panicle; receptacle naked; involucre 3.5–5 mm. high, with linear bracts; ray-flowers 7–14, pistillate, yellow; disk-flowers several, perfect, yellow. Achenes about 1.5 mm. long, similar to 558.

Control.—The same as for 558.

558. *Solidago graminifolia* Salisb. var. *Nuttallii* Fern. Narrow-leaved goldenrod, Bushy goldenrod, Creeping yellow-weed.

Perennial; reproducing by seeds and creeping rootstocks. Common in old fields and pastures; on dry or moist, gravelly or clay soils. Common in eastern North America, westward to South Dakota and Missouri. Native. August–September.

Description.—Stems erect, simple or branched above, rigid, glabrous, 5–10 dm. high; from long creeping rootstocks. Leaves alternate, simple, sessile, entire, linear-lanceolate, glabrous or nearly so, the lower scarce or wanting. Heads numerous, in a dense terminal flat-topped corymb, obovoid-cylindric, sessile, 20–30-flowered; involucral bracts closely appressed, in 3–many series, straw-colored or yellowish-green; receptacle flat, naked; ray-flowers 6–20, yellow, pistillate; disk-flowers several, perfect, yellow. Achenes about 2.5 mm. long, obovoid, apex truncate, with a central tubercle, ribbed, pubescent, dull, light brown; pappus capillary, usually deciduous.

Control.—Plow infested fields shallow in autumn and plant a cultivated crop the following year.

559. *Solidago rugosa* Mill. Rough goldenrod.

Perennial; reproducing by seeds and short rootstocks. Fields, pastures, and waste places; mostly on dry, stony or gravelly soil. Widespread and locally abundant throughout eastern North America. Native. August–September.

Description.—Stems erect, simple or branched, villous, 5–20 dm. high. Leaves alternate, simple, sessile, lanceolate, mostly sharply serrate, thin, loosely veiny but with one main vein, pubescent. Heads in spreading racemes forming a broad panicle; receptacle naked; involucre with appressed bracts, 3–4 mm. high; ray-flowers 6–9, pistillate, yellow; disk-flowers 4–7, perfect, yellow. Achenes about 1 mm. long, similar to 558.

Control.—The same as for 558.

Several other species of Solidago, native to various sections of North America, are sometimes abundant in old pastures and neglected fields. These do not persist under cultivation or if the tops are mowed about twice a year before seeds are formed.

560. *Sonchus arvensis* L. Perennial sow thistle, Field sow thistle, Creeping sow thistle, Gutweed. Fig. 133.

Perennial; reproducing by seeds and creeping roots. Grain fields, cultivated fields, grasslands, and waste places; mostly on rich soils.

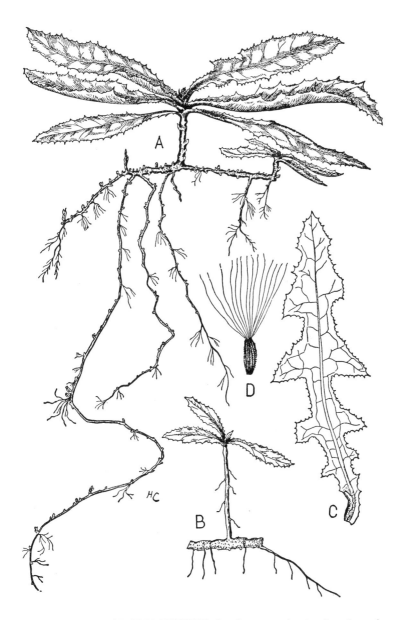

Fig. 133—**PERENNIAL SOW THISTLE,** *Sonchus arvensis:* **A,** plant in spring, showing spreading root system with new shoots starting from buds ($\times \frac{1}{4}$); **B,** new plant starting from a small piece of root ($\times \frac{1}{2}$); **C,** a deeply lobed basal leaf common in some forms ($\times \frac{1}{3}$); **D,** seed (achene), ($\times 6$).

Widespread throughout the northern United States and southern Canada; most troublesome in the grain sections of the Northwest [20] and in Manitoba and Saskatchewan.[21] Introduced from Europe. June–September.

Description.—Stems from creeping roots [20] which are fleshy and which sometimes produce tuberous enlargements, erect, simple, or branched, stout, ridged, glabrous and often glaucous, 6–15 dm. high; juice milky. Leaves alternate, simple, clasping, prickly-dentate, runcinate-pinnatifid, the base cordate, glabrous; the lower with margined petioles. Heads about 4 cm. in diameter, in large open corymbose clusters, with many perfect bright yellow ligulate flowers; peduncles glandular-pubescent or glabrous; involucral bracts subulate, somewhat imbricated, green, glandular-pubescent or glabrous; receptacle naked. Achenes 2.5–4 mm. long, narrowly ovate with truncate base and the apex with a central tubercle, somewhat flattened, with 5 prominent ribs on each side, transversely wrinkled, reddish-brown; pappus capillary, simple, persistent.

Control.—Keep all sow thistles mowed frequently enough to prevent seed formation. Summer fallow throughout the growing season (19). Avoid a continuous grain crop and use a short rotation (20). Plant early crops such as millet, barley or oats. After harvest plow and cultivate until time to sow winter rye. When the rye is removed follow with late summer cultivation. The next spring sow sweet clover with the spring grain. After the grain is harvested pasture the sweet clover in late summer and the next year.

This description includes the form of the perennial sow thistle with glabrous peduncle and receptacle usually considered as *Sonchus arvensis* var. *glabrescens* Wimm. and Grab., but by some authors also designated as a distinct species, *Sonchus uliginosus* Bieb.[22, 23]

561. *Sonchus asper* (L.) Hill. Spiny-leaved sow thistle. Fig. 134, a–c.

Annual; reproducing by seeds. Cultivated fields, gardens, grain fields, and waste places. Widespread throughout North America. Introduced from Europe. July–September.

Description.—Stems from a tap-root, erect, branched, stout, 3–20 dm. high; juice milky. Leaves similar to 560 but more prickly, the base auricled and rounded, the lower often spatulate and undivided. Heads similar to 560, 1.2–2.5 cm. in diameter; ligulate flowers pale

[20] Stevens, 1924, 1926. [21] Tullis, 1924.
[22] Long, 1922. [23] Pretz, 1923.

yellow. Achenes similar to 560, 2–3 mm. long, with 3 small ribs on each side, slightly wrinkled, orange-brown.

Control.—Clean cultivation followed by hand hoeing and weeding of scattered plants missed by the cultivator (6, 5). Infested grain fields should be harrowed while the grain is small so as to destroy the sow thistle seedlings (8). Mow waste places before seeds are formed (11).

562. *Sonchus oleraceus* L. Sow thistle, Annual sow thistle, Hares lettuce, Colewort, Milk thistle. Fig. 134, d–e.

Annual; reproducing by seeds. Cultivated fields, gardens and grain fields. Widespread throughout North America. Introduced from Europe. July–September.

Description.—Stems from a tap-root, erect, branched, stout, glabrous and glaucous, 3–20 dm. high; juice milky. Leaves similar to 560 but less prickly-dentate, auricles small and acute. Heads similar to 560, in cymose panicles, 1.2–2.5 cm. in diameter; ligulate flowers pale yellow. Achenes similar to 560, 2–3 mm. long, the ribs less prominent.

Control.—The same as 561.

563. *Tanacetum vulgare* L. Tansy, Bitter buttons, Hind-head, Parsley fern.

Perennial; reproducing by seeds and rootstocks. Along roadsides, waste places, and neglected fields; dry, sandy or gravelly soils. Widespread in eastern North America; infrequent westward to the Pacific Coast; most common in the northeastern states. Introduced from Europe as a medicinal herb. July–September.

Description.—Stems erect, simple, rather stout, glabrous, 3–9 dm. high; usually forming dense colonies. Leaves alternate, 1–3-pinnately divided, dentate, glabrous, petioles winged. Heads numerous, in dense terminal corymbose clusters, erect, less than 1 cm. in diameter, with many disk-flowers; receptacle convex, chaffy; involucral bracts numerous, imbricated, dry, scarious; disk-flowers yellow, perfect, except the marginal ones are chiefly pistillate. Achenes about 1.5 mm. long, linear, with truncate apex, 5-angled, yellow-brown; pappus a short 5-toothed crown.

Control.—Tansy does not persist under cultivation. In grasslands and along roadsides mow the tansy close to the ground several times during the summer (11). Small patches can be destroyed by digging the rootstocks.

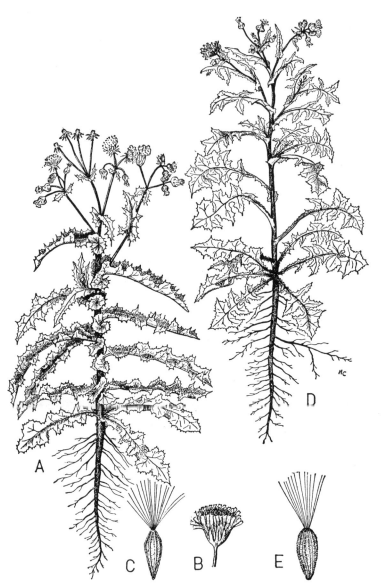

Fig. 134—**SPINY-LEAVED SOW THISTLE,** *Sonchus asper:* **A,** plant showing general habit ($\times \frac{1}{6}$); **B,** head of flowers ($\times \frac{1}{2}$); **C,** seed (achene), ($\times 6$).
SOW THISTLE, *Sonchus oleraceus:* **D,** plant showing general habit ($\times \frac{1}{6}$); **E,** seed (achene), ($\times 6$).

564. *Taraxacum erythrospermum* Andrz. (*T. laevigatum* DC.) Redseeded dandelion.

Perennial; reproducing by seeds. Lawns, grasslands, and waste places. Widespread but rather local, mostly in the northeastern and north central states. Introduced from Europe. May–June.

Description.—Very similar to 565 in general habit and appearance but the leaves more deeply divided. Heads smaller, 2–3 cm. in diameter, sulfur-yellow; involucral bracts glaucous. Achenes about 3 mm. long, oblong-obovate, 4–5 ribbed, terminating in a beak about 6 mm. long, red or reddish-brown; pappus capillary, copious, about 5 mm. long.

Control.—The same as for 565.

565. *Taraxacum officinale* Weber. Dandelion, Lions-tooth, Blowball, Cankerwort. Fig. 135, a–d.

Perennial; reproducing by seeds and by new shoots from roots. Lawns, meadows, and pastures. Widespread throughout North America. Introduced from Eurasia; according to some authorities, also native in northern North America. May–June.

Description.—Stems very short, from a long thick taproot, bearing a rosette of leaves and several large, hollow, glabrous scapes; juice milky. Leaves basal, varying in outline from oblong to spatulate, runcinate, pinnatifid, or rarely nearly entire, usually pubescent. Heads solitary on the scape, 3–5 cm. broad, with many perfect bright yellow ligulate flowers; involucral bracts in two rows, the outer elongated and reflexed, the inner linear and erect; receptacle flat or convex, naked. Achenes similar to 564 but yellow-brown; pappus capillary, 3–4 mm. long, persistent.

Control.—Spudding (12). Cut off tops before seeds form.

566. *Tragopogon major* Jacq. (*T. dubius* Am. auth., not Scop.).

Biennial or perennial; reproducing by seeds. Fields, roadsides, and waste places. New York to Washington, south to Virginia, Illinois, Texas, and California. Introduced from Europe. May–July.

Description.—Similar to *T. pratensis* but peduncle enlarged upwardly below the head; involucre in fruit 4–7 cm. high; achenes 2–4 cm. long.

Control.—The same as for 567.

567. *Tragopogon porrifolius* L. Salsify, Vegetable-oyster, Oysterplant, Goats-beard, Noon-plant, Jerusalem star.

Biennial; reproducing by seeds. Old meadows, roadsides, and

waste places. Widespread throughout the United States and eastern Canada; locally common. Introduced from Europe; grown in gardens for the edible root, frequently escaping and spreading. June–July.

Description.—Stem from a fleshy tap-root, erect, branched, 6–12 dm. high, glabrous; juice milky. Leaves alternate, simple, entire, linear or tapering from a clasping base, keeled, glabrous. Heads solitary, terminal, 7–10 cm. in diameter, with many purple ligulate flowers closing by noon; peduncle dilated and hollow for several cms. below the head; involucral bracts lanceolate-attenuate, erect, in one row, equal, extending beyond the rays; receptacle convex, naked. Achenes 10–15 mm. long, narrowly fusiform, terminating in a long beak, straight or curved, with 5–10 rough-scaly ribs, dull, light brown; pappus of numerous yellow plumose bristles, 2–3 cm. long.

Control.—Badly infested fields should be plowed and cultivated for a year. Meadows and waste places should be mowed as soon as the first flower-heads appear and again later.

568. *Tragopogon pratensis* L. Yellow goats-beard, Meadow salsify, Noon-flower, Morning sun. Fig. 127, a–c.

Biennial; reproducing by seeds. Old meadows, roadsides and waste places. Locally common from New Brunswick to Manitoba, southward to Georgia and Colorado; infrequent westward. Introduced from Europe. June–July.

Description.—A plant with the same general appearance as 567 but with yellow flowers and the peduncle slightly, or not at all, dilated below the head; heads smaller, 4–5 cm. in diameter; achenes 5–10 mm. long.

Control.—The same as for 567.

569. *Tussilago farfara* L. Coltsfoot, Cough-wort, Ginger-root, Clayweed, Dove-dock, Horse-hoof. Fig. 135, e–k.

Perennial; reproducing by seeds and rootstocks. Moist clay banks, along ditches and becoming more common in pastures and fields. Newfoundland to Minnesota, southward to Pennsylvania and Ohio. Introduced from Europe. April–May.

Description.—Stems appearing in early spring, from horizontal rootstocks, erect, scapose, scaly-bracted, woolly when young. Leaves basal, simple, palmately veined, petioled, dentate, angular or cordate, glabrous above, white-woolly beneath, thick, appearing near the end of the flowering season. Heads solitary, terminal, many-flowered, 2–3

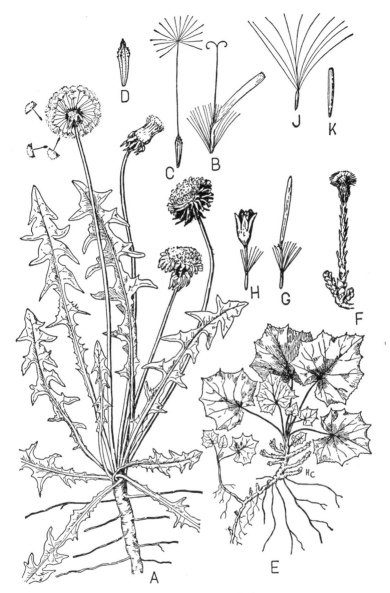

Fig. 135—**DANDELION**, *Taraxacum officinale:* **A**, plant showing general habit ($\times \frac{1}{3}$); **B**, flower ($\times 2$); **C**, seed (achene) with pappus ($\times 2$); **D**, seed (achene) without pappus ($\times 3$).
COLTSFOOT, *Tussilago farfara:* **E**, plant showing habit ($\times \frac{1}{4}$); **F**, head of flowers ($\times \frac{1}{3}$); **G**, ray-flower ($\times 2$); **H**, disk-flower ($\times 2$); **J**, seed (achene) with pappus ($\times 2$); **K**, seed (achene) without pappus ($\times 3$).

cm. in diameter; involucral bracts in one row, oblong, obtuse, often purplish; receptacle flat, naked; ray-flowers yellow, in several rows, pistillate, fertile; disk-flowers yellow, perfect but sterile, numerous. Achenes 3–4 mm. long, cylindric, with many vertical ribs, glossy, yellow or reddish-brown; pappus capillary, often breaking and leaving a crown of scales.

Control.—Improved drainage (21). Clean cultivation.

570. *Xanthium orientale* L. (*X. pensylvanicum* Wallr., *X. canadense* Mill., *X. commune* Britton). Cocklebur, Clotbur, Sheep bur, Button-bur, Ditch-bur. Fig. 126, a–e.

Annual; reproducing by seeds. Fields, roadsides, and waste places; lowlands, especially on river bottom land subject to overflow. Widespread from southern Canada throughout the United States to Mexico; native and most abundant and troublesome in the Mississippi Valley; mostly introduced eastward; by some authors considered to be a native of Eurasia and Central America. August–September.

Description.—Stems erect, with spreading branches, coarse, rough-pubescent, stout, angled and often red-spotted. Leaves alternate, simple, long-petioled, broadly ovate-cordate, dentate or somewhat lobed, rough-pubescent. Heads unisexual, the staminate small, in short terminal spikes, several-flowered, the pistillate in axillary clusters; staminate involucre top-shaped, with 7–12 separate bracts, pistillate involucre ovoid, coriaceous, closed, hairy and spiny, bearing 2 pistillate flowers developing into a hard prickly bur. Bur oblong, light brown, about 2 cm. long; prickles numerous, crowded, 3–6 mm. long, hooked at the summit, hairy; beaks 2, stout, incurved; achenes 1–1.5 cm. long, 2 in each bur, oblong, flattened, with pointed apex, dark brown.

Control.—Clean cultivation followed by hand hoeing and weeding. Mow waste places before the cockleburs have formed seeds. Mature plants should be burned after mowing.

The cocklebur is represented in the United States by several very similar forms based largely on variations in the shape, hairiness and spininess of the mature burs. Since these characteristics of the burs do not appear to be constant, it has seemed best for the present purpose to follow Wiegand and Eames [24] and treat all the forms under one species.

[24] Wiegand and Eames, 1926.

The spiny burs are easily scattered by hooking on clothing and the fur of animals. Usually only one of the pair of seeds in each bur germinates in the first season. The cocklebur seedlings are poisonous to livestock, especially when they are grazed in the cotyledon stage. Swine appear to be more susceptible to poisoning than cattle or sheep.[25]

571. *Xanthium spinosum* L. Spiny cocklebur, Spiny clotbur, Clotweed, Spanish thistle, Dagger cocklebur.

Annual; reproducing by seeds. Waste places and pastures, on alluvial soil along streams. Widespread in the eastern United States. Introduced from tropical America. August–September.

Description.—Stems erect, 3–9 dm. high, slender, with spreading branches, yellowish, hoary-pubescent, armed at the axils with triple yellow spines. Leaves alternate, short-petioled, lanceolate or ovate-lanceolate, pinnately lobed, dark gray-green above with white veins, white-downy beneath. Heads unisexual, similar to 570; the pistillate involucre developing into a bur containing two achenes. Bur oblong, very slightly flattened, 10–13 mm. long, light brown, prickles hooked, glabrous, 3–4 mm. long; beak short, straight; achenes similar to 570 but smaller.

Control.—The same as for 570.

Xanthium strumarium L. is listed (1954) as a weed in the northeastern states.

[25] Marsh, Roe, and Clawson, 1923.

GLOSSARY

Accrescent. Growing larger with age.
Accumbent (cotyledons). Having the edges against the radicle.
Achene. Small, dry and hard, one-seeded indehiscent fruit.
Acuminate. Gradually tapering to a long point.
Acute. Sharp pointed.
Adnate. Congenital union of two different organs.
Adventitious. Appearing out of usual order or place.
Aggregate fruit. One formed by the coherence of several pistils that were distinct in the flower.
Albuminous seed. One containing an endosperm or albumen, the material surrounding the embryo.
Alternate. One (leaf or bud) at a node; placed singly at different heights on the stem.
Anatomosing. Forming a network of cross-veins.
Angiospermae. Plants with seeds borne in an ovary.
Annual. Of one season's duration from seed to maturity and death.
Anther. The pollen-bearing part of the stamen.
Apetalous. Having no petals.
Apiculate. Ending in a short pointed tip.
Appressed. Lying close and flat against.
Aril. An appendage or covering of a seed, growing from the hilum or base of the seed.
Ascending. Rising obliquely upwards.
Attenuate. Becoming very narrow.
Auriculate. With ear-shaped appendages.
Awl-shaped. Narrow and sharp-pointed; gradually tapering from the base to a slender or stiff point.
Awn. A bristle-like part or appendage.
Axil. Upper angle formed where the leaf joins the stem.
Axile. Situated on the axis.
Axillary. Situated in an axil.
Axis. The main or central line of development of a plant or organ; the stem.
Barbed. Furnished with rigid points or short bristles, usually reflexed like the barb of a fish-hook.
Beak. A long, prominent and substantial point, usually applied to prolongations of fruits and pistils.
Bearded. Bearing a long awn, or furnished with long or stiff bristles.
Berry. Fleshy fruit, soft throughout.
Bi-. A Latin prefix signifying two, twice or doubly.
Biennial. Of two seasons' duration from seed to maturity and death.
Blade. The expanded part of a leaf.

Bract. A much reduced leaf, particularly the small or scale-like leaves of a flower-cluster or associated with the flowers.
Bracteate. Having bracts.
Bracteose. Bracts numerous or conspicuous.
Bractlet. A secondary bract, as one borne on the pedicel of a flower.
Branchlet. An ultimate division of a branch, not including the last season's growth.
Bristle. A stiff strong hair.
Bulb. An underground leaf-bud with fleshy scales and a short axis.
Bulblet. A small bulb, borne on the stem or inflorescence.
Bulbous. Having the character of a bulb.
Calyx. The outer whorl of floral envelopes; the outer perianth whorl.
Campanulate. Bell-shaped; cup-shaped with a broad base.
Canescent. Hoary with gray pubescence.
Capillary. Hair-like; very slender.
Capitate. Shaped like a head; aggregated into a dense or compact cluster, a head.
Capsule. Dry dehiscent fruit of a compound pistil.
Carinate. Having a keel or projecting longitudinal medial line on the lower or outer surface.
Carpel. One of the foliar units of a compound pistil. A simple pistil has one carpel.
Caruncle. An appendage at or about the hilum of a seed.
Catkin. A scaly-bracted spike, with imperfect flowers.
Cauline. Pertaining or belonging to the stem.
Cell. A structure containing a cavity, as the cells of an ovary or anther.
Chaff. A small thin scale or bract, becoming dry or membranous; in particular the bracts in the flower-heads of composites.
Chlorophyll. The green coloring matter within the cells of plants.
Ciliate. Fringed with hairs on the margin.
Cinereous. Ash-color; light gray.
Circumscissile. Dehiscing by a regular transverse circular line of division, the top usually separating as a lid.
Clasping. Leaf partly or completely surrounding a stem.
Claw. The long, narrow, petiole-like base of a petal or sepal in some flowers.
Cleft. Lobed, with the incisions extending halfway or more to the midrib of a leaf.
Comose. Furnished with or like a tuft of hairs.
Compound leaves. Those in which the blade consists of two or more separate parts (leaflets).
Compound pistil. One composed of two or more fused carpels.
Conduplicate. Two parts folded lengthwise along the midrib, particularly cotyledons or leaves in a bud.
Cone. A dense and usually elongated collection of flowers or fruits that are borne beneath scales, or a collection of sporophylls on an axis, the whole mass forming a fruit-like body.
Connate. Congenital union of like structures.

Glossary

Connivent. Coming together or converging but not organically connected.
Cordate. Heart-shaped, with the point away from the base.
Coriaceous. Of the texture of leather.
Corm. A solid bulb-like part usually subterranean; the enlarged, fleshy, solid base of a stem.
Corolla. The inner whorl of floral envelopes; the inner perianth whorl.
Corymb. A flat or convex flower-cluster with the outer flower opening first.
Cotyledon. Seed-leaf; the primary leaf or leaves in the embryo.
Creeping. A trailing shoot that takes root throughout its length.
Crenate. With rounded teeth, scalloped.
Crested. With an elevated and irregular or toothed ridge.
Culm. The stem of grasses or sedges.
Cuneate. Wedge-shaped; triangular with the acute angle downward.
Cuspidate. Tipped with a sharp and rigid point.
Cyme. A broad more or less flat-topped flower-cluster, with central flowers opening first.
Deciduous. Falling off in autumn (as leaves); falling early.
Decompound. More than once compound or divided.
Decumbent. Reclining at base but the summit ascending.
Decurrent. Extending down the stem below the insertion.
Deflexed. Bent or turned abruptly downward.
Dehiscent. Opening regularly by valves or slits, as a capsule.
Deltoid. Triangular; delta-like.
Dentate. Toothed, teeth pointing outward.
Denticulate. Furnished with minute teeth.
Depressed. Somewhat flattened from the end.
Di-. A Greek prefix signifying two or twice.
Diadelphous (stamens). Combined in two sets or groups.
Dichotomous. Forking regularly by pairs.
Dicotyledon. A plant of the subdivision of Angiospermae bearing two cotyledons or seed-leaves.
Diffuse. Loosely branching or spreading; of open growth.
Digitate. Hand-like; compound with the members arising from a common point.
Dioecious. Unisexual, with staminate and pistillate flowers on different plants.
Discoid. Disk-like; in Compositae, having only disc flowers.
Disk. A more or less fleshy or elevated development of the receptacle about the pistil; receptacle in the heads of composites.
Disk-flowers. The tubular flowers in the center of heads of composites, as distinguished from the marginal or ray-flowers.
Dissected. Divided into many slender segments.
Divaricate. Widely divergent or spreading.
Divergent. Spreading away from each other.
Divided. Separated to the base.
Dorsal. Back; relating to the back or outer surface of a part of an organ.

Drupe. Fleshy indehiscent fruit with endocarp stony and exocarp soft; stone-fruit.
Ellipsoid. Solid body with elliptical outline.
Elliptical. A flat part or body that is oval or oblong with the ends rounded.
Emarginate. With a shallow notch at the apex.
Empty glume. One of the two basal bracts of the grass spikelet, a glume without a flower in its axil.
Endocarp. The inner layer of a pericarp.
Endosperm. Nutritive tissue or material about the embryo in a seed.
Entire. Margin even, not toothed, notched or divided.
Ephemeral. Short-lived; persisting for one day only.
Epigynous. Borne on the ovary; used of floral parts when ovary is inferior and flower not perigynous.
Erose. With margin as if gnawed or eroded.
Evergreen. Remaining green in its dormant season or throughout the year.
Exalbuminous seed. Without an albumen or endosperm.
Exocarp. The outer layer of a pericarp.
Exserted. Projecting beyond an envelope.
Falcate. Scythe-shaped or sickle-shaped.
Fascicle. A condensed or close cluster.
Fertile. Said of pollen-bearing stamens and seed-bearing flowers.
Filiform. Thread-like, long and very slender.
Fimbriate. Fringed.
Flexuous. Having a more or less zigzag wavy form.
Floret. Individual flower of grasses and composites; also other very small flowers that form a part of a dense inflorescence.
Floriferous. Flower-bearing.
Flowering glume. One of the bracts of the grass spikelet bearing a flower in its axil.
Follicle. A fruit consisting of a single carpel splitting along the inner or upper suture only.
Free central (placentation). Placenta projecting from the bottom of the ovary but otherwise free.
Frond. The leaf of ferns.
Fruit. The ripened ovary or ovaries with the adnate parts; the seed-bearing organ.
Fusiform. Spindle-shaped, swollen in the middle and tapering toward each end.
Gamopetalous. Petals united; corolla of one piece.
Gibbous. Swollen on one side.
Glabrate. Somewhat glabrous, or becoming glabrous with maturity.
Glabrous. Without hairs.
Gland. A secreting part or appendage, but term often used for small swellings or projections on various organs.
Glandular. Furnished with glands, or gland-like.
Glaucous. Covered or whitened with a bloom.
Globose. Spherical in form or nearly so.

Glossary

Globular. Nearly globose.
Glomerate. In a compact cluster or clusters.
Glomerule. A compact head-like cluster.
Glume. A small chaff-like bract; in particular, those of the grass spikelet.
Glutinous. Sticky, gluey.
Grain. Like an achene but the seed-coat and thin pericarp fused throughout into one body, particularly the fruit of grasses.
Granular, granulose. Composed of, or appearing as if covered by, minute grains.
Habit. The general aspect of a plant, or its mode of growth.
Habitat. The home of a plant; the situation in which it grows wild.
Halberd-shaped. Shaped like an arrowhead but with the basal lobes spreading.
Hastate. Halberd-shaped like an arrowhead, but with basal lobes widely spreading.
Head. A short compact flower-cluster of more or less sessile flowers.
Herb. Plant naturally dying to the ground; without persistent stem above ground; without a definite woody structure.
Herbaceous. Not woody; of the texture of an herb.
Hilum. In the seed the scar or mark indicating the point of attachment.
Hirsute. With rather coarse or rough hairs.
Hispid. Provided with stiff or bristly hairs.
Hyaline. Transparent or translucent.
Hypogynous. Borne on the axis or under the ovary, said of stamens or petals and sepals.
Imbricated. Overlapping like the shingles on a roof.
Imperfect (flower). One lacking either stamens or carpels.
Included. Not protruded, as stamens not projecting from the corolla; not exserted.
Incumbent cotyledons. The back of one cotyledon lying against the radicle.
Indehiscent. Not regularly opening, as a seed-pod or anther.
Indurated. Hardened.
Indusium. The covering of the sorus or fruit-dot in ferns.
Inferior ovary. One in which the sepals appear attached on the top.
Inflexed. Bent inward.
Inflorescence. Flower-cluster; mode of flower-bearing.
Internode. The part of the stem between two nodes.
Involucel. An involucre about a part of a flower-cluster; a secondary involucre.
Involucre. A whorl of small leaves or bracts standing close below a flower or flower-cluster.
Involute. Rolled inward toward upper surface.
Irregular flower. The parts of a whorl or series not all alike.
Keeled. Ridged like the bottom of a boat; also the two united petals of a papilionaceous flower form a keel.
Lacerate. With margin appearing as if torn.
Laciniate. With narrow irregular lobes.

Lanceolate. Several times longer than wide, broadest near the base and narrowed to the apex.

Lateral. On or at the side.

Leaflet. One part of a compound leaf.

Legume. Dehiscent dry fruit of a simple pistil normally splitting along two sides.

Lemma. Lower bract inclosing grass flower.

Lenticular. Lens-shaped, with two convex surfaces.

Ligulate. Strap-shaped; particularly as applied to the ray-flowers of composites.

Ligule. A projection or outgrowth from the top of the sheath in grasses and similar plants.

Linear. Long and narrow with parallel margins.

Lobed. Divided into segments about to the middle.

Loculicidal (dehiscence). Splitting through the back of each cell or carpel of a capsule.

Lyrate. Pinnatifid, but with an enlarged terminal lobe and smaller lateral lobes.

Maculate. Spotted.

Marginal (flowers). The outer flowers, particularly in the head of composites, either ligulate or tubular.

Membranaceous. Thin, rather soft, and somewhat translucent.

Mericarp. One of the two carpels of the fruit of a member of the Parsley family.

-Merous. In composition, referring to the number of parts; as flowers 5-merous, in which the parts of each kind or series are five or in fives.

Micropyle. The point on the seed which represents the closed orifice of the ovule.

Monadelphous. Stamens united into one group by their filaments.

Moniliform. Like a string of beads.

Monocotyledon. A plant of the subdivision of Angiospermae bearing one cotyledon or seed-leaf.

Monoecious. Staminate and pistillate flowers on the same plant.

Mucronate. Furnished with an abrupt minute point.

Multiple fruit. One formed by the coherence of pistils and associated parts of the several flowers of an inflorescence.

Naked. Without calyx or corolla (flower); without chaff (receptacle of composites).

Nerved. With simple veins or slender ribs.

Netted-venation, -veined. The principal veins of a leaf forming a network.

Neutral (flower). Without stamens or carpels.

Node. A joint or place where leaves are attached to a stem.

Nut. A hard, indehiscent, one-celled and one-seeded fruit, usually resulting from a compound ovary.

Nutlet. A small nut.

Ob-. A Latin syllable usually indicating inversion; as obovate, inverted ovate.

Glossary

Oblanceolate. Like lanceolate, but with the narrow end towards the stem.
Oblique (leaves). Unequal sided.
Oblong. Longer than broad, and with the sides nearly parallel most of their length.
Obtuse. Blunt or rounded at the end.
Ocrea. Sheath formed by the fused stipules, as in some Polygonaceae.
Opposite. Two (leaves or buds) at a node.
Orbicular. Circular.
Ovary. The part of the pistil bearing the ovules.
Ovate. With an outline like that of a hen's egg, with the broad end toward the base.
Ovoid. A solid that is oval in flat outline.
Ovule. Undeveloped seed.
Palea. Upper of two bracts inclosing grass flower.
Palet. In the grass flower, the upper of the two inclosing bracts, the low one being the flowering glume or lemma.
Palmate. Radiating fan-like from approximately one point.
Palmately compound (leaf). Leaflets radiating from one point.
Panicle. An elongated, irregularly branched, raceme-like inflorescence.
Papilionaceous corolla. Butterfly-like; pea-like flower with standard, wings and keel.
Papillate. Bearing minute protuberances.
Pappus. The modified calyx-limb in composites, forming a crown of bristles, scales, etc., at the summit of the achene.
Parallel venation, -veined. The principal veins of a leaf running parallel or nearly so.
Parasitic. Growing on, and deriving nourishment from, another plant.
Parietal. On the inner wall or surface of a capsule.
Parted. Cleft or cut not quite to the base.
Pectinate. Comb-like, pinnatifid with very narrow close divisions or parts.
Pedicel. Stem of an individual flower in a cluster.
Peduncle. Stem of a solitary flower or of a flower-cluster.
Pellucid. Clear, nearly transparent.
Peltate. Attached to its stalk inside the margin; shield-shaped.
Perennial. Of three or more seasons' duration.
Perfect (flower). Having both stamens and carpels.
Perfoliate. The stem appearing to pass through the leaf.
Perianth. The floral envelope considered together, mostly used for flowers in which there is no clear distinction between calyx and corolla, as in lilies.
Pericarp. The matured ovary.
Perigynous. Borne around the ovary and not beneath it, as when calyx, corolla and stamens are borne on the edge of a cup-shaped receptacle.
Persistent. Remaining attached; leaves not all falling off at the same time.
Petal. One of the divisions of a corolla.
Petaloid. Petal-like.
Petiole. The stalk of a leaf.

Pilose. Shaggy, with soft hairs.
Pinna. A primary division or leaflet of a compound leaf.
Pinnate. Feather-formed; with the leaflets of a compound leaf arranged on each side of a common axis.
Pinnately compound. Leaflets arranged on each side of a common axis of a compound leaf.
Pinnatifid. Pinnately cleft or parted.
Pinnule. A secondary pinna or leaflet in a pinnately decompound leaf.
Pistil. The ovule-bearing and seed-bearing organ with style and stigma.
Pistillate. Having pistils and no stamens; female.
Placenta. Part or place in the ovary where ovules are attached.
Plicate. Folded into plaits, usually lengthwise, like a closed fan.
Plumose. Plumy, feather-like; with fine hairs as in the pappus bristles of some composites.
Plumule. The bud or growing point of an embryo plant.
Pod. A general term to designate a dry dehiscent fruit.
Pollen. The spores or grains borne by the anther, containing the male element.
Pollinium. A coherent mass of pollen, as in milkweeds.
Poly-. A Greek syllable meaning many.
Polygamous. Bearing perfect and imperfect flowers on the same plant.
Polypetalous. Having separate petals.
Pome. Fruit of apple, hawthorn, etc.; fleshy fruit with bony or leathery several-celled core and soft outer part.
Prickle. A small sharp outgrowth from the bark, rind or epidermis.
Prismatic. Shaped like a prism.
Procumbent. Trailing or lying flat upon the ground but not rooting.
Prostrate. Lying flat upon the ground.
Pteridophyte. A fern or related plant; a spore-bearing vascular plant.
Puberulent. Minutely pubescent.
Pubescent. Covered with soft short hairs.
Punctate. With translucent or colored dots, depressions or pits.
Raceme. A simple flower-cluster of pedicelled flowers on a common elongated axis.
Rachilla. A diminutive or secondary axis; in particular, in the grasses and sedges, the axis of the spikelet.
Rachis. The axis of a spike or compound leaf.
Radical (leaves). Appearing to come from the root or from the base of the stem near the ground.
Radicle. The part of the embryo below the cotyledons.
Ramose. With many branches.
Ray-flower. The modified flowers of the outer part of the heads of some composites, with a strap-like extension of the corolla.
Receptacle. The more or less enlarged or elongated end of the peduncle or flower axis.
Recurved. Curved outward or backward.
Reflexed. Bent outward or backward.

Glossary

Regular flower. All the parts of a whorl or series alike.
Reniform. Kidney-shaped.
Repand. Wavy-margined.
Reticulate. The veins forming a network.
Retrorse. Bent or curved over, back or downward.
Retuse. With shallow notch at otherwise rounded apex.
Revolute. Margin rolled backward or under.
Rhizome. Underground stem, rootstock.
Rhombic. With outline of equilateral oblique-angled parallelogram.
Rib. A primary vein or nerve in a leaf or similar organ; any prominent elevated line along a body.
Rootstock. Underground stem, rhizome.
Rosette. A very short stem or axis bearing a dense cluster of leaves.
Rotate. Wheel-shaped.
Rudimentary. Imperfectly developed or in an early state of development.
Rugose. Wrinkled; generally due to the depressions of the veins in the upper surface of the leaf.
Runcinate. Coarsely toothed or cut, the pointed teeth turned toward the base of the leaf.
Runner. A slender trailing stem taking root at the nodes.
Sagittate. Shaped like an arrowhead; triangular with the basal lobes pointing downward.
Salverform. With a slender tube and an abruptly spreading border.
Samara. An indehiscent winged fruit.
Scabrous. Rough to the touch.
Scales. Dry and appressed, modified or reduced leaves or bracts.
Scape. A leafless peduncle arising from the ground.
Scapose. Resembling a scape.
Scarious. Thin, dry and membranaceous.
Scurfy. Covered with small bran-like scales.
Sepal. One of the divisions of a calyx.
Septicidal (dehiscence). Splitting through the partitions of a capsule.
Serrate. Having sharp teeth pointing forward.
Serrulate. Finely serrate.
Sessile. Without a stalk.
Setose. Full of bristles.
Sheath. A long or more or less tubular structure surrounding an organ or part.
Silique. Capsule with two valves separating from a thin longitudinal partition.
Simple leaves. Those in which the blade is all in one piece.
Sinuate. Wavy-margined.
Sinus. Spaces between two lobes.
Sorus. The fruit-dot or -cluster of ferns (plural sori).
Spadix. A thick or fleshy spike, surrounded or subtended by a spathe.
Spathe. The bract or leaf surrounding or subtending a flower-cluster or a spadix.

Spatulate. Gradually narrowed downward from a rounded summit.
Spermatophyte. A seed-bearing plant.
Spike. A flower-cluster like a raceme, but with sessile or nearly sessile flowers.
Spikelet. A secondary spike; a unit of the inflorescence of the grasses.
Spine. A sharp, rather slender, rigid outgrowth.
Spinulose. Provided with small spines.
Sporangium. A spore-case.
Spore. A simple reproductive body usually consisting of a single detached cell, and containing no embryo, particularly in the ferns and lower plants.
Spur. A tubular or horn-like projection from the calyx or corolla; it usually secretes nectar.
Stamen. The pollen-bearing or male organ of a flower.
Staminate. Having stamens and no pistils; male.
Standard. The upper and broad, more or less erect, petal in a papilionaceous flower.
Stellate. Star-shaped.
Sterile. Infertile, barren.
Stigma. The part of the pistil which receives the pollen.
Stipe. The stalk of a pistil; the petiole of a fern leaf.
Stipel. Stipule of a leaflet.
Stipule. A basal appendage of a petiole (usually two).
Stolon. A shoot that bends to the ground and takes root at its tip and gives rise to a new plant.
Stoloniferous. Bearing runners or shoots that take root.
Striate. Marked with fine longitudinal lines or ridges.
Strict. Very straight and upright.
Strigose. With appressed, sharp, straight and stiff hairs.
Style. The part of the pistil connecting the ovary and stigma, usually more or less elongated.
Stylopodium. A disk-like expansion of the base of the style, particularly in the Parsley family.
Sub-. A Latin prefix usually signifying somewhat or slightly.
Subtended. To stand below or close to, as a bract below a flower.
Subulate. Awl-shaped; broad at base, narrow and tapering from the base to a sharp rigid point, the sides generally concave.
Succulent. Juicy; fleshy; soft and thickened in texture.
Superior ovary. One that is free from the calyx or perianth.
Tap-root. One with a stout tapering body, usually vertical.
Tendril. A thread-shaped organ, modified leaf or stem part, by which a plant clings to a support.
Terete. Circular in cross-section.
Terminal. At the end of a stem or branch.
Ternate. In threes.
Thorn. A degenerated sharp-pointed branch.
Throat. The opening of a gamopetalous corolla or gamopetalous calyx.
Tomentose. Densely hairy with matted wool.
Translucent. Partially transparent.

Glossary

Tri-. Three or three times.
Trifoliate. Of three leaflets.
Trimorphic. Occurring in three forms.
Truncate. Ending abruptly as if cut off transversely.
Tuber. A thickened part; usually an enlarged end of a subterranean stem, as a rootstock.
Tubercle. A small tuber; a rounded protruding body.
Tubular. Hollow and of an elongated or pipe-like form.
Turbinate. Shaped like a top; inversely conical.
Twig. A young shoot: used to denote the growth of the past season only.
Umbel. An umbrella-like flower-cluster.
Umbellet. A secondary umbel.
Unarmed. Destitute of spines, prickles or thorns, etc.
Uncinate. Hooked at the tip.
Undulate. With a wavy margin or surface.
Unisexual. Of one sex; staminate or pistillate.
Utricle. A small, bladdery, one-seeded fruit.
Valve. The units or pieces into which a capsule splits; a separable part of a pod.
Veins. The branches of the fibro-vascular bundles forming the framework in leaves.
Ventral. Front; relating to the inner face of an organ; opposite of dorsal.
Vesicle. A small bladder or air-cavity.
Villous. Long and soft-hairy; shaggy.
Viscid. Sticky.
Whorled. Three or more (leaves or buds) at a node.
Wing. A thin, dry, membranaceous expansion or flat extension or appendage of an organ; the lateral petals of a papilionaceous flower.
Woolly. Provided with long, soft, more or less matted hairs.

READY REFERENCE DATA

UNITS OF WEIGHT AND CAPACITY

1 ounce	= 28.35 grams
1 pound	= 453.6 grams or 16 ounces
1 gram	= 1 cubic centimeter or $\frac{1}{28}$ of an ounce
1 liter	= 1.0567 quarts liquid measure
1 gallon	= 3.735 liters or 4 quarts or 231 cubic inches
1 gallon water	= 8 pounds
1 cubic foot	= 6 gallons

UNITS OF LENGTH AND AREA

1 millimeter (mm.)	= 0.0394 inch or $\frac{1}{25}$ inch
1 centimeter (cm.)	= 0.3937 inch or $\frac{2}{5}$ inch
1 decimeter (dm.)	= 3.937 inches or about 4 inches
1 meter (m.)	= 3.28 feet or $39\frac{1}{3}$ inches
1 kilometer	= 1,000 meters or 3,280 feet 10 inches or 0.62 mile (approximately)
1 inch	= 2.54 centimeters or 25.4 millimeters
1 foot	= 30.48 centimeters
1 yard	= 0.914 meter
1 rod	= 5.5 yards
1 mile	= 1.609 kilometers or 320 rods or 5,280 feet
1 square rod	= $272\frac{1}{4}$ square feet or $\frac{1}{160}$ of an acre
1 acre	= 43,560 square feet or 160 square rods or 0.4 hectare (approximately)
1 hectare	= 2.471 acres or 10,000 square meters
20 by 55 feet	= $\frac{1}{40}$ acre (approximately)

TEMPERATURE

To change C.° to F.° multiply by $\frac{9}{5}$ and then add 32. To change F.° to C.° subtract 32 and then multiply by $\frac{5}{9}$, or by formula, $F = C/5 \times 9 + 32$ and $C = F - 32 \times \frac{5}{9}$.

MISCELLANEOUS

1 pound applied to 100 square feet = 435 pounds per acre (approximately)
1 pound applied to 1,000 square feet = 43 pounds per acre (approximately)
1 pound applied to 1 square foot = $21\frac{3}{4}$ tons per acre

LITERATURE REFERENCES

ABRAMS, LEROY. Illustrated Flora of the Pacific States. Vol. 1, 1940; vol. 2, 1944; vol. 3, 1951.

AHLGREN, GILBERT A., GLENN C. KLINGMAN, and DALE E. WOLF. Principles of Weed Control. 368 pp. Wiley, New York. 1951.

ANONYMOUS. Sulphate of iron and its relation to the farmer and the weed. In Bull. of Amer. Steel and Wire Co. 30 Church St. N.Y.C.

ARNY, A. C. Quack grass eradication. Minn. Agr. Exp. Sta. Bull. 151. 1915.

— Quack grass control. Minn. Agr. Exp. Sta. Ext. Circ. 25. 1928.

— Variations in the organic reserves in underground parts of five perennial weeds from late April to November. Minn. Agr. Exp. Sta. Techn. Bull. 84. 1932.

ARNY, A. C., R. O. BRIDGFORD, and R. S. DUNCAN. Eradicating perennial weeds with chlorates. Minn. Agr. Exp. Sta. Ext. Circ. 32. 1930.

ASLANDER, ALFRED. Chlorates as plant poisons. Jour. Amer. Soc. Agron. 18: 12. 1926.

— Sulphuric acid as a weed spray. Jour. Agr. Research 34:1065–1091. 1927.

— Experiments on the eradication of Canada thistle, Cirsium arvense, with chlorates and other herbicides. Jour. Agr. Research 36:11. 1928.

ATKESON, F. W., H. W. HULBERT, and T. R. WARREN. Effect of bovine digestion and manure storage on the viability of weed seeds. Jour. Amer. Soc. Agron. 26:390–399. 1934.

BAKKE, A. L. European Bindweed. Iowa Agr. Exp. Sta. Circ. 124. 1930.

— Leafy spurge, *Euphorbia Esula* L. Iowa Agr. Exp. Sta. Research Bull. 198. 1936.

— Control of Leafy Spurge, *Euphorbia Esula* L. Iowa Agr. Exp. Sta. Research Bull. 222. 1937.

— Experiments on the control of European Bindweed. Iowa Agr. Exp. Sta. Research Bull. 259. 1939.

BAKKE, A. L., W. G. GAESSLER, and W. E. LOOMIS. Relation of Root Reserves to control of European Bindweed, *Convolvulus arvensis* L. Iowa Agr. Exp. Sta. Research Bull. 254. 1939.

BALL, W. S. White horse nettle. Mon. Bull. Dept. of Agr., Calif. 21:348–349. 1932.

— Weed control. Mon. Bull. Dept. of Agr., Calif. 22:252–257. 1933.

— Poverty weed (Iva axillaris Pursh). Mon. Bull. Dept. of Agr. Calif. 22:305. 1933.

BALL, W. S., AND MARGARET K. BELLUE. "Kelp" or Swamp Knotweed. Mon. Bull. Dept. Agr. Calif. 25. 273–275. 1936.

BALL, W. S., A. S. CRAFTS, B. A. MADSON, and W. W. ROBBINS. Weed Control. Calif. Agr. Ext. Serv. Circ. 97. 1936.

BALL, W. E., and O. C. FRENCH. Sulfuric Acid for Control of Weeds. Calif. Agr. Exp. Sta. Bull. 596. 1935.

BALL, W. S., B. A. MADSON, and W. W. ROBBINS. The control of weeds. Calif. Agr. Ext. Serv. Circ. 54. 1931.

BALL, W. S., and W. W. ROBBINS. Russian Knapweed. Mon. Bull. Dept. of Agr., Calif. 20:666–668. 1931.

— Camel thorn. Alhagi camelorum. Mon. Bull. Dept. of Agr., Calif. 22:258–259. 1933.

— Spiny clotbur (Xanthium spinosum L.). Mon. Bull. Dept. of Agr., Calif. 22:278. 1933.

— Heliotrope. Mon. Bull. Dept. of Agr., Calif. 22:379–380. 1933.

BALL, W. S., W. W. ROBBINS, and M. K. BELLUE. The star thistles (Centaurea spp.). Mon. Bull. Dept. of Agr., Calif. 22:294–298. 1933.

BARNETT, H. L., and H. C. HANSON. Control of leafy spurge and review of literature on chemical weed control. N. Dak. Agr. Exp. Sta. Bull. 277. 1934.

BARR, C. G. Organic Reserves in the Roots of bindweed. Journ. Agr. Research 60:391–413. 1940.

BATHO, GEO. Leafy Spurge. Manitoba Dept. Agr. and Imm. Circ. 106. 1932.

BEACH, C. L. Viability of weed seeds in feeding stuffs. Vt. Agr. Exp. Sta. Bull. 138. 1908.

BEAL, W. J. Seeds of Michigan weeds. Mich. Agr. Exp. Sta. Bull. 260. 1910.

— Michigan weeds. Mich. Agr. Exp. Sta. Bull. 267 (ed. 2). 1915.

BEAUMONT, A. B. Toxicity of several chemicals to a species of moss common to old pastures in the New England States. Journ. Amer. Soc. Agron. 27:134–137. 1935.

BECKWITH, C. S., and JESSIE G. FISKE. Weeds of cranberry bogs. N. J. Agr. Exp. Sta. Circ. 171. 1925.

BELLUE, M. K. Weeds of California seed rice. Mon. Bull. Dept. of Agr., Calif. 21:290–296. 1932.

— New weeds confused with hoary cress. Mon. Bull. Dept. of Agr., Calif. 22:288–293. 1933a.

— Austrian Field Cress—new and noxious. Mon. Bull. Dept. of Agr., Calif. 22:385–386. 1933b.

— Garden Rocket, *Eruca sativa* Mill., a "new" flax weed. Mon. Bull. Dept. Agr. Calif. 25:280–282. 1936.

— Perennial pepper-cress, *Lepidium latifolium* L. Mon. Bull. Dept. Agr. Calif. 27:296–300. 1938.

— Silver-sheathed Knotweed as a pest in southwestern alfalfa. Mon. Bull. Dept. Agric. Calif. 24:238–241. 1935.

Bibliography of Weed Investigations. Division of Weed Investigations. U.S. Dept. of Agr. Reprint from Weeds.

BIOLETTI, FREDERIC T. The extermination of morning-glory. Calif. Agr. Exp. Sta. Circ. 69. 1911.

BLAKE, S. F. Two new western weeds. Science, LV:1426. 455–456. 1922.

— Erucastrum Pollichii in West Virginia. Rhodora 26:22. 1924.

— Potentilla intermedia in the Boston District. Rhodora 30:107–108. 1928.

BLATCHLEY, W. S. The Indiana weed book. 191 pp. Nature Publ. Co. Indianapolis. 1912.

BOHMONT, D. W. Weeds of Wyoming. Wyo. Agr. Exp. Sta. Bull. 325. July, 1953.

BOLLEY, H. L. Distribution of weed seeds by winter winds. N. Dak. Agr. Exp. Sta. Bull. 17:102–105. 1895.
— Studies upon weeds in 1900. N. Dak. Agr. Exp. Sta. Ann. Rept. 11:48–56. 1901.
— Weeds and means of eradication. N. Dak. Agr. Exp. Sta. Bull. 80. 1908.
BOTTEL, A. E. Introduction and control of camel thorn. Alhagi camelorum Fisch. Mon. Bull. Dept. of Agr., Calif. 22:261–263. 1933.
BOWSER, W. E., and J. D. NEWTON. Decomposition and movement of herbicides in soils, and effects on soil microbiological activity and subsequent crop growth. Canadian Jour. Research 8:73–100. 1933.
BOYD, G. W., and C. L. CORKINS. Wyoming weeds and their control. Wyo. Agr. Ext. Serv. Circ. 33. 1940.
BRENCHLEY, W. E. Weeds of farm land. pp. x + 239, pls. 2, figs. 38. 1920.
— Spraying for weed eradication. Jour. Bath and West and Southern Counties Society Ser. 5. 19:1–20. 1925.
BRITTON, N. L. The leafy spurge becoming a pest. Jour. N. Y. Bot. Garden 22:73–75. 1921.
BRITTON, NATHANIEL LORD, and ADDISON BROWN. An illustrated flora of the northern United States and adjacent Canada and the British Possessions. 1913. (Revised edition: The new Britton and Brown illustrated flora of the northeastern United States and adjacent Canada. 3 Vols. Revised by Henry Allen Gleason. New York. New York Botanical Garden. 1952.
BROWN, D. E., and J. E. MCMURTREY, JR. Value of natural weed fallow in the cropping system for tobacco. Maryland Agr. Exp. Sta. Bull. 363. 1934.
BROWN, J. G., and R. B. STREETS. Sulphuric acid spray: a practical means for control of weeds. Ariz. Agr. Exp. Sta. Bull. 128. 1928.
BROWN, W. S. The cranberry in Oregon. Oregon Agr. Exp. Sta. Bull. 225. 1927.
BURGESS, J. L., and C. H. WALDRON. Farm weeds of North Carolina and methods for their control. N. C. Dept. of Agr. Bull. 40:8. No. 259. 1919.
CAMERON, W. H., and H. P. SMITH. Prickly pear eradication and control. Texas Agr. Exp. Sta. Bull. 575. 1939.
CAMPBELL, E. G. Value of weeds to the farmer. Jour. Amer. Soc. Agron. 16:91–96. 1924.
CATES, J. S. The eradication of quack-grass. U. S. Dept. Agr. Farmers Bull. 464. 1911.
CATES, J. S., and H. R. COX. The weed factor in the cultivation of corn. U. S. Dept. Agr. Bur. Plant Ind. Bull. 257. 1912.
CATES, J. S., and W. J. SPILLMAN. A method of eradicating Johnson grass. U. S. Dept. Agr. Farmers Bull. 279. 1907.
CLARK, G. H., and J. FLETCHER. Farm weeds of Canada. 56 Col. Pl. 103 pp. Ottawa. 1906.
CLAUSEN, R. T. Rapistrum in northern North America. Rhodora 42:201–202. 1940.
CLAWSON, A. B. Larkspur or Poisonweed. U. S. Dept. Agr. Farmers Bull. 988. 1918. (Revised 1933.)

COCKERELL, T. D. A. Lamium purpureum in Colorado. Rhodora 28:112. 1926.
COOK, W. H. Fire hazards in the use of oxidizing agents as herbicides. Canadian Jour. Research 8:509–544. 1933.
COOK, W. H., and A. C. HALFERDAHL. Chemical weed killers (a review). Nat. Research Council Canada. Bull. 18. 1937.
COOLEY, L. M. Wild Bramble Eradication. N. Y. (Geneva) Agr. Exp. Sta. Bull. 674. 1936.
CORKINS, C. L., and A. B. ELLEDGE. Continuous burning to eradicate noxious weeds. The Reclamation Era. 30:140–142. 1940.
CORMANY, C. E. Chicory growing in Michigan. Mich. Agr. Exp. Sta. Spec. Bull. 167. 1927.
COUCH, J. F. Trembles (or milk sickness). U. S. Dept. Agr. Circ. 306. 1933.
COULTER, JOHN MERLE, and AVEN NELSON. New manual of botany of the central Rocky Mountains. American Book Company, New York. 1909.
COX, H. R. The eradication of bindweed or wild morning-glory. U. S. Dept. Agr. Farmers Bull. 368. 1909.
— Controlling Canada thistles. U. S. Dept. Agr. Farmers Bull. 545. 1913.
— Eradication of ferns from pasture lands in the eastern U. S., U. S. Dept. Agr. Farmers Bull. 687. 1926.
CRAFTS, A. S. Progress in chemical weed control. Mon. Bull. Dept. of Agr., Calif. 22:264–268. 1933.
— The use of arsenical compounds in the control of deep-rooted perennial weeds. Hilgardia 7:361–372. 1933a.
— Sulfuric acid as a penetrating agent in arsenical sprays for weed control. Hilgardia 8:125–147. 1933b.
— Factors influencing the effectiveness of Sodium Chlorate as a herbicide. Hilgardia 9:437–457. 1935.
— The Toxicity of Sodium arsenite and Sodium chlorate in four California Soils. Hilgardia 9:458–498. 1935.
— Plot tests with chemical soil sterilants in California. Univ. Calif. Exp. Sta. Bull. 648. 1941.
CRAFTS, A. S., and P. B. KENNEDY. The physiology of Convolvulus arvensis (Morning-glory or bindweed) in relation to its control by chemical sprays. Plant Physiol. 3:329–344. 1930.
CRAFTS, A. S., and R. N. RAYNOR. The herbicidal properties of boron compounds. Hilgardia 10:343–374. 1936.
CROWLEY, D. J. Cranberry growing in Washington. Wash. Agr. Exp. Sta. Bull. 230. 1929.
DARLINGTON, H. T. Dr. Beal's seed experiment in Michigan. Amer. Jour. of Bot. 9:266–269. 1922.
DARLINGTON, H. T., E. A. BESSEY, and C. R. MEGEE. Michigan weeds. Mich. Agr. Exp. Sta. Special Bull. 304. 1940.
DARLINGTON, WM. American weeds and useful plants. 1859.
DARROW, G. M. Managing cranberry fields. U. S. Dept. Agr. Farmers Bull. 1401. 1924.
DEATRICK, E. P. The spotting method of weed eradication. Science LXXI: 487–488. 1930.

DETLING, L. E. A revision of the North American Species of Descurainia. Amer. Midl. Nat. 22:481–520. 1939.

DETMERS, FREDA. Canada thistle, Cirsium arvense Tourn. Ohio Agr. Exp. Sta. Bull. 414. 1927.

DURRELL, L. W. Common weeds of Colorado lawns. Colo. Exp. Sta. Bot. Sect. Bull. 310. 1926.

Eastern Section National Weed Committee of Canada Proceedings. Published annually. Vol. 1, 1947+

ECKART, C. F. How thermogen enhances the growth of plants. 29 pp. Honolulu, Hawaii. 1923.

EDMOND, F. B. Mulch paper for vegetable crops. Mich. Agr. Exp. Sta. Quart. Bull. 9:3. 1929.

EGGINTON, G. E. Colorado weed seeds. Colo. Agr. Exp. Sta. Bull. 260. 1921.

EGGINTON, G. E., and W. W. ROBBINS. Irrigation water as a factor in the dissemination of weed seeds. Colo. Agr. Exp. Sta. Bull. 253. 1920.

ESSARY, S. H. Control of Dodder in Lespedeza. Univ. of Tenn. Agr. Exp. Sta. Circ. 22. 1928.

ESSER, P. Die Giftpflanzen-Deutschlands. 212 pp. Braunschweig. 1910.

FERNALD, M. L. *Urtica gracilis* and some related North American species. Rhodora 28:191–199. 1926.

FERTIG, STANFORD N. 1954 Weed Control in Field Crops. Cornell Ext. Circ. 821, Revised Febr. 1954. Ithaca, N.Y.

FISKE, JESSIE G. Weeds of New Jersey. N. J. Agr. Exp. Sta. Circ. 161. 1924. (Revised 1931.)

FLEMING, C. E., and N. F. PETERSON. Don't feed fox-tail hay to lambing ewes. Nev. Agr. Exp. Sta. Bull. 97. 1919.

FLINT, L. H. Crop-plant stimulation with paper mulch. U. S. Dept. Agr. Techn. Bull. 75. 1928.

— Suggestions for paper-mulch trials. U. S. Dept. Agr. Circ. 77. 1929.

FRANZKE, C. J., and A. N. HUME. Field Bindweed. S. Dak. Agr. Exp. Sta. Bull. 305. 1936.

FREED, VIRGIL. Oregon Weed Control Recommendations. Readers Weeders, 28. Ore. State College. 1953. Corvallis, Ore.

GAHN, BESSIE W. How to control ragweed, the principal cause of autumn hay fever. U. S. Dept. Agr. Leaflet 95. 1933.

GARMAN, H. Some Kentucky weeds and poisonous plants. Ky. Agr. Exp. Sta. Bull. 183. 1914.

GARNER, E. S., and S. C. DAMON. The persistence of certain lawn grasses as affected by fertilization and competition. R. I. Agr. Exp. Sta. Bull. 217. 1929.

GARRETT, A. C. Some introduced plants of Utah. Torreya 21:76–79. 1921.

GEORGIA, ADA. A manual of weeds, 593 pp. Macmillan, New York. 1914.

GILBERT, B. E. An analytical study of the putting greens of Rhode Island golf courses. R. I. Agr. Exp. Sta. Bull. 212. 1928.

GILKEY, HELEN M. The most important noxious weeds of Oregon. Ore. Agr. Exp. Sta. Ext. Bull. 412. 1929.

GLEASON, HENRY ALLEN. *See* BRITTON, N. L. and A. BROWN.

GODFREY, G. H. Control of Nutgrass with Chloropicrin. Soil Science 47: 391–394. 1939.

GOSS, W. L. The vitality of buried seeds. Jour. Agr. Research 29:349–362. 1924.

GRANT, C. V., and A. A. HANSEN. Poison ivy and poison sumac and their eradication. U. S. Dept. Agr. Farmers Bull. 1166. 1929.

GRAU, FRED V. Use of sodium chlorate and other chemicals for controlling turf weeds. Bull. U. S. Golf Assoc., Green Sect. 13:154–179. 1933.

GRAY, ASA. Manual of Botany. American Book Company, New York. 8th ed. (Revised by M. L. Fernald) 1950.

GRAY, G. P. Spraying for the control of wild morning-glory within the fog belt. Calif. Agr. Exp. Sta. Circ. 168. 1917.

— Herbicides. Mon. Bull. Dept. of Agr., Calif. 11:263–269. 1922.

— Weed control along fencerows and roadways. Mon. Bull. Dept. of Agr., Calif. 8:599–603. 1919.

GRESS, E. M. Pennsylvania weeds. Penn. Dept. of Agr. Gen. Bull. 416. 1925.

GROH, HERBERT. Some recently noticed mustards. Scientific Agriculture 13: 722–727. 1933.

— Hoary Cresses in Canada. Scient. Agric. 20:750–756. 1940.

— *Rapistrum* spp., in some recently noticed mustards. Scient. Agric. 13: 726–727. 1933.

— Turkestan alfalfa as a medium of weed introduction. Scient. Agric. 21: 36–43. 1940.

HAGAN, W. A., and A. ZEISSIG. Experimental bracken poisoning of cattle. Cornell Veterinarian 16:194–208. 1927.

HANSEN, A. A. Canada thistle and methods of eradication. U. S. Dept. Agr. Farmers Bull. 1002. 1918.

— The hawkweeds or paintbrushes. U. S. Dept. Agr. Circ. 130. 1920.

— The toll of weeds in Indiana. Ind. Acad. Sci. Proc., pp. 105–109. 1921.

— Lawn pennywort: a new weed. U. S. Dept. Agr. Circ. 165. 1921.

— Austrian field cress. Torreya 22:73–77. 1922.

— Nineteen noxious weeds of Indiana. Purdue Univ. Agr. Exp. Sta. Circ. 106. 1922.

— Recent Indiana weeds. Ind. Acad. Sci. Proc. 38:293–296. 1923.

HANSON, HERBERT C., and VELVA E. RUDD. Leafy spurge, life history and habits. N. Dak. Agr. Exp. Sta. Bull. 266. 1933.

HARMON, G. W., and F. D. KEIM. The percentage and viability of weed seeds recovered in the feces of farm animals and their longevity when buried in manure. Jour. Amer. Soc. Agron. 26:762–767. 1934.

HARPER, H. J. The use of sodium chlorate in the control of Johnson grass. Jour. Amer. Soc. Agron. 22:417–422. 1930.

HARTUNG, W. J. The functions of paper mulch in pineapple culture. 31 pp. Honolulu, Hawaii. 1926.

HARVEY, R. B. The action of toxic agents used in the eradication of noxious plants. Jour. Amer. Soc. Agron. 23:481–489. 1931a.

— Ammonium thiocyanate as a weed eradicant. Jour. Amer. Soc. Agron. 23:944–946. 1931b.

HAYDEN, ADA. Distribution and reproduction of Canada thistle. Amer. Jour. Bot. 21:355–373. 1934.

HELGESON, E. A. Russian Knapweed and Perennial Peppergrass. N. Dak. Agr. Exp. Sta. Bull. 292. 1940.

HENKEL, ALICE. Weeds used in medicine. U. S. Dept. Agr. Farmers Bull. 188. 1917.

HENRY, HELEN H. The seeds of quack grass and certain wheat grasses compared. Jour. Agr. Research 35:537–546. 1927.

HILLMAN, F. H. Dodder in relation to farm seeds. U. S. Dept. Agr. Farmers Bull. 306. 1907.

HILLS, J. L., and C. H. JONES. Commercial feeding stuffs. Vt. Agr. Exp. Sta. Bull. 131. 1907.

HILLS, J. L., C. H. JONES, and P. A. BENEDICT. Commercial feeding stuffs, principles and practice of stock feeding. Vt. Agr. Exp. Sta. Bull. 152. 1910.

HOWITT, J. EATON, and JOHN D. MACLEOD. The Weeds of Ontario. Ontario Dept. of Agr. Bull. 409. Reprinted June, 1949. Toronto, Canada.

HUFFAKER, C. B., and C. E. KENNETT. Ecological tests in biological control of Klamath weed. J. Econ. Entom. 45:1060–4. Dec., 1952.

HULBERT, H. W., R. S. BRISTOL, and L. V. BENJAMIN. Methods affecting the efficiency of chlorate weed killers. Idaho Agr. Exp. Sta. Bull. 189. 1931.

HULBERT, H. W., J. D. REMSBERG, and H. L. SPENCE. Controlling perennial weeds with chlorates. Jour. Amer. Soc. Agron. 22:423–433. 1930.

HULBERT, H. W., H. L. SPENCE, and L. V. BENJAMIN. The eradication of Lepidium Draba. Jour. Amer. Soc. Agron. 26:858–864. 1934.

HUME, A. N., and S. L. SLOAN. Quack grass and western wheat grass. S. Dak. Agr. Exp. Sta. Bull. 170. 1916.

HUTCHINS, A E. Mulch paper in vegetable production. Minn. Agr. Exp. Sta. Bull. 298. 1933.

INCE, J. W. Fertility and weeds. N. Dak. Agr. Exp. Sta. Bull. 112. 1915.

JACKMAN, E. R., et al. Control of Perennial Weeds in Oregon. Ore. State College Ext. Bull. 510. 1938.

JENKINS, E. H. Feeds, seeds and weeds. Conn. (New Haven) Exp. Sta. Bull. 161. pp. 3–6. 1909.

JEPSON, WILLIS LINN. A manual of the flowering plants of California. Univ. Calif. Press. 2nd printing, 1951.

JOHNSON, E. Emulsification of Diesel oil for puncture vine control. Mon. Bull. Dept. of Agr., Calif. 16. 1927.

— Recent developments in the use of herbicides in California. Mon. Bull. Dept. of Agr., Calif. 17:7–16. 1928.

— The puncture vine in California. Calif. Agr. Exp. Sta. Bull. 528:1–42. 1932.

JOHNSTON, C. O. Goat grass, a new wheat-field weed growing troublesome. U. S. Dept. Agr. Yearbook. 277–279. 1931.

JOHNSTONE-WALLACE, D. B. The soils and crop production in Genesee County, N. Y. Part II. Pastures. N. Y. (Cornell) Agr. Exp. Sta. Bull. 567. 1933.

JONES, J. W. Progress in experiments on water-grass control at the Biggs Rice Field Station, 1922–1923. Calif. Agr. Exp. Sta. Bull. 375. 1924.

JOSSELYN, JOHN. New Englands Rareties discovered: in birds, beasts, fishes, serpents and plants of that country. p. 85. London. 1672.

KAINS, MAURICE G. Chicory growing. U. S. Dept. Agr. Div. of Botany Circ. 29. 1900.

KALTER, GRACE M. The weeds of the Miami Valley. The Miami Bulletin, Miami Univ. IX:2. 1910.

KARPER, R. E. The blueweed and its eradication. Tex. Agr. Exp. Sta. Bull. 292. 1922.

KEIM, F. D., and A. L. FROLIK. Common grass weeds of Nebraska. Nebr. Agr. Exp. Sta. Bull. 288. 1934.

KEMPTON, F. E., and N. F. THOMPSON. The common barberry and how to kill it. U. S. Dept. Agr. Circ. 356. 1925.

KENNEDY, P. B. Observations on some rice weeds in California. Calif. Agr. Exp. Sta. Bull. 356:465–494. 1923.

KENNEDY, P. B., and A. S. CRAFTS. The application of physiological methods to weed control. Plant Physiol. 2:503–506. 1927.

KEPHART, L. W. Quackgrass. U. S. Dept. Agr. Farmers Bull. 1307. 1923.

KIESSELBACH, T. A., N. F. PETERSEN, and W. W. BURR. Bindweeds and their control. Nebr. Agr. Exp. Sta. Bull. 287. 1934.

KINCH, R. C., and F. D. KEIM. Eradication of bindweed in bluegrass lawns. Jour. Amer. Soc. Agron. 29:30–39. 1937.

KORSMO, E. Unkräuter im Ackerbau der Neuzeit. 580 pp. Julius Springer, Berlin. 1930.

LATSHAW, W. L., and J. W. ZAHNLEY. Experiments with sodium chlorate and other chemicals as herbicides for field bindweed. Jour. Agr. Research 35:757–767. 1927.

— Killing field bindweed with sodium chlorate. Kans. Agr. Exp. Sta. Circ. 136. 1928.

LEACH, B. P. The weed problem with suggestions for control. Bull. U. S. Golf Assoc. Green Sect. 7:206–209. 1927. 8:28–33; 218–221. 1928.

LINDSEY, A. A. Unpublished thesis.

LONG, BAYARD. Sonchus uliginosus in Philadelphia area. Torreya 22:91–98. 1922.

LONG, H. C. Suppression of weeds by fertilizers and chemicals. 57 pp. The author. Surbiton, Surrey, England. 1934.

— Weeds of arable land. Misc. Publ. 61. Ministry of Agr. and Fisheries, 10. London, S. W. I. 1929.

LOOMIS, W. E., E. V. SMITH, R. BISSEY, and L. E. ARNOLD. The absorption and movement of sodium chlorate when used as an herbicide. Jour. Amer. Soc. Agron. 24:724–739. 1933.

MAGILL, W. W. Geese work overtime in strawberry fields. N.J. State Hort. Soc. N. 33:2476. Feb., 1952.

MAGRUDER, ROY. Paper mulch for the vegetable garden. Ohio Agr. Exp. Sta. Bull. 447. 1930.

MARSH, C. D. The Loco-weed disease. U. S. Dept. Agr. Farmers Bull. 1054. 1919.

Literature References

MARSH, C. D., and A. B. CLAWSON. Cicuta or water hemlock. U. S. Dept. Agr. Bull. 69. 1914.
— The stock-poisoning death camas. U. S. Dept. Agr. Farmers Bull. 1273. 1922.
— Toxic effect of St. Johnswort (Hypericum perforatum) on cattle and sheep. U. S. Dept. Agr. Techn. Bull. 202. 1930.
MARSH, C. D., A. B. CLAWSON, J. F. COUCH, and W. W. EGGLESTON. The whorled milkweed (Asclepias galioides) as a poisonous plant. U. S. Dept. Agr. Bull. 800. 1920.
MARSH, C. D., A. B. CLAWSON, and H. MARSH. Lupines as poisonous plants. U. S. Dept. Agr. Bull. 405. 1916.
— Larkspur poisoning of livestock. U. S. Dept. Agr. Bull. 365. 1916.
MARSH, C. D., G. C. ROE, and A. B. CLAWSON. Livestock poisoning by cocklebur. U. S. Dept. Agr. Circ. 283. 1923.
— Nuttall's Death Camas (Zygadenus Nuttallii) as a poisonous plant. U. S. Dept. Agr. Bull. 1376. 1926.
MAY, W. I. Whorled milkweed. Colo. Agr. Exp. Sta. Bull. 255. 1920.
MCCALL, M. A., and H. M. WANSER. The principles of summer-fallow tillage. Wash. Agr. Exp. Sta. Bull. 183. 1924.
MCNAIR, J. B. Rhus dermatitis from Rhus Toxicodendron, radicans and diversiloba. Univ. of Chicago Press. 298 pp. 1923.
MONTEITH, J. JR. Controlling weeds in putting greens. Bull. U. S. Golf Assoc., Green Sect. 10:142–155. 1930.
MONTEITH, JOHN JR., and A. E. RABBITT. Killing weed seed in soil with chloropicrin (Tear gas). Turf Culture 1:63–79. 1939.
MOORE, R. A., and A. L. STONE. The eradication of farm weeds with iron sulphate. Wisc. Agr. Exp. Sta. Bull. 179. 1909.
MUENSCHER, W. C. Some changes in the weed flora of Whatcom County, Washington. Torreya 30:130–134. 1930a.
— Leafy spurge and related weeds. N. Y. (Cornell) Ext. Bull. 192. 1930b.
— Perennial sow thistle and related weeds. N. Y. (Cornell) Ext. Bull. 195. 1930c.
— Lead arsenate experiments on the germination of weed seeds. N. Y. (Cornell) Agr. Exp. Sta. Bull. 508. 1930d.
— Killing perennial weeds with chlorates during winter. N. Y. (Cornell) Agr. Exp. Sta. Bull. 542. 1932.
— Wild mustard and related weeds. N. Y. (Cornell) Ext. Bull. 168 (revised). 1934a.
— Poison ivy and poison sumac. N. Y. (Cornell) Ext. Bull. 191 (revised). 1934b.
MUENSCHER, W. C., and BASSETT MAGUIRE. Notes on some New York plants. Rhodora 33:165–167. 1931.
MUNN, M. T. Spraying lawns with iron sulphate to eradicate dandelions. N. Y. (Geneva) Agr. Exp. Sta. Bull. 466:421–459. 1919.
MUSSELMAN, H. H. Essentials of a mulch paper laying machine. Mich. Agr. Exp. Sta. Circ. 126. 1929.

NEEL, L. R. Control of broom sedge. Tenn. Agr. Exp. Sta. Circ. 57. 1939.
NELSON, JAMES C. A new weed from Oregon. Torreya 22:86–88. 1922.
NORRIS, ELVA L. Nebraska weeds. Nebr. Dept. of Agr. Bull. 101. 1929.
— Ecological study of the weed population of eastern Nebraska. Univ. of Nebr., Univ. Studies 39:29–91. 1939.
North Central States Weed Control Conference Proceedings and Research Report. Published annually. Vol. 1, 1944+
Northeastern States Weed Control Conference Proceedings. Published annually. Vol. 1, 1947+
OFFORD, H. R. The chemical eradication of Ribes. U. S. Dept. Agr. Techn. Bull. 240. 1931.
OLIVE, E. W. The killing of mustard and other noxious weeds in grain fields by the use of iron sulphate. S. Dak. Agr. Exp. Sta. Bull. 112. 1909.
OSWALD, E. I. The effect of animal digestion and fermentation of manures on the vitality of seeds. Md. Agr. Exp. Sta. Bull. 128. 1908.
OSWALD, W. L., and ANDREW BOSS. Minnesota weeds, descriptions, identifications, and eradication. Minn. Agr. Exp. Sta. Bull. 139. 1914.
PAMMEL, L. H. Manual of poisonous plants. p. 363. 977 pp. Cedar Rapids, Iowa. 1911.
— Weeds of Farm and garden. 281 pp. Orange Judd. Co. 1911a.
— Some weedy grasses injurious to livestock, especially sheep. Iowa Agr. Exp. Sta. Circ. 116. 1929.
PAMMEL, L. H., and C. M. KING. Some new weeds of Iowa. Iowa Agr. Exp. Sta. Circ. 98. 1925.
— The weed flora of Iowa. Iowa Geol. Surv. Bull. 4: (revised). 1926.
PAMMEL, L. H., C. M. KING, and A. HAYDEN. Marsh cress, a bad weed. Iowa Agr. Exp. Sta. Circ. 120. 1929.
PAVLYCHENKO, T. K., and J. B. HARRINGTON. Competitive efficiency of weeds and cereal crops. Canadian Jour. Research 10:77–94. 1934.
PEITERSEN, A. K., and R. T. BURDICK. Perennial peppergrass, a noxious weed in Colorado. Colo. Agr. Exp. Sta. Bull. 264. 1920.
PHILLIPS, CLAUDE E. 100 weeds—aids to their identification by basal leaf characteristics. Northeastern Weed Control Conference, Misc. Paper 184. Jan., 1954. Del. Agr. Exp. Sta.
PIEMEISEL, R. L. Weedy abandoned lands and the weed hosts of the beet leaf hopper. U. S. Dept. Agr. Circ. 229. 1932.
PIPAL, F. J. White top and its control. Purdue Agr. Exp. Sta. Circ. 85. 1918.
— Wild garlic and its eradication. Purdue Agr. Exp. Sta. Bull. 176. 1914.
POPE, W. T. Manual of wayside plants of Hawaii. Advertiser Publishing Co., Ltd. Honolulu, Hawaii, U. S. A. 1929.
POWELL, H. O. Russian thistle, its use and control. Dominion Agricultural Credit Co., Regina, Sask. 19 pp. 1933.
PRETZ, H. W. Additional notes on Sonchus uliginosus. Torreya 23:79–85. 1923.
PRINCE, A. H. Weeds and grasses of Arkansas rice fields. Univ. of Ark. Ext. Circ. 253. 1927.
REA, H. E. Control of Early Weed and Grass Seedlings in Cotton. Texas Agr. Exp. Sta. Progress Report 1508. Nov. 14, 1952.

Literature References

RIDLEY, H. N. The dispersal of plants throughout the world. 744 pp. L. Reeve and Co., Ltd. Kent. 1930.

ROBBINS, W. W. Alien Plants growing without cultivation in California. Calif. Agr. Exp. Sta. Bull. 637. 1940.

ROBBINS, W. W., A. S. CRAFTS, and R. N. RAYNOR. Weed Control. 503 pp. Mc-Graw-Hill, New York. 2nd. ed. 1952.

ROBINSON, B. L. Erucastrum Pollichii adventive in America. Rhodora 13:10–12. 1911.

ROGERS, C. F. Canada thistle and Russian knapweed and their control. Colo. Agr. Exp. Sta. Bull. 348. 1928.

— Zygophyllum fabago in Colorado. Science LXIX:600–601. 1929.

ROGERS, C. F., L. W. DURRELL and L. B. DANIELS. Three important perennial weeds of Colorado. Colo. Agr. Exp. Sta. Bull. 313 (ed. 2). 1927.

ROGERS, C. F., and I. HATFIELD. Carbon disulphide for the eradication of perennial weeds. Colo. Exp. Sta. Bull. 347. 1929.

ROLLINS, R. C. On two weedy Crucifers. Rhodora 302–306. 1940.

RUNNELS, H. A., and J. H. SCHAFFNER. Manual of Ohio weeds. Ohio (Wooster) Agr. Exp. Sta. Bull. 475. 1931.

RYDBERG, PER AXEL. Flora of the prairies and plains of Central North America. New York Botanical Garden, New York. 1932.

SAMPSON, A. W., and K. W. PARKER. St. Johns wort on range lands in California. Calif. Agr. Exp. Sta. Bull. 503. 1930.

SCHAFER, E. G. The bindweed. Wash. Agr. Exp. Sta. Pop. Bull. 137.2–19. 1927.

SCHAFER, E. G., O. C. LEE and J. R. NELLER. Eradicating the bindweed with sodium chlorate. Wash. Agr. Exp. Sta. Bull. 235. 1929.

SCHULZ, E. R., and N. F. THOMPSON. Some effects of sodium arsenite when used to kill the common barberry. U. S. Dept. Agr. Bull. 1316. 1925.

SCHWEINITZ, LEWIS D. VON. Remarks on the plants of Europe which have become naturalized in the United States. Annals Lyceum of Natural History of N. Y. 3:148–155. (1832) 1828–1836.

SELBY, A. D. Spraying to kill weeds—some useful methods. Ohio (Wooster) Agr. Exp. Sta. Circ. 102. 1910.

SIEVERS, A. F. Peppermint and spearmint as farm crops. U. S. Dept. Agr. Farmers Bull. 1555. 1929.

— American medicinal plants of commercial importance. U. S. Dept. Agr. Misc. Publication 77. 1930.

SMALL, JOHN K. Another Sonchus for America. Torreya 21:100. 1921.

— The Austrian field cress again. Torreya 23:23–25. 1923.

— Manual of the southeastern flora. Judd, New York. 1933.

SMILEY, F. J., collaborator. Weeds of California and methods of control. Mon. Bull. Dept. of Agr., Calif. 11. 1922.

SMITH, ALFRED. Effect of paper mulch on soil temperature, soil moisture, and yield of crops. Hilgardia. 6:159–201. 1931.

SMITH, E. V., and E. L. MAYTON. Nut grass eradication studies. II. The eradication of nut grass, *Cyperus rotundus* L., by certain tillage treatments. Jour. Amer. Soc. Agron. 30:18–21. 1938.

Southern Weed Control Conference Proceedings. Published annually. Vol. 1, 1948+

SPENCE, H. L., and H. W. HULBERT. Idaho Perennial Weeds. Univ. Idaho Ext. Bull. 98. 1935.

SPENCER, E. R. Just weeds, 317 pp. 1940. Chas. Scribner's Sons, N. Y.

STANDLEY, PAUL C. A new United States weed: Hymenophysa pubescens. Science, N. S. LXII No. 1614. pp. 509–510. 1925.

State noxious-weed seed requirements recognized in the administration of the federal seed act. U.S. Dept. Agr. Production and Marketing administration, Washington, D.C. Revised Oct., 1951.

STEVENS, O. A. Perennial sow thistle, growth and reproduction. N. Dak. Agr. Exp. Sta. Bull. 181. 1924.

— The sow thistle. N. Dak. Agr. Exp. Sta. Circ. 32:3–16. 1926.

— North Dakota weeds. N. Dak. Agr. Exp. Sta. Bull. 243. 1930.

— The number and weight of seeds produced by weeds. Amer. Jour. Bot. 19: 784–794. 1932.

— Plants of the mustard family in North Dakota. N. Dak. Agr. Exp. Sta. Bimonth. Bull. 1 (5):7–10. 1939.

— North Dakota Weeds. N. Dak. Agr. Ext. Serv. Circ. 156. 1937.

STEWART, GEORGE, and D. W. PITTMAN. Ridding the land of wild morningglory. Utah Agr. Exp. Sta. Bull. 189. 1924.

STEWART, R. T., R. G. REEVES, and L. G. JONES. The spurge nettle. Jour. Amer. Soc. Agron. 28:907–913. 1936.

STITT, R. E. Dodder control in annual lespedezas. Jour. Amer. Soc. Agron. 31:338–343. 1939.

SWINGLE, D. B., and ALFRED ATKINSON. A warning against fan weed. Mont. Agr. Exp. Sta. Circ. 12. 1911.

SWINGLE, D. B., H. E. MORRIS, and E. W. JAHNKE. Fifty important weeds of Montana. Mont. Ext. Bull. 45. 1920.

TALBOT, M. W. Johnson grass as a weed. U. S. Dept. Agr. Farmers Bull. 1537. 1928.

— Wild garlic and its control. U. S. Dept. Agr. Leaflet 43. 1929.

THOMPSON, H. C. Experimental studies of cultivation of certain vegetable crops. N. Y. (Cornell) Agr. Exp. Sta. Memoir 107. 1927.

THOMPSON, H. C., and HANS PLATENIUS. Results of mulch paper with vegetable crops. Proc. Amer. Soc. Hort. Sci. (1931) 28:304–308. 1932.

THOMPSON, H. C., P. H. WESSELS, and H. S. MILLS. Cultivation experiments with certain vegetable crops on Long Island. N. Y. (Cornell) Agr. Exp. Sta. Bull. 521:1–14. 1931.

THOMPSON, N. F. Kill the common barberry with chemicals. U. S. Dept. Agr. Circ. 268. 1923.

THOMSON, R. B., and H. B. SIFTON. A guide to the poisonous plants and weed seeds of Canada and the Northern United States. Univ. Toronto Press. 169 pp., 40 figs. 1922.

THORNTON, B. J., and L. W. DURRELL. Colorado weeds. Colo. Agr. Exp. Sta. Bull. 403. 1933.

— Weeds of Colorado. Colo. Exp. Sta. Bull. 466. 1941.

Literature References

TICE, C. Weeds and their control. British Columbia Dept. of Agr. Bull. 106. 1932.

TIDESTROM, I. Botanical Notes. Rhodora 36:309–312. 1934.

— Flora of Utah and Nevada. Washington Govt. Printing Office, Washington, D.C. 1925.

TRACY, S. M. Bermuda grass. U. S. Dept. Agr. Farmers Bull. 814. 1917.

TUKEY, H. B. Plant Regulators in Agriculture. New York, John Wiley and Sons, Inc. 1954.

TULLIS, M. P. Weeds, their identification and control. Prov. of Saskatchewan, Dept. of Agr. Bull. 57 (ed. 3). 1923.

— The control of sow thistle. Saskatchewan Dept. of Agr. Bull. 58. 1924.

VINALL, H. N. Johnson grass: its production for hay and pasturage. U. S. Dept. Agr. Farmers Bull. 1476. 1926.

WAHLENBERG, W. G. Investigations in weed control by zinc sulphate and other chemicals. U. S. Dept. Agr. Techn. Bull. 156. 1930.

Weeds, Journal of the Association of Regional Weed Conferences. W. F. Humphrey Press, Inc. Geneva, N. Y. Published quarterly. Vol. 1, 1951+

Weeds of the North Central states. Univ. Ill. Agr. Exp. Sta. Circ. 718. Feb. 1954.

WELTON, F. A. Sodium chlorate as a lawn weed killer. Ohio Agr. Exp. Sta. Bimonthly Bull. 141. 1929.

WELTON, F. A., and J. C. CARROLL. Control of lawn weeds and the renovation of lawns. Ohio Agr. Exp. Sta. Bull. 619. 1941.

Western Canadian Weed Control Conference Proceedings. Published annually. Vol. 1, 1947+

Western Weed Control Conference Proceedings. Published annually. Vol. 1. 1938+

WESTGATE, W. A., and R. N. RAYNOR. A new selective spray for the control of certain weeds. Calif. Agr. Exp. Sta. Bull. 634. 1940.

WESTOVER, H. L. Planting and care of lawns. U. S. Dept. Agr. Farmers Bull. 1677. 1931.

WESTVELD, R. H. Preliminary results in eradicating weeds with zinc sulphate and by burning in forest nursery seed beds. Mich. Agr. Exp. Sta. Quart. Bull. 15:254–261. 1933.

WHEELER, LOUIS C. The Names of three species of Brassica. Rhodora 40: 306–309. 1938.

WIEGAND, K. M., and A. J. EAMES. The flora of the Cayuga Lake Basin, New York, N. Y. (Cornell) Agr. Exp. Sta. Memoir 92. 1926.

WIMMER, FR., and H. GRABOWSKI. Flora Silesiae Vratislaviae 3:82–92. 1829.

YOUNGKEN, H. W. A comparative study of the seeds and spikes of certain caulescent species of Plantago. Amer. Jour. of Pharm. 106:No. 5. 1–9. 1934.

YUNCKER, T. G. Revision of the North American and West Indian species of Cuscuta. Univ. of Ill. Biological Monographs VI:1–142. 1921.

— The genus Cuscuta. Mem. Torr. Bot. Club 18:113–331. 1932.

APPENDIX I

Changes in Botanical Nomenclature

MUENSCHER NOMENCLATURE	CURRENT NOMENCLATURE
Acnida altissima Riddell	*Amaranthus tuberculatus* (Moq.) Sauer
Aegilops cylindrica Host	*Triticum cylindricum* (Host) Ces., Pass. and Gib.
Alhagi camelorum Fisch.	*Alhagi pseudalhagi* (Bieb.) Desv.
Alliaria officinalis Andrz.	*Alliaria petiolata* (Bieb.) Cavara and Grande
Ampelamus albidus (Nutt.) Britt.	*Cynanchum laeve* (Michx.) Pers.
Anaphalis margaritacea (L.) C. B. Clarke	*Anaphalis margaritacea* (L.) Benth. and Hook.
Anthemis arvensis L. var. *agrestis* (Wallr.) DC.	*Anthemis arvensis* L.
Arctium minus (Hill) Bernh.	*Arctium minus* Bernh.
Asclepias galioides HBK.	*Asclepias subverticillata* (Gray) Vail
Astragalus diphysus Gray	*Astragalus lentiginosus* Dougl. var. *diphysus* (Gray) Jones
Bidens comosa Wiegand	*Bidens comosa* (Gray) Wiegand
Brassica rapa L.	*Brassica rapa* L. subsp. *sylvestris* (L.) Janchen
Camelina dentata Pers.	*Camelina alyssum* (Miller) Thell.
Camelina sativa Crantz	*Camelina sativa* (L.) Crantz
Cardaria pubescens (C. A. Mey.) Rollins	*Cardaria pubescens* (C. A. Meyer) Jarmolenko
Cassia tora L.	*Cassia obtusifolia* L.
Centaurea vochinensis Bernh.	*Centaurea nigrescens* Willd. subsp. *nigrescens*
Chorispora tenella (Willd.) DC.	*Chorispora tenella* (Pall.) DC.
Chrysanthemum leucanthemum L. var. *pinnatifidum* Lecoq and Lamotte	*Chrysanthemum leucanthemum* L.
Cirsium arvense Scop.	*Cirsium arvense* (L.) Scop.
Convolvulus sepium L.	*Calystegia sepium* (L.) R. Br.
Coronopus procumbens Gilibert	*Coronopus squamatus* (Forssk.) Aschers.
Cytisus scoparius L.	*Cytisus scoparius* (L.) Link
Dactyloctenium aegyptium (L.) Richter	*Dactyloctenium aegyptium* (L.) Willd.

Appendix I

Muenscher Nomenclature	Current Nomenclature
Digitaria ischaemum (Schreb.) Muhl.	*Digitaria ischaemum* (Schreb.) Schreb. ex Muhl.
Dipsacus sylvestris Huds.	*Dipsacus fullonum* L.
Echinochloa pungens (Poir.) Rydb.	*Echinochloa muricata* (Beauv.) Fern. var. *muricata*
Eleusine indica Gaertn.	*Eleusine indica* (L.) Gaertn.
Eragrostis megastachya (Koel.) Link	*Eragrostis cilianensis* (All.) E. Mosher
Eragrostis poaeoides Beauv.	*Eragrostis minor* Host
Erechtites prenanthoides (A. Rich.) DC.	*Erechtites minima* (Poir.) DC.
Eremocarpus setigerus Benth.	*Eremocarpus setigerus* (Hook.) Benth.
Erigeron canadensis L.	*Conyza canadensis* (L.) Cronq.
Erigeron divaricatus Michx.	*Conyza ramosissima* Cronq.
Eruca sativa Mill.	*Eruca vesicaria* (L.) Cav. subsp. *sativa* (Mill.) Thell.
Franseria discolor Nutt.	*Ambrosia tomentosa* Nutt.
Franseria tomentosa Gray	*Ambrosia grayi* (Nels.) Shinners
Franseria tenuifolia Torr. and Grey	*Ambrosia confertifolia* DC.
Galinsoga ciliata (Raf.) Blake	*Galinsoga quadriradiata* Ruiz and Pavon
Gaura coccinea Pursh	*Gaura coccinea* Nutt. ex Pursh
Gaura odorata Lag.	*Gaura odorata* Sessé ex Lag.
Glycyrrhiza lepidota (Nutt.) Pursh	*Glycyrrhiza lepidota* Pursh
Gnaphalium macounii Greene	*Gnaphalium viscosum* HBK.
Halogeton glomeratus (C. A. Meyer) Coult.	*Halogeton glomeratus* (Bieb.) C. A. Meyer
Hedeoma pulegioides Pers.	*Hedeoma pulegioides* (L.) Pers.
Helenium nudiflorum Nutt.	*Helenium flexuosum* Raf.
Helenium tenuifolium Nutt.	*Helenium amarum* (Raf.) Rock
Hieracium florentinum All.	*Hieracium piloselloides* Vill.
Hieracium praealtum Gochnat	*Hieracium praealtum* Vill. ex Gochnat
Hieracium pratense Tausch	*Hieracium caespitosum* Dumort.
Hordeum nodosum L.	*Hordeum brachyantherum* Nevski
Lactuca scariola L.	*Lactuca serriola* L.
Lepidium repens Boiss.	*Lepidium repens* (Schrenk) Boiss.
Linaria canadensis (L.) Dumont var. *texana* (Scheele) Pennell	*Linaria texana* Scheele
Lippia cuneifolia Steud.	*Phyla cuneifolia* (Torr.) Greene
Melilotus alba Desr.	*Melilotus alba* Medic.
Melilotus officinalis (L.) Desr.	*Melilotus officinalis* (L.) Pall.
Navarretia intertexta Hook.	*Navarretia intertexta* (Benth.) Hook.

Changes in Botanical Nomenclature

MUENSCHER NOMENCLATURE	CURRENT NOMENCLATURE
Navarretia squarrosa H. & A.	*Navarretia squarrosa* (Esch.) H. & A.
Neslia paniculata Desv.	*Neslia paniculata* (L.) Desv.
Nicandra physalodes (L.) Pers.	*Nicandra physalodes* (L.) Gaertn.
Oxalis europaea Jord.	*Oxalis stricta* L.
Oxalis florida Salisb.	*Oxalis dillenii* Jacq. subsp. *filipes* (Small) Eiten
Oxalis stricta L.	*Oxalis dillenii* Jacq.
Paspalum ciliatifolium Michx.	*Paspalum setaceum* Michx. var. *ciliatifolium* (Michx.) Vasey
Polanisia graveolens Raf.	*Polanisia dodecandra* (L.) DC.
Polygonum coccineum Muhl.	*Polygonum amphibium* L.
Polygonum sachalinense F. Schmidt	*Polygonum sachalinense* F. Schmidt ex Maxim.
Polygonum setaceum Baldw.	*Polygonum setaceum* Baldw. ex Ell.
Rhus diversiloba T. and G.	*Toxicodendron diversilobum* (T. and G.) Greene
Rhus microcarpa Steud.	*Toxicodendron radicans* (L.) Kuntze
Rhus radicans L.	*Toxicodendron radicans* (L.) Kuntze
Rhus radicans L. var. *rydbergii* Small	*Toxicodendron rydbergii* (Small) Greene
Rhus toxicodendron L.	*Toxicodendron toxicarium* (Salisb.) Gillis
Rhus vernix L.	*Toxicodendron vernix* (L.) Kuntze
Roemeria refracta	*Roemeria refracta* DC.
Rorippa islandica (Oeder) Borbas	*Rorippa palustris* (L.) Besser
Rorippa islandica var. *hispida*	*Rorippa palustris* (L.) Besser subsp. *hispida* (Desv.) Jonsell
Rorippa sylvestris Bess.	*Rorippa sylvestris* (L.) Besser
Rudbeckia serotina Nutt.	*Rudbeckia hirta* L. var. *pulcherrima* Farwell
Rumex hastatulus Baldw.	*Rumex hastatulus* Baldw. ex. Ell.
Rumex mexicanus Meissn.	*Rumex triangulivalvis* (Dans.) Rech. f.
Saponaria vaccaria L.	*Vaccaria pyramidata* Medic.
Sedum purpureum (L.) Link	*Sedum telephium* L.
Sida hederacea Dougl.	*Malvella leprosa* (Ortega) Krapov.
Silene cserei Baumg.	*Silene csereii* Baumg.
Silene cucubalus Wibel	*Silene vulgaris* (Moench) Garcke
Solanum villosum Mill.	*Solanum sarrachoides* Sendt.
Solidago graminifolia (L.) Salisb. var. *Nuttallii* Fern.	*Solidago graminifolia* (L.) Salisb.
Sonchus arvensis L. var. *glabrescens* Wimm. and Grab.	*Sonchus arvensis* L. subsp. *uliginosus* (Bieb.) Nyman

Muenscher Nomenclature	Current Nomenclature
Sonchus uliginosus Bieb.	*Sonchus arvensis* L. subsp. *uliginosus* (Bieb.) Nyman
Sorgum halepense (L.) Pers.	*Sorghum halepense* (L.) Pers.
Sorgum vulgare Pers.	*Sorghum bicolor* (L.) Moench
Specularia perfoliata (L.) A. DC.	*Triodanis perfoliata* (L.) Nieuw.
Sporobolus poiretii (R. & S.) Hitchc.	*Sporobolus indicus* (L.) R. Br.
Sporobolus vaginiflorus Wood	*Sporobolus vaginiflorus* (Torr.) Wood
Stachys annua L.	*Stachys annua* (L.) L.
Stachys arvensis L.	*Stachys arvensis* (L.) L.
Stellaria media Cyrill	*Stellaria media* (L.) Vill.
Swainsonia salsula (Pall.) Taub.	*Sphaerophysa salsula* (Pall.) DC.
Tragopogon major Jacq.	*Tragopogon dubius* Scop.
Urtica holosericea Nutt.	*Urtica dioica* L. subsp. *gracilis* (Ait.) Seland. var. *holosericea* (Nutt.) C. L. Hitchc.
Valerianella olitoria (L.) Poll.	*Valerianella locusta* (L.) Laterrade
Veronica hederaefolia L.	*Veronica hederifolia* L.
Vicia angustifolia Reichard	*Vicia sativa* L. subsp. *nigra* (L.) Ehrh.
Vicia tetrasperma (L.) Moench	*Vicia tetrasperma* (L.) Schreber
Xanthium orientale L.	*Xanthium strumarium* L.
Zigadenus gramineus Rydb.	*Zigadenus venenosus* S. Wats. var. *gramineus* (Rydb.) Walsh
Zigadenus paniculatus S. Wats.	*Zigadenus paniculatus* (Nutt.) S. Wats.

APPENDIX II

Weeds Mentioned in This Book
(Arranged Alphabetically by Scientific Name Followed by Standardized Common Name)

SCIENTIFIC NAME	STANDARDIZED COMMON NAME
Abutilon theophrasti Medic.	velvetleaf
Acalypha rhomboidea Raf.	rhombic copperleaf
virginica L.	Virginia copperleaf
Achillea millefolium L.	common yarrow
Acnida altissima Riddell (see *Amaranthus tuberculatus*)	
Adonis annua L.	pheasants-eye
Aegilops cylindrica Host (see *Triticum cylindricum*)	
Aegopodium podagraria L.	bishops goutweed
Aeschynomene virginica (L.) B.S.P.	northern jointvetch
Agrimonia gryposepala Wallr.	agrimony
Agropyron repens (L.) Beauv.	quackgrass
smithii Rydb.	western wheatgrass
Agrostemma githago L.	corn cockle
Aira caryophyllea L.	silver hairgrass
Alhagi camelorum Fisch. (see *A. pseudalhagi*)	
pseudalhagi (Bieb.) Desv.	camelthorn
Alliaria officinalis Andrz. (see *A. petiolata*)	
petiolata (Bieb.) Cavara & Grande	garlic mustard
Allium canadense L.	wild onion
tricoccum Ait.	wild leek
vineale L.	wild garlic
Alopecurus geniculatus L.	water foxtail
pratensis L.	meadow foxtail
Amaranthus albus L.	tumble pigweed
graecizans L.	prostrate pigweed
hybridus L.	smooth pigweed
lividus L.	livid amaranth
palmeri S. Wats.	Palmer amaranth
powellii S. Wats.	Powell amaranth
retroflexus L.	redroot pigweed
spinosus L.	spiny amaranth

Appendix II

Scientific Name	Standardized Common Name
Amaranthus (cont.)	
tuberculatus (Moq.) Sauer	tall waterhemp
Ambrosia artemisiifolia L.	common ragweed
bidentata Michx.	lanceleaf ragweed
confertifolia DC.	slimleaf bursage
grayi (Nels.) Shinners	woollyleaf bursage
psilostachya DC.	western ragweed
tomentosa Nutt.	skeletonleaf bursage
trifida L.	giant ragweed
Ammannia coccinea Rottb.	purple ammannia
Ampelamus albidus (Nutt.) Britt. (see *Cynanchum laeve*)	
Amphicarpum purshii Kunth	annual goobergrass
Amsinckia intermedia Fisch. & Mey.	coast fiddleneck
lycopsoides Lehm.	tarweed fiddleneck
Anagallis arvensis L.	scarlet pimpernel
Anaphalis margaritacea (L.) Benth. & Hook.	pearly everlasting
Andropogon gerardii Vitman	big bluestem
scoparius Michx.	little bluestem
virginicus L.	broomsedge
Angelica atropurpurea L.	purplestem angelica
Antennaria plantaginifolia (L.) Hook.	plantainleaf pussytoes
Anthemis arvensis L.	corn chamomile
cotula L	mayweed
Anthoxanthum odoratum L.	sweet vernalgrass
Apocynum androsaemifolium L.	spreading dogbane
cannabinum L.	hemp dogbane
Arabidopsis thaliana (L.) Heynh.	mouseearcress
Arctium lappa L.	great burdock
minus Bernh.	common burdock
Arenaria serpyllifolia L.	thymeleaf sandwort
Argemone intermedia Sweet	bluestem pricklepoppy
mexicana L.	Mexican pricklepoppy
Aristida dichotoma Michx.	churchmouse threeawn
oligantha Michx.	prairie threeawn
Arrhenatherum elatius (L.) Mert. & Koch var. *bulbosum* (Willd.) Spenner	tuber oatgrass
Artemisia absinthium L.	absinth wormwood
annua L.	annual wormwood
biennis Willd.	biennial wormwood
tridentata Nutt.	big sagebrush
vulgaris L.	mugwort

Weeds Mentioned in This Book 539

Scientific Name	Standardized Common Name
Asclepias galioides HBK. (see *A. subverticillata*)	
labriformis Jones	labriform milkweed
speciosa Torr.	showy milkweed
subverticillata (Gray) Vail	poison milkweed
syriaca L.	common milkweed
tuberosa L.	butterfly milkweed
verticillata L.	eastern whorled milkweed
Aster pilosus Willd.	white heath aster
simplex Willd.	white field aster
Astragalus diphysus A. Gray (see *A. lentiginosus* var.)	
lentiginosus Dougl. var. *diphysus* (Gray) Jones	blue loco
mollisimus Torr.	woolly loco
Atriplex patula L.	spreading orach
patula L. var. *hastata* (L.) Gray	halberdleaf orach
rosea L.	red orach
Avena fatua L.	wild oat
Axyris amaranthoides L.	Russian pigweed
Barbarea verna (Mill.) Aschers.	early wintercress
vulgaris R. Br.	yellow rocket
Bassia hyssopifolia (Pall.) Ktze.	fivehook bassia
Bellis perennis L.	English daisy
Berberis vulgaris L.	European barberry
Berteroa incana (L.) DC.	hoary alyssum
Bidens bipinnata L.	spanishneedles
cernua L.	nodding beggarticks
comosa (Gray) Wieg.	leafybract beggarticks
connata Muhl.	purplestem beggarticks
frondosa L.	devils beggarticks
polylepis Blake	coreopsis beggarticks
vulgata Greene	tall beggarticks
Brassica hirta Moench	white mustard
juncea (L.) Coss.	Indian mustard
kaber (DC.) L. C. Wheeler var. *pinnatifida* (Stokes) L. C. Wheeler (see *B. kaber*)	wild mustard
nigra (L.) Koch	black mustard
rapa L. (see *B. rapa* subsp.)	
rapa L. subsp. *sylvestris* (L.) Janchen	birdsrape mustard
Bromus commutatus Schrad.	hairy chess
japonicus Thunb.	Japanese brome
mollis L.	soft chess

Appendix II

Scientific Name	Standardized Common Name
Bromus (cont.)	
racemosus L.	upright brome
rubens L.	red brome
secalinus L.	cheat
sterilis L.	poverty brome
tectorum L.	downy brome
Calamovilfa brevipilis (Torr.) Scribn.	riverbank sandreed
Calystegia sepium (L.) R. Br.	hedge bindweed
Camelina alyssum (Mill.) Thell.	flatseed falseflax
dentata Pers. (see *C. alyssum*)	
microcarpa Andrz.	smallseed falseflax
sativa (L.) Crantz	largeseed falseflax
Campanula rapunculoides L.	creeping bellflower
Campsis radicans (L.) Seem.	trumpetcreeper
Cannabis sativa L.	hemp
Capsella bursa-pastoris (L.) Medic.	shepherdspurse
Cardamine pratensis L.	cuckoo bittercress
Cardaria draba (L.) Desv.	hoary cress
pubescens (C. A. Meyer) Jarmolenko	hairy whitetop
Carduus acanthoides L.	plumeless thistle
crispus L.	welted thistle
nutans L.	musk thistle
Carex bullata Schkuhr	button sedge
lupulina Muhl.	hop sedge
Carum carvi L.	caraway
Cassia fasciculata Michx.	partridgepea
marilandica L.	wild senna
nictitans L.	sensitive partridgepea
obtusifolia L.	sicklepod
occidentalis L.	coffee senna
tora L. (see *C. obtusifolia*)	
Cenchrus longispinus (Hack.) Fern.	longspine sandbur
Centaurea calcitrapa L.	purple starthistle
cyanus L.	cornflower
diffusus Lam.	diffuse knapweed
iberica Trevir.	Iberian starthistle
jacea L.	brown knapweed
maculosa Lam.	spotted knapweed
melitensis L.	Malta starthistle
nigrescens Willd. subsp. *nigrescens*	Vochin knapweed
repens L.	Russian knapweed
solstitialis L.	yellow starthistle

Weeds Mentioned in This Book 541

SCIENTIFIC NAME	STANDARDIZED COMMON NAME
Centaurea (cont.)	
vochinensis Bernh. (see	
C. nigrescens subsp.)	
Cerastium arvense L.	field chickweed
viscosum L.	sticky chickweed
vulgatum L.	mouseear chickweed
Chaenorrhinum minus (L.) Lange	dwarf snapdragon
Chamaedaphne calyculata (L.) Moench	leatherleaf
Chelidonium majus L.	greater celandine
Chenopodium album L.	common lambsquarters
ambrosioides L.	mexicantea
var. *anthelminticum* (L.) Gray	drug wormseed goosefoot
bonus-henricus L.	perennial goosefoot
botrys L.	jerusalemoak goosefoot
capitatum (L.) Aschers.	blite goosefoot
glaucum L.	oakleaf goosefoot
hybridum L.	mapleleaf goosefoot
murale L.	nettleleaf goosefoot
paganum Reichenb.	pigweed goosefoot
polyspermum L.	manyseeded goosefoot
rubrum L.	red goosefoot
urbicum L.	city goosefoot
Chondrilla juncea L.	rush skeletonweed
Chorispora tenella (Pall.) DC.	blue mustard
Chrysanthemum leucanthemum L.	oxeye daisy
var. *pinnatifidum* Lecoq & Lamotte (see *C. leucanthemum*)	
Cichorium intybus L.	chicory
Cicuta maculata L.	spotted waterhemlock
Cirsium altissimum (L.) Spreng.	tall thistle
arvense (L.) Scop.	Canada thistle
pumilum (Nutt.) Spreng.	pasture thistle
vulgare (Savi) Tenore	bull thistle
Cleome serrulata Pursh	Rocky Mountain beeplant
Cnicus benedictus L.	blessed thistle
Cnidoscolus stimulosus (Michx.) Gray	bullnettle
Commelina communis L.	dayflower
Comptonia peregrina (L.) Coult.	sweetfern
Conium maculatum L.	poison hemlock
Conringia orientalis (L.) Dumort.	haresear mustard
Convallaria majalis L.	lily-of-the-valley
Convolvulus arvensis L.	field bindweed
sepium L. (see *Calystegia sepium*)	

542   Appendix II

SCIENTIFIC NAME	STANDARDIZED COMMON NAME
Conyza canadensis (L.) Cronq.	horseweed
ramosissima Cronq.	dwarf fleabane
Corispermum hyssopifolium L.	hyssopleaf tickweed
Coronopus didymus (L.) Sm.	swinecress
procumbens Gilibert (see *C. squamatus*)	
squamatus (Forssk.) Aschers.	creeping watercress
Crataegus spp.	hawthorns
Crepis biennis L.	rough hawksbeard
capillaris (L.) Wallr.	smooth hawksbeard
tectorum L.	narrowleaf hawksbeard
Crotalaria sagittalis L.	rattlebox
Croton capitatus Michx.	woolly croton
glandulosus L.	tropic croton
Cucurbita foetidissima HBK.	buffalo gourd
Cuphea carthagenensis (Jacq.) Macbr.	tarweed cuphea
petiolata (L.) Koehne	clammy cuphea
Cuscuta campestris Yunck.	field dodder
coryli Engelm.	hazel dodder
epilinum Weihe	flax dodder
epithymum Murr.	clover dodder
glomerata Choisy	composite dodder
gronovii Willd.	swamp dodder
indecora Choisy	largeseed dodder
pentagona Engelm.	fiveangled dodder
planiflora Tenore	smallseed dodder
polygonorum Engelm.	polygonum dodder
suaveolens Serr.	Chile dodder
Cyclolloma atriplicifolium (Spreng.) Coult.	winged pigweed
Cymbalaria muralis Gaertn., Mey. & Scherb.	Kenilworth ivy
Cynanchum laeve (Michx.) Pers.	honeyvine milkweed
nigrum (L.) Pers.	black swallowwort
vincetoxicum (L.) Pers.	white swallowwort
Cynodon dactylon (L.) Pers.	bermudagrass
Cynoglossum officinale L.	houndstongue
Cyperus diandrus Torr.	low flatsedge
esculentus L.	yellow nutsedge
iria L.	rice flatsedge
rotundus L.	purple nutsedge
strigosus L.	false nutsedge
Cytisus scoparius (L.) Link	Scotch broom
Dactyloctenium aegyptium (L.) Willd.	crowfootgrass

Weeds Mentioned in This Book 543

SCIENTIFIC NAME	STANDARDIZED COMMON NAME
Danthonia spicata (L.) Beauv.	poverty oatgrass
Datura stramonium L.	jimsonweed
tatula L. (see *D. stramonium*)	
Daucus carota L.	wild carrot
pusillus Michx.	southwestern carrot
Delphinium menziesii DC.	Menzies larkspur
strictum A. Nels.	strict larkspur
Dennstaedtia punctilobula (Michx.) Moore	hayscented fern
Descurainia longipedicellata O. E. Schulz	slimstem tansymustard
pinnata (Walt.) Britt.	tansymustard
richardsonii (Sweet) O. E. Schulz	Richardson tansymustard
sophia (L.) Webb	flixweed
Desmodium canadense (L.) DC.	hoary tickclover
Dianthus armeria L.	Deptford pink
Digitalis purpurea L.	common foxglove
Digitaria filiformis (L.) Koel.	slender crabgrass
ischaemum (Schreb.) Schreb. ex Muhl.	smooth crabgrass
sanguinalis (L.) Scop.	large crabgrass
Diodia teres Walt.	poorjoe
Diplotaxis muralis (L) DC.	stinking wallrocket
tenuifolia (L.) DC.	slimleaf wallrocket
Dipsacus fullonum L.	teasel
laciniatus L.	cutleaf teasel
sylvestris Huds. (see *D. fullonum*)	
Draba verna L.	whitlowwort
Dracocephalum parviflorum Nutt.	American dragonhead
Duchesnea indica (Andr.) Focke	India mockstrawberry
Dulichium arundinaceum (L.) Britt.	dulichium
Dyssodia papposa (Vent.) Hitchc.	fetid marigold
Echinochloa colonum (L.) Link	junglerice
crus-galli (L.) Beauv.	barnyardgrass
muricata (Beauv.) Fern. var. *muricata*	prickly barnyardgrass
pungens (Poir.) Rydb. (see *E. muricata* var.)	
Echinocystis lobata (Michx.) T. & G.	wild cucumber
Echium vulgare L.	blue thistle
Eleocharis palustris (L.) R. & S.	creeping spikerush
Eleusine indica (L.) Gaertn.	goosegrass
Ellisia nyctelea L.	waterpod
Epilobium angustifolium L.	fireweed
hirsutum L.	hairy willowweed
paniculatum Nutt.	panicle willowweed

Appendix II

Scientific Name	Standardized Common Name
Equisetum arvense L.	field horsetail
hyemale L.	scouringrush
palustre L.	marsh horsetail
sylvaticum L.	sylvan horsetail
telmateia Ehrh.	giant horsetail
Eragrostis cilianensis (All.) E. Mosher	stinkgrass
frankii C. A. Meyer	sandbar lovegrass
minor Host	little lovegrass
pectinacea (Michx.) Nees	tufted lovegrass
poaeoides Beauv. (see *E. minor*)	
Erechtites hieracifolia (L.) Raf.	American burnweed
minima (Poir.) DC.	Australian burnweed
prenanthoides (A. Rich.) DC. (see *E. minima*)	
Eremocarpus setigerus (Hook.) Benth.	turkey mullein
Erigeron annuus (L.) Pers.	annual fleabane
canadensis L. (see *Conyza canadensis*)	
divaricatus Michx. (see *Conyza ramosissima*)	
philadelphicus L.	Philadelphia fleabane
pulchellus Michx.	robinsplantain
strigosus Muhl.	rough fleabane
Erodium cicutarium (L.) L'Her.	redstem filaree
Eruca sativa Mill. (see *E. vesicaria* subsp.)	
vesicaria (L.) Cav. subsp. *sativa* (Mill.) Thell.	garden rocket
Erucastrum gallicum (Willd.) O. E. Schulz	dog mustard
Erysimum cheiranthoides L.	wormseed mustard
repandum L.	bushy wallflower
Eupatorium capillifolium (Lam.) Small	dog fennel
maculatum L.	joepyeweed
perfoliatum L.	boneset
rugosum Houtt.	white snakeroot
serotinum Michx.	late eupatorium
Euphorbia corollata L.	flowering spurge
cyparissias L.	cypress spurge
dentata Michx.	toothed spurge
esula L.	leafy spurge
helioscopia L.	sun spurge

Weeds Mentioned in This Book

Scientific Name	Standardized Common Name
Euphorbia (cont.)	
lucida Waldst. & Kit.	shining spurge
maculata L.	spotted spurge
marginata Pursh	snow-on-the-mountain
peplus L.	petty spurge
supina Raf.	prostrate spurge
vermiculata Raf.	hairy spurge
Fimbristylis autumnalis (L.) R. & S.	slender fimbristylis
Franseria discolor Nutt.	
(see *Ambrosia tomentosa*)	
tenuifolia Torr. & Gray	
(see *Ambrosia confertifolia*)	
tomentosa Gray	
(see *Ambrosia grayi*)	
Fumaria officinalis L.	fumitory
Galeopsis tetrahit L.	hempnettle
Galinsoga ciliata (Raf.) Blake	
(see *G. quadriradiata*)	
quadriradiata Ruiz & Pavon	hairy galinsoga
Galium aparine L.	catchweed bedstraw
asprellum Michx.	rough bedstraw
boreale L.	northern bedstraw
mollugo L.	smooth bedstraw
verum L.	yellow bedstraw
Gaultheria shallon Pursh	salal
Gaura biennis L.	biennial gaura
coccinea Nutt. ex Pursh	scarlet gaura
odorata Sessé ex Lag.	scented gaura
parviflora Dougl.	smallflower gaura
sinuata Nutt.	wavyleaf gaura
villosa Torr.	hairy gaura
Genista tinctoria L.	dyers greenweed
Geranium carolinianum L.	Carolina geranium
columbinum L.	longstalk geranium
maculatum L.	spotted geranium
molle L.	dovefoot geranium
pratense L.	meadow geranium
pusillum L.	small geranium
Geum aleppicum Jacq. var. *strictum* (Ait.) Fern.	yellow avens
macrophyllum Willd.	largeleaf avens
Glechoma hederacea L.	ground ivy
Glycyrrhiza lepidota Pursh	wild licorice
Gnaphalium macounii Greene	
(see *G. viscosum*)	

Scientific Name	Standardized Common Name
Gnaphalium (cont.)	
obtusifolium L.	fragrant cudweed
uliginosum L.	low cudweed
viscosum HBK.	clammy cudweed
Grindelia squarrosa (Pursh) Dunal	gumweed
Gutierrezia dracunculoides (DC.) Blake	common broomweed
Hackelia floribunda (Lehm.) I. M. Johnston	western stickseed
virginiana (L.) I. M. Johnston	Virginia stickseed
Halogeton glomeratus (Bieb.) C. A. Meyer	halogeton
Hedeoma pulegioides (L.) Pers.	American pennyroyal
Helenium amarum (Raf.) Rock	bitter sneezeweed
autumnale L.	common sneezeweed
flexuosum Raf.	purplehead sneezeweed
nudiflorum Nutt. (see *H. flexuosum*)	
tenuifolium Nutt. (see *H. amarum*)	
Helianthus annuus L.	sunflower
ciliaris DC.	Texas blueweed
maximiliani Schrad.	Maximilian sunflower
petiolaris Nutt.	prairie sunflower
tuberosus L.	Jerusalem artichoke
Heliotropium curassavicum L.	seaside heliotrope
Hemerocallis fulva L.	tawny daylily
Hibiscus trionum L.	Venice mallow
Hieracium aurantiacum L.	orange hawkweed
caespitosum Dumort.	yellow hawkweed
florentinum All. (see *H. piloselloides*)	
floribundum Wimm. & Grab.	yellowdevil hawkweed
pilosella L.	mouseear hawkweed
piloselloides Vill.	kingdevil hawkweed
praealtum Vill. ex Gochnat	tall kingdevil hawkweed
pratense Tausch (see *H. caespitosum*)	
Holcus lanatus L.	velvetgrass
Hordeum brachyantherum Nevski	meadow barley
jubatum L.	foxtail barley
murinum L.	wall barley
nodosum L. (see *H. brachyantherum*)	
pusillum Nutt.	little barley

Weeds Mentioned in This Book

SCIENTIFIC NAME	STANDARDIZED COMMON NAME
Hydrocotyle sibthorpioides Lam.	lawn pennywort
Hypericum perforatum L.	St. Johnswort
punctatum L.	spotted St. Johnswort
spathulatum (Spach) Steud.	shrubby St. Johnswort
Hypochoeris radicata L.	spotted catsear
Inula helenium L.	elecampane
Ipomoea coccinea L.	scarlet morningglory
hederacea (L.) Jacq.	ivyleaf morningglory
pandurata (L.) G. F. W. Mey.	bigroot morningglory
purpurea (L.) Roth	tall morningglory
Iresine celosia L.	Jubasbush bloodleaf
Isatis tinctoria L.	dyers woad
Iva axillaris Pursh	povertyweed
xanthifolia Nutt.	marshelder
Juncus bufonius L.	toad rush
effusus L.	soft rush
gerardii Loisel.	saltmeadow rush
tenuis Willd.	slender rush
Kalmia angustifolia L.	sheep laurel
latifolia L.	mountain laurel
polifolia Wang.	pale laurel
Kickxia elatine (L.) Dumort.	sharppoint fluvellin
Knautia arvensis (L.) Duby	field scabious
Kochia scoparia (L.) Schrad.	kochia
Lactuca biennis (Moench) Fern.	biennial lettuce
canadensis L.	tall lettuce
muralis (L.) Gaertn.	wall lettuce
pulchella (Pursh) DC.	blue lettuce
scariola L. (see *L. serriola*)	
serriola L.	prickly lettuce
Lamium amplexicaule L.	henbit
maculatum L.	spotted deadnettle
purpureum L.	red deadnettle
Lappula echinata Gilib.	European sticktight
Lapsana communis L.	nipplewort
Lathyrus palustris L.	marsh peavine
pratensis L.	meadow peavine
Leersia oryzoides (L.) Swartz	rice cutgrass
Leontodon autumnalis L.	fall hawkbit
Leonurus cardiaca L.	motherwort
Lepidium campestre (L.) R. Br.	field pepperweed
densiflorum Schrad.	greenflower pepperweed
latifolium L.	perennial pepperweed
perfoliatum L.	yellowflower pepperweed
repens (Schrenk) Boiss.	lens pepperweed

548 Appendix II

Scientific Name	Standardized Common Name
Lepidium (cont.)	
ruderale L.	narrowleaf pepperweed
virginicum L.	Virginia pepperweed
Leptochloa fascicularis (Lam.) Gray	bearded sprangletop
Lespedeza striata (Thunb.) H. & A.	common lespedeza
violacea (L.) Pers.	violet lespedeza
Linaria canadensis (L.) Dumont	oldfield toadflax
canadensis var. *texana* (Scheele) Pennell (see *L. texana*)	
texana Scheele	Texas toadflax
vulgaris Mill.	yellow toadflax
Lippia cuneifolia (Torr.) Steud. (see *Phyla cuneifolia*)	
Lithospermum arvense L.	corn gromwell
officinale L.	gromwell
Lobelia inflata L.	indiantobacco
Lolium temulentum L.	darnel
Lonicera japonica Thunb.	Japanese honeysuckle
Lupinus perennis L.	perennial lupine
Lychnis alba L.	white cockle
dioica L.	red campion
floscuculi L.	meadow campion
Lycopsis arvensis L.	small bugloss
Lycopus americanus Muhl.	American bugleweed
uniflorus Michx.	oneflower bugleweed
Lygodesmia juncea (Pursh) D. Don	skeletonweed
Lysimachia nummularia L.	moneywort
terrestris (L.) BSP.	swampcandle loosestrife
Lythrum salicaria L.	purple loosestrife
Madia glomerata Hook.	cluster tarweed
sativa Mol. var. *congesta* Torr. & Gray	Chilean tarweed
Malva moschata L.	musk mallow
neglecta Wallr.	common mallow
Malvella leprosa (Ortega) Krapov.	alkali sida
Marrubium vulgare L.	white horehound
Matricaria matricarioides (Less.) Porter	pineappleweed
Medicago lupulina L.	black medic
sativa L.	alfalfa
Melilotus alba Medic.	white sweetclover
altissima Thuill.	tall yellow sweetclover
officinalis (L.) Pall.	yellow sweetclover
Mentha arvensis L.	field mint
gentilis L.	red mint

Weeds Mentioned in This Book

Scientific Name	Standardized Common Name
Mentha (cont.)	
piperita L.	peppermint
spicata L.	spearmint
Mirabilis hirsuta (Pursh) MacM.	hairy four-o'clock
linearis (Pursh) Heimerl	narrowleaf four-o'clock
nyctaginea (Michx.) MacM.	wild four-o'clock
Mollugo verticillata L.	carpetweed
Muhlenbergia frondosa (Poir.) Fern.	wirestem muhly
schreberi J. F. Gmel.	nimblewill
Navarretia intertexta (Benth.) Hook.	woolly gilia
squarrosa (Esch.) Hook. & Arn.	skunkweed gilia
Nepeta cataria L.	catnip
Neslia paniculata (L.) Desv.	ball mustard
Nicandra physalodes (L.) Gaertn.	apple-of-Peru
Oenothera biennis L.	common eveningprimrose
Onoclea sensibilis L.	sensitive fern
Opuntia spp.	pricklypears
Origanum vulgare L.	wild marjoram
Ornithogalum umbellatum L.	star-of-Bethlehem
Orobanche ludoviciana Nutt.	Louisiana broomrape
minor J. E. Smith	clover broomrape
ramosa L.	hemp broomrape
Oryza sativa L.	rice
Osmunda cinnamomea L.	cinnamon fern
Oxalis dillenii Jacq.	southern yellow woodsorrel
subsp. *filipes* (Small) Eiten	yellow woodsorrel
europaea Jord. (see *O. stricta*)	
florida Salisb. (see *O. dillenii* subsp.)	
stricta of authors, non L. (see *O. dillenii*)	
stricta L.	common yellow woodsorrel
Oytropis lambertii Pursh	Lambert crazyweed
Panicum capillare L.	witchgrass
dichotomiflorum Michx.	fall panicum
gattingeri Nash	Gattinger panicum
virgatum L.	switchgrass
Papaver dubium L.	field poppy
rhoeas L.	corn poppy
Parthenium hysterophorus L.	ragweed parthenium
Paspalum ciliatifolium Michx. (see *P. setaceum* var.)	
dilatatum Poir.	dallisgrass
distichum L.	knotgrass
floridanum Michx.	Florida paspalum

Scientific Name	Standardized Common Name
Paspalum (cont.)	
laeve Michx.	field paspalum
setaceum Michx. var. *ciliatifolium* (Michx.) Vasey	fringeleaf paspalum
Passiflora incarnata L.	maypop passionflower
Pastinaca sativa L.	wild parsnip
Penstemon digitalis Nutt.	digitalis penstemon
gracilis Nutt.	slender penstemon
Phacelia purshii Buckl.	Pursh phacelia
Phalaris brachystachya Link	shortspike canarygrass
canariensis L.	canarygrass
paradoxa L.	hood canarygrass
Phyla cuneifolia (Torr.) Greene	wedgeleaf frogfruit
Physalis alkekengi L.	Chinese lanternplant
heterophylla Nees	clammy groundcherry
ixocarpa Brotero	tomatillo groundcherry
lobata Torr.	purpleflower groundcherry
longifolia Nutt.	longleaf groundcherry
pubescens L.	downy groundcherry
subglabrata Mack. & Bush	smooth groundcherry
Phytolacca americana L.	common pokeweed
Picris echioides L.	bristly oxtongue
hieracioides L.	hawkweed oxtongue
Plantago aristata Michx.	bracted plantain
indica L.	whorled plantain
lanceolata L.	buckhorn plantain
major L.	broadleaf plantain
media L.	hoary plantain
purshii R. & S.	woolly plantain
rugelii Dcne.	blackseed plantain
virginica L.	paleseed plantain
Poa annua L.	annual bluegrass
Polanisia dodecandra (L.) DC.	clammyweed
graveolens Raf. (see *P. dodecandra*)	
Polygonum amphibium L.	water smartweed
argyrocoleon Steud.	silversheath knotweed
aviculare L.	prostrate knotweed
coccineum Muhl. (see *P. amphibium*)	
convolvulus L.	wild buckwheat
cuspidatum Sieb. & Zucc.	Japanese knotweed
erectum L.	erect knotweed
hydropiper L.	marshpepper smartweed
hydropiperoides Michx.	mild smartweed

Weeds Mentioned in This Book

Scientific Name	Standardized Common Name
Polygonum (cont.)	
lapathifolium L.	pale smartweed
orientale L.	princesfeather
pensylvanicum L.	Pennsylvania smartweed
persicaria L.	ladysthumb
sachalinense F. Schmidt ex Maxim.	Sakhalin knotweed
scandens L.	hedge smartweed
setaceum Baldw. ex Ell.	bristly smartweed
Portulaca oleracea L.	common purslane
Potentilla anserina L.	silverweed cinquefoil
argentea L.	silvery cinquefoil
canadensis L.	common cinquefoil
fruticosa L.	shrubby cinquefoil
intermedia L.	downy cinquefoil
norvegica L.	rough cinquefoil
recta L.	sulphur cinquefoil
reptans L.	creeping cinquefoil
simplex Michx.	oldfield cinquefoil
Prunella vulgaris L.	healall
Prunus virginiana L.	common chokecherry
Pteridium aquilinum (L.) Kuhn	
var. *latiusculum* (Desv.) Underw.	eastern bracken
var. *pubescens* Underw.	western bracken
Ranunculus abortivus L.	smallflower buttercup
acris L.	tall buttercup
arvensis L.	corn buttercup
bulbosus L.	bulbous buttercup
repens L.	creeping buttercup
sceleratus L.	celeryleaf buttercup
testiculatus Crantz	testiculate buttercup
Raphanus raphanistrum L.	wild radish
sativus L.	garden radish
Ratibida pinnata (Vent.) Barnh.	pinnate coneflower
Reseda lutea L.	yellow mignonette
Rhus diversiloba T. & G. (see *Toxicodendron diversilobum*)	
microcarpa Steud. (see *Toxicodendron radicans*	
radicans L. (see *Toxicodendron radicans*)	
radicans L. var. *rydbergii* Small (see *Toxicodendron rydbergii*)	
toxicodendron L. (see *Toxicodendron toxicarium*)	

Appendix II

Scientific Name	Standardized Common Name
Rhus (cont.)	
vernix L. (see *Toxicodendron vernix*)	
Roemeria refracta DC.	Roemer poppy
Rorippa austriaca (Crantz) Bess.	Austrian fieldcress
islandica (Oeder) Borbas (see *R. palustris*)	
islandica var. *hispida* (see *R. palustris* subsp.)	
palustris (L.) Bess.	marshcress
subsp. *hispida* (Desv.) Jonsell	hispid marshcress
sylvestris (L.) Bess.	yellow fieldcress
Rosa arkansana Porter	Arkansas rose
eglanteria L.	sweetbrier rose
Rubus allegheniensis Porter	Allegheny blackberry
flagellaris Willd.	northern dewberry
Rudbeckia hirta L.	hairy coneflower
hirta L. var. *pulcherrima* Farwell	blackeyedsusan
laciniata L.	cutleaf coneflower
serotina Nutt. (see *R. hirta* var.)	
triloba L.	browneyedsusan
Rumex acetosa L.	sorrel
acetosella L.	red sorrel
altissimus Wood	pale dock
crispus L.	curly dock
hastatulus Baldw. ex Ell.	heartwing sorrel
mexicanus Meissn. (see *R. triangulivalvis*)	
obtusifolius L.	broadleaf dock
patientia L.	patience dock
triangulivalvis (Dans.) Rech. f.	willowleaf dock
Salix spp.	willows
Salsola kali L. var. *tenuifolia* Tausch	Russian thistle
Salvia reflexa Hornem.	Rocky Mountain sage
Sanguisorba minor Scop.	small burnet
Saponaria officinalis L.	bouncingbet
vaccaria L. (see *Vaccaria pyramidata*)	
Satureja vulgaris (L.) Fritsch	wild basil
Scirpus americanus Pers.	American bulrush
atrovirens Willd.	green bulrush
cyperinus (L.) Kunth	woolgrass bulrush
Scleranthus annuus L.	knawel

Weeds Mentioned in This Book 553

Scientific Name	Standardized Common Name
Scrophularia lanceolata Pursh	lanceleaf figwort
marilandica L.	Maryland figwort
Sedum acre L.	mossy stonecrop
purpureum (L.) Link (see *S. telephium*)	
telephium L.	liveforever
Senecio aureus L.	golden ragwort
jacobaea L.	tansy ragwort
vulgaris L.	common groundsel
Sesbania exaltata (Raf.) Cory	hemp sesbania
Setaria faberi Herrm.	giant foxtail
glauca (L.) Beauv.	yellow foxtail
verticillata (L.) Beauv.	bristly foxtail
viridis (L.) Beauv.	green foxtail
Sherardia arvensis L.	field madder
Sibara virginica (L.) Rollins	sibara
Sicyos angulatus L.	burcucumber
Sida hederacea (Dougl.) Torr. ex Gray (see *Malvella leprosa*)	
spinosa L.	prickly sida
Silene antirrhina L.	sleepy catchfly
csereii Baumg.	biennial campion
cucubalus Wibel (see *S. vulgaris*)	
dichotoma Ehrh.	hairy catchfly
gallica L.	English catchfly
noctiflora L.	nightflowering catchfly
vulgaris (Moench) Garcke	bladder campion
Silphium perfoliatum L.	cupplant
Sisymbrium altissimum L.	tumble mustard
irio L.	London rocket
officinale (L.) Scop.	hedge mustard
Sium suave Walt.	waterparsnip
Smilax glauca Walt.	cat greenbrier
Solanum carolinense L.	horsenettle
dulcamara L.	bitter nightshade
elaeagnifolium Cav.	silverleaf nightshade
nigrum L.	black nightshade
rostratum Dunal	buffalobur
sarrachoides Sendt.	hairy nightshade
triflorum Nutt.	cutleaf nightshade
villosum Mill. (see *S. sarrachoides*)	
Solidago canadensis L.	Canada goldenrod
gigantea Ait.	giant goldenrod
graminifolia (L.) Salisb.	narrowleaf goldenrod

Appendix II

SCIENTIFIC NAME	STANDARDIZED COMMON NAME
Solidago (cont.)	
var. *nuttallii* Fern.	
(see *S. graminifolia*)	
nemoralis Ait.	gray goldenrod
rugosa Mill.	rough goldenrod
Sonchus arvensis L.	perennial sowthistle
arvensis L. subsp. *uliginosus*	swamp sowthistle
(Bieb.) Nyman	
arvensis var. *glabrescens* Wimm. &	
Grab. (see *S. arvensis* subsp.)	
asper (L.) Hill	spiny sowthistle
oleraceus L.	annual sowthistle
uliginosus Bieb.	
(see *S. arvensis* subsp.)	
Sorghum bicolor (L.) Moench	shattercane
halepense (L.) Pers.	johnsongrass
vulgare Pers. (see *S. bicolor*)	
Specularia perfoliata (L.) A. DC.	
(see *Triodanis perfoliata*)	
Spergula arvensis L.	corn spurry
Spergularia rubra (L.) J. & C. Presl.	red sandspurry
Sphaerophysa salsula (Pall.) DC.	swainsonpea
Spiraea alba DuRoi	narrowleaf meadowsweet
douglasii Hook.	Douglas spirea
latifolia (Ait.) Borkh.	meadowsweet
tomentosa L.	hardhack
Sporobolus indicus (L.) R. Br.	smutgrass
neglectus Nash	annual dropseed
poiretii (R. & S.) Hitchc.	
(see *S. indicus*)	
vaginiflorus (Torr.) Wood	poverty dropseed
Stachys annua (L.) L.	hedgenettle betony
arvensis (L.) L.	fieldnettle betony
palustris L.	marsh betony
Stellaria graminea L.	little starwort
media (L.) Vill.	chickweed
Stipa comata Trin. & Rupr.	needleandthread
spartea Trin.	porcupinegrass
Succisa australis (Wulf.) Reichenb.	devilsbit
Swainsonia salsula (Pall.) Taub.	
(see *Sphaerophysa salsula*)	
Symphytum officinale L.	common comfrey
Tanacetum vulgare L.	tansy
Taraxacum erythrospermum Andrz.	redseed dandelion
officinale Weber	common dandelion

Weeds Mentioned in This Book

SCIENTIFIC NAME	STANDARDIZED COMMON NAME
Thlaspi arvense L.	field pennycress
perfoliatum L.	thoroughwort pennycress
Thymus serpyllum L.	creeping thyme
Torilis japonica (Houtt.) DC.	Japanese hedgeparsley
Toxicodendron diversilobum (T. & G.) Greene	Pacific poison oak
radicans (L.) Kuntze	poison ivy
rydbergii (Small) Greene	western poison ivy
toxicarium (Salisb.) Gillis	poison oak
vernix (L.) Kuntze	poison sumac
Tragopogon dubius Scop.	western salsify
major Jacq. (see *T. dubius*)	
porrifolius L.	common salsify
pratensis L.	meadow salsify
Tribulus terrestris L.	puncturevine
Trichostema dichotomum L.	bluecurls
Trifolium agrarium L.	hop clover
arvense L.	rabbitfoot clover
dubium Sibth.	small hop clover
fimbriatum Lindl. (see *T. wormskjoldii*)	
procumbens L.	low hop clover
wormskjoldii Lehm.	Columbia clover
Triodanis perfoliata (L.) Nieuw.	Venus lookingglass
Triticum cylindricum (Host) Ces., Pass. & Gib.	jointed goatgrass
Tussilago farfara L.	coltsfoot
Typha latifolia L.	common cattail
Ulex europaeus L.	gorse
Urtica dioica L.	stinging nettle
dioica L. subsp. *gracilis* (Ait.) Seland. var. *holosericea* (Nutt.) C. L. Hitchc.	California slender nettle
holosericea Nutt. (see *U. dioica* subsp.)	
lyallii S. Wats.	Lyall nettle
procera Muhl.	tall nettle
urens L.	burning nettle
Vaccaria pyramidata Medic.	cow cockle
Valerianella locusta (L.) Laterrade	European cornsalad
olitoria (L.) Poll. (see *V. locusta*)	
Veratrum viride Ait.	white hellebore
Verbascum blattaria L.	moth mullein
lychnitis L.	white mullein
phlomoides L.	clasping mullein

Scientific Name	Standardized Common Name
Verbascum (cont.)	
thapsus L.	common mullein
Verbena bracteata Lag. & Rodr.	prostrate vervain
hastata L.	blue vervain
officinalis L.	European vervain
stricta Vent.	hoary vervain
urticifolia L.	white vervain
Vernonia altissima Nutt.	tall ironweed
baldwinii Torr.	western ironweed
Veronica agrestis L.	field speedwell
arvensis L.	corn speedwell
chamaedrys L.	germander speedwell
filiformis Sm.	creeping speedwell
hederifolia L.	ivyleaf speedwell
officinalis L.	common speedwell
peregrina L.	purslane speedwell
persica Poir.	birdseye speedwell
polita Fries	wayside speedwell
serpyllifolia L.	thymeleaf speedwell
Vicia angustifolia L.	
(see *V. sativa* subsp.)	
cracca L.	bird vetch
sativa L.	common vetch
sativa L. subsp. *nigra* (L.) Ehrh.	narrowleaf vetch
tetrasperma (L.) Schreb.	fourseed vetch
villosa Roth	hairy vetch
Vinca minor L.	periwinkle
Viola arvensis Murr.	field violet
Woodwardia virginica (L.) Sm.	Virginia chainfern
Xanthium orientale L.	
(see *X. strumarium*)	
spinosum L.	spiny cocklebur
strumarium L.	heartleaf cocklebur
Zigadenus elegans Pursh	mountain deathcamas
gramineus Rydb.	
(see *Z. venenosus* var.)	
nuttallii Gray	Nuttall deathcamas
paniculatus (Nutt.) S.	foothill deathcamas
venenosus S. Wats.	meadow deathcamas
var. *gramineus* (Rydb.) Walsh	grassy deathcamas
Zygophyllum fabago L.	Syrian beancaper

APPENDIX III

Bibliography for Current Nomenclature and Common Names

American Joint Committee on Horticultural Nomenclature. Standardized Plant Names. 2d ed. J. Horace McFarland Co., Harrisburg, Pa. 1942.

BAILEY, LIBERTY H., and ETHEL Z. BAILEY, revised and expanded by the staff of the L. H. Bailey Hortorium. Hortus Third. Macmillan Publishing Co., New York. 1976.

BARNEBY, RUPERT C. Atlas of North American *Astragalus*. New York Botanical Garden. (Memoirs of the New York Botanical Garden, vol. 13). 2 vols., 1964.

CANNE, JUDITH M. A revision of the genus *Galinsoga* (Compositae: Heliantheae). Rhodora 79:319–389. 1977.

CORRELL, DONOVAN S., and HELEN B. CORRELL. Aquatic and wetland plants of southwestern United States. Stanford University Press, Stanford, Calif. 2 vols. 1975.

CORRELL, DONOVAN S., and MARSHALL C. JOHNSTON. Manual of the vascular plants of Texas. Texas Research Foundation, Renner, Texas. 1963.

EITEN, GEORGE. Taxonomy and regional variation of *Oxalis* section *Corniculatae*. I. Introduction, keys and synopsis of the species. Amer. Midl. Nat. 69:257–309. 1963.

FRYXELL, PAUL A. The North American Malvellas (Malvaceae). Southwest. Nat. 19:97–103. 1974.

GILLIS, WILLIAM T. The systematics and ecology of poison-ivy and the poison-oaks (*Toxicodendron*, Anacardiaceae). Rhodora 73:72–159, 161–237, 370–443, 465–540. 1971.

GLEASON, HENRY A., and ARTHUR CRONQUIST. Manual of vascular plants of northeastern United States and adjacent Canada. D. Van Nostrand Co., New York. 1963.

GOULD, FRANK W. The grasses of Texas. Texas A&M University Press, College Station, Texas. 1975.

HITCHCOCK, C. LEO and ARTHUR CRONQUIST. Flora of the Pacific Northwest. University of Washington Press, Seattle, Wash. 1973.

KOCH, STEVEN D. Notes on the genus *Eragrostis* (Gramineae) in the southeastern United States. Rhodora 80:390–403. 1978.

KOMAROV, V. L., et al. (ed.) Flora of the U.S.S.R. (Flora SSSR). Translated from the Russian. Published for the National Science Foundation, Washington, D.C., by the Israel Program for Scientific Translations, Jerusalem. 1963–1972.

MITCHELL, RICHARD S., and K. KENNETH DEAN. Polygonaceae (Buckwheat family) of New York State. Contributions to a Flora of New York State. I. New York State Museum Bulletin 431. 1978.

MUNZ, PHILIP A. A California flora. University of California Press, Berkeley, Calif. 1959.

PAYNE, WILLARD W. A re-evaluation of the genus *Ambrosia* (Compositae). Jour. Arn. Arboretum 45:401–438. 1964.

RADFORD, ALBERT E., HARRY E. AHLES, and C. RITCHIE BELL. Manual of the vascular flora of the Carolinas. The University of North Carolina Press, Chapel Hill, N.C. 1968.

SEYMOUR, FRANK C. The flora of New England. Charles E. Tuttle Co., Rutland, Vt. 1969.

STUCKEY, RONALD L. Taxonomy and distribution of the genus *Rorippa* (Cruciferae) in North America. Sida 4:279–430. 1972.

Terminology Committee of the Weed Science Society of America. Report of the Subcommittee on Standardization of Common and Botanical Names of Weeds. Weed Sci. 19:435–476. 1971.

TUTIN, THOMAS G., et al. (ed.). Flora Europaea. 4 vols. (to date). Cambridge University Press, Cambridge. 1964–1976.

INDEX

Synonyms are in italics. Page numbers in italics refer to illustrations.

Absinthe 430
Abutilon theophrasti 310, *312*
Acalypha rhomboidea 296
  virginica 297
  virginica 296
Achillea millefolium 422, *492*
Acnida altissima 196
  tuberculata 196
Adonis annua 217
Aegilops cylindrica 146
Aegopodium podagraria 321, *322*
Aeschynomene virginica 47, 290
Agents of dissemination 14
Agricultural seed impurities 16
Agrimonia gryposepala 264
Agrimony 264
Agropyron repens 111, *112*
  smithii 114
Agrostemma githago 201, *202*
Ague-weed 463
Aira caryophyllea 114
Aizoaceae 199
Alfalfa 285
Alfilaria 292
Alhagi camelorum 276
Alkali-grass 160
Alliaria officinalis 229
Allionia nyctaginea 196
Allium 157
  canadense 154, *155*
  tricoccum *155*, 156
  vineale *155*, 156
Almond, earth 148
Alopecurus geniculatus 115
  pratensis 115
Alsine media 217
Alum-root 295
Alyssum, hoary 232, *257*
Amaranth, green *193*, 194, 195
  mat 192
  palmers 194
  red 194
  spiny *193*, 196
  spleen 194

Amaranth, thorny 196
Amaranth family 192
Amaranth pigweed *193*, 194, 195
Amaranthaceae 192
Amaranthus albus 192, *193*
  blitoides 192
  graecizans 192, *193*
  graecizans 192
  hybridus *193*, 194
  lividus 194
  palmeri 194
  powellii 195
  retroflexus *193*, 195
  spinosus *193*, 196
Amber 314
Ambrosia, tall 425
Ambrosia artemisiifolia 423, *424*
  bidentata 425
  elatior 423
  psilostachya 423, *424*
  trifida *424*, 425
American cancer 197
Ammania coccinea 47
Ampelamus albidus 339, *342*
Amphicarpum purshii 46
Amsinckia intermedia 356, *362*
  lycopsoides 357
Anacardiaceae 306
Anagallis arvensis 334, *335*
Anaphalis margaritacea 425, *467*
Andropogon furcatus 115
Andropogon gerardii 115
  scoparius 115
  virginicus 116
Angelica, American 323
  purple-stem 323
Angelica atropurpurea 323
Animal digestion, effect on seeds 22
Animals as agents of dissemination 20
Annuals 9
Antennaria 426
  plantaginifolia 426, *467*
Anthemis arvensis var. agrestis 426, *427*
  cotula *427*, 428

Anthoxanthum odoratum 116
Apargia autumnale 484
Apocynaceae 336
Apocynum androsaemifolium 336, *337*
  cannabinum *337*, 338
Apple-of-Peru 383
Apple-of-Sodom 386
Arabidopsis thaliana 230
Arctium lappa 428, *429*
  minus *429*, 430
Arenaria serpyllifolia 201, *214*
Argemone intermedia 227
  mexicana 226
Aristida, few-flowered 117, *147*
Aristida dichotoma 117, *147*
  oligantha 117, *147*
Arrhenatherum elatius var. bulbosum 148
Artemisia absinthium 430
  annua 431, *432*
  biennis 431, *432*
  tridentata 433
  vulgaris *432*, 433
Artichoke, Jerusalem 472, *473*
Asclepiadaceae 339
Asclepias galioides 343
  labriformis 343
  speciosa 339
  syriaca 339, *340*
  tuberosa *340*, 341
  verticillata 341
Asparagus, wild 486
Aspris caryophyllea 114
Asses-ears 364
Aster, white field 433
  white heath 425
Aster paniculatus 433
Aster pilosus 425
  simplex 433
Asthma-weed 421
Astragalus diphysus 290
  mollissimus 290
Atriplex, tumbling 180
Atriplex patula 180, *181*
  patula var. hastata 180, *181*
  rosea 180
Auger-seed 146
Austrian field cress 254
Austrian field pea 290
Avena fatua 117, *118*
Avens, yellow 265
Axyris amaranthoides 182

Bachelors-button 440
Backwort 364
Bacon-weed 183
Ball mustard 252
Ballogan 484
Balm, field 369
  stinking 371
Balsam, old field 468
Balsam-apple 418
Bamboo, Mexican 167
Barbarea verna 230
  vulgaris 230, *231*
Barberry 226, *266*
Barberry, European 226
Barberry family 226
Barilla *178*, 191
Barley, little 134
  wall 133
  wild 131, 133
Barley-grass 133
Barn-grass 127
Barnyard-grass 127, *128*, 129
Barren strawberry 271
Basil, field 381
  stone 381
  wild *380*, 381
Basil-weed 381
Bassia hyssopifolia 182
Beans, Indian 283
Bear-bind 166, 344
Bear corn 159
Bear-grass 123
Beard-grass 115, 116
Beard-grass, broom 115
Beard-tongue 394, *403*
Beaver poison 323
Bedflower 414
Bedstraw, ladys 414
  northern *412*, 413
  rough *412*, 413
  white hedge 413
  yellow *412*, 414
Bee nettle 369, 371
Bee-plant, Rocky Mountain 228
Beetles 53
Beggar-ticks 264, *435*, 436, 437, *438*
Beggar-ticks, swamp 437
  tall 437
Beggars-buttons 428
Bellflower, 419, *420*
Bellflower, creeping, 419
Bellis perennis 434

Index

Bellwort, clasping 421
Benefits from weeds 32
Berberidaceae 226
Berberis vulgaris 226, *266*
Bergamot, wild 377
Bermuda-grass 123
Berteroa incana 232, *257*
Betony, Pauls 401
Bidens bipinnata 434, *435*
 cernua *435*, 436
 comosa 437
 connata 437
 frondosa *435*, 436
 involucrata 437
 polylepis 437, *438*
 vulgata *435*, 437
Biennials 9
Big taper 398
Bignonia radicans 404
Bignonia family 404
Bignoniaceae 404
Bindweed 217, 344, *345*
Bindweed, black *165*, 166
 bracted 346
 climbing 166
 European 344
 great 346
 hedge *165*, 172, 346
 ivy 166
 knot 166
Biological methods 55
Bird-grass 164
Bird-seed 493
Bird-seed-grass 139
Birds as agents of dissemination 20
Birds-eye, 217, 400
Birds-eye, red 206
Birds-nest 325, 329
Birds pepper 250
Bitter buttons 499
Bittersweet, European 387, *388*
Bitterweed 423, 425, 431, 459, 471
Bivonea stimulosa 297
Black brush 269
Black-eyed Susan 491, *492*
Black-grass 153
Black-jacks 409
Black medic *280*, 283
Blackberry 46
Blackberry, running 273
Blackseed-grass 145
Blackweed 423

Bladder-pod 421
Blister flower 222
Blister plant 220
Blite, strawberry 186
Blite mulberry 186
Blitum capitatum 186
Blood-leaf 196
Blood stanch 459
Bloodwort 422
Bloodwort, mouse 475
Blow-ball 501
Blue-bottle 440
Blue bur 360
Blue curls 382
Blue devil 357
Blue joint-grass 114
Blue sailors 447
Blue-stem, Colorado 114
Blue thistle 357, *358*
Blue-weed, 357, 474
Bluebell family 419
Boebera papposa 456
Boneset 463
Boneset, purple 463
 tall 463
Bonnet-grass 124
Borage family 356
Boraginaceae 356
Bottle-brush 106
Bottle-grass 142
Boulder fern 110
Bouncing Bet *204*, 206
Bouquet-violet 318
Bowmans root 338
Bracken 110, 111
Brake, hog 110, 111
 meadow 110
 polypod 110
Brake fern *109*, 110, 111
Bramble, trailing 273
Brambles 274
Brassica alba 232
 arvensis 234
 campestris 236
Brassica hirta 232, *233*
 juncea *233*, 234
 kaber var. pinnatifida 234, *235*
 nigra 236, *237*
 rapa 236, *237*
Brier, green 160
 sand 386
 saw 160

Weeds

Bristly foxtail *141*, 142
Brome, red 121
 soft 119
Brome-grass, barren 122
 downy *120*, 122
 smooth 119
Bromus commutatus 119
 hordeaceus 119
 japonica 123
 mollis 119, *120*
 racemosus 121
 rubens 121
 secalinus *120*, 121
 sterilis 122
 tectorum *120*, 122
Broom beard-grass 115
Broom brush 316
Broom-rape 406
Broom-rape, clover 406
 hemp 406
 Louisiana 406
 tobacco 406
Broom-rape family 406
Broom sedge 116
Broomweed 425
Bruisewort 364
Bubble poppy 211
Buckhorn, western 407
Buckthorn-weed 356
Buckwheat, climbing false 172
 hedge 172
 wild 166
Buckwheat family 164
Buffalo bur *388*, 390
Bugleweed 375
Bugleweed, cutleaf 375
Bugloss 489
Bugloss, small 364
 vipers 357
Bugseed 190
Bull nettle 386, 389
Bull rattle 205
Bulrush 47
Bulrush, dark green 151
Bunk 447
Bur, blue 360
 sheep 360
Bur-grass 123
Bur-marigold 436
Bur-marigold, smaller 436
Bur-ragweed 465
Burdock, great 428, *429*

Burdock, smaller *429*, 430
Burnet, garden 274
 salad 274
Burning of weeds 55
Burning-bush 191
Burnt-weed 319
Burnut, ground 291
Burnweed, Australian 457
 yellow 356
Burweed 360
Butter-and-eggs *392*, 394
Butter flower 220
Butter print 310
Butter-rose 220
Butter-weed 310
Buttercup 224, *225*
Buttercup, bog 224
 bulbous 222, *223*
 creeping 222, *223*
 field 222, *225*
 meadow 220
 small-flowered 218, *221*
 tall field 220, *221*
Butterfly-weed *340*, 341
Button-bur 504
Button-weed 411

Cactaceae 317
Cactus, Russian 191
Cactus family 317
Cadlock 236, 252
Calamint 381
Calamovilfa brevipilis 16
Calfkill 331
Calico bush 333
Caltrop 291
Caltrop family 291
Caltrops 440
Camas, *see* Zigadenus 160
Camel thorn 276
Camelina dentata 238
 microcarpa 236, *251*
 sativa 238
Campanula rapunculoides 419, *420*
Campanulaceae 419
Campion 209, *212*
Campion, bladder *210*, 211
 corn 201
 meadow 206
 red 206
 white 205
Campsis radicans 404, *405*

Index

Canada thistle 448, *449*
Canary-grass 139
Cancer jalap 197
Candle-wick 398
Candy-grass 130
Cankerwort 501
Cannabis sativa 162
Caper, Syrian bean 291
Caper family 228
Capparidaceae 228
Caprifoliaceae 415
Capriola dactylon 123
Capsella bursa-pastoris 238, *240*
Carara didyma 241
Caraway 323
Cardamine pratensis 238
Cardaria draba 239, *240*
  pubescens 239
Carduus acanthoides 439
  arvensis 448
  crispus 440
  nutans 439
Careless weed 195
Careless weed, prickly 196
Carex 46
  bullata 46
  lupulina 152
Carpenters-weed 379
Carpet-weed 199, *200*
Carpet-weed family 199
Carrot, wild 325, *328*
Carum carvi 323
Caryophyllaceae 201
Case weed 238
Cashew family 306
Cassia chamaecrista 277
Cassia fasciculata 277
  marilandica 277
  nictitans 278
  occidentalis 278
  tora 278
Cat-tail, broad-leaved 47
Catch-weed 411
Catchfly, English 211
  forked 211
  hairy 211
  night-flowering *210*, 213
  sleepy 209, *210*
Catmint 378
Catnip *374*, 378
Cats-ear 477
Cats-foot 369

Cats-milk 303
Celandine 227
Celandine, great 227
Celery seed 179
Cenchrus carolinianus 123
Cenchrus longispinus 123, *132*
  pauciflorus 123
Centaurea calcitrapa 440
  cyanus 440
  diffusa 441
  iberica 441
  jacea 442, *443*
  maculosa 442, *443*
  melitensis 442, *443*
  picris 444
  repens 444
  solstitialis *443*, 444
  vochinensis 445
Centaury, brown 442
  rayless 442
Cerastium arvense 203, *216*
  viscosum 203
  vulgatum 205, *216*
Chaenorrhinum minus 391, *392*
Chaetochloa lutescens 140
  verticillata 142
  viridis 142
Chafe-weed 468
Chamaedaphne calyculata 46
Chamaenerion angustifolium 319
Chamomile, corn 426, *427*
  dogs 428
  fetid 428
  field 426
  rayless 487
Charlock 232, 234
Charlock, jointed 252
  white 252
Cheat 121
Cheese-rennet 414
Cheeses 313
Chelidonium majus 227
Chemical weed control 56, 61
Chenopodiaceae 180
Chenopodium album 183, *184*
  ambrosioides 183, *188*
  ambrosioides var. anthelminticum 185
  bonus-henricus 185, *188*
  botrys 185, *188*
  capitatum 186
  glaucum *184*, 186
  hybridum *184*, 186

Chenopodium (continued)
  murale 187
  paganum 187, *188*
  polyspermum 189
  rubrum 189
  urbicum *184*, 189
Chess, *120*, 121
Chess, early 122
  Japanese 123
  slender 122
  soft 119, *120*
  upright 121
Chicken-weed 413
Chickweed 216, 217
Chickweed, field 203, *216*
  germander 404
  Indian 199
  meadow 203
  mouse-ear 203, 205, *216*
  poison 334
  red 334
  sandwort 201
  whorled 199
Chicory 447, *485*
Childrens-bane 323
Chinamens-greens 195
Chinese lantern plant 385
Choctaw root 338
Choke-cherry 271
Chondrilla juncea 445
Chorispora tenella 260
Chrysanthemum leucanthemum var. pinnatifidum *446*, 447
Chufa 148
Cichorium intybus 447, *485*
Cicuta maculata 323, *324*
Cinnamon fern 108
Cinquefoil 267, *268*, 271
Cinquefoil, downy 269
  hoary 267
  rough *268*, 269
  rough-fruited 270
  shrubby 269, *272*
  silvery 267, *272*
  sulfur *268*, 270
Cirsium altissimum 448
  arvense 448, *449*
  arvense, disease of 56
  lanceolatum 452
  pumilum 452, *453*
  vulgare 452, *453*
Clammy-weed 228

Weeds

Classification of weeds 12
Clay-weed 502
Clean machinery 49
Clean seed 49
Cleavers 411, *412*, 413
Cleavers, yellow 414
Cleome serrulata 228
Clotbur 430, 504
Clotbur, spiny 505
Clotweed 505
Clover, black 283
  bush 282
  honey 285
  hop 286
  little hop *280*, 287
  low hop *280*, 287
  old-field 286
  rabbit-foot *280*, 286
  stinking 228
  stone 286
  tree 285
  white sweet 285
  yellow 286
  yellow sweet 286
Clover choker 407
Cnicus benedictus 454, *455*
Cnidoscolus stimulosus 297
Coakum 197
Coast-blite 189
Cock-grass 121
Cockle, clammy 213
  corn 201, *202*
  cow *204*, *208*
  pink 208
  purple 201
  spring 208
  sticky 213
  tarry 209
  white 205, *207*
Cockle-button 428, 430
Cocklebur *481*, 504
Cocklebur, dagger 505
  spiny 505
Cocksfoot panicum 127
Cockspur-grass 127
Coco 148
Coco-grass 150
Coco sedge 148, 150
Coffee plant 320
Coffee senna 278
Coffee-weed 278, 447
Colewort 499

Index

Colorado blue-stem 114
Colorado bur 390
Coltsfoot 502, *503*
Comfrey 364
Comfrey, wild 359
Commelina communis 152
Commelinaceae 152
Compass plant 482
Compositae 422
Composite family 422
Comptonia peregrina 162, *266*
Coneflower 491
Coneflower, tall 425
Conium maculatum 325, *326*
Conringia orientalis 241, *251*
Control methods 48
Convallaria majalis 157, *158*
Convolvulaceae 344
Convolvulus arvensis 344, *345*
  sepium 346, *347*
Conyza canadensis 459
Copper-leaf 296
Coreopsis tinctoria 425
Corispermum hyssopifolium 190
Corn, bear 159
  chicken 145
  wild 145
Corn-bind 166, 344
Corn campion 201
Corn cockle 201, *202*
Corn-flower 440
Corn mullein 201
Corn rose 201
Corn-salad 415, *416*
Coronopus didymus 241, *259*
  procumbens 241
Cotton-grass 46
Cotton-weed 310, 339, 425
Couch-grass 111
Cough-wort 502
Cow-bell 211
Cow cockle *204*, 208
Cow-herb 208
Cow-itch 404
Cow-poison 218
Cowbane, spotted 323
Cowthwort 373
Crab-grass, large *126*, 127
  purple 127
  slender 125
  small 125, *126*
  sprouting 137

Cranesbill *293*, 295
Cranesbill, Carolina 294
  long-stalked 294
  small-flowered 296
  spotted 295
  wild, *293*, 295
Crassulaceae 262
Crataegus 264
Craw pea 282
Creeping charlie 334, 369
Creeping jenny 334, 344
Crepis biennis 456
  capillaris 454
  tectorum 456
Cress, Austrian field 254, *255*
  bastard 260
  bitter 230
  carpet 241
  cow 249
  creeping yellow 256
  field 249
  globe-podded hoary 239
  hoary 239, *240*
  marsh 254, *255*
  mouse-ear 230
  penny 260
  perennial pepper 252
  rocket 230
  St. Barbaras 230
  swine 241, *259*
  wart 241
  winter 230, *231*
  yellow *255*, 256
  yellow water 254
Crotalaria sagittalis 278, *284*
Croton capitatus 297
  glandulosus 298
Crow flower 206
Crowfoot, bulbous 222, *223*
  celery-leaved 224
  creeping 222, *223*
  cursed *223*, 224
  smooth-leaved 218
  tall 220
Crowfoot family 217
Crowfoot-grass 124, 127, 129
Crown-of-the-field 201
Crown-weed 425
Cruciferae 229
Cuckold 434
Cuckoo, snake 205
Cuckoo-button 430

Weeds

Cuckoo flower 206, 238
Cucumber, one-seeded bur 419
  wild 418
Cucurbita foetidissima 419
Cucurbitaceae 418
Cudweed 468
Cudweed, clammy 466
  low *467*, 468
  marsh 468
Cultivated plants becoming weeds 15
Cultivation 51, 57, 62
Culver-foot 295
Cup plant 425
Cuphea petiolata 318
Curdwort 414
Cuscuta 350, 352
  arvensis 351
  campestris 346
  coryli 348, *350*
  epilinum 348, *350*
  epithymum 348, *350*
  glomerata 349
  gronovii 349, *350*
  indecora 349
  pentagona *350*, 351
  planifera *350*, 351
  polygonorum 352
  racemosa var. *chiliana* 349
  suaveolens 352
Cut-grass 46
Cycloloma atriplicifolium 190
Cymbalaria muralis 391, *392*
Cynanchum nigrum 343
  vincetoxicum *337*, 343
Cynodon dactylon 123
Cynoglossum officinale 357, *358*
Cyperaceae 148
Cyperus diandrus 148
  esculentus 148, *149*
  iria 150
  rotundus 150
  strigosus *149*, 150
Cypress spurge 300, *302*
Cytisus scoparius 279

Dactylotenium aegyptium 124
Daisy, blue 447
  blue spring 459
  cone-headed 491
  English 434
  field 447
  lawn 434

ox-eye *446*, 447, 491
  stinking 428
  white 447
  yellow 491
Dallis-grass 138
Dandelion 501, *503*
Dandelion, coast 477
  fall 484, *490*
  false 477
  red-seeded 501
Danthonia spicata *118*, 124
Darky-head 491
Darnel 134, *161*
Datura stramonium 383, *384*
  tatula 383
Daucus carota 325, *328*
  pusillus 327
Day-lily 157
Dayflower 152
Deadly-hemlock 325
Death camas, foothill 160
  grassy, 160
  meadow 160, *161*
  mountain 160
  Nuttalls 160
Death-weed 479
Deerwort 463
Delphinium menziesii 218, *219*
  simplex 218
  strictum 218, *219*
Dennstaedtia punctilobula *109*, 110
Deptford pink 205, *214*
Descurainia pinnata *235*, 242
  richardsonii 242
  sophia 242
Desmodium 281
  canadense 279
Devil-weed 360, *479*
Devils-bit 418
Devils boot-jack 436
Devils cabbage 239
Devils-grass 111
Devils-grip 199
Devils-gut 213
Devils-hair 348
Devils paintbrush 477
Devils plague 325
Devils shoestring 486
Devils-vine 346
Dewberry 273
Dianthus armeria 205, *214*
Dicksonia, hairy 110

Dicksonia punctilobula 110
Digging, control by 57
Digitalis purpurea 393
Digitaria humifusa 125
Digitaria filiformis 125
  ischaemum 125, *126*
  sanguinalis *126*, 127
Dill-weed 428
Diodia teres 411
Diplotaxis muralis 244
  tenuifolia 244, *245*
Dipsacaceae 417
Dipsacus laciniatus 417
  sylvestris *416*, 417
Diseases of crops, weeds as hosts for 26
Disking 53
Dissemination 14
Dissemination, animals as agents of 20
  fruits modified for *19*
  seeds modified for *19*
  water as agent of 18
  wind as agent of 18
Ditch-bur 504
Dock, bitter 179
  blunt-leaved 179
  broad-leaved *177*, 179
  curly 176, *177*
  dove 502
  narrow-leaved 176
  pale 174, 179
  patience *175*, 179
  smooth 174
  sour 172, *175*, 176
  succory 484
  velvet 398
  water 174
  white 179
  willow-leaved 179
  yellow 176
Dodder, *350*, 352
Dodder, Chilean 349
  clover 348, *350*
  common 349, *350*
  field *350*, 351
  flax 348, *350*
  hazel 348, *350*
  large-seeded 351
  large-seeded alfalfa 349
  onion 349
  small-seeded alfalfa *350*, 351
  thyme 348
  western field 346

Dog bur 357
Dog fennel *427*, 428, 461, *462*
Dog nettle 369
Dogbane, hemp 338
  spreading 336, *337*
Dogbane family 336
Dogs-tooth-grass 123
Door-weed 164
Dove-dock 502
Dove-weed 298
Doves-foot 295
Downy brome-grass *120*, 122
Draba verna 244
Dracocephalum parviflorum 368
Dragonhead 368
Drainage 59
Drop-seed 135, 145
Drop-seed, Mexican 135
Drouth-weed, woolly white 298
Duchesnea indica 265
Dulichium arundinaceum 46
Duration of weeds 11
Dyers broom 281
Dyers greenwood 281
Dyeweed 281
Dyssodia papposa 456, *462*

Earth almond 148
Earth-apple 472
Echinochloa colonum 47
  crusgalli 127, *128*
  muricata 129
  pungens 129
Echinocystis lobata 418
Echium vulgare 357, *358*
Edible weeds 33
Eggs-and-bacon 394
Eglantine 273
Egyptian-grass 124, 142
Elder, marsh 480, *481*
  small-flowered marsh 479
Elecampane *470*, 479
Eleocharis palustris 47
Eleusine indica *128*, 129
Elf-dock 479
Elf-wort 479
Ellisia 354, *355*
Ellisia nyctelea 354, *355*
Emetic-weed 421
Eola-weed 314
Epilobium angustifolium 319
  hirsutum 319

Weeds

Equisetaceae 106
Equisetosis 106
Equisetum arvense 106, *107*
  hyemale 106, *107*
  palustre *107*, 108
  sylvaticum *107*, 108
  telmateia 108
Eradication of weeds 48
Eragrostis cilianensis 130
Eragrostis frankii 130
  major 130
  megastachya *128*, 130
  minor 131
  pectinacea 130
  pilosa 130
  poaeoides 131
Erechtites hieracifolia 456, *483*
  prenanthoides 457
Eremocarpus setigerus 298
Ericaceae 331
Erigeron annuus 457, *458*
  canadensis 459, *460*
  divaricatus 425
  philadelphicus *458*, 459
  pulchellus 459, *460*
  ramosus 461
  strigosus *458*, 461
Eriophorum 46
Erodium cicutarium 292, *293*
Eruca sativa 244, *245*
Erucastrum gallicum *245*, 246
  pollichii 246
Erysimum cheiranthoides *213*, 216
  repandum *243*, 247
Eupatorium capillifolium 461, *462*
  maculatum 463
  perfoliatum 463
  purpureum 463
  rugosum 463, *464*
  serotinum 425
  urticaefolium 463
Euphorbia 298
  corollata 298
  cyparissias 300, *302*
  dentata 301
  esula 301, *302*
  helioscopia *299*, 303
  hirsuta 305
  lucida 303
  maculata *299*, 304
  maculata 305
  marginata 304

  nutans 304
  peplus *299*, 304
  preslii 304
  supina *299*, 305
  vermiculata *299*, 305
  virgata 301
Euphorbiaceae 296
Evening lychnis 205
Evening primrose 320
Evening primrose family 319
Everlasting, clammy 466, *467*
  early 426
  fragrant *467*, 468
  pearly 425, *467*
  plantain-leaved 426
Eye-bright 304, 334, 421

Faitours-grass 301
Fallow 58
False flax 236
Fan-weed 253, 260
Fat-hen 180, 183
Feather-geranium 185
Federal seed act 32
Felon herb 475
Fennel 471
Fennel, dog *427*, 428, 461, *462*
  hogs 428
  yellow dog 471
Fern, boulder 110
  brake *109*, 110, 111
  chain 46
  cinnamon 108
  eagle 111
  hay-scented *109*, 110
  meadow 162
  sensitive *109*, 110
  shrubby 162
  upland 111
Fern family 110
Fern gale 162
Fern-wort 162
Fetticus 415
Fever plant 320
Fever-weed 463
Feverfew 264
Fiddle-necks 356
Field chickweed 203, *216*
Field kale 234
Field mustard 234
Figwort 395
Figwort family 391

Fimbristylis autumnalis 47
Finger-grass 125, 127
Finger-weed 356
Fireball 191
Fireweed 319, 356, 456, *483*
Fireweed, Mexican 191
Five-finger 271
Five-finger, tall 269
Flannel-leaf 398
Flatweed 477
Flax, Dutch 236
  false 236
  flat-seeded false 238
  large-seeded false 238
  small-seeded false 236, *251*
  western 236
Flax-grass 117
Flaxweed 394
Fleabane 459, *460*
Fleabane, daisy 457, 459
  dwarf 425
  Philadelphia *458*, 459
  rough daisy *458*, 461
Flicker-tail-grass 131
Flixweed 242
Flower-of-an-hour 310, *312*
Flowering fern family 108
Fluellin 401
Fluellin, sharp-pointed 393
Fog-fruit 368
Fool-hay 135
Forget-me-not, yellow 356
Four-o'clock, wild 196
Four-o'clock family 196
Foxglove 393
Foxglove, wild 394
Foxtail, bent 115
  bristly *141*, 142
  golden 140
  green *141*, 142
  marsh 115
  nodding 142
  red 121
  water 115
  yellow 140, *141*
Foxtail-grass 131
Foxtail-grass, meadow 115
Foxtail-rush 106
Franseria, white-leaved 465
Franseria discolor 465
  tenuifolia 466
  tomentosa 465

French-weed 260, 466
Frost-blite 183
Frost flower 433
Fruits modified for dissemination *19*
Fullers herb 206
Fumaria officinalis 228
Fumariaceae 228
Fumitory 228
Furze 287

Gaertneria discolor 465
Gagroot 421
Galeopsis tetrahit 369, *380*
Galingale 148
Galingale, edible 148
Galinsoga 466, *494*
  ciliata 466, *494*
  parviflora 466
Galium aparine 411, *412*
  asprellum *412*, 413
  boreale *412*, 413
  mollugo *412*, 413
  verum *412*, 414
Gallow-grass 162
Garden rocket 244, *245*
Garget 197
Garlic, crow 156
  field 156
  meadow 154
  wild *155*, 156
Garlic-mustard 229
Gaultheria shallon 46
Gaura, biennial 320
Gaura biennis 320
  coccinea 320
  odorata 320
  parviflora 320
  sinuata 320
  villosa 320
Geese 53
Genista tinctoria 281
Geraniaceae 292
Geranium, wild 295
Geranium carolinianum 294
  columbinum 294
  maculatum *293*, 295
  molle *293*, 295
  pratense 296
  pusillum 296
Geranium family 292
Germination of buried seeds 6

570   Weeds

Geum aleppicum var. strictum 265
  macrophyllum 267
  strictum 265
Gill-over-the-ground 369
Ginannia lanata 131
Ginger-root 502
Gipsy-combs 417
Gipsy-weed 401
Girasole 472
Glasswort, prickly 191
Glechoma hederacea 369, 370
Glycyrrhiza lepidota 281
Gnaphalium decurrens 466
Gnaphalium macounii 466, 467
  obtusifolium 467, 468
  polycephalum 468
  uliginosum 467, 468
Goat-grass 146
Goat-weed 314
Goats-beard 501
Goats-beard, yellow 483, 502
Gold cup 220
Gold-thread vine 349
Goldenrod 495
Goldenrod, bushy 496
  Canada 495
  gray 425
  narrow-leaved 496
  rough 496
  tall 495
Gonolobus laevis 339
Good King Henry 185, 188
Goose-grass 128, 199, 164, 411
Goosefoot 184, 189
Goosefoot, many-seeded 189
  maple-leaved 184, 186
  nettle-leaved 187
  oak-leaved 184, 186
  perennial 185
  red 189
  white 183
Goosefoot family 180
Gorse 287
Gosmore 446, 477
Gourd, wild 419
Gourd family 418
Goutweed 321, 322
Gramineae 111
Grass family 111
Grass pink 205
Graveyard-weed 300
Gray mile 363
Grazing 53

Green brier 160
Green-flowered pepper-grass 249
Green-vine 344
Grimsel 493
Grindelia squarrosa 469, 470
Grip-grass 411
Gromwell 361, 363
Ground-cherry 384, 385, 386
Ground-cherry, perennial 386
  purple-flowered 386
Ground-heal 401
Ground ivy 369, 370
Groundsel 493, 494
Guinea-grass, false 142
Gum-plant 469
Gum succory 445
Gumweed 469, 470
Gutierrezia dracunculoides 425
Guttiferae 314
Gutweed 496

Hackbrush 276
Hackelia floribunda 359
  virginiana 359
Hair-grass, silver 114
Hairweed 348
Halogeton glomeratus 178, 191
Hard-heads 442
Hardhack 276
Hardhack, yellow 269
Hares-ear mustard 241
Harrowing 52, 62
Harts-eye 329
Hawkbit 484
Hawks-beard 454, 456
Hawkweed 475
Hawkweed, field 477
  mouse-ear 475, 476
  orange 474, 476
Hawthorn 264
Hay-fever weed 423
Heal-all 372, 379
Healing herb 364
Hearts-ease 170, 316
Heartweed 170
Heath family 331
Hedeoma pulegioides 370, 371
Hedge mustard 248, 258
Hedgehog-grass 123
Helenium autumnale 469
  nudiflorum 471
  tenuifolium 471

Index 571

Helianthus annuus 472, *473*
  ciliaris 474
  maximiliani 425
  petiolaris 474
  tuberosus 472, *473*
Heliotrope, alkali 360
  seaside 360
  wild 360, *362*
Heliotropium curassavicum 360, *362*
Hellebore, American *158*, 159
  false 159
  green 159
  white 159
Hemerocallis fulva 157
Hemp 162
Hemp, American 338
  Indian 310, *337*, 338
  water 196
  wild 369, 425
Hemp nettle 369
Henbit 371
Herb sherard 414
Heronsbill 292
Hibiscus trionum 310, *312*
Hieracium auranticum 474, *476*
  florentinum 475, *476*
  floribundum 475
  pilosella 475, *476*
  praealtum 477
  pratense *476*, 477
Highwater shrub 480
Hillwort 382
Hind-head 499
Hoary alyssum 232, *257*
Hoary cress 239, *240*
Hoeing 51
Hog brake 111
Hog-rush 153
Hog-weed 423, 459
Hogwort 297
Holcus halepensis 142
Holcus lanatus *118*, 131
Honey-bloom 336
Honeysuckle, Japanese 415
Honeysuckle family 415
Honeyvine 339, *342*
Hordeum jubatum 131, *132*
  murinum 133
  nodosum 133
  pusillum 134
Horehound 376
Horehound, water 375
Horse-elder 479

Horse-hoof 502
Horse-knobs 442
Horse nettle 386, *388*
Horse-pipes 106
Horse-weed 425, 459
Horseheal 479
Horsetail *107*, 108
Horsetail, field 106, *107*
  giant 108
  ivory 108
  shade 108
  wood *107*, 108
Horsetail family 106
Horsetail fern 106
Hounds-tongue 357, *358*
Houndsbane 376
Hunger weed 222
Husk-tomato 385, 386
Hydrocotyle rotundifolia 327
Hydrocotyle sibthorpioides 327
Hydrophyllaceae 354
Hymenophysa pubescens 239
Hypericum perforatum 314, *315*
  prolificum 316
  spathulatum 316
Hypochoeris radicata *446*, 477
Hyssop, wild 365

Impurities of agricultural seed 16, 31
Indian beans 283
Indian chickweed 199
Indian hemp 310, *337*, 338
Indian mallow 310, 313
Indian mustard *233*, 234
Indian physic 338
Indian pink 206
Indian poke 159
Indian tobacco *420*, 421
Indian wickup 319
Indigo, curly 290
  tall 290
Indigo weed 47
Inkberry 197
Insects on weeds 27
Inula helenium *470*, 479
Ipecac, milk 336
Ipomoea coccinea 352
  hederacea 353
  pandurata 353
  purpurea 353
Iresine celosia 196
  paniculata 196

572 Weeds

Ironweed 365, 369
Ironweed, tall 425
  western 425
Isatis tinctoria 247, *259*
Iva axillaris *478*, 479
  xanthifolia 480, *481*
Ivray 134
Ivy 333
Ivy, coliseum 391
  ground 369, *370*
  Kenilworth 391, *392*
  poison 306, *307*
  three-leaved 306

Jacobs-ladder 394
Jacobs-staff 398
Jalap, cancer 197
Jamestown-weed 383
Japanese knotweed 167, *173*
Jatropha stimulosa 297
Jerusalem artichoke 472, *473*
Jerusalem-oak 185, *188*
Jerusalem star 501
Jerusalem-tea 183
Jim Hill mustard 258
Jimson-weed 383, *384*
Joe-pye weed 463
Johnson-grass 142, *143*
Joint-grass 47
Joint-grass, blue 114
Jubas bush 196
Juncaceae 152
Juncus 46
  bufonius 152
  effusus 153, *161*
  gerardii 153
  tenuis 154

Kale, field 234
  wild 252
Kalmia angustifolia 331, *332*
  latifolia *332*, 333
  polifolia *332*, 333
Kedlock 232, 234
Ketmia, bladder 310
Key to weeds 64
Kickxia elatine *392*, 393
Kidney-vine 413
King devil 475, *476*, 477
Kinghead 425
Klamath-weed 314
Klinkweed 241

Knapweed, brown 442
  rayed 442
  Russian 444
  spotted 442, *443*
Knautia arvensis *416*, 417
Knawel *200*, 208
Knot-grass 111, 164
Knot-grass, German 208
Knot-root-grass 135
Knotweed 164, *168*, 169
Knotweed, biting 167
  erect 167, *168*
  Japanese 167, *173*
  silver-sheathed 164
  spotted 170
  swamp 166
Kochia 191
  scoparia 191

Labiatae 368
Lactuca biennis 484
  canadensis 484
  muralis 480
  pulchella 482
  serriola 482, *485*
Ladys-thumb *168*, 170
Lambkill 331
Lambs-quarters 180, 183, *184*
Lambs-tongue 410
Lamium amplexicaule 371, *372*
  maculatum *372*, 373
  purpureum *372*, 373
Lantern-plant, Chinese 385
Lappula echinata 360, *361*
  floribunda 359
  virginiana 359
Lapsana communis *483*, 484
Larkspur 218, *219*
Larkspur, low 218, *219*
Lathyrus palustris 282
  pratensis 282
Laurel, alpine 333
  dwarf 331
  mountain *332*, 333
  narrow-leaved 331
  pale 333
  poison 333
  sheep 331, *332*
  swamp *332*, 333
Laws, weed 31
Leafy spurge 301, *302*
Leather-leaf 46

Leek, wild *155*, 156
Leersia oryzoides 46
Leguminosae 276
Lentil tare 290
Leontodon autumnalis 484, *490*
Leonurus cardiaca 373, *374*
Lepachys pinnata 491
Lepidium apetalum 249
Lepidium campestre 249, *253*
  densiflorum *240*, 249
  draba 239
  latifolium 252
  perfoliatum *240*, 250
  repens 239
  ruderale 250
  virginicum *240*, 250
Leptilon canadense 459
Leptochloa fascicularis 47
Lespedeza 283
  striata 283
  violacea 282
Lettuce, blue 482
  hares 499
  lambs 415
  prickly 482
  showy 482
  wild 482, 484, *485*
Licorice, American 281
  wild 281
Liliaceae 154
Lily, tawny orange 157
Lily family 154
Lily-of-the-valley 157, *158*
Linaria canadensis 393, *405*
  canadensis var. texana 394
  cymbalaria 391
  elatine 393
  minor 391
  vulgaris *392*, 394
Lions-ear 373
Lions-tail 373
Lions-tooth 484, 501
Lippia cuneifolia 368
Lithospermum arvense *361*, 363
  officinale *361*, 363
Little wale 363
Live-forever 262, *263*
Lobelia 160, 421
  inflata *420*, 421
Lobelia family 421
Lobeliaceae 421
Loco, blue 290

Loco, purple 290
  white 290
Loco weeds 290
Lolium temulentum 134, *161*
London pride 206
Longevity of seeds 4, 5
Lonicera japonica 415
Loosestrife, clammy 318
  creeping 334
  purple 318
Loosestrife family 318
Love-grass 130
Love-grass, low 131
  strong-scented 130
  tufted 130
Love-vine 351
Lovers-pride 170
Lupine 283, *284*
Lupinus perennis 283, *284*
Lychnis, evening 205
Lychnis alba 205, *207*
  dioica 206
  flos-cuculi 206, *207*
Lycopsis arvensis 364
Lycopus americanus 375
  uniflorus 375
Lygodesma juncea 486
Lysimachia nummularia 334, *335*
  terrestris 46
Lythraceae 318
Lythrum salicaria 318

Machinery for weed control 49
Mad-apple 383
Madder, blue field *412*, 414
  wild 413
Madder family 411
Madderwort 430
Madia, common 486
Madia glomerata 486
  sativa 486
Madnip 329
Mahonia 226
Maize thorn 440
Mallow, alkali 314
  false 313
  Indian 310, 313
  low 313
  musk 311
  round-leaved *312*, 313
  thistle 313
  Venice 310

574 Weeds

Mallow (continued)
  white 314
Mallow family 310
Malva moschata 311
  neglecta *312*, 313
  rotundifolia 313
Malvaceae 310
Mares-tail 459
Marguerite 447
Marigold, fetid 456, *462*
Marijuana 162
Marjoram, pot 378
  wild *370*, 378
Marrube 376
Marrubium vulgare 376
Marsh elder 480, *481*
Marsh elder, small flowered 479
Marvel 376
Mat-grass 164
Matricaria matricarioides *427*, 487
  suaveolens 487
May-pop 317
Mayweed 423, 428
Mayweed, false 456
Meadow brake 110
Meadow foxtail-grass 115
Meadow-grass 130
Meadow-pine 106
Meadowsweet 274, *275*, 276
Meadowsweet, woolly 276
Mealweed 183
Means-grass 142
Mechanical injury from weeds 30
Medic, black *280*, 283
  hop 283
Medicago lupulina *280*, 283
  sativa 285
Medicinal weeds 33
Melilotus alba 285
  altissima 286
  officinalis 286
Mentha arvensis 376
  gentilis 377
  piperita 377
  spicata 378
Mercury 306
Mercury, three-seeded 296
Mercury-weed 296
Mesquite-grass 131
Metric system 518
Mexican bamboo 167
Mexican fireweed 191

Mexican-tea 183
Miami mist 356
Mignonette, cut-leaved 262
  yellow 262, *263*
Mignonette family 262
Milfoil 422
Milk, tainted by weeds 30
Milk-grass 415
Milkweed 339, 482
Milkweed, climbing 343
  common *340*
  orange 341
  poison 298, 343
  showy 339
  wandering 336
  white-flowered 298
  whorled 341, 343
Milkweed family 339
Millet, double-seeded 46
  Morocco 142
  Polish 127
  wild 140, 142
Millet-grass 142
Mint, brandy 377
  corn 376
  creeping whorled 377
  dog 381
  field 376
  garden 378
  lamb 377, 378
  our ladys 378
  squaw 371
  wild 376
Mint family 368
Mirabilis hirsuta 197
  linearis 197
  nyctaginea 196
Mock-apple 418
Modesty 310
Mollugo verticillata 199, *200*
Moneywort 334, *335*
Moonshine 425
Morning glory 352, 353
Morning glory, ivy-leaved 353
  small-flowered 344
  wild 344, 346, *347*
Morning glory family 344
Morning sun 502
Mosquito plant 371
Motherwort 373, *374*
Mouse-ear 426, 468
Mowing 54

Index

Mugwort *432*, 433
Muhlenbergia frondosa 135
  mexicana 135
  schreberi 135
Mulch paper 60
Mullein *397*, 398
Mullein, corn 201
  moth 396, *397*
  turkey 298
  white 396
Musk 311
Musk mallow 311
Musk plant 311
Muskrat-weed 323
Musquash root 323
Mustard, ball 252, *257*
  black 236, *237*
  blue 260
  brown 236
  dog *245*, 246
  field 234
  hares-ear 241, *251*
  hedge *248*, 258
  Indian *233*, 234
  Jim Hill 258
  leaf 234
  spreading 247
  tall hedge 258
  tansy 166, *225*, 242
  treacle 241, *243*, 246, 247
  tumble *248*, 258
  water 230
  white 232, *233*
  wild 234, *235*
  wormseed *243*, 246
Mustard family 229
Myrica asplenifolia 162
Myricaceae 162
Myrtle 338
Myrtle, yellow 334

Navarretia intertexta 354
  squarrosa 354
Neckweed 162, 401
Needle-and-thread-grass 146
Needle-grass 146
Nepeta cataria *374*, 378
  hederacea 369
Neslia paniculata 252, *257*
Nettle, bee 369, 371
  blind 371
  bull 386, 389
  dead 371, *372*
  dog 369
  flowering 369
  hedge *380*, 381
  hemp 369, *380*
  horse 386, *388*
  low 381
  red dead *372*, 373
  slender 162, 163
  small 163
  spotted dead *372*, 373
  spurge 297
  stinging 162, 163
  tall 162, 163
  white horse 389
Nettle family 162
Nicandra physalodes 383
Nigger-head 491
Nightshade, beaked 390
  black *388*, 389
  blue 387
  climbing 387
  cut-leaved 390
  deadly 389
  garden 389
  hairy 391
  silver-leaf 389
  three-flowered 390
  woody 387
Nightshade family 383
Nimble Kate 419
Nimblewill 135
Nipple-wort *483*, 484
None-such 283
Noon-flower 502
Noon-plant 501
Norta altissima 258
Notholcus lanatus 131
Noxious weeds 31
Nut-grass 150
Nut-grass, northern 148
  yellow 148, *149*
Nut sedge 148, 150
Nyctaginaceae 196

Oat-grass 117
Oat-grass, tuber 148
  wild *118*, 124
Oats, wild 117, *118*
Oenothera biennis 320
Oil seed, Siberian 236
Old fog 124

Weeds

Old maids pink 201, 206
Old witch-grass 135, *136*
Onagraceae 319
Onion, wild 154, *155*
Onoclea sensibilis *109*, 110
Opium, wild 482
Opuntia 317
Orache 180, *181*
Orache, halberd-leaved 180, *181*
 red 180
 thorn 182
Orange hawkweed 474, *476*
Orange milkweed 341
Orange paintbrush 474
Orange root 341
Organy 378
Origanum vulgare *370*, 378
Ornithogalum umbellatum *158*, 159
Orobanchaceae 406
Orobanche ludoviciana 406
 minor 406
 ramosa 406
Orpine, garden 262
Orpine family 262
Oryza sativa 148
Osmunda cinnamomea 108
Osmundaceae 108
Ox-eye daisy *446*, 447
Ox-tongue 489, *490*
Ox-tongue, bristly *478*, 489
Oxalidaceae 291
Oxalis corniculata 291
Oxalis europaea 289, 291
 filipes 292
 florida 292
 stricta 292
Oxybaphus hirsutus 197
 linearis 197
 nyctagineus 196
Oxytropis lambertii 290
Oyster-plant 501

Paintbrush, devils 477
 orange 474
 yellow *476*, 477
Panic-grass 127
Panicum, cocksfoot 127
Panicum capillare 135, *136*
 dichotomiflorum *136*, 137
 gattingeri *136*, 137
 virgatum 137
Pansy, field 316

Pansy, wild 316
Papaver dubium 228
 rhoeas 227
Papaveraceae 226
Paper mulch 60
Parsley, Japanese hedge 331
 poison 325
Parsley family 321
Parsley fern 499
Parsnip, water 329, *330*
 wild *328*, 329
Parsonsia petiolata 318
Parthenium 487, *488*
 hysterophorus 487, *488*
Partridge pea 277, 278
Paspalum 138
 angustifolium 139
 ciliatifolium 138
 dilatatum 138
 distichum 139
 floridanum 47
 laeve 139
Passiflora incarnata 317
Passifloraceae 317
Passion-flower 317
Passion-flower family 317
Pastinaca sativa *328*, 329
Pasturing 53
Patience dock *175*, 179
Pauls betony 401
Pea, Austrian field 290
 blue 283
 cat 288
 craw 282
 meadow 282
 wild 278, 282, **288**
Pear, prickly 317
Pearl plant 363
Peco 218
Penny-cress 260
Penny mountain 382
Pennyroyal, American 371, *370*
 bastard 382
 mock 371
Pennywort, lawn 327
Penstemon digitalis 394, *403*
 gracilis 395
 laevigatus var. *digitalis* 394
Pepper, birds 250
 mild water 169, *171*
 poor mans 250
 water 167

Index

Pepper-grass 240, 250
Pepper-grass, clasping-leaved 240, 250
  downy 249
  field 249, 253
  green-flowered 240, 249
  perennial 239, 252
Pepper plant 167, 238
Peppermint 377
Pepperwort, hoary 239
Perennials 9
Periwinkle 338
Persicaria, pale 169
Persicary 170
Persicary, glandular 170
  swamp 170
Phacelia purshii 356
Phalaris brachystachya 47
  canariensis 139
  paradoxa 47
Pheasants eye 217
Phlox family 354
Physalis alkekengi 385
  heterophylla 384, 385
  ixocarpa 386
  lobata 386
  longifolia 386
  pubescens 386
  subglabrata 384, 386
Phytolacca americana 197, 198
  decandra 197
Phytolaccaceae 197
Pick-purse 213, 238
Picris echioides 478, 489
  hieracioides 489, 490
Picry 306
Pigeon berry 197
Pigeon-foot 295
Pigeon-grass 127, 142
Pigeon-weed 363
Pigweed 183, 187, 188
Pigweed, amaranth 193, 194, 195
  prostrate 192, 193
  rough 194, 195
  Russian 182
  spreading 192
  strawberry 186
  strong-scented 183
  tumbling 192
  white 192
  winged 190
Pilewort 395, 456
Pimpernel, scarlet 334, 335

Pimpernelle 274
Pin-grass 292
Pin-weed 292
Pine-grass 106
Pineapple-weed 427, 487
Pink, Deptford 205, 214
  French 440
  grass 205
  hedge 206
  Indian 206
  meadow 206
  old maids 201, 206
  rush 486
  windmill 211
Pink family 201
Pink-weed 164
Pitchfork-weed 436
Pitchforks 436
Plantaginaceae 407
Plantago arenaria 407
Plantago aristata 407, 408
  indica 407, 408
  lanceolata 408, 409
  major 408, 409
  media 410
  purshii 410
  rugelii 408, 410
  virginica 411
Plantain, bracted 407, 408
  broad-leaved 408, 409, 410
  buck 409
  buckhorn 408, 409
  dooryard 409
  English 409
  hoary 410, 411
  narrow-leaved 409
  pale 410
  purple-stemmed 410
  Purshs 410
  rat-tail 407
  robins 459, 460
  Rugels 408, 410
  sandwort 407
  white 426
  whorled 407, 408
  woolly 410
Plantain family 407
Pleurisy root 341
Plowing 52, 62
Poa annua 139
Poinsettia dentata 301
Poison ash 308

Poison berry 387, 389
Poison creeper 306
Poison dogwood 308
Poison elder 308
Poison-hemlock 325, *326*
Poison ivy 306, *307*
Poison oak 306
Poison parsley 325
Poison stinkweed 325
Poison sumac 308, *309*
Poison-weed 218
Poisonous weeds 28
Poke, Virginia 197
Pokeberry 197
Pokeweed 197, *198*
Pokeweed family 197
Polanisia graveolens 228
Polemoniaceae 354
Polygonaceae 164
Polygonum 172
Polygonum argyrocoleon 164
  aviculare 164, *168*
  coccineum 166
  convolvulus *165*, 166
  cuspidatum 167, *173*
  erectum 167, *168*
  hydropiper 167, *171*
  hydropiperoides 169, *171*
  lapathifolium 169
  orientale 170
  pensylvaticum 170, *171*
  persicaria *168*, 170
  sachalinense 167
  scandens *165*, 172
  setaceum 172
Polypod brake 110
Polypodiaceae 110
Poorland flower 447
Poormans weather-glass 334
Poppy, bubble 211
  corn 227
  field 227, 228
  Mexican 226
  prickly 226, 227
  red 227
  roemeria 228
Poppy family 226
Porcupine-grass 146, *147*
Portulaca, wild 199
Portulaca oleracea 199, *200*
Portulacaceae 199
Pot-herb 230

Potentilla anserina 46
  argentea 267, *272*
  canadensis 267, *268*
  canadensis 271
  fruticosa 269, *272*
  intermedia 269
  monspeliensis 269
  norvegica *268*, 269
  pumila 267
  recta *268*, 270
  reptans 270
  simplex *268*, 271
Poverty-grass 115, 117, 124, *147*
Poverty-weed 425, 466, *478*, 479
Poverty-weed, silver-leaf 465
  woolly-leaved 465
Prairie weed 269
Prevention of seed production 50
Prevention of weeds 48
Prickly pear 317
Primrose, evening 320
  field 320
  tree 320
Primrose family 334
Primrose family, evening 319
Primulaceae 334
Princes-feather 170
Propagation by roots *13*
  by stems *10*
Prunella vulgaris *372*, 379
Prunus virginiana 271
Psyllium seed 409
Pteridium aquilinum var. latiusculum *109*, 111
  aquilinum var. pubescens 110
Pteris aquilina 111
Puccoon 363
Puke-weed 421
Pulling of weeds 50
Pulse family 276
Puncture vine *289*, 291
Purple-head 170
Purslane 199, *200*
Purslane, milk *299*, 305
  winter 401
Purslane family 199
Purslane speedwell *399*, 401
Pursley 199
Purvain 365
Pusley 199
Pusley, Chinese 360
Pussy-toes 426

Index

Quack-grass 111, *112*
Quack-grass, western 114
Quakers bonnets 283
Queen Annes lace 325
Quitch-grass 111

Rabbit-ears 241
Radicula austriaca 254
  palustris 254
  sylvestris 256
Radish 254, *261*
Radish, jointed 252
  wild 252, *261*, 254
Ragged jade 206
Ragged robin 206, *207*
Ragweed 423, *424*
Ragweed, false 480
  giant 425
  great *424*, 425
  lance-leaved 425
  perennial 423, *424*
  western 423
Ragwort 493
Ragwort, golden, 425
Ramsted 394
Ranunculaceae 217
Ranunculus abortivus 218, *221*
  acris 220, *221*
  arvensis 222, *225*
  bulbosus 222, *223*
  repens 222, *223*
  sceleratus *223*, 224
  testiculatus 224, *225*
Raphanus raphanistrum 252, *261*
  sativus 254, *261*
Ratibida pinnata 491
Rattle-box 278, *284*
Rattlesnake weed 327
Rattleweed 278
Red brome 121
Red-ink plant 197
Red-river-weed 480
Red robin 206
Red-root 162, 195, *363*
Red scale 180
Red shanks 167, *170*
Red stem 318
Red-weed 174
Reed-grass 46
Reproduction of weeds 9
Reseda lutea 262, *263*
Resedaceae 262

Rheumatism-weed 338
Rhus diversiloba 306
  microcarpa 306
  radicans 306, *307*
  radicans var. rydbergii 306
  toxicodendron 306
  toxicodendron 306
  vernix 308, *309*
Rib-grass 409
Ribwort 409
Rice, jungle 47
  red 148
Richweed 463
Ripple-grass, western 407
Rocket, Dyers 262
  garden 244, *245*
  large sand 244, *245*
  wall 244
  yellow 230
Rocket cress 230
Rocket salad 244
Roemeria refracta 228
Roemeria poppy 228
Rorippa austriaca 254, *255*
  islandica 254, *255*
  islandica var. hispida 254, *255*
  palustris 254
  sylvestris *255*, 256
Rosa arkansana 273
  eglanteria *266*, 273
  rubiginosa 273
Rosaceae 264
Rose, prairie 273
  sweetleaf 273
  wild 273
Rose family 264
Rose petty 459
Rosin rose 314
Rosin-weed 469
Rotation of crops 58
Rubiaceae 411
Rubus 46, 274
  flagellaris 273
  villosus 273
Rudbeckia hirta 491
Rudbeckia laciniata 425
  serotina 491,*492*
  triloba 491
Rumex 180
  acetosa 172, *175*
  acetosella 174, *175*
  altissimus 174

Rumex (continued)
crispus 176, *177*
hastatulus 176, *178*
mexicanus 179
obtusifolius *177*, 179
patientia *175*, 179
Rush 153
Rush, bog 153, *161*
club 151
field 154
hog- 153
meadow 151
path 154
poverty 154
shore 46
slender 154
slender yard 154
soft 153
spike 47
toad 152
Rush family 152
Rush-grass 145
Rush nut 148
Russian cactus 191
Russian pigweed 182
Russian thistle *181*, 191
Russian tumble weed 191
Rutabaga 236, *237*
Rye-grass, poison 134

Sage, lance-leaved 383
Sage-brush 433
Sainfoil 279
St. Barbaras cress 230
St. Johns-wort 314, *315*
St. Johns-wort, shrubby 316
St. Johns-wort family 314
Salal 46
Salix 46
Salsify 501
Salsify, meadow, 502
Salsoli kali var. tenuifolia *181*, 191
pestifer 191
Saltbush 180
Saltwort 191
Salvers-grass, quack 300
Salvia reflexa 383
Sand brier 386
Sandbur 123, *132*, 390
Sandbur-grass 123
Sandvine 339
Sandweed 213

Sandwort, thyme-leaved 201, *214*
Sanguisorba minor 274
Sanicle, Indian 463
white 463
Saponaria officinalis *204*, 206
vaccaria *204*, 208
Satin flower 217
Satin-grass 135
Satureja vulgaris *380*, 381
Scabiosa arvensis 417
Scabious, field *416*, 417
southern 418
sweet 457
Scabwort 479
Scale-grass 47
Scarlet berry 387
Scirpus 47
americanus 46
atrovirens 151
cyperinus 151
Scleranthus annuus *200*, 208
Scoke 197
Scorpion weed 356
Scotch broom 279
Scouring-rush 106, *107*
Scouring-rush, tall 106
Scourwort 206
Scratch-grass 411
Scrophularia lanceolata 396
marilandica 395
Scrophulariaceae 391
Scurvy 236
Scurvy-grass 290
Scutch-grass 111, 123
Sedge 148, *149*, 150
Sedge, coco 148, 150
hop 152
nut 148, 150
Sedge family 148
Sedge-grass 116
Sedum acre 262
purpureum 262, *263*
triphyllum 262
Seed impurities 16, 31
Seed production 4
Seeds, effect of animal digestion on 22
germination of buried 6
longevity of 4, 5
viability of 4, 6, 22
Seeds in soil 6, 8
Seeds modified for dissemination *19*
Self-heal 379

Index

Senecio aureus 425
  jacobaea 493
  vulgaris 493, *494*
Senna, coffee 278
  wild 277
Sensitive fern *109*, 110
Sensitive plant, large-flowered 277
Senvil 236
Senvre 232
Sesbania exaltata 47, 290
Setaria faberi 142
  glauca 140, *141*
  lutescens 140
  verticillata *141*, 142
  viridis *141*, 142
Sheep bur 504
Sheep lice 357
Sheep poison 331
Shelly-grass 111
Shepherds-bag 238
Shepherds-clock 334
Shepherds-purse 238, *240*
Sherard, herb 414
Sherardia arvensis *412*, 414
Shoo-fly 310
Shrubby St. Johns-wort 316
Sibara virginica 256
Siberian mustard 239
Siberian oil seed 236
Sicklepod 278
Sicyos angulatus 419
Sida, prickly *312*, 313
  spiny 313
Sida hederacea 314
  spinosa *312*, 313
Silene anglica 211
Silene antirrhina 209, *210*
  csereii 209, *212*
  cucubalus *210*, 211
  dichotoma 211
  gallica 211
  inflata 211
  latifolia 211
  noctiflora *210*, 213
Silk plant 410
Silkweed 339
Silphium perfoliatum 425
Silver button 425
Silver leaf 425
Silver-weed 276
Simson 493
Sinapis alba 232

Sinapis arvensis 234
Sisymbrium altissimum *248*, 258
  incisum 242
  irio 260
  officinale *248*, 258
  sophia 242
Sium cicutaefolium 329
Sium suave 329, *330*
Skeleton-weed 445, 486
Skevish 459
Skunk-tail-grass 131
Skunkweed 354
Slobber-weed 304
Smartweed 167, *171*
Smartweed, mild 169
  pale 169
  Pennsylvania 170, *171*
  spotted 170
  water 166, 167
Smilax 160
  glauca 160
Smother crops 59
Smut-grass 145
Snake cuckoo 205
Snake flower 357
Snake-grass 106
Snake-grass, meadow 130
Snake-weed 325
Snakeroot, white 463, *464*
Snapdragon, small 391, *392*
  wild 394
Sneeze-weed 469, 471
Sneeze-weed, purple-headed 471
Snow-on-the-mountain 304
Soap plant 160
Soapwort 206
Soapwort, white 206
Solanaceae 383
Solanum carolinense 386, *388*
  dulcamara 387, *388*
  elaeagnifolium 389
  nigrum *388*, 389
  rostratum *388*, 390
  triflorum 390
  villosum 391
Soldier weed 196
Solidago canadensis 495
  gigantea 495
  graminifolia var. nuttallii 496
  nemoralis 425
  rugosa 496
  serotina 495

Sonchus arvensis 496, *497*
    arvensis var. glabrescens 498
    asper 498, *500*
    oleraceus 499, *500*
    uliginosus 498
Sorgum halepense 142, *143*
    vulgare 145
Sorrel, cow 174
    field 174
    garden 172
    green 172
    horse 174
    ladys 291
    meadow 172
    mountain 174
    red-top 174
    sheep 174, *175*
    tall 172
    wild 176, *178*
    yellow wood *289*, 291
Sour dock 172, *175*
Sour-grass 174, 292
Sour-weed 174
Sow-bane 187
Sow thistle 499, *500*
Sow thistle, annual 499
    creeping 496
    field 496
    perennial 496, *497*
    spiny-leaved 498, *500*
Spanish needles 434, *435*
Spanish-tea 183
Spear-grass, annual 139
    dwarf 139
    tufted 130
Spearmint 378
Specularia perfoliata *420*, 421
Speedwell *399*, 401, 402
Speedwell, corn *399*, 400
    creeping 404
    field *399*, 402
    garden 404
    germander 400, 404
    ivy-leaved 401
    purslane *399*, 401
    rock 400
    thyme-leaved *399*, 404
    wall 400
Spergula arvensis 213, *214*
Spergularia rubra *212*, 215
Spiderwort family 152
Spiraea alba 274, *275*

Spiraea douglasii 276
    latifolia 276
    salicifolia 274
    tomentosa *275*, 276
Spleenwort bush 162
Spoon-wood 333
Sporobolus neglectus 145
    poiretii 145
    vaginiflorus 145
Spotted-hemlock 323
Spudding 54
Spurge, broad-leaved 303
    cypress 300, *302*
    flowering 298
    hairy *299*, 305
    leafy 301, *302*
    nodding 304
    petty *299*, 304
    salvers 300
    shining 303
    spotted *299*, 304
    stubble 304
    sun *299*, 303
    toothed 301
    wart 303
Spurge family 296
Spurge-nettle 297
Spurry 213, *214*
Spurry, red sand- *212*, 215
Spurwort 414
Squaw-weed 463
Squirrel-tail-grass 131, *132*
Stachys annua 381
    arvensis *380*, 381
    palustris 382
Staggerweed 218
Staggerwort 469
Star-of-Bethlehem *158*, 159
Star-thistle 442, *443*
Star-thistle, Iberian 441
    purple 440
    yellow 444
Starweed 217
Starwort 217
Starwort, grassy 215
    yellow 479
Stavesacre 218
Steam for killing weeds 55
Steeplebush *275*, 276
Stellaria graminea 215, *216*
    media *216*, 217
Stick-tights *435*, 436, 437

Index

Stickseed 359, 360, *361*
Stickweed 264
Sticky stem 318
Stink-grass *128*, 130
Stinking willie 493
Stinkweed 228, 260, 428, 456
Stinkweed, poison 325
Stinkwort 383
Stipa, western 146
Stipa comata 146
  spartea 146, *147*
Stitchwort, grass-leaved 215, *216*
Stone-grass 164
Stone seed 363
Stonecrop, mossy 262
Storksbill 292, *293*
Straw mulch 60
Strawberry, barren 271
  Indian 265
  mock 265
  spinach 186
Strawberry blite 186
Succisa australis 418
Succory 447
Succory, gum 445
Succory dock 484
Sumac, poison 308, *309*
  swamp 308
Summer-cypress 191
Summer fallow 58
Summer-grass 140
Sun dial 283
Sunflower 425, 472, *473*
Sunflower, false 480
  prairie 474
  swamp 469
  wild 472
Swainsonia salsula 290
Swallow-wort 227
Swallow-wort, black 343
  white *337*, 343
Swamp laurel *332*, 333
Swamp sumac 308
Sweating plant 463
Sweet betty 206
Sweet brier *266*, 273
Sweet bush 162
Sweet-fern 162, *266*
Sweet Gale family 162
Sweet root 281
Sweet vernal-grass 116
Sweet william, wild 206

Swine-bane 187
Swine cress 241, *259*
Switch-grass *136*, 137
Symphytum officinale 364
Syntherisma ischaemum 125
  sanguinale 127
Syrian bean caper 291

Tackweed 291
Tanacetum vulgare 499
Tansy 499
Tansy, false 431
  wild 423
Taraxacum erythrospermum 501
  laevigatum 501
  officinale 501, *503*
Tar-fitch 282
Tare, lentil 290
  smooth 290
  wild 288
  yellow 282
Tarweed 318, 357, *362*, 469, 486
Teasel *416*, 417
Teasel, card 417
Teasel family 417
Tecoma radicans 404
Thistle, Barnabys *443*, 444
  blessed 454, *455*
  blue 357, *358*
  bull 452, *453*
  Canada 448, *449*
  card 417
  creeping 448
  fragrant 452
  green 448
  horse 482
  milk 482, 499
  musk 439
  Napa 442, *443*
  nodding 439
  pasture 452, *453*
  perennial 448
  plume 452
  plumeless 439
  Russian *181*, 191
  small-flowered 448
  Spanish 505
  spear 452
  star- 442, *443*
  tall 448
  Texas 390
  tumbling 191

Weeds

Thistle, Barnabys (*continued*)
  Turkestan 444
  welted 440
Thlaspi arvense *253*, 260
  perfoliatum 260
Thorn, maize 440
  white 264
Thorn-apple 264, 383
Thoroughwort 425, 463
Thousand-leaf 422
Thunder flower 205
Thunderwood 308
Thyme, creeping *372*, 382
  horse 381
  wild 382
Thymus serpyllum *372*, 382
Tick trefoil 279
Tick trefoil, Canada 279
Tickle-grass 131, 135
Tickseed, leafy-bracted 437
Timothy, water 115
Tipton-weed 314
Tithymalus cyparissias 300
Toadflax *392*, 393
Toadflax, old field 393, *405*
  yellow 394
Tobacco, Indian *420*, 421
  ladies 426, *467*
Tocalote 442
Tomatilla 386
Tomato, wild 386, 390
Tongue-grass 217, 250
Tongue grass, wild 219
Torches 398
Torilis japonica 331
Tormentil 270
Toxicodendron radicans 306
  vernix 308
Tragopogon dubius 501
Tragopogon major 501
  porrifolius 501
  pratensis *483*, 502
Treacle mustard 241, *243*, 246, 247
Tread-softly 297
Trefoil 283
Trefoil, Canada tick 279
  hop 286, 287
  tick 279
Trembles 465
Tribulus terrestris *289*, 291
Trichostema dichotomum 382
Trifolium agrarium 286

Trifolium arvense *280*, 286
  dubium *280*, 287
  fimbriatum 46
  procumbens *280*, 287
Triple-awn 117
Trompillo 389
Trumpet-creeper 404, *405*
Trumpet-weed 463
Tumble mustard *248*, 258
Tumble weed 190, 192, *193*
Tumble weed, Russian 191
Tumble weed-grass 135
Turnip, wild 252
Turnpike-geranium 185
Tussilago farfara 502, *503*
Typha latifolia 47

Ulex europaeus 287
Umbelliferae 321
Umbrella plant 47
Umbrella-wort 196
Urtica dioica 162
  gracilis 163
  holosericea 164
  lyallii 164
  procera 163
  urens 163
Urticaceae 162
Uses of weeds 32

Valerian family 415
Valerianaceae 415
Valerianella locusta 415
Valerianella olitoria 415, *416*
Vegetable-oyster 501
Velvet-grass *118*, 131
Velvet-leaf 310, *312*
Velvet-weed 310, 320
Venus-cup 417
Venus looking-glass *420*, 421
Veratrum viride *158*, 159
Verbascum blattaria 396, *397*
  lychnitis 396
  phlomoides 398
  thapsus *397*, 398
Verbena bracteata 365, *366*
  bracteosa 365
  hastata 365, *366*
  officinalis 367
  stricta 367
  urticifolia *366*, 367
Verbenaceae 365

Vernal-grass, sweet 116
Vernonia altissima 425
  baldwinia 425
Veronica 398
Veronica, creeping *355*
Veronica agrestis 404
  arvensis *399*, 400
  chamaedrys 400
  filiformis *355*, 400
  hederaefolia 401
  officinalis *399*, 401
  peregrina *399*, 401
  persica *399*, 402
  polita *399*, 402
  serpyllifolia *399*, 404
  tournefortii 402
Vervain, blue 365, *366*
  bracted 365, *366*
  European 367
  hoary 367
  mullein-leaved 367
  nettle-leaved 367
  prostrate 365
  white *366*, 367
  woolly 367
Vervain family 365
Vetch, bird 288
  cow 288
  four-seeded 290
  hairy 290
  narrow-leaved 288
  slender 290
  spring 290
  tufted 288
  wild 288
  winter 290
Vetchling, marsh 282
  yellow 282
Viability of seeds 4, 6, 22
Vicia angustifolia 288
  cracca 288
  sativa 290
  tetrasperma 290
  villosa 290
Vinca minor 338
Vincetoxicum nigrum 343
  officinale 343
Viola arvensis 316
Violaceae 316
Violet family 316
Vipers bugloss 357
Virginia poke 197

Warlock 236
Warmot 430
Wart-grass 303
Wartweed 227, 303
Wartwort 468
Water as disseminating agent 18
Water cress, yellow 254
Water-grass 127
Water-hemlock 323, *324*
Water parsnip 329, *330*
Water pepper 167
Water pepper, mild 169, *171*
Waterleaf family 354
Wax balls 296
Waxweed, blue 318
Way-grass 164
Waybent 133
Weather-grass 146
Weed seeds in soil 6, 8
Weeds, benefits derived from 32
  classification of 12
  dissemination of, see Dissemination
  duration of 11
  early introductions of 23
  edible 33
  habits of 3
  keys to 64
  losses caused by 24
  medicinal 33
  poisonous 28
  reproduction of 9
  seed production of 4
  source of 21, 23
Weeds as hosts for crop diseases 26
  declared noxious by law 31
  harboring insect pests 27
  of cranberry bogs 45, 46
  of cultivated fields 43
  of gardens 43
  of grain fields 44, 45
  of hay fields 41
  of lawns 36, 37
  of meadows 41
  of pastures 38, 39
  of rice fields 46, 47
  of special habitats 36
  of turfs 36, 37
  producing mechanical injury 30
Wheat-grass 111
Wheat-grass, western 114
Wheat oats 117
Wheat-thief 121, 363

586 Weeds

Whip-tongue 413
White bottle 211
White cap 276
White horse 124
White mans foot 410
White robin 205
White root 341
White thorn 264
White-top 239, 457, *458*, 461
White-weed 239, 360, 457, 465
Whitlow-grass 244
Wickup, Indian 319
Wicky 331
Willow 46
Willow, flowering 319
Willow-herb 319
Willow-weed 169, 170
Wind 18
Wind witch 191
Windmill pink 211
Winged pigweed 190
Winter annuals 9
Winter cress 230, *231*
Winter sweet 378
Winter-weed 217
Wire-grass 117, 123, 124, 129, 135, 139, 154
Witch-grass 111
Witch-grass, Gattingers *136*, 137
 old, 135, *136*
 spreading *136*, 137
Witches-hair 135
Woad 247, 250
Woad waxen 281
Wode whistle 325
Wolf-grass 115
Wood-grass 135
Wood sorrel, upright yellow 292

Wood sorrel, yellow *289*, 291, 292
Wood Sorrel family 291
Woodwardia virginica 46
Wool-grass 151
Woolmat 357
Wormseed 183, *188*
Wormwood 430, 433
Wormwood, annual 431, *432*
 biennial 431, *432*
Woundwort 382

Xanthium canadense 504
 commune 504
Xanthium orientale 481, 504
 pensylvanicum 504
 spinosum 505
 strumarium 505

Yard-grass 129
Yarr 213
Yarrow 422, *492*
Yellow cress *255*, 256
Yellow devil 475, 477
Yellow mignonette 262, *263*
Yellow nut-grass 148, *149*
Yellow paintbrush *476*, 477
Yellow rocket 230
Yellow star 469
Yellow-weed 222, 230
Yellow-weed, creeping 496

Zigadenus elegans 160
 gramineus 100
 nuttallii 160
 paniculatus 160
 venenosus 160, *161*
Zygophyllaceae 291
Zygophyllum fabago 291

Library of Congress Cataloging in Publication Data

Muenscher, Walter Conrad Leopold, 1891–1963.
    Weeds.

    Reprint of the 2d ed., 1955, published by Macmillan, New York.
    Bibliography: p.
    Includes index.
    1. Weeds—United States—Identification. 2. Weeds—Canada—Identification. 3. Weed control. I. Title.
SB612.A2M8   1980   632'.58'097   79-48017
ISBN 0-8014-1266-8